W9-ATJ-792

DATA HANDLING IN SCIENCE AND TECHNOLOGY — VOLUME 2

Chemometrics: a textbook

DATA HANDLING IN SCIENCE AND TECHNOLOGY

Volume 1 Microprocessor Programming and Applications for Scientists and Engineers by R.R. Smardzewski

Volume 2 Chemometrics: A Textbook by D.L. Massart, B.G.M. Vandeginste, S.N. Deming, Y. Michotte and L. Kaufman

Volume 3 Experimental Design: A Chemometric Approach by S.N. Deming and S.L. Morgan

Volume 4 Advanced Scientific Computing in BASIC with Applications in Chemistry, Biology and Pharmacology by P. Valkó and S. Vajda

DATA HANDLING IN SCIENCE AND TECHNOLOGY — VOLUME 2

Advisory Editors: B.G.M. Vandeginste and L. Kaufman

Chemometrics: a textbook

D.L. MASSART
Farmaceutisch Instituut, Vrije Universiteit Brussel, Belgium

B.G.M. VANDEGINSTE
Laboratorium voor Analytische Chemie, Katholieke Universiteit Nijmegen, The Netherlands

S.N. DEMING
Department of Chemistry, University of Houston, Texas, U.S.A.

Y. MICHOTTE and L. KAUFMAN
Farmaceutisch Instituut, Vrije Universiteit Brussel, Belgium

ELSEVIER

Amsterdam — Oxford — New York — Tokyo 1988

ELSEVIER SCIENCE PUBLISHERS B.V.
Sara Burgerhartstraat 25
P.O. Box 211, 1000 AE Amsterdam, The Netherlands

Distributors for the United States and Canada:

ELSEVIER SCIENCE PUBLISHING COMPANY INC.
655, Avenue of the Americas
New York, NY 10010, U.S.A.

First edition 1988
Second impression 1989
Third impression 1990

Library of Congress Cataloging-in-Publication Data

Chemometrics : a textbook.

(Data handling in science and technology ; no. 2)
Includes bibliographies and index.
1. Chemistry, Analytic--Statistical methods.
I. Massart, Desiré L. (Desiré Luc), 1941- .
II. Series.
QD75.4.S8C46 1987 543'.001'5195 87-9011
ISBN 0-444-42660-4

ISBN 0-444-42660-4

Printed in The Netherlands

CONTENTS

Introduction

Most chemists, whether they are biochemists, organic chemists, pharmaceutical or clinical chemists and most medical doctors, pharmacists, and biologists who apply a chemical discipline need to carry out chemical determinations, i.e. perform chemical analysis. This book is addressed to all these scientists who carry out analytical determinations. Its purpose is to give an introduction in the science of chemometrics, a discipline of analytical chemistry.

The word chemometrics is defined in Chap. 1. This definition indicates that chemometrics is concerned with formal methods for the selection and optimization of analytical methods and procedures and for the interpretation of data. About ten years ago, no book existed that discussed these problems. This led Auke Dijkstra and two of the present authors (D.L.M. and L.K.) to write the first book on chemometrics [1]. The subject was then very new but it has evolved considerably since then. For this reason (and also because the book, although reprinted three times, was sold out), the need was felt for what, at first, we thought would be a new, updated edition of the first book. However, it has finally become a completely new book (not more than 10–20% of the original text has been used unchanged in this book).

In some respects, the spirit in which the book was written has remained unchanged so that we can cite the introduction of the original to explain some aspects of the philosophy underlying this volume.

"The trend towards a more formal approach of the selection of analytical methods is not really new, but it has definitely grown stronger in the last few years. It is not only felt among those who make this their research field in general analytical chemistry, but also by analytical chemists who are concerned more directly with analytical practice, such as clinical chemists and by those who need the results of analytical determinations, such as physicians using laboratory tests for medical diagnosis. At about the same time that concepts such as information were introduced into general analytical chemistry, clinical chemists began to use multivariate data analysis techniques to investigate which analytical methods yield the most diagnostic information. If one looks at the literature cited by analytical chemists and clinical chemists, one finds that they cite different literature and that, in general, there seems to be very little communication between the two groups.

In some applications, formal methods for the investigation of the performance of analytical methods were introduced many years ago. This is the case, for example, with official analytical chemists, who have developed methods for the evaluation of errors likely to occur in analytical procedures. Many analytical chemists from other

Reference p. 3

specialities, but who are also concerned with the evaluation of analytical methods, seem, however, to ignore the existence of such methods.

We have tried to combine the knowledge stored in these (and other) different specialities in the hope of stimulating a more systematic application of formal selection methods in analytical chemistry. Our first idea was to limit this book to newer methods or concepts, such as information theory and operational research; but it was soon clear that it would be meaningless to try and make a synthesis and not include classical statistical concepts. Therefore a number of chapters on classical statistical methods were added.

Unfortunately, most chemists are daunted by the task of learning how these mathematical methods function and this is not helped by the difficulty of establishing a link between the formal mathematics found in most books on this subject and analytical problems.

We have tried to treat the mathematical topics as lucidly as possible and to illustrate the text with examples, in some instances abandoning a rigorous mathematical treatment. In those chapters where the subject matter is probably new to most chemists, only the most elementary explanations are given, often in words, because we think it more important to emphasize the underlying philosophy than to explain the mathematics. In doing so, we hope we have removed the barriers of applying formal methods to optimization problems in analytical chemistry. One major difficulty encountered when writing this book was the mathematical symbolism. We have tried to present a coherent set of symbols throughout the book but, because of the diversity of the methods described, we have not been entirely successful in this respect. Nevertheless, we think that the symbols used should be sufficiently clear."

Several other things have changed. One of these was the team of authors. Unfortunately, Professor Auke Dijkstra was unable to participate in rewriting this book. His ideas are, however, present everywhere in the volume. Auke Dijkstra was one of the very first chemometricians with an overview of the whole field and his influence on the work of the European authors of this book is of an everlasting nature.

Bernard Vandeginste and Yvette Michotte, who were associated with the first book having written parts of a chapter, have now become full authors and Stan Deming joined the team to rewrite the chapters on optimization.

A very important change is that the first book was mainly concerned with the optimization and selection of methods and, to a much lesser degree, with data acquisition and interpretation. Since many of the formal methods proposed here are used both for method development and for data evaluation, this has not resulted in the discussion of many other methods, but in a shift of emphasis. Less weight is given to information theory, systems theory, and operations research and more to time series, correlation, and transformation methods, filtering, smoothing, etc.

Finally, our aim, which went beyond that of the original book, was to write a tutorial book. Since, in didactical texts, too many references are irritating, we have chosen to cite in the text only when strictly necessary, but to give additional references at the end of most chapters.

Reference

1 D.L. Massart, A. Dijkstra and L. Kaufman, Evaluation and Optimization of Laboratory Methods and Analytical Procedures, Elsevier, Amsterdam, 1978.

Chapter 1

Chemometrics and the Analytical Process

1. Definition of chemometrics

The subject matter of this book is chemometrics, a term coined in 1972, which can be defined as the chemical discipline that uses mathematical, statistical, and other methods employing formal logic (a) to design or select optimal measurement procedures and experiments, and (b) to provide maximum relevant chemical information by analyzing chemical data.

Chemometrics has found widespread application in analytical chemistry and therefore that, essentially, is what this book is about. At the same time, it is also a book about the *essentials* of analytical chemistry. If one leaves out the words mathematical, etc. from the definition, one observes that chemometrics is really about what all analytical chemists try to do, namely to design optimal analytical procedures and to try and obtain as much information as possible from the results. Since chemometrics does this with the help of mathematical methods, it has evolved to the theoretical cornerstone of what we will call the analytical process, i.e. the reasoning followed by the analyst to select and optimize procedures, to carry them out in an efficient way, and to interpret the results correctly with a maximum of relevant information as the end product.

2. The intelligent laboratory concept

Figure 1 gives a general picture of the analytical process. Further detail is given in Fig. 2.

The analytical process starts with a problem that has to be solved and to solve it one needs chemical information. It is the task of the analytical chemist to provide

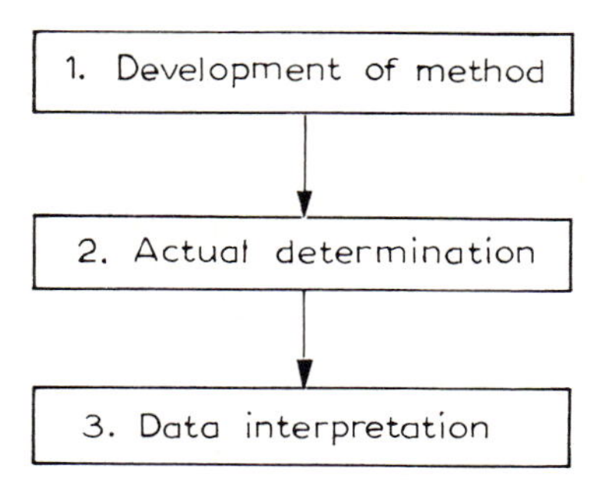

Fig. 1. The main steps in the analytical process. Chemometrics is concerned only with steps 1 and 3.

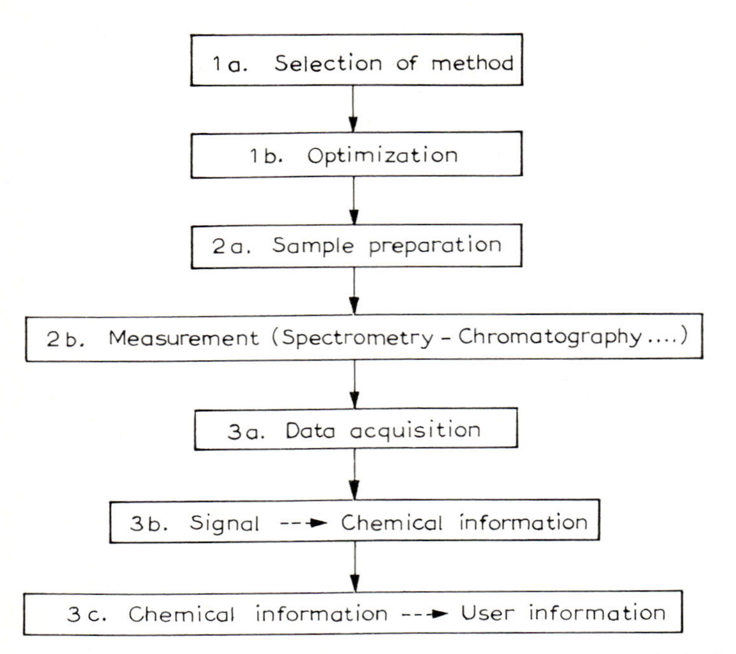

Fig. 2. A more detailed view of the analytical process. The numbers in the boxes relate to Fig. 1.

this. His first step will be to select a method. Let us suppose he has to know the quality of a certain foodstuff and to do this he needs to determine the trace element content. He then has to decide whether he will use atomic absorption spectrometry (AAS), neutron activation analysis, inductively coupled plasma emission, or some other method. If he chooses AAS, his next decision will then be to decide on a flame or flameless method and if he selects the latter, what ashing temperature or gradient would be indicated. He will also have to decide on the pretreatment of the sample (wet ashing or low-temperature ashing, extraction or not, what kind of extraction, etc). All this will lead him to an initial procedure. Very probably, this procedure will not be the final one. The analytical chemist then tries to optimize the initial procedure by experimental optimization. He changes the pH of the buffer to obtain a more complete extraction and the drying temperature in the oven to obtain more reproducible results.

The procedure is now available and the analysis can begin. This is usually divided into two parts, the pretreatment and the actual determination. The pretreatment consists of operations such as weighing, extracting, drying, centrifugation, etc. This step is often the most difficult and time-consuming and determines the quality and efficiency of the method.

The result of the determination is a (usually electrical) signal and nowadays it is retrieved from the instrument by a computer. Very often, this signal is first treated to make it more useful by, for instance, reducing noise and it is then translated into chemical information. This means that a list of chemical identities and concentrations is now obtained. To achieve this translation, models describing the relationship between signal and concentration or identity, such as calibration models, are required. Some analysts finish here but one should remember, of course, that the analysis was carried out to solve a problem. This means that the chemical information should be translated into user or diagnostic information. Is the foodstuff acceptable for consumption?; does the analysis of an air sample indicate that a certain industry is responsible for air pollution at the collection point?; does the result of a patient's blood tests indicate a certain disease?, etc. While the answer may be simple in some cases (for example, the foodstuff contains chemical X in excess of dose Y and therefore violates legal rules) it may be much more complex in certain cases and necessitate the application of certain mathematical techniques.

The analytical process described in this way can be considered as a system regulated by two feedback loops (Fig. 3). The first, internal to the laboratory, is the quality evaluation loop. Its purpose is to verify whether the performance of the method is good enough to achieve the analytical purpose for which it was developed and carried out. This loop requires the definition and evaluation of performance criteria (is the method "good" enough?) and the development of quality control schemes (does the method remain good enough when it is carried out repeatedly or continuously over a period of time?).

The second loop (the decision loop in Fig. 3) is the interaction with the outside world. The analytical results (hopefully) serve to solve a problem for, or to make a decision by, the person or organisation that asked for the results in the first place. This usually leads to new questions or, when the results did not bring the expected solution to the problem, to a better formulation of the question. In many instances, the analytical results also serve to control some process and the characteristics of the process determine the required characteristics of the analytical method.

Chemometrics is involved in steps 1a, 1b, 3a, 3b and 3c and in both control loops. Practical chemometrics is a matter of carrying out computations and this means that, in each of these steps, a computer is involved. This is certainly true also for steps 2a and 2b. More and more instruments are now attached to a computer (2a) and robots (2b), which are really computers with a hand, are often used to carry out the pretreatment step. In fact, we conclude that all the steps of Fig. 2 are computer-compatible. It is our belief that the separate functions of Figs. 2 and 3 will slowly be integrated and controlled by a central laboratory information system. When this integration has been achieved, an intelligent laboratory will have been developed. It will be able to select and optimize a procedure by itself, carry it out, extract the relevant information, check its own good functioning, and help in making decisions.

The integrated intelligent laboratory described in this way will rely heavily on software and the purpose of this book is to give the formal and mathematical background of the algorithms and techniques used. In fact, an alternative definition of chemometrics in analytical chemistry could be that it is the chemical discipline that studies mathematical, statistical and other methods employing formal logic to

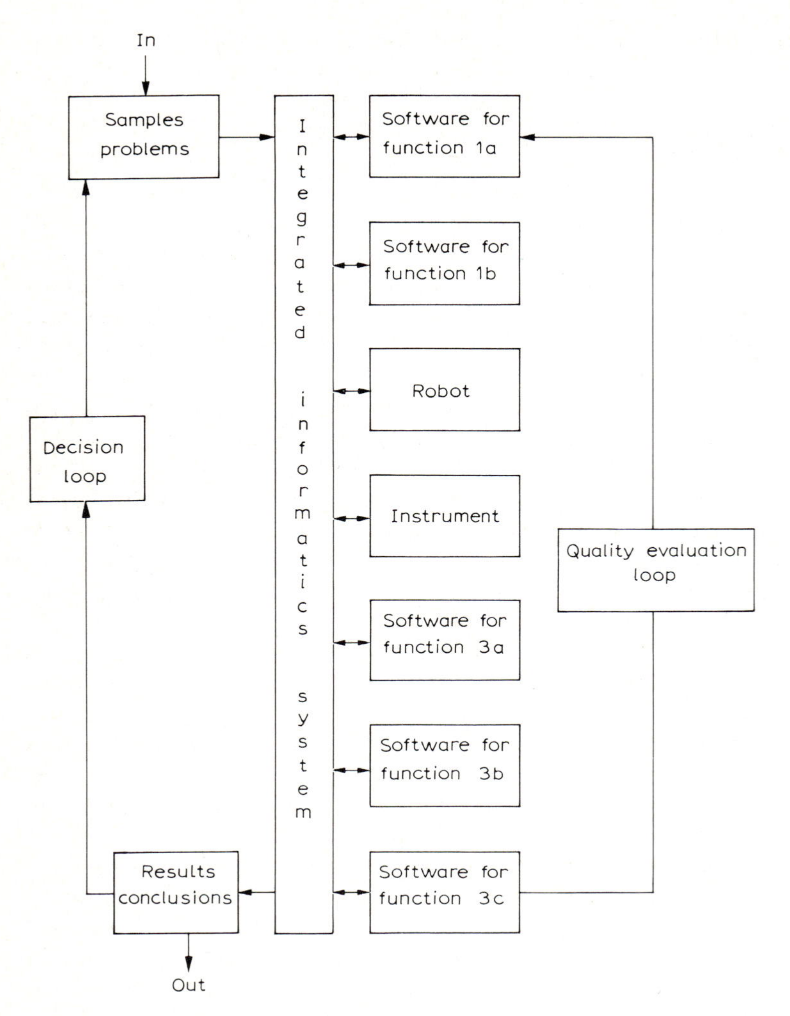

Fig. 3. The analytical process and its environment. The numbers and letters in the boxes relate to Fig. 2.

achieve the development of an integrated intelligent laboratory as described in Fig. 3.The technical software problems are not discussed. For instance, although we consider that robotics may become an important part of the intelligent laboratory, its development is mainly a question of hardware and software technology and therefore there will be no chapter on robotics in this book. There is also a lot of interest nowadays in expert systems. These will certainly be of use in those steps where the analytical chemist uses expertise, such as in the development of the analytical procedure or in the interpretation of spectroscopic data (structural analysis). However, again, this is mainly a problem of software and knowledge engineering, which we believe to be beyond the scope of a textbook on chemometrics today.

3. Organization of the book

The mathematical techniques and the formal concepts which must be introduced, unfortunately cannot be ordered in the same way as in Figs. 1–3. Because the first thing one really does when one needs to develop a new procedure is to define its desired characteristics, we started the book with a discussion on performance characteristics. Indeed, before it is possible to make a selection of a method or to carry out an optimization, one must have criteria according to which this may be done. Consequently, the performance of analytical procedures has to be evaluated by determining one or more performance characteristics of the procedure. The set of criteria has to be defined for each problem and will include quantities such as precision, accuracy, limits of detection and interferences. Up to now, most of these criteria have been used in quantitative analysis and it is probable that another set of characteristics will be required for qualitative analysis. Measures of information may be used for this purpose and for that reason a chapter is devoted to information theory.

Chapters 1–11 also introduce some of the basic statistical concepts (hypothesis testing, analysis of variance, straight line calibration, etc.) that form the basis for the more sophisticated applications of Chaps. 12–27.

With the methods and statistics from Chaps. 1–11, one can now select methods on a formal basis. However, they do not yet permit us to go to the next stage in the analytical process, the optimization step, because one of the techniques needed to achieve this applies multiple regression. For this reason, Chaps. 12–15 are devoted to a series of mathematically related methods concerned with regression, correlation and autocorrelation, and transformations. These techniques give the necessary background to carry out operations related with data acquisition, such as signal enhancement by a reduction of the noise level (filtering, smoothing, Chap. 15), signal restoration (deconvolution, pseudo-deconvolution), characterization and modelling of the signal, for instance to resolve or describe spectra by regression (Chap. 13), or time-dependent processes by time series methodology (Chap. 14). The step following data acquisition in Fig. 2, namely translation to chemical information, requires calibration and calibration requires carrying out a regression. Simple one-component straight-line calibration is discussed in Chap. 5 and the generalization to polynomials and multicomponent problems is discussed in Chap. 13. Multiple regression permits multicomponent analysis (Chap. 13).

With the mathematical knowledge obtained in Chaps. 1–15, it is possible to study experimental optimization methods, which is the subject of Chaps. 16–19. They treat response surface methodology, the use of analysis of variance and regression applied to optimization and introduce some methods, such as the Simplex method, that are used specifically for optimization purposes.

The last chapter of this part is devoted to some optimization methods that have been developed specifically for use in chromatography, the area of analytical chemistry where optimization has been most successful.

Chapters 20–23 are concerned with multivariate data analysis, or at least to an aspect that has been widely studied in recent years, namely pattern recognition. In the scheme of Fig. 2, it is to be placed among the methods that are used for translating chemical results into user results. Indeed, these methods are used mostly for categorizing the materials that were subjected to the analysis. The fundamental idea is that, when a sample is characterized by many different analytical results, these results form a pattern. This pattern is used to answer three questions, namely

(a) how to display multidimensional data in a lower, preferably two-dimensional space (Chap. 21), without loss of significant information,

(b) how to detect groups of samples with similar patterns (Chap. 22), and

(c) how to classify an object in one of two or more known classes on the basis of its pattern (Chap. 23).

Chapters 24–27 have to do mainly with decision processes. Several formal techniques have been developed outside analytical chemistry to help make certain decisions. Some of them have found their way into analytical or clinical chemistry. Chapter 25 discusses some operations research methods. These methods have been developed to help managers take decisions, for instance about routing or queueing problems, and can be applied by chemists for the same purpose. Chapter 26 is about making decisions on the basis of analytical results when there is an element of uncertainty: does a certain number of analytical results outside some specification mean that a batch of samples is defective, taking into account the distribution of defective samples and analytical precision?; does a certain clinical result mean that a patient is ill, taking into account the distribution of results for these tests in a population?; and, again, the analytical precision? In Chap. 27, aspects of control theory are used to derive a model for the selection of the sampling scheme and the analytical method to control a process (industrial, ecological, etc.).

Recommended reading

In addition to the book to which this volume is the successor and which appeared first in 1978 (see Introduction), there is one other general book on chemometrics:

M.A. Sharaf, D.L. Illman and B.R. Kowalski, Chemometrics, Wiley, New York, 1986.

There are also two books with a general view on chemometrics, although the subject is somewhat more specific, namely:

S.T. Balke, Quantitative Column Liquid Chromatography. A Survey of Chemometric Methods, Elsevier, New York, 1984.

G. Kateman and F.W. Pijpers, Quality Control in Analytical Chemistry, Wiley, New York, 1981.

General information can also be obtained from the following journals, which are specifically directed towards this field.

Journal of Chemometrics and Intelligent Laboratory Systems (published by Elsevier)
Journal of Chemometrics (published by Wiley)

and from the two yearly reviews on chemometrics in Analytical Chemistry or the section on chemometrics from Analytica Chimica Acta.

There is a Chemometrics Society, with a Newsbulletin (write to Professor W. Wegscheider, Technical University of Graz, Institute for Analytical Chemistry, Technikerstrasse 4, A-8010 Graz, Austria.). The definition of chemometrics was derived by us from this bulletin (No. 7, 1981). The intelligent laboratory concept is explained in more detail in

D.L. Massart and P.K. Hopke, Chemometrics and distributed software, J. Chem. Inform. Comput. Sci., 25 (1985) 308.

Chapter 2

Precision and Accuracy

1. General discussion of concepts

1.1 Introduction

The purpose of carrying out a determination is to obtain a valid estimate of a "true" value. When one considers the criteria according to which an analytical procedure is selected, precision and accuracy are usually the first to come to mind and most textbooks concerned with analytical chemistry discuss and define these terms. One would therefore expect that there are universally accepted definitions and methods for determining these quantities. Unfortunately, this is not the case.

Of the many definitions proposed, we prefer those given in ref. 1 because these seem to be the most appropriate from both the analytical and statistical points of view (see Sects. 1.4 and 1.5).

1.2 Categories of errors in analytical chemistry

In analytical chemistry, several types of deviation from the true value are encountered when carrying out a determination. One can call these deviations uncertainties or errors and both terms will be used. Roughly, the following categories may be considered: random (indeterminate) errors cause imprecise measurements and are therefore assessed by means of the precision (or imprecision, as preferred by some workers), while systematic errors cause inaccurate (incorrect) results and are referred to in terms of accuracy (or inaccuracy as preferred by some workers). If the true value is 100 and a set of measurements yields the results 98, 101, 99, and 100, random errors occur. Results such as 110, 108, 109, and 111 would indicate the occurrence of systematic errors. The total error observed is the sum of the systematic and random errors. Usually, the precision is studied first because systematic errors can be determined only when random errors are sufficiently small and their size is known.

When an analyst carries out a number of replicate determinations on the same sample using the same procedure, apparatus, reagents, etc., results that are subject to random and normally distributed errors are obtained. The results of replicate determinations are considered to be a random sample from a normal population of results. The standard deviation of this distribution is generally called the precision of a procedure. It is often obtained under favourable conditions and is not what could be called the "real-life" precision. When the procedure is to be applied as a routine method, other sources of error will be introduced and the precision will decrease. For example, it is often observed that the precision calculated for samples analysed in several batches or on several days is worse than that for samples analysed in one batch or on the same day. The latter is sometimes called the day-to-day or between-days precision, while the former is the within-day precision.

These additional sources of variation are not necessarily random. When they are caused by unstable reagents or by the ageing of parts of the apparatus (for example, the pump tubes in a continuous automatic analyser), they are systematic. Such a time-dependent error is sometimes known as drift and is discussed in more detail in Chap. 6.

In the same manner, when a procedure is carried out on the same sample by several laboratories, each with their own personnel, apparatus, reagents, etc., one usually observes a normal distribution of errors broader than that obtained when a single analyst carries out the determinations or when they are carried out in a single laboratory. This effect results from the fact that each laboratory makes some systematic errors or introduces its own bias owing to, for example, impurity of the reagents or incomplete directions for carrying out the procedure. Laboratory biases themselves may be normally distributed. Thus, the distribution obtained when the sample is analysed using the same method by several laboratories is also normal. The dispersion around the mean can again be considered to be a measure of precision and, in other definitions, this is called the reproducibility. In Chap. 4, it is shown how to assess this measure of precision by inter-laboratory comparison.

Procedures are also subject to inherent, systematic (and therefore not normally distributed) errors. Systematic errors are generally said to influence the accuracy, although there is some divergence of opinion and terminology on this point (see Sect. 1.5).

Systematic errors may be constant (absolute) or proportional (relative). A constant error refers to a systematic error independent of the true concentration of the substance to be determined and is expressed in concentration units. A proportional error is a systematic error that depends on the concentration of the analyte and is expressed in relative units, such as a percentage.

The main sources of constant error are

(a) insufficient selectivity (interference), which is caused by another component that also reacts so that falsely high values are obtained; measures of selectivity are discussed in Chap. 8;

(b) matrix effects; this source of error is due to the presence of a component which does not by itself produce a reading but which inhibits or enhances the measurement (these interferences also cause an insufficient selectivity); and

(c) inadequate blank corrections.

Proportional errors are caused by errors in the calibration and, more particularly, by different slopes of the calibration lines for the sample and standard. The incorrect assumption of linearity over the range of analysis will also cause errors related to the concentration to be determined.

Systematic errors can be studied by a variety of methods, which are discussed in Chap. 3.

There are other sources of error which cannot be classified easily in one of these categories. An example in automatic continuous analysis concerns the contamination caused by previous samples and is called the carry-over error. This occurs when successive samples take a common path in an automated system. Because of its dependence on the parameters of the method, it can be considered as a systematic error. On the other hand, it is not constant, nor is it proportional to the concentration of the sample analysed but to that of the previous sample.

1.3 Precision and accuracy as criteria

Precision and accuracy together determine the error of an individual determination. They are among the most important criteria for judging analytical procedures by their results. Many workers consider these quantities to describe the state of the art and their improvement is regarded as the only possible aim of optimization studies. However, analysts proposing a method for a particular procedure should ask themselves whether an increase in the precision and accuracy of the determination is really important or even useful. All sources of variation must be taken into account. For example, if sampling is to be regarded as part of the analysis, then sampling errors must also be considered. In some instances, these errors are very large and can dominate the total error. An example is a potassium determination carried out routinely in an agricultural laboratory [2]. It was found that 87.8% of the error was due to sampling errors (84% for sampling in the field and 3.8% due to inhomogeneity of the laboratory sample), 9.4% to between-laboratory error, 1.4% to the sample preparation, and only 1.4% to the precision of the measurement. It is clear that, in this instance, an increase in the precision of measurement is of little interest. A comparable situation is found in clinical chemistry where the purpose of the analysis is to investigate whether the values fall in the normal range or not. Because of biological variability, this range can be very large. There have been interesting studies of the effect of analytical error on normal values (the values considered to be normal for a population) and clinical usefulness. Whatever the results of these studies, it seems evident that there is no sense in trying to obtain a method with 0.01% precision and accuracy when the normal range is of the order of 20%. These aspects are discussed in more detail in Chaps. 24–27, in which the relationship between analytical chemistry and its environment is considered. Therefore, this and the two following chapters are essentially descriptive in the sense that the assessment of precision and accuracy (or their components) is discussed without considering requirements for their magnitude.

1.4 Definition and measurement of precision (repeatability, reproducibility)

Different definitions of the above three terms have been proposed and we shall restrict ourselves to four of them. The first was given in ref. 1: "Precision refers to the reproducibility of measurement within a set, that is, to the scatter or dispersion of a set about its central value". The term "set" is defined as referring to a number (n) of independent replicate measurements of some property. Readers are urged to use this definition with an understanding of its limitations, such as the fact that the values obtained are usually based on a small number of observations and should therefore be regarded as an estimate of the parameter. By adding this comment, the definition of ref. 1 conforms with statistical usage. Statisticians make a careful distinction between a population quantity and its estimate, the sample quantity. In statistics, one studies the population on the basis of samples drawn from the population in a random way. The random sample, which should be representative of the studied population, is then used to check assumptions about the population (statistical tests) or to estimate parameters of the population (mean, variance, etc.). Consider the following example. An analyst is asked to measure the sodium concentration of a water sample. n replicates are required. This means that the analyst will draw n independent samples from the population which, here, is the water sample given to him. He will produce n results x_i ($i = 1, \ldots, n$). The mean, $\bar{x}$, of the x_i values is an estimate of μ, the mean sodium content of the population from which the samples were drawn.

The definition given in ref. 1 does not specify whether or not the set of measurements is carried out by a single operator. As will be seen in Chap. 4, this is of great practical importance. On the other hand, the International Organization for Standardization [3] prefers the following definitions. Reproducibility: the closeness of agreement between individual results obtained with the same method or identical test material but under different conditions (different operator, different apparatus, different laboratory and/or different time). Repeatability: closeness of agreement between successive results obtained with the same method or identical test material and under the same conditions (same operator, same apparatus, same laboratory and same time).

According to Youden and Steiner [4], the precision is composed of random within-laboratory errors and unidentified systematic errors in individual laboratories (laboratory bias). These errors are also normally distributed. In this instance, precision is considered to be identical with the reproducibility as defined above, with repeatability as a component. Other terms such as scatter and analytical variability are also used occasionally. According to the International Union of Pure and Applied Chemistry (IUPAC) [5], precision "relates to the variations between variates, i.e. the scatter between variates". Some workers prefer the term imprecision to precision in order to avoid the linguistic difficulty that a procedure becomes more precise when its measure, the precision, decreases. In our view, collaboration between statisticians and analytical chemists is so important that semantic difficulties should be avoided. Terms such as reproducibility, repeatability, and imprecision, which are not used by statisticians, should not be used by chemists except in a colloquial sense, i.e. when there is no need to attach a precise meaning to them. The analytical chemist should clearly specify what he means by precision if there can be a doubt.

The following measures of precision within a set (as defined above) are proposed in ref. 1.

"Standard deviation is the square root of the quantity (sum of squares of deviations of individual results from the mean, divided by one less than the number of results in the set). The standard deviation, s, is given by

$$s = \sqrt{\frac{1}{n-1} \sum_{i=1}^{n} (x_i - \bar{x})^2}$$

Standard deviation has the same units as the property being measured. It becomes a more reliable expression of precision as n becomes large. When the measurements are independent and normally distributed, the most useful statistics are the mean for the central value and the standard deviation for the dispersion." One observes that the symbol s is used for the estimate of the true standard deviation, σ. This is correct statistical practice. Recent rules approved by IUPAC [5] state that, when the number of replicates is smaller than 10, s should be used instead of σ. In our view, it is preferable always to use s for an estimate, even a good one, and to reserve σ for the "true" value. It should be noted here that statisticians make a distinction between a biased and an unbiased estimate. The standard deviation as defined above is an unbiased estimate and should therefore be represented by $\hat{\sigma}$, where the "hat" on σ indicates that it is unbiased. We would really prefer to use this symbolism throughout this book but, as we do not want to introduce or create symbolism and terminology that would be unfamiliar to analytical chemists, we shall refrain from doing so, except occasionally when some distinction is important.

"Variance, s^2, is the square of the standard deviation."

"Relative standard deviation is the standard deviation expressed as a fraction of the mean, i.e. $s/\bar{x}$. It is sometimes multiplied by 100 and expressed as a percentage. Relative standard deviation is preferred over coefficient of variation."

Two other quantities are defined, although they are not to be recommended as measures of precision except when the set consists of only a few measurements. These quantities are the mean (or average) deviation, given by

$$\frac{1}{n} \sum_{i=1}^{n} |x - \bar{x}|$$

and the range, given by the difference in magnitude between the largest and smallest results in a set.

One should observe that the often used standard error, which is the standard deviation of the mean and is equal to $s/\sqrt{n}$, is of no direct interest in evaluating the precision as a criterion for evaluating a procedure. It does give, however, an idea of the confidence which one can have in the mean value, $\bar{x}$, obtained for the analysis of a particular sample. Indeed, by performing, for example, n replicate analyses on the same sample, one obtains n results with a mean of $\bar{x}$ and standard deviation s. The standard error, $s/\sqrt{n}$, only tells us which standard deviation of the mean $\bar{x}$ can be expected when repeating the procedure, i.e. again performing n replicate analyses on the same sample.

References p. 31

1.5 Definitions of bias and accuracy

When analytical determinations are carried out, they yield different results, x_i. A result x_i can differ from the true value, μ_0, which is unknown and, in statistical terminology, this difference is referred to as the error of x_i

$$e_i = x_i - \mu_0$$

If enough measurements are made, a stable mean $\bar{x}$ is obtained, where $\bar{x}$ is an estimate of the mean, μ, of an unlimited number of determinations (the population). The absolute difference between μ as represented by $\bar{x}$ and the true value, μ_0, is called the bias or systematic error.

e_i can thus be written as

$$e_i = \underbrace{x_i - \mu}_{\text{random error}} + \underbrace{\mu - \mu_0}_{\text{bias}}$$

Consider the following example. Five replicate measurements with method A yield the results 2.8, 2.7, 3.0, 3.2 and 3.3. $\bar{x}_1 = 3$. Method B yields the results 4.8, 4.7, 5.0, 5.2 and 5.3. $\bar{x}_2 = 5$. $\bar{x}_1$ estimates μ_1, the mean one expects for an infinite number of replicates (the population mean); $\bar{x}_2$ estimates μ_2. Suppose the true value is 3 ($\mu_0 = 3$). The error e_{11} of the first result of method 1 can be written as

$$e_{11} = \underbrace{2.8 - 3}_{\text{random deviation}} + \underbrace{3 - 3}_{\text{bias} = 0}$$

The error e_{12} of the first result of method 2 is written as

$$e_{12} = \underbrace{4.8 - 5}_{\text{random deviation}} + \underbrace{5 - 3}_{\text{bias} = 2}$$

It should be noted that the bias obtained by experimentation is only an estimate of the true bias, as it is calculated by using $\bar{x}$, which is itself an estimate.

It is necessary to consider, at this stage, the terms "laboratory bias" and "method bias". The former, as seen in the preceding section, contributes to the precision of a method (inter-laboratory precision), while the latter constitutes the systematic error due to the method as such. The accuracy for an inter-laboratory trial is identical with the method bias. An inter-laboratory trial consists of distributing an aliquot of a homogenized sample to a number of randomly chosen laboratories in order to evaluate an analytical procedure. From the point of view of an individual laboratory, however, the systematic error is the sum of the method bias (common to all laboratories using the method) and the laboratory bias (for the laboratory in question). This, too, is called the accuracy and the meaning of the term accuracy is therefore not always clear.

Let us now turn to the definition given in ref. 1. It states that "Accuracy normally refers to the difference (error or bias) between the mean, $\bar{x}$, of the set of results and the value, $\hat{x}$, which is accepted as the true or correct value for the quantity measured. It is also used as the difference between an individual value x_i and $\hat{x}$. The absolute accuracy of the mean is given by $\bar{x} - \hat{x}$ and of an individual value by $x_i - \hat{x}$". In this definition, $\hat{x}$ has the same meaning as the symbol μ_0 used by us and in most statistical books. The definition of ref. 1 is ambiguous because it consists, in fact, of two different sub-definitions. The first relates to the mean obtained with a particular method and is a synonym of systematic error, while the other relates to individual results and is therefore made up of a combination of the systematic error and the random error.

This introduces a new difficulty as this definition allows the use of the word accuracy for the sum of the errors due to systematic and random causes. The combination of both has been called total error by some workers.

The definition of IUPAC [5] is very similar and states that "accuracy relates to the difference between a result (or mean) and the true value".

In this book, we shall use the word accuracy in the general sense, i.e. in a colloquial way. When distinctions are important, we shall use the following terms. (i) Laboratory bias (see the section on precision), being the systematic error introduced by a laboratory. This bias is considered to be part of the inter-laboratory precision. (ii) Method bias, being the systematic error introduced by the use of a particular method. It is the same for all laboratories. (iii) Total error, for combinations of errors due to method bias and random errors (inter- and intra-laboratory precision).

1.6 Laboratory bias

Analytical chemists developing new methods should realize that these methods will be used by chemists in other analytical laboratories who may not have the same fundamental knowledge of the method and may therefore simply follow the procedure proposed with their own apparatus, reagents, etc. Very often a new method, when it is used under actual working conditions, gives poor results. To describe this frequently observed phenomenon in terms of precision, we have stated that the overall precision of a method, s, is composed of two terms, namely an intra-laboratory precision (s_r) and an inter-laboratory precision or laboratory bias (s_b). It is known that s_b is usually larger than s_r.

A typical example was given by Wernimont [6] (see Fig. 1) in an article concerned with a study of sources of variation (laboratories, analysts within laboratories, different days for the same analyst, replicate determinations).

It can be seen that the total variation for single tests carried out in any laboratory, on any day and by any analyst ($s = 0.27$) can be explained by the variation among laboratories ($s_b = 0.25$). The other sources of error are much less important.

Many analytical chemists consider that this effect is due to imperfect or incomplete descriptions of procedures and that, provided that procedures are described in sufficient detail, every laboratory should obtain results with the same precision and

References p. 31

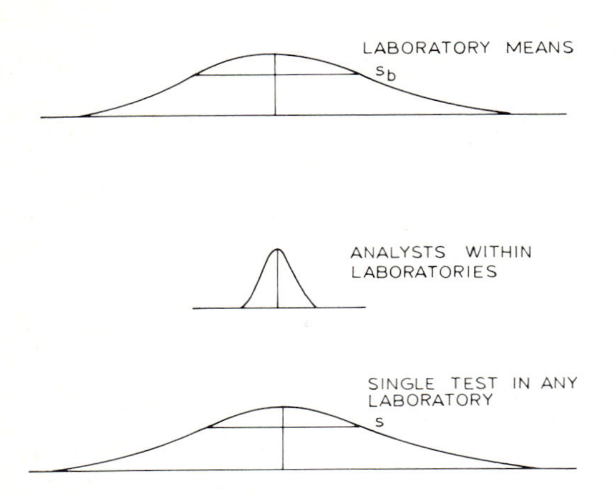

Fig. 1. A comparison of sources of variation in the determination of acetyl (adapted from ref. 6).

accuracy. In fact, this is not true. Very interesting work in this respect has been carried out under the auspicies of the Association of Official Analytical Chemists and their conclusions were given in their statistical manual [4].

The only way of reaching a conclusion about the inter-laboratory precision of an analytical method under actual working conditions is to homogenize a sample and to distribute it to a number of laboratories for analysis, i.e. to carry out an inter-laboratory trial. Intercomparisons are carried out in two situations

(a) when a method has been tested by one or a few laboratories, shown to be free of method bias and proved sufficiently precise in the laboratory of the promotors to warrant an examination of its general usefulness and

(b) when several methods are in use for a certain determination and one wants to know whether they yield the same or significantly different results.

Situation (b) is discussed in Chap. 4. In this section, only situation (a) is considered, in particular when one wants to demonstrate that laboratory biases are present.

The frequent occurrence of systematic errors in user laboratories may appear surprising. However, it can be demonstrated easily by using a two-sample chart or Youden plot. If the participants in collaborative studies are asked to analyse two samples of more or less analogous constitution and there is no systematic laboratory (or method) bias, the probability of finding a high result (+) should be equal to the probability of obtaining a low result (−) for each participating laboratory. This also means that the combination of two high results (+ +), two low results (− −) and both possibilities of obtaining one low and one high result (+ − or − +) are equally probable. By plotting the result for sample A against the result for sample B for each laboratory, one should obtain a diagram similar to Fig. 2(a). In fact, one nearly always obtains a result such as that shown in Fig. 2(b), i.e. a significantly high prevalence of + + and − − results, showing that more laboratories than expected deliver either two high or two low results.

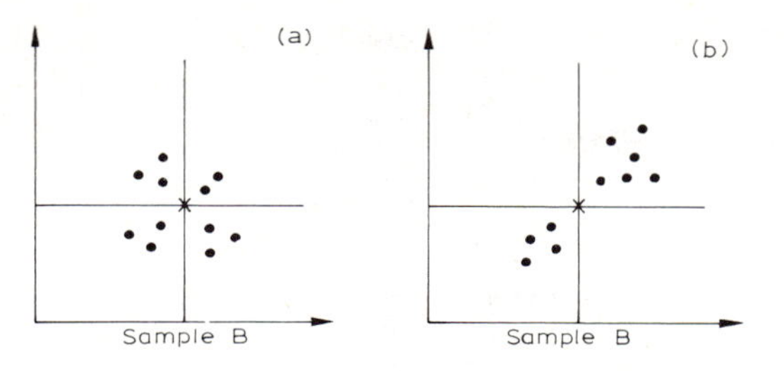

Fig. 2. The two-sample method for the detection of laboratory bias. (a) No laboratory bias; (b) laboratory bias.

Further, as Youden and Steiner's two-sample test does not distinguish method and laboratory bias, there is little reason to carry out intercomparison of different methods in this way. Method bias can be evaluated with the analysis of variance technique (see Chap. 4).

Horwitz et al. [7] showed that there is a relationship between the concentration levels determined and the attainable precision. This attainable precision is independent of the determination method. Expressed as the relative standard deviation (RSD) and containing the between- and within-laboratory components, it was shown to decrease by a factor 2 when the level of concentration decreases by a

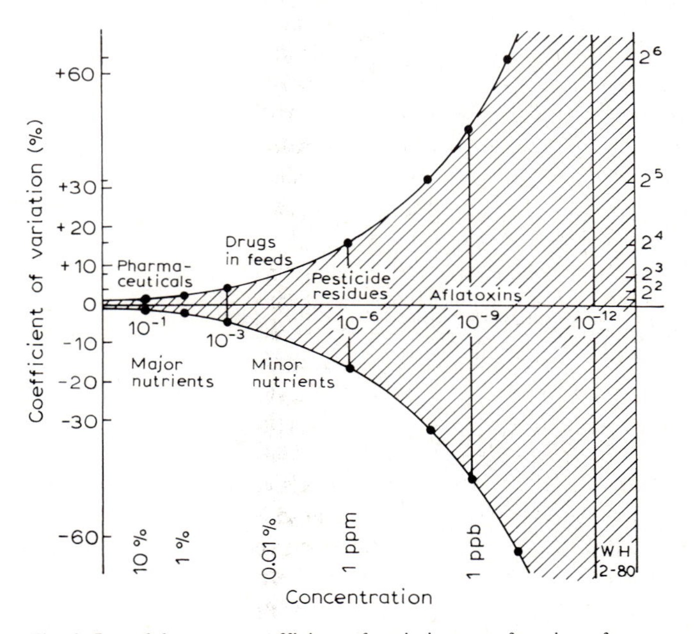

Fig. 3. Inter-laboratory coefficient of variation as a function of concentration [7].

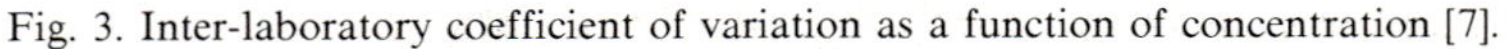

References p. 31

factor 100. Typical values are: active components of a drug in a drug formulation at a concentration of 1% (RSD 2%), drugs in feeds at a 0.01% concentration (RSD 8%), pesticide residues at the 1 ppm level (RSD 16%), aflatoxins in foodstuffs at the 10 ppb level (RSD 32%), etc. Mathematically, this relation is given by

$$\mathrm{RSD} = 2^{(1-0.5 \log x)}$$

where x is the concentration in μg g^{-1} expressed as negative powers of 10. It is illustrated by Fig. 3. It was also shown that the within-laboratory component is usually between half and two-thirds of the precision as defined above.

These results probably reflect the best that can be expected of a method or a laboratory and in most practical instances it is probably of little avail to try and develop methods with higher precision.

2. Mathematical

It is beyond the scope of this book to give a detailed review of statistical principles which can be found in many statistical handbooks. Nevertheless, a few basic concepts and equations should be given.

2.1 Discrete and continuous random variables

A variable is called random when a certain probability is attached to each value. If only a finite number of values can be taken, the variable is known as discrete. It is also regarded as discrete if the number of points is infinite but can be arranged in a countable sequence. If this is not the case, the variable is called continuous. Examples of continuous variables are temperature, length, concentration; on the other hand, the number of peaks in a chromatogram is a discrete variable.

2.2 Probability functions. Cumulative frequency distribution functions

2.2.1 Continuous variables

When a large number of measurements must be represented, it is useful to group the data into classes and to count the number of measurements belonging to each class. This can be represented graphically by a histogram in which the measurements are represented by rectangles, the heights being proportional to the (relative) frequencies and the widths representing the class width (Fig. 4). The total area under the curve equals 1. Suppose that samples of larger and larger size are taken and measured to finer and finer intervals. One would obtain, in the limit, a smooth curve as indicated in Fig. 4. This is called a continuous probability density function for which

$$\int_{-\infty}^{\infty} \mathrm{f}(x)\,\mathrm{d}x = 1$$

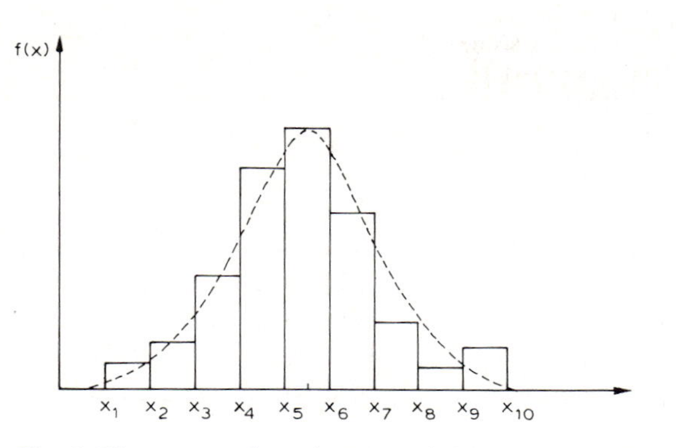

Fig. 4. Histogram and continuous probability (density) function.

2.2.2 Discrete variables

The probability function of x can be represented graphically by drawing a vertical line on the x axis, the height of which corresponds to the probability at each value of x (see Fig. 5 as an example). The sum over all terms $P(x)$ equals 1.

The cumulative distribution function is defined as

$$P(X \leqslant x_n) = \sum_{i=0}^{n} P(x_i) = P(x_0) + P(x_1) + \ldots + P(x_n)$$

2.3 Mean and variance in samples and populations

The mean and variance in samples and population can be calculated according to the formulas given in Table 1.

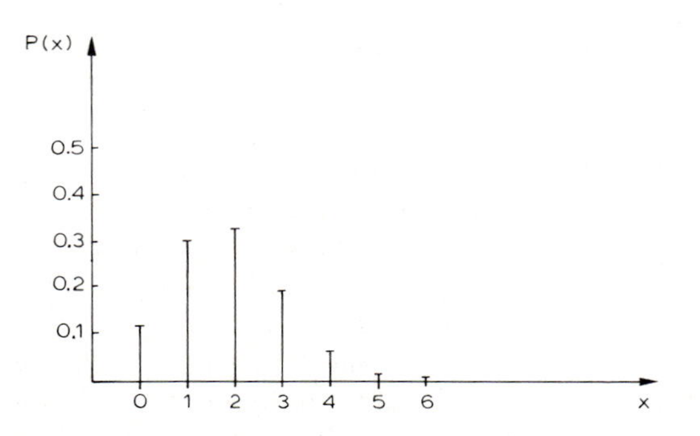

Fig. 5. Probability function for a discrete variable. $P(x_0) = 0.1176$; $P(x_1) = 0.3025$; $P(x_2) = 0.3241$; $P(x_3) = 0.1852$; $P(x_4) = 0.0595$; $P(x_5) = 0.0102$; $P(x_6) = 0.0007$.

$$\sum_{i=0}^{n} P(x_i) = 1$$

References p. 31

TABLE 1

Mean and variance in samples and population

	Mean	Variance
Sample of size n n measurements	$\bar{x} = \sum_{i=1}^{n} \frac{x_i}{n}$	$s^2 = \frac{\sum_{i=1}^{n} (x_i - \bar{x})^2}{n-1}$
Population Discrete variable	$\mu = \sum_{i=0}^{\infty} x_i P(x_i)$	$\sigma^2 = \sum_{i=0}^{\infty} (x_i - \mu)^2 P(x_i)$
Continuous variable	$\mu = \int_{-\infty}^{\infty} x \, f(x) \, dx$	$\sigma^2 = \int_{-\infty}^{\infty} (x - \mu)^2 \, f(x) \, dx$

2.4 Mean and variance after transformation of the variable x

Sometimes, the variable x with known mean and variance is transformed into the new variable y. The mean and variance of y can easily be calculated as shown in Table 2.

The last transformation of Table 2 is called "scaling" and will be discussed in Sect. 2.6.2.

Important transformations are the log and the square root transformations. They are used, for example, in the modelling of calibration lines (see the example in Chap. 5) and in pattern recognition (see Chap. 23).

2.5 Mean and variance of a linear function of random variables

Situations often arise where a new variable, y, is obtained which is a linear function of other random variables, x_i.

$$y = a_1 x_1 + a_2 x_2 + \ldots + a_n x_n$$

TABLE 2

Mean and variance after transformation of x

Transformation	Mean	Variance
$y = x + b$	$\bar{y} = \bar{x} + b$	$s_y^2 = s_x^2$
$y = ax$	$\bar{y} = a\bar{x}$	$s_y^2 = a^2 s_x^2$
$y = x - \bar{x}$	$\bar{y} = 0$	$s_y^2 = s_x^2$
$y = (x - \bar{x})/s_x$	$\bar{y} = 0$	$s_y^2 = 1$

The mean and variance of y can be calculated from

$$\mu_y = \sum_{i=1}^{n} a_i \mu_{x_i} \tag{1}$$

$$\sigma_y^2 = \sum_{i=1}^{n} a_i^2 \sigma_{x_i}^2 + 2 \sum_{i<j}^{n} a_i a_j \sigma_{x_i} \sigma_{x_j} \rho_{ij} \tag{2}$$

where ρ_{ij} is the correlation coefficient between x_i and x_j (see also Chap. 14).

When the variables are independent, the correlation is zero and eqn. (2) reduces to

$$\sigma_y^2 = \sum_{i=1}^{n} a_i^2 \sigma_{x_i}^2$$

A simple example of an application in medicine can be given. The mean blood pressure (y) is defined as the sum of 1/3 of the systolic blood pressure (x_1) and 2/3 of the diastolic blood pressure (x_2).

$$y = \tfrac{1}{3}x_1 + \tfrac{2}{3}x_2$$

For men aged 40–60 years, $\bar{x}_1$ is found to be 149 with standard deviation $s_1 = 11.4$; $\bar{x}_2 = 92$ with $s_2 = 9.9$. The correlation coefficient between x_1 and x_2 is 0.65.

Applying eqns. (1) and (2), one calculates

$$\bar{y} = \tfrac{1}{3} \times 149 + \tfrac{2}{3} \times 92 = 111$$

and

$$s_y^2 = \left(\tfrac{1}{3}\right)^2 \times 11.4^2 + \left(\tfrac{2}{3}\right)^2 \times 9.9^2 + 2 \times \left(\tfrac{1}{3}\right) \times \left(\tfrac{2}{3}\right) \times 11.4 \times 9.9 \times 0.65 = 90.6$$

$$s_y = 9.52$$

An analytical example is given in Chap. 5, Sect. 6 on calibration.

2.6 The normal distribution

2.6.1 Statistical description

The probability function of a normal distribution is given by

$$f(x) = \frac{1}{\sigma\sqrt{2\pi}} \exp\left\{-\frac{1}{2}\left(\frac{x-\mu}{\sigma}\right)^2\right\}$$

where μ and σ are the mean value and standard deviation, respectively, of this probability function. The cumulative frequency distribution function of the normal distribution is given by

$$F(x) = \int_{-\infty}^{x} \frac{1}{\sigma\sqrt{2\pi}} \exp\left\{-\frac{1}{2}\left(\frac{x-\mu}{\sigma}\right)^2\right\} dx$$

References p. 31

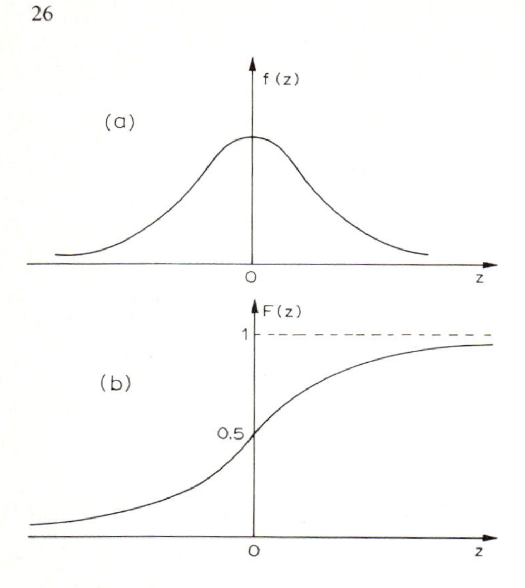

Fig. 6. The standardized normal probability function (a) and the resulting cumulative frequency distribution function (b).

When a variable x has a normal distribution with a mean value μ and variance σ^2, this is written as

$$x \sim N(\mu, \sigma^2)$$

2.6.2 The standardized normal distribution

An important particular case arises when the mean value is zero and the variance is unity. Such a variable is called a standardized or reduced or unit normal variable (or variate)

$$z \sim N(0, 1)$$

The z score is known as a "standardized" variable because its units are standard deviations. The probability function is given by

$$f(z) = \frac{1}{\sqrt{2\pi}} e^{-z^2/2}$$

and its cumulative frequency distribution is given by

$$F(z) = \int_{-\infty}^{x} \frac{1}{\sqrt{2\pi}} e^{-z^2/2} dx$$

The functions are illustrated in Fig. 6.

It can be shown that if a variable x has an $N(\mu, \sigma^2)$ distribution, the variable $z = (x - \mu)/\sigma$ has an $N(0, 1)$ distribution; z is called the reduced variable of x.

The transformation $(x - \mu)/\sigma$ is called scaling, autoscaling or z transformation (see also Sect. 2.4). Values of the distribution function of z are given in most

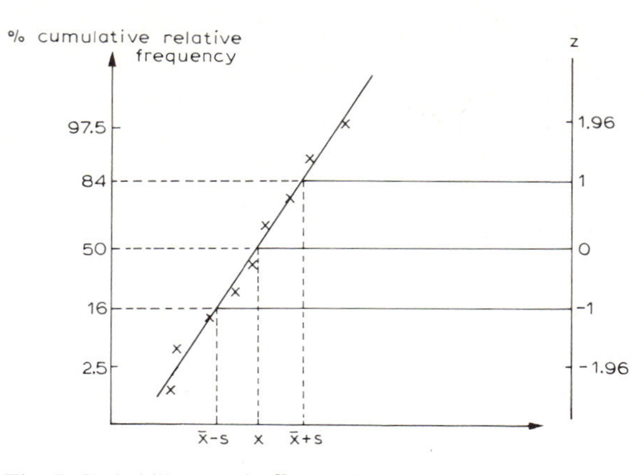

Fig. 7. Probability graph. % cumulative relative frequencies plotted against the upper class boundaries.

statistical handbooks (see also Appendix, Table 1). Notice that a two-sided table is given (for a definition, see Chap. 3). Tables of the cumulative distribution function of z can be found in statistical handbooks.

Scaling is a very important operation: any normal distribution, with a certain mean μ and variance σ^2 can be reduced to the same normal distribution with mean 0 and variance 1. The effect of scale, measurement units is thus removed.

In order to check whether a frequency distribution can be approximated by a normal distribution, one can use probability graph paper. The given frequency distribution is then converted into a cumulative frequency distribution. The cumulative relative frequencies are plotted against the upper class boundaries on probability graph paper. If a straight line is obtained, one can state that a normal distribution closely fits the data (see Fig. 7). This of course is not a test; tests are given in Chap. 3.

2.6.3 The central limit theorem

An extremely important theorem in mathematical statistics is the *central limit theorem*. Consider the sum

$$y = x_1 + x_2 + \ldots + x_n$$

of n independent variables x_i with mean μ_i and variance σ_i^2 $(i = 1, 2, \ldots, n)$. The central limit theorem states that, for large values of n, the distribution of y is approximately normal with mean $\Sigma\mu_i$ and variance $\Sigma\sigma_i^2$ whatever the distributions of the independent variables x_i might be.

This result is also important to analytical chemists because it explains why distributions of errors often tend to be approximately normal. Indeed, in analytical chemistry, the overall error is often a function of many component errors and can be approximated as a linear function of independently distributed component errors. An important condition, of course, is that the component errors have the same order of magnitude and that no single source of error dominates all the others. As will be seen in Chap. 5, this condition is not always fulfilled.

An important application of the central limit theorem is the distribution of the mean; in this context, the theorem can be stated as follows. If all possible random samples, each of size n, are taken from any population with mean μ and standard deviation σ, the distribution of the sample means will have a mean $\mu_{\bar{x}} = \mu$, a variance $\sigma_{\bar{x}}^2 = \sigma^2/n$ and will be normally distributed when the parent population is normally distributed or will be approximately normally distributed for large samples ($n \geqslant 30$) regardless of the shape of the parent population.

2.7 Other distributions

2.7.1 The binomial distribution

If a population consists of items belonging to two mutually exclusive categories, it is said to be a binomial population. Consider a series of independent trials and suppose that the outcome of each trial is either A, with probability $P(A) = \pi$, or $\bar{A}$ (not A) with probability $P(\bar{A}) = 1 - \pi$. If n trials are carried out, one can calculate the probability of exactly i times the event A

$$P(i) = C_n^i \pi^i (1 - \pi)^{n-i}$$

with

$$C_n^i = \frac{n!}{i!(n-i)!}$$

A simple example from quality control: the probability, estimated by the relative frequency, of finding a defective sample in a batch is 0.02. One can use the binomial distribution to calculate the chance of finding 1 defective piece out of a random sample of 5, viz.

$$P(1) = C_5^1 (0.02)^1 (0.98)^4 = 0.0922$$

In Chap. 26 on decision making, a more detailed example of this kind is given.

According to Sect. 2.3, the mean and variance of the binomial distribution can be calculated as

$$\mu = n\pi$$

$$\sigma^2 = n\pi(1 - \pi)$$

It can be shown that if $n\pi$ and $n(1 - \pi)$ become larger than 5, the binomial distribution tends to normality.

2.7.2 The Poisson distribution

The Poisson distribution describes a discrete variable related to discrete events in a continuous interval (e.g. of time). One approach is to consider a limiting case of the binomial distribution in which (a) n tends to infinity; (b) π tends to zero; (c) $n\pi$

tends to a finite number λ. The probability of obtaining a result i, $P(i)$, can then be calculated as

$$P(i) = \frac{e^{-\lambda}\lambda^i}{i!}$$

According to Sect. 2.3, the mean and variance can be calculated as

$$\mu = \sigma^2 = \lambda$$

In situations where the outcome of an experiment is a count (e.g. radio immuno assay, RIA, or neutron activation analysis), a Poisson distribution is obtained and the standard deviation of the response is calculated as the square root of the mean value. An application of the Poisson distribution is given in Sect. 1 of Chap. 25.

For a value of λ larger than 10, the Poisson distribution tends to normality. The relationship between the normal, binomial and Poisson distributions can be summarized as follows

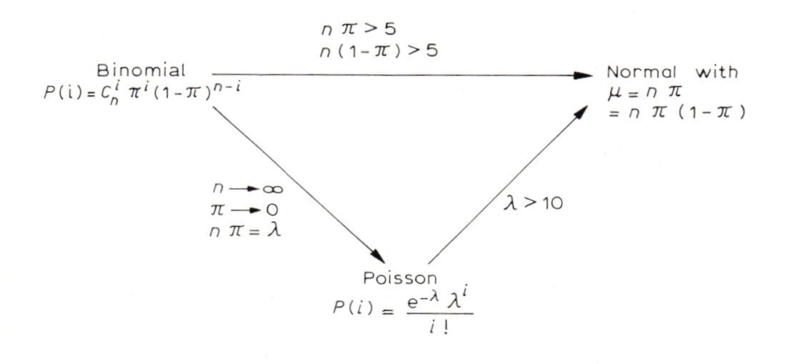

2.7.3 The chi-square distribution

If $z_1, z_2, \ldots, z_k$ are independent unit normal variables, then the variable

$$\chi_k^2 = z_1^2 + z_2^2 + \ldots + z_k^2$$

is said to have a chi-square distribution with k degrees of freedom. Values for the chi-square distribution are given in statistical tables and are depicted in Fig. 8.

The chi-square distribution is used in the well-known chi-square test to find out whether an observed distribution is drawn from a population with a particular theoretical distribution. An example will be given in Chap. 3.

2.7.4 The t-distribution

Let us consider an $N(0, 1)$ normal variable z and a χ_k^2 variable independent of z. The variable t, given by

$$t_k = \frac{z}{\sqrt{\chi_k^2/k}}$$

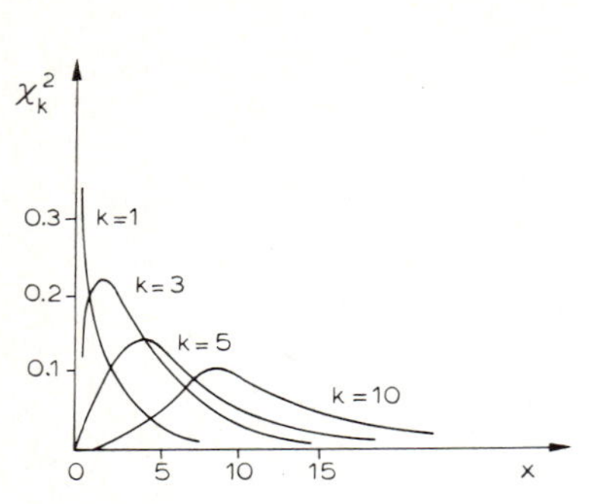

Fig. 8. Chi-square probability distribution function.

is said to have a t-distribution with k degrees of freedom. It is also called a Student distribution (see Fig. 9). Critical values of Student distributions are given in Table 2 of the Appendix. Notice that a two-sided table is given (for a definition, see Chap. 3). One-sided tables and tables of the cumulative distribution function of t can also be found in statistical handbooks. Care should be taken not to confuse these tables.

As $k \to \infty$, the t-distribution tends to normality. Many statistics used in Chap. 3 to test the equivalence of two analytical methods have a t-distribution.

2.7.5 The F-distribution

The ratio

$$F_{k,m} = \frac{s_k^2}{s_m^2}$$

is said to have an F- or Fisher–Snedecor distribution. The values of this distribution are tabulated in Table 3 of the Appendix. The F-distribution, which is illustrated in

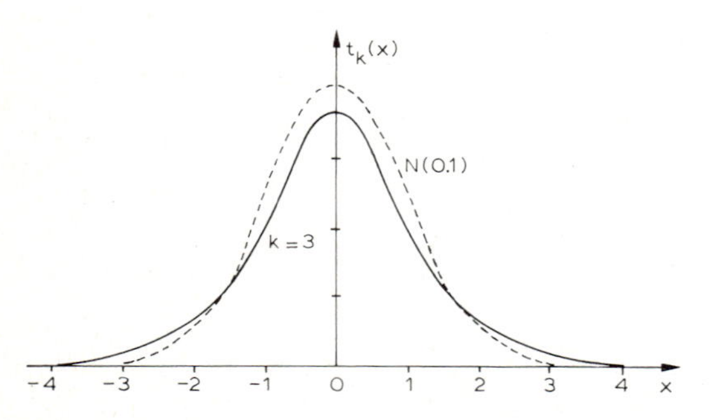

Fig. 9. The t-distribution for $k = 3$ compared with the $N(0, 1)$ distribution.

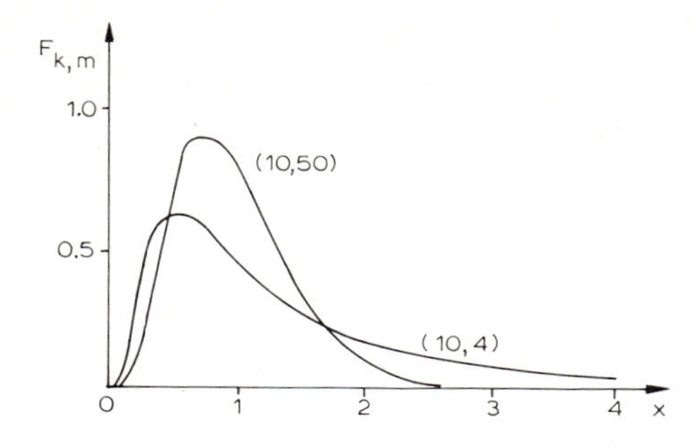

Fig. 10. The *F*-distribution.

Fig. 10, will be used in Chap. 3 where the precisions of two methods are compared; it is also used as the test statistic in the analysis of variance technique (Chap. 4).

References

1 Guide for use of terms in reporting data, Anal. Chem., 54 (1982) 157.
2 F.H.B. Vermeulen, Neth. J. Agric. Sci., 5 (1957) 221.
3 International Organization for Standardization, ISO/TC, 69 (1966).
4 W.J. Youden and E.H. Steiner, Statistical Manual of the Association of Official Analytical Chemists, Washington, DC, 1975.
5 IUPAC, Compendium of Analytical Nomenclature, Pergamon Press, Oxford, 1978.
6 G. Wernimont, Anal. Chem., 23 (1951) 1572.
7 W. Horwitz, L.R. Kamps and K.W. Boyer, J. Assoc. Off. Anal. Chem., 63 (1980) 1344.

Recommended reading

J.O. Westgard, Precision and accuracy: concepts and assessment by method evaluation testing, CRC Crit. Rev. Clin. Lab. Sci., April (1981) 283.

W. Horwitz, Evaluation of analytical methods used for regulation of foods and drugs, Anal. Chem., 54 (1982) 67A.

G.T. Wernimont, in W. Spendley (Ed.), Use of Statistics to Develop and Evaluate Analytical Methods, The Association of Official Analytical Chemists, Arlington, VA, 1985.

E.D. Schall, Collaborative procedures of the Association of Official Analytical Chemists, Anal. Chem., 50 (1978) 337A.

J. Mandel and T.W. Lashof, Interpretation and generalization of Youden's two-sample diagram, J. Qual. Technol., 6 (1974) 22.

R.S. Wayne, Collaborative study of gas–liquid chromatographic determination of malathion in formulations and technical materials, J. Assoc. Off. Anal. Chem., 62 (1979) 292.

A.J. Sheppard, A.E. Waltking, H. Zmachinski and S.T. Jones, Two lipoxidase methods for measuring *cis*, *cis*-methylene interrupted polyunsaturated fatty acids in fats and oils: collaborative study, J. Assoc. Off. Anal. Chem., 61 (1978) 1419.

J. Buttner, R. Borth, J.H. Boutwel and P.M.G. Broughton, International federation of clinical chemistry provisional recommendation of quality control in clinical chemistry. I. General principles and terminology, Clin. Chim. Acta, 63 (1975) F25.

J.O. Westgard, R.N. Carey and S. Wold, Criteria for judging precision and accuracy in method development and evaluation, Clin. Chem., 20 (1974) 825.

W. Horwitz, Evaluation of analytical methods used for regulation of foods and drugs, Anal. Chem., 54 (1982) 67A.

W. Horwitz, Evaluation of analytical methods used for regulation, J. Assoc. Off. Anal. Chem., 65 (1982) 525.

Chapter 3

Evaluation of Precision and Accuracy. Comparison of Two Procedures

1. Introduction

One way of selecting an analytical method is to compare several methods according to their performance characteristics. Often, one procedure is already being used and one can consider replacing it by a cheaper or faster procedure or, in general, by a procedure with more desirable characteristics. A prerequisite for doing this is that the new method should be sufficiently accurate, i.e. sufficiently free from method bias, and this aspect is the main concern of this section.

The simplest means of determining the accuracy of a method is to analyse a standard or reference material for which the concentration of the analyte is known with high accuracy and precision. The difference between the known true value and the mean of replicate determinations with the "test" method is due to the sum of method bias and random errors. It is therefore necessary to estimate the proportion of each type of error and the strategy used for this purpose is to investigate first whether the deviation can be explained by random errors alone. This is done with a t-test and, when this shows that the mean of the determinations does not significantly differ from the true value, the deviation between the given and the experimental values is considered to be due to random errors and the method is considered accurate. If not, the deviation is considered to be a measure of the bias. Often, it is simply stated that the deviation is equal to the bias, whereas it is, of course, only an estimate of the method bias.

This procedure of investigating accuracy has the disadvantage that the result is valid only for the particular reference material used. Often, no standard material of the particular type needed is available. In this case, one often compares the method being investigated or "test method" with an existing method, called the "reference method", for which it is usually assumed that there is no method or laboratory bias. In view of our discussion of the components of inter-laboratory precision in the previous chapter, this is a hazardous assumption and, for this reason, the final evaluation of a method should preferably be carried out as an inter-laboratory study.

When one assumes that the reference method is accurate, both the reference and test methods are used to carry out a number of determinations. Sometimes, one analyses replicates from the same sample, but in this instance one will learn only whether the method is accurate for the particular material being analysed. It is therefore preferable to analyse a range of samples with both methods. The results obtained can be used in several ways.

(1) Ideally, the results obtained with both methods should be completely correlated, i.e. the correlation coefficient, r, (see Chap. 14) should be equal to unity. The

References p. 57

correlation coefficient, however, cannot be interpreted directly in terms of accuracy. For example, does $r = 0.95$ mean that the method should be considered accurate or not? Therefore, the value of the correlation coefficient will serve only as a preliminary indication and it will not be discussed further.

(2) Tests can be applied to investigate whether the differences obtained are significant or not. According to whether one assumes a normal distribution of errors or does not make any assumptions, a t-test (Sect. 3) or a non-parametric test (Sect. 4) will be carried out.

(3) If one plots the results from one method against those from the other, the regression line should ideally pass through the origin and have a slope of unity. The intercept on the ordinate is therefore a measure of method bias (at least when the measurements are made over a sufficiently large concentration range), while the deviation of the slope from 1 is a measure of proportional error. The application of regression analysis to method comparison is discussed in Sect. 5. In Sect. 6, the application of this technique to recovery experiments (standard addition techniques), used to detect proportional errors, is also considered.

(4) The variances for the replicate analysis of one sample by two methods can be compared using the F-test (Sect. 7). This is a means of comparing the precision of the respective methods.

When comparing more than 2 methods, techniques such as the analysis of variance can be used (see Chap. 4).

2. Statistics

One of the most important aspects of applied science is the experimental examination of hypotheses. The rationalization of this examination requires an objective technique for accepting or rejecting a hypothesis. Such a technique must be based on quantification of the available information; it must take into account the risk a scientist is willing to take of making a wrong decision. This risk or uncertainty is due to the fact that the entire population is not measured; measurements are only carried out on a sample taken from the population. The difference between characteristics of the sample and those of the population can lead to erroneous conclusions. The following procedure, which is a model for statistical decision making, will be used throughout this book. The procedure consists of several steps, which are considered in the following sub-sections.

2.1 The elaboration of a test

The elaboration of a test consists of a number of consecutive steps.

(1) Clearly formulate the question to be answered.

(2) Select the appropriate test. When several tests are available, the conditions for using each of them must be examined.

(3) Decide which level of significance will be given to the selected test. This level, denoted by α, is defined as the probability of rejecting the null hypothesis when it is true. α is usually given an a priori value of 0.01 or 0.05 (see Sect. 2.2).

(4) State the hypotheses. Two types of hypothesis are encountered in statistics. The null hypothesis, H_0, is always a hypothesis of no difference and is the negation of an effect or a difference which has been measured by the scientist. In other words, the observed effect or difference is said to be due to chance. The existence of this effect or difference is called the research or alternative hypothesis and is denoted by H_1.

(5) The calculation of the test statistic (e.g. a *t*-, *z*- or *F*-value).

(6) The comparison of the calculated test statistic with the theoretical tabulated value at the chosen significance level, taking into account the number of degrees of freedom.

(7) Make the decision. This depends on the test chosen. For instance, for the *t*-tests (see Sect. 3), one accepts the null hypothesis if the calculated value is smaller than the tabulated one and rejects H_0 (and accepts H_1) if the calculated value exceeds the tabulated value. The probability of having rejected H_0 when, in fact, it was true is called the *P*-value (i.e. the value of α after having performed the test). For the Mann–Whitney test (see Sect. 4.1), one accepts H_0 when the calculated value is larger than the tabulated value.

2.2 Types of error

α is the risk which is given as tolerable before the test is performed and is defined as the probability of rejecting H_0 when, in fact, it is true. This error is called the error of the first type or type 1 error.

As said before, α is usually given an a priori value of 0.01 or 0.05. However, circumstances may arise where very different α values are to be preferred. For instance, a hypothesis may have such excessive consequences, that one is unwilling to reject it, even when there is only a 0.05 probability that it is true. The choice of α also depends on the β risk one is willing to take. In addition to the α error, it is also possible that the null hypothesis will be accepted when, in fact, it is false. This error is called the error of the second type and its probability is denoted by β. The importance of β can be illustrated in the following case. A reference and a test method are compared. One takes the risk α, e.g. 0.05, to falsely tell that both methods yield different results and the risk β, e.g. 0.1, to falsely tell that both methods yield no different results. Making the β error is of relevance when the difference between the two methods is of sufficient magnitude to yield systematically wrong results when using the test method. Unlike the α error, the β error is not a single value, but a series of values depending on the α level, sample size and, in the case of a comparison between two means, of the differences between μ_1 and μ_2. The relationship between both kinds of error is also illustrated in Fig. 1 where distributions of the mean are shown. One notes that decreasing α increases β [Fig. 1(a) and (b)]. If an observation x falls at the point indicated in Fig. 1(b), it lies in the region of acceptance of H_0. However, x could also be sampled from the alternative distribution and, in this case, an error of size β is committed when accepting H_0. The usual procedure is to choose a small value of α because α errors

References p. 57

are usually quite serious. However, β will often be larger than one would consider to be a satisfactorily small value, e.g. 0.1. In this case, conclusions have to be formulated carefully. While one accepts H_0, one does not rule out the possibility that H_1 is true. The customary procedure is to accept H_0 until additional data force one to a different conclusion. The only way to reduce both α and β is to increase the sample size. Indeed, as the standard deviation of the mean is $\sigma/\sqrt{n}$ (see Chap. 2, Sect. 6.3), an increase in n will reduce the spread of μ, concentrating its distribution more closely around its central value and thus reducing both α and β [see Fig. 1(c) and (d)].

By giving a value to α and β, one can compute the minimum sample size to discover a certain difference. For the comparison of two sample means, the number of observations necessary in each group is given by

$$n = \frac{2(z_\alpha + z_\beta)^2}{D^2} \tag{1}$$

where z_α and z_β are the unit normal deviates for given values of α and β (see

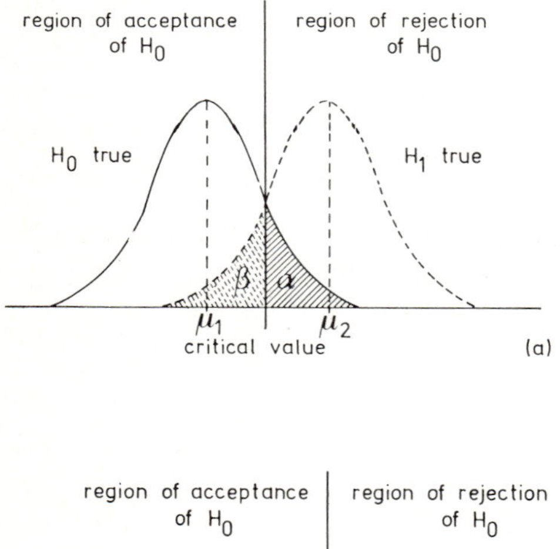

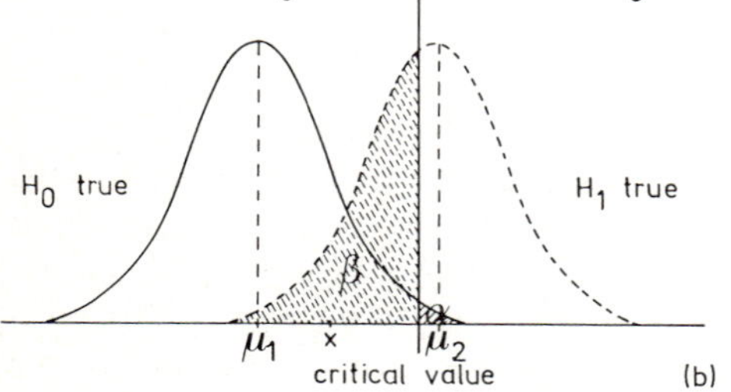

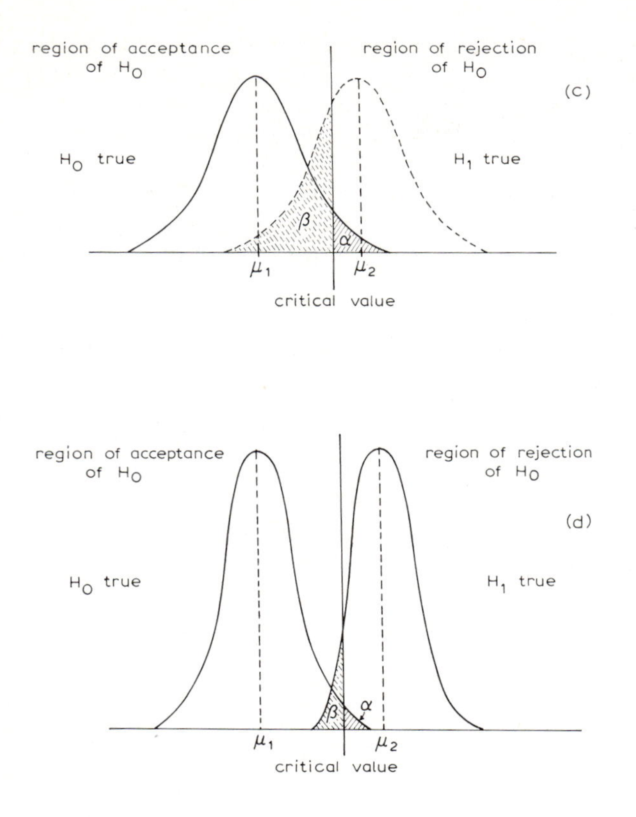

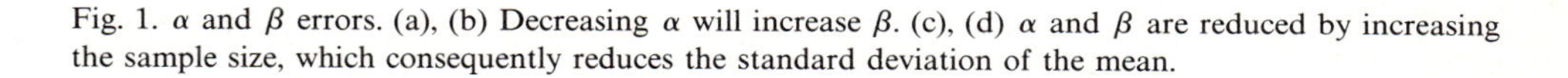
Fig. 1. α and β errors. (a), (b) Decreasing α will increase β. (c), (d) α and β are reduced by increasing the sample size, which consequently reduces the standard deviation of the mean.

Appendix, Table 1).

$$D = \frac{\delta}{\sqrt{(\sigma_1^2/2) + (\sigma_2^2/2)}}$$

or, if $\sigma_1 = \sigma_2 = \sigma$

$$D = \frac{\delta}{\sigma}$$

where δ is the difference between the means which it is considered important to detect. The use of this formula can be illustrated by the following example. A test method and a reference method for the determination of Pb in blood are compared. A difference of the mean concentration of 10 ppb is considered important. The precision of both methods is supposed to be the same; σ is estimated to be 20 ppb.

References p. 57

TABLE 1

The four possible outcomes in a hypothesis test

Decision	True situation in the population	
	H_0 true; H_1 false	H_0 false; H_1 true
Accept H_0	No error; probability $= 1 - \alpha$	Type 2 error; propability $= \beta$
Reject H_0	Type 1 error; probability $= \alpha$	No error; probability $= 1 - \beta =$ power of the test

What size of sample is required to detect an increase in the mean of 10 ppb with the test method, setting $\alpha = 0.05$ and $\beta = 0.1$? By applying eqn. (1), one calculates $n = 69$. As a one-sided test is used, notice that $z_\alpha = 1.645$ (see Sect. 2.3). As β is always taken one-sided, $z_\beta = 1.282$. Tables to obtain sample sizes as a function of α, β and the difference of μ_1 and μ_2 are available [1].

As a conclusion, one can say that the analyst performing an experiment should balance α and β errors and consider the "cost" of each kind of error. He should also decide whether or not the additional work and cost occasioned by increasing the sample size are worthwhile.

The definitions of type 1 and type 2 errors may be summarized in Table 1.

2.3 One- and two-sided tests

The formulation of the research hypothesis can lead to a one- or a two-sided test. Suppose the means of two independent samples are to be compared. H_0 is then $\mu_1 = \mu_2$ and, if the question to be answered is "are the two means different from each other?", H_1 is $\mu_1 \neq \mu_2$. In this case, a two-sided test will be performed. If the question is "does sample 1 yield a smaller mean than sample 2?", H_1 will be $\mu_1 < \mu_2$ and a one-sided test will be performed.

When a two-sided statistical table is used and the test is two-sided, the calculated value of the test statistic will be compared with the tabulated value at a significance level α and, in the case of a one-sided test, to the tabulated value at a significance

TABLE 2

The formulation of H_0 in one- and two-sided tests

Sign in the alternative hypothesis	$<$	$\neq$	$>$
Region of rejection of H_0 (critical region)	One region left side	Two regions one on each side	One region right side
Test performed	One-sided	Two-sided	One-sided

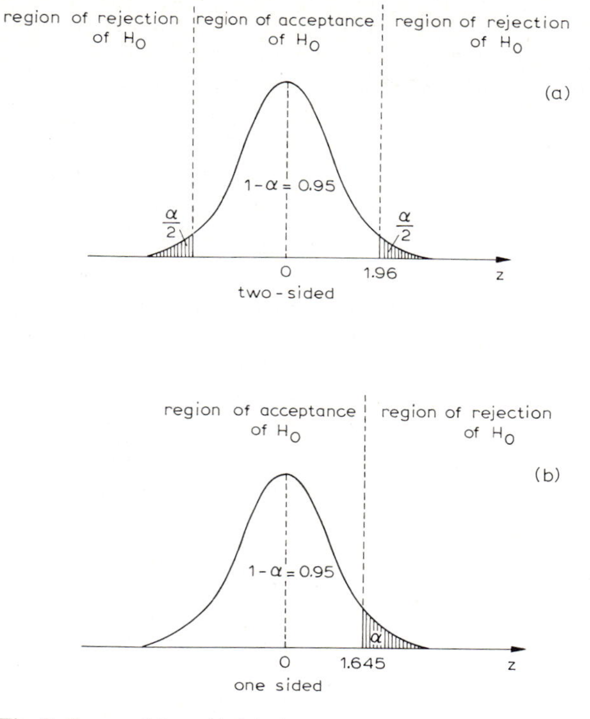

Fig. 2. One- and two-sided tests.

level of 2α. The difference between the two kinds of test is illustrated by the following example. For the comparison of two means in the case of large samples, the test statistic z is calculated (see Sect. 3.2.1). z is a unit normal variable, the distribution of which is given in Fig. 2. The difference between one- and two-sided tests is summarized in Table 2.

In the case of a two-sided test, the test statistic will be compared with 1.96, the value of z for $\alpha = 0.05$ which can be found in a two-sided table of the unit normal distribution. For a one-sided test carried out at the same significance level of 5%, however, the test statistic is compared with 1.645, i.e. the value of z for 2α when the same two-sided table is used.

2.4 Statistical scales

When selecting a test for solving a statistical problem, various factors must be taken into account, such as the nature of the population being studied, the way the sample was or will be drawn, and the type of test to be used. The way in which the measurements are made forms the basis of the mathematical operations necessary for carrying out a test. The types of measurement used are called statistical or measurement scales and four types can be distinguished, viz. the nominal, ordinal, interval and ratio scales. These scales, together with the mathematical operations associated with them, are discussed below.

2.4.1 The nominal scale

The nominal scale, which is mathematically the weakest scale, is used when the only information known about the elements of a sample is its classification into groups. The symbols used for describing the groups or the names of the classes form the nominal scale. For example, when studying the results of a determination of glucose in blood, it is possible to classify the results into two groups: the values outside the normal range (abnormal values) and those within the normal range (normal values). This classification constitutes a nominal scale.

2.4.2 The ordinal scale

It may be possible, in addition to the classification of the elements of a sample into classes, to compare the different classes and to define an order for them. If this order is complete, i.e. if each pair of classes can be compared, the scale of classification is called an ordinal scale.

If one again considers a series of glucose results, one can make a classification according to whether the results are below the normal range (low values), within the normal range (normal values) or above the normal range (high values) and an ordinal scale is defined.

A classification does not imply a distance between the classes but only a sequence according to which "low values" are situated below "normal values" and normal values below "high values".

Conventionally, statistical tests using parameters such as the arithmetic mean or standard deviation may not be valid for data in an ordinal scale since the distances between groups have no real meaning. Most statistical tests used in an ordinal scale are of the non-parametric type. Some of these tests will be described further in this chapter.

2.4.3 The interval scale

An interval scale has the same properties as the ordinal scale but, in addition, the distance between any two points of the scale can be measured. In an interval scale, it is necessary to choose a zero point and a unit of measurement. An example is the Celsius temperature scale, which originally referred all temperatures to the melting point of ice. Each temperature measurement is then located a number of degrees above or below this level.

2.4.4 The ratio scale

A ratio scale has the same properties as the interval scale except that the zero point cannot be chosen but is defined by nature. Examples of ratio-scaled variables are absolute temperature, weight, the concentration of glucose in the blood.

This scale is the strongest statistical scale available and all tests can be carried out under it. In practice, all statistical tests can be applied on both the interval and ratio scales. The differentiation between the two scales has no consequences for the present book.

3. Evaluation of method bias using tests on the mean (t-test)

When one analyses a standard or reference sample [e.g. those proposed by organizations such as The American Society of Testing and Materials (ASTM), The National Bureau of Standards (NBS), The International Atomic Energy Agency (IAEA), and Le Bureau Communautaire de Reference (BCR)] by a new method, one will have to decide whether the result obtained differs significantly from the stated concentration. This concentration is the result of a large number of careful determinations by the organization issuing the sample, while the result obtained with the new ("test") method is the mean of a number of replicate determinations. Statistically, one therefore compares the means of two populations. In practice, the only population parameter given for the reference material is the mean. Often, no standard deviation is given.

Reference samples of this type have real value only when they have been certified with sufficient care. An example of how this should be done is the certification procedure used by the NBS, which uses three principal modes of measurement for reference samples: measurement with a method of known accuracy by at least two analysts working independently; measurement with at least two independent methods, the estimated accuracies of which are good compared with the accuracy required for certification; and measurement according to a collaborative scheme, incorporating qualified laboratories.

Therefore, although there is an obvious need for standard materials, one should recommend individual laboratories and organizations not to issue their own standard materials except when unavoidable, but to leave it to the few organizations that have long and established experience in this field. The reservations in the preceding paragraphs do not mean that individual laboratories should not test their methods by comparison with a reference method or a standard material with incomplete information. It is better to make a study of the accuracy of a proposed method with the available reference materials, however imperfect the statistical data may be, rather than make no study of the accuracy at all. However, it is necessary to bear in mind the limitations in the conclusions that can be drawn from such comparisons.

After these introductory and cautionary remarks, we can turn to the statistical methodology. One has to decide whether or not there is a significant difference between the stated value which is accepted as the true value, μ_0, and the experimental estimate, $\bar{x}$ of μ obtained with the test method. This includes a definition of an acceptable α level.

References p. 57

3.1 The one-sample case

Let us first investigate the case where μ_0 has been determined with high precision so that the standard deviation can be considered to be approximately zero. We state here that the value given for the reference sample is, by definition, equal to the "true" value.

Statistically, one can state the problem as follows: "can the sample of size n with mean $\bar{x}$ and variance s^2 be considered as coming out of the population with mean μ_0?". If certain conditions are fulfilled in the case of small samples, that question can be answered by performing a Student t-test which tests the difference between μ (estimated by $\bar{x}$) and μ_0.

If several random samples of size n are drawn from the population, $\bar{x}$ will take different values. It can be shown that, for large samples (most statisticians call a sample large when $n \geqslant 30$), $\bar{x}$, which is also a random variable, is normally distributed with mean μ and variance σ^2/n

$$\bar{x} \sim N(\mu, \sigma^2/n)$$

It should be noted that x does not need to be normally distributed in the population (see also Chap. 2, Sect. 6.3).

(a) $n \geqslant 30$. One calculates the statistic z as

$$z_{\text{cal}} = \frac{\bar{x} - \mu_0}{s/\sqrt{n}}$$

If the null hypothesis is true, z has an $N(0, 1)$ distribution; H_0 is accepted if the calculated absolute z value is smaller than 1.96 (two-sided test) or 1.645 (one-sided test) at the 0.05 level of significance.

(b) $n < 30$. x has to be normally distributed in the case of small samples. One calculates

$$t_{\text{cal}} = \frac{\bar{x} - \mu_0}{s/\sqrt{n}}$$

If the null hypothesis is true, t has the Student distribution with $n-1$ degrees of freedom. The calculated t-value is compared with the theoretical value at $n-1$ degrees of freedom and at the chosen significance level α.

Example. Analysis of an international standard material with known Pb concentration ($\mu_0 = 0.340$ μg g^{-1}) to investigate a newly developed method. 15 replicate analyses are performed and the following results obtained

0.380; 0.346; 0.291; 0.278; 0.404; 0.331; 0.409;

0.285; 0.361; 0.268; 0.306; 0.243; 0.316; 0.223; 0.299

Is the mean obtained by the analyst different from μ_0 ($\alpha = 0.05$)? Mean and standard deviation are calculated as follows.

$$\bar{x} = \frac{\Sigma x_i}{n} = 0.316$$

$$s = \sqrt{\frac{\Sigma(x_i - \bar{x})^2}{n-1}} = \sqrt{\frac{\Sigma x_i^2 - n\bar{x}^2}{n-1}} = 0.056$$

$H_0 : \mu = \mu_0$ (μ is estimated by $\bar{x}$).
$H_1 : \mu \neq \mu_0$, as a two-sided test will be performed.

As the sample size is small, the condition of normality of x has to be fulfilled (normal distribution of Pb levels in the standard material). The test statistic t is calculated from

$$t_{cal} = \frac{0.316 - 0.340}{0.056/\sqrt{15}} = -1.660$$

The theoretical t value for a two-tailed test and 14 degrees of freedom is 2.145. As $|t_{cal}|$ is smaller than t_{theor}, the null hypothesis is accepted and it is concluded that, at the chosen significance level and with this sample size, no significant difference between $\bar{x}$ and μ_0 can be shown.

3.2 The two-sample case

The second case to be considered is when the standard deviation on the reference sample is not negligible. Statistically, one asks: "could the two independent samples with means $\bar{x}_1$ and $\bar{x}_2$, variances s_1^2 and s_2^2, and sample sizes n_1 and n_2, come out of a population with mean $\mu = \mu_1 = \mu_2$?". A t-test for the comparison of the means of two independent samples can be performed after having checked certain conditions (see Sect. 3.2.1).

This evaluation procedure only enables one to conclude that the method is accurate (or not) for the analysis of a sample of that particular concentration. In order to obtain more general conclusions, one determination can be carried out with each method on n different samples, which should preferably include a sufficient variety of matrices and a range of concentrations. The question now is whether the differences, d_i, between the results of the two methods are significantly different from zero. If this is not so, the methods will be considered to give the same result. Another consequence is that the differences between d_i and zero are then due to random errors. One could say that $\bar{d}$, the mean value of d_i, is compared with the reference value zero. Formally, one compares the difference between the means of two related (or paired) samples. This can be done by performing a t-test for paired measurements (see Sect. 3.2.2).

Of all the evaluation procedures mentioned, the last one is the most desirable. One should be aware, however, of its limitations. If the requirements for its proper use are not met, the t-test will yield erroneous results. In practice, this will occur in the following situations.

(a) If a systematic error is caused in only one or a few of the samples by an interferent present only in those samples, the random error in the samples can mask the systematic error, or else the systematic error in one sample may lead to such a high t-value that it is concluded that the method is generally inaccurate.

(b) The t-test is valid for a constant systematic error or proportional errors in a

References p. 57

very restricted concentration range but not for proportional errors over a wider range, as the research hypothesis is that the difference between both procedures (populations in statistical terminology) is independent of the concentration. Proportional errors depend on the concentration so that neither the t-test nor the non-parametric tests discussed in Sect. 4 are valid.

In Sect. 1, it was argued that method comparison studies should preferably be carried out on an inter-laboratory basis, so as to take into consideration the effect of laboratory bias on test and reference methods. In some situations, however, laboratory bias may be considered of little importance.

One situation of this nature sometimes occurs in clinical laboratories. Clinical chemists are often less concerned with comparisons of their data with those from other laboratories than with the internal consistency of their data. A clinical laboratory which carries out statistical quality control will determine its own normal values for a particular test, thereby eliminating the effect of laboratory bias, or else adjust the values by the analysis of control sera. Therefore, a concept such as the inter-laboratory precision or laboratory bias is of less importance than it is to official analysts. To be fair, it should be noted that there is a strong trend towards more inter-laboratory quality control (proficiency testing) in the clinical laboratory.

On the other hand, it is vital for the validity of the statistical evaluation of clinical chemistry data to take into account the day-to-day precision, as clinical laboratories carry out the same tests daily for long periods. Barnett and Youden [2] suggest the following procedure. Samples from up to five patients are collected and analysed with both the standard and the investigated method on successive days until a total of 40 samples has been analysed. In this way, the standard deviation used in the t-test will be representative for between-day precision, which is more meaningful than within-day precision, and it will also incorporate the effect of interfering substances (such as drugs) which affect patient values.

3.2.1 Comparison of the means of two independent samples

For the comparison of the means of two independent samples, two cases have again to be considered.

(a) Large samples (n_1 and $n_2 \geqslant 30$).

For large samples and for any distribution of x, one has

$\bar{x}_1 : N(\mu_1, \sigma_1^2/n_1)$

$\bar{x}_2 : N(\mu_2, \sigma_2^2/n_2)$ (see also Chap. 2, Sect. 6.3)

The difference $\bar{x}_1 - \bar{x}_2$ is then also normally distributed with mean $\mu_1 - \mu_2$ and variance

$$\sigma^2 = \frac{\sigma_1^2}{n_1} + \frac{\sigma_2^2}{n_2}$$

For that reason, the statistic

$$z = \frac{\bar{x}_1 - \bar{x}_2}{\sqrt{(s_1^2/n_1) + (s_2^2/n_2)}}$$

is a unit normal variable, i.e. a normally distributed variable with mean 0 and variance 1. The calculated z value is compared with the theoretical value at the chosen significance level. For $\alpha = 0.05$, the theoretical value is 1.96 for a two-tailed test and 1.645 for a one-tailed test.

Example. A reference sample analysed in an inter-comparison study yields for the analysis of compound A a mean $\bar{x}_1 = 32.6$ with variance $s_1^2 = 6.55$ ($n_1 = 36$). The analyst analyses the reference sample with his own method and finds for 30 replicate analyses ($n_2 = 30$) a mean $\bar{x}_2 = 31.6$ with variance $s_2^2 = 4.05$. Can it be shown that the investigated method yields lower results than those obtained in the inter-comparison study ($\alpha = 0.05$)?

$H_0: \mu_1 = \mu_2$

$H_1: \mu_1 > \mu_2$

as a one-sided test will be performed.

In the case of large samples, one calculates

$$z = \frac{32.6 - 31.6}{\sqrt{(6.55/36) + (4.05/30)}} = 1.776$$

As the calculated value is larger than 1.645, the theoretical value of z at the signficiance level $\alpha = 0.05$ for a one-tailed test, H_0 is rejected and one concludes that, indeed, lower results are obtained with the test method. The probability of the type 1 error, i.e. having drawn a wrong conclusion, is $0.035 \leqslant P < 0.04$.

(b) Small samples (n_1 and/or $n_2 < 30$).

For small samples, two conditions have to be fulfilled for performing a t-test.

(1) The variable x is normally distributed in the population. This condition can be checked with appropriate tests if sufficient data are available. In some cases, however, the distribution of x is known to be normal (e.g. the literature or previous experiments) and one can thus consider the assumption of normality to be valid. It is always useful to construct a histogram of the data from which shape some information about the distribution of x can be obtained. Visual inspection of the histogram gives a rough idea about the symmetry of the distribution, about the mean and the standard deviation, about the occurrence of outliers etc. It should, however, be emphasized that, although the histogram is a useful help to the scientist, it does not have the value of a test, so one should always be careful in drawing conclusions.

(2) The variances σ_1^2 and σ_2^2 of the two populations from which the samples are drawn should be equal. This condition is tested with the F-test of Fisher–Snedecor, which consists in calculating the statistic F as the ratio of the sample variances s_1^2 and s_2^2. This ratio is described by the F-distribution (see Chap. 2). By convention, one calculates the F ratio by dividing the largest variance by the smallest. One generally sets a significance level of $\alpha = 0.05$ and has to compare the calculated F ratio with the theoretical value at $(n_1 - 1)$ and $(n_2 - 1)$ degrees of freedom which is tabulated or calculated from the theoretical distribution function.
H_0 is formulated as $\sigma_1^2 = \sigma_2^2$ and H_1 as $\sigma_1^2 \neq \sigma_2^2$. H_0 is accepted if the calculated F-value is smaller than the theoretical value. When both conditions of normality

and homogeneity of variances are fulfilled, a t-test can be performed by calculating the statistic

$$t = \frac{\bar{x}_1 - \bar{x}_2}{\sqrt{s^2[(1/n_1) + (1/n_2)]}}$$

which is distributed as Student's t with $n_1 + n_2 - 2$ degrees of freedom. A pooled variance, s^2, is calculated as

$$s^2 = \frac{(n_1 - 1)s_1^2 + (n_2 - 1)s_2^2}{n_1 + n_2 - 2}$$

The calculated t-value is then compared with the theoretical value at the chosen significance level, α, and at $n_1 + n_2 - 2$ degrees of freedom. If, for small samples, the condition of normality is not fulfilled, one can perform a non-parametric test such as the U-test of Mann and Whitney (see Sect. 4). If only the condition of homogeneity of variances (also called homoscedasticity) is not fulfilled, one can carry out Cochran's test (see BALANCE software).

Example. Consider the same problem as above (large samples), but with two smaller samples available.

$\bar{x}_1 = 32.6 \quad s_1^2 = 6.55 \quad n_1 = 11$

$\bar{x}_2 = 31.6 \quad s_2^2 = 4.05 \quad n_2 = 13$

As the sample sizes are small, two conditions must be met for the validity of the t-test. Let us assume the condition of normality to be fulfilled. The condition of homogeneity of variances can be checked with the F-test (see also Sect. 7).

F-test $\quad H_0: \sigma_1^2 = \sigma_2^2$
($\alpha = 0.05$)

$\quad \mathrm{H}_1: \sigma_1^2 = \sigma_2^2$

$$F = \frac{s_1^2}{s_2^2} = \frac{6.55}{4.05} = 1.62$$

The calculated F-value is compared with the theoretical value at 10 and 12 degrees of freedom, i.e. 3.37. H_0 is thus accepted; no difference in variance can be shown.

As both conditions are fulfilled, one can now perform a t-test with

$\mathrm{H}_0: \mu_1 = \mu_2$

and

$\mathrm{H}_1: \mu_1 > \mu_2$

$$t = \frac{32.6 - 31.6}{\sqrt{5.19(1/11 + 1/13)}} = 1.071$$

$$s^2 = \frac{10 \times 6.55 + 12 \times 4.05}{22} = 5.19$$

The theoretical value for a one-tailed test and 22 degrees of freedom is 1.717. H_0 is then accepted and one concludes that, in this experiment with this sample size and at the chosen significance level, no difference in means can be shown.

3.2.2 Comparison of the means of two related (paired) samples

(a) Large samples (n, the number of pairs ⩾ 30).

In the case of large samples, the quantity

$$z = \frac{\bar{d}}{s_d/\sqrt{n}}$$

is a unit normal variable in which $\bar{d}$ is the mean of the differences d_i between pairs and s_d^2 is the variance of the d_i.

For a significance level of $\alpha = 0.05$, the z-value is compared with 1.96 (two-sided test) or 1.645 (one-sided test) in the usual way.

(b) Small samples.

If the number of pairs, n, is smaller than 30, the condition of normality of x is required or at least the normality of the differences d_i. If this is the case, the quantity

$$t = \frac{\bar{d}}{s_d/\sqrt{n}}$$

has a Student t-distribution with $(n - 1)$ degrees of freedom.

Example. The content of compound A is analysed in 10 samples using the reference method, R, and a test method, T. Do both methods yield different results ($\alpha = 0.05$)?

Sample	x_R	x_T	$d_i = x_R - x_T$
1	114	116	−2
2	49	42	+7
3	100	95	+5
4	20	10	+10
5	90	94	−4
6	106	100	+6
7	100	96	+4
8	95	102	−7
9	160	150	+10
10	110	104	+6
			$\Sigma d_i = +35$

The null hypothesis is that both methods yield the same results or, in other words, that the observed differences between R and T are due to chance. H_0 is formulated as

$H_0: \mu_d = 0$

$H_1: \mu_d \neq 0$

as a two-tailed test is required. The validity of the test requires a normal distribution of the x_i, or at least of the d_i in the population (normal distribution of the levels of compound A).

$\Sigma d_i = 35$

$$\bar{d} = \frac{\Sigma d_i}{n} = 3.50$$

$$s_d = \sqrt{\frac{\Sigma d_i^2 - \left[(\Sigma d_i)^2/n\right]}{n-1}} = 5.85$$

$$t = \frac{\bar{d}}{s_d/\sqrt{n}} = \frac{3.50}{5.85/\sqrt{10}} = 1.89$$

References p. 57

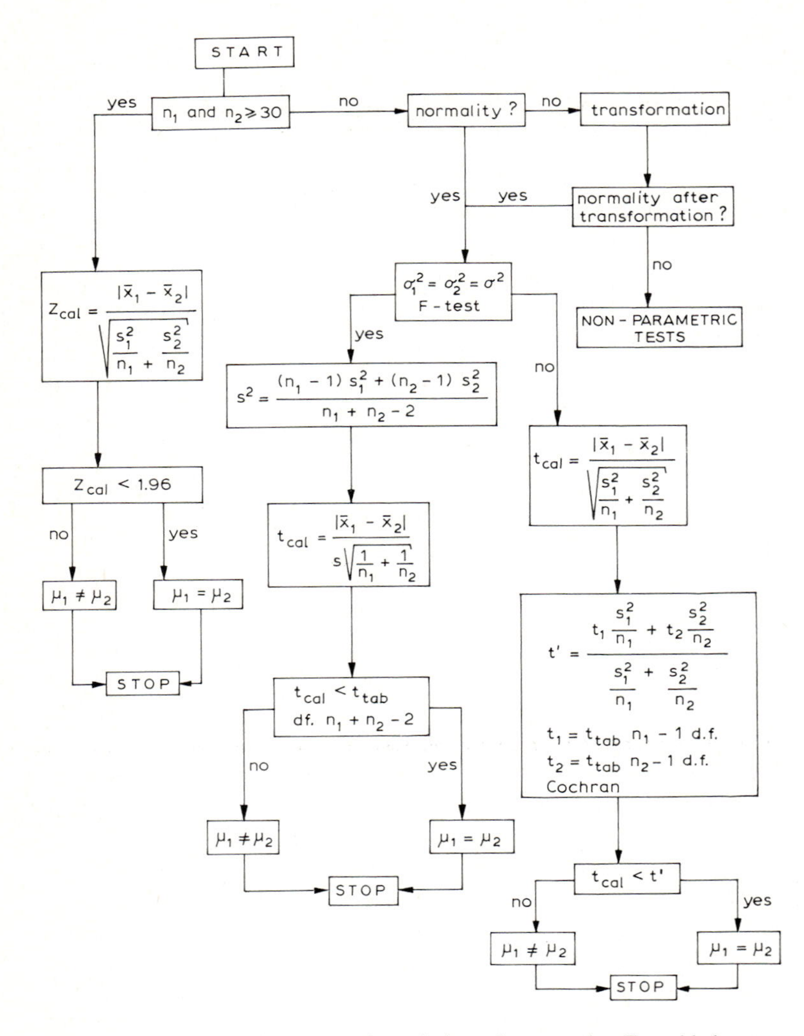

Fig. 3. Comparison of the means of two independent samples. Two-sided tests; $\alpha = 0.05$.

The calculated t-value is compared with the theoretical value at $\alpha = 0.05$ and 9 degrees of freedom, i.e. 2.262. As the calculated value is smaller than the theoretical value, H_0 is accepted. This means that, with this sample size and at the chosen significance level, no difference between the two methods can be shown.

The flow schemes of Figs. 3 and 4 summarize how to select a test correctly.

4. Non-parametric tests for the comparison of methods

In the previous sections, the comparison of methods by the use of the t-test was discussed. When using such tests for small samples ($n < 30$), one implicitly assumes

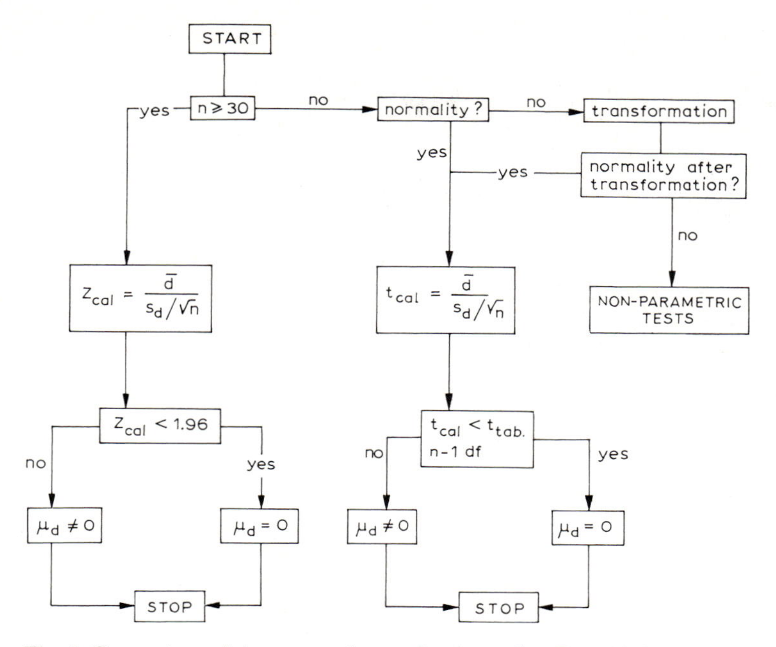

Fig. 4. Comparison of the means of two related samples. Two-sided tests; $\alpha = 0.05$.

that the results for each method are normally distributed, but often this is not so, or at least it cannot be proved conclusively. In fact, according to some studies, it seems that normal distributions are obtained in fewer instances than is generally accepted. Reasons why non-gaussian distributions are obtained are, for example, heterogeneity of samples, rounding off (producing a discontinuous distribution) and measurements near the detection limit (with sub-zero readings set to zero) [3].

The use of tests based on a normal distribution can then lead to erroneous conclusions. A detailed discussion of transformations to obtain normal distributions for use in clinical chemistry was given by Harris and De Mets [4] and by Martin et al. [5], the most common being the log–normal distribution (the logarithm of the data are normally distributed). When no gaussian distribution can be obtained, one can use methods which do not require assumptions about the distribution (so-called distribution-free methods). These methods do not require calculations of the usual parameters, such as the mean or standard deviation, and are therefore also called non-parametric. They have the advantage of always being valid and only require very simple calculations. Therefore, once one knows these methods, one is tempted to use them on every occasion. However, it should be stressed that they are less efficient and require more replicate measurements than the parametric methods for a certain level of α. Non-parametric tests have in common that they do not use the values of the quantitative variables but their ranks and that they are based on counting. For nominal and ordinal scales, only

References p. 57

non-parametric tests are valid (see Sect. 2.4). Parametric tests require quantitative variables on interval or ratio scales.

For those who want to know more about non-parametric testing, we recommend the excellent book by Siegel [6]. We have selected the Mann–Whitney U-test for the comparison of independent samples and the Wilcoxon matched-pairs signed-ranks test, also called the Wilcoxon T-test, for paired measurements. Although both tests are not subject to the suppositions inherent to the parametric t-tests, they are still almost as efficient as the t-tests. When normality of the data is doubtful, one should always resort to these tests in the case of small samples.

4.1 The Mann–Whitney U-test (independent samples)

Example. Two groups of measurements are to be compared; the data are

A: 11.2; 13.7; 14.8; 11.1; 15.0; 16.1; 17.3; 10.9; 10.8; 11.7

B: 10.9; 11.2; 12.1; 12.4; 15.5; 14.6; 13.5; 10.8

One first ranks all the data, taking groups A and B together, and gives rank 1 to the lowest result, rank 2 to the second etc. Calling n_1 and n_2 the number of data in the group with the smallest and largest number of results, respectively, and R_1 and R_2 the sums of ranks in these two groups, then

$$U_1 = n_1 n_2 + \frac{n_1(n_1+1)}{2} - R_1$$

$$U_2 = n_1 n_2 + \frac{n_2(n_2+1)}{2} - R_2$$

The smaller of the two U values is used. When ties are present, the avarage of the ranks is given. In the example given, the lowest result, 10.8, is given the rank 1. As 10.8 is obtained twice, they are both given the rank $(1+2)/2 = 1.5$.

The Mann–Whitney test in fact compares the median of two samples. The smaller the difference between the medians, the smaller the difference between U_1 and U_2.

$H_0: U_1 = U_2$

$H_1: U_1 \neq U_2$ (two-sided test)

	Result	Rank
A	10.8	1.5
B	10.8	1.5
A	10.9	3.5
B	10.9	3.5
A	11.1	5
A	11.2	6.5
B	11.2	6.5
A	11.7	8
B	12.1	9
B	12.4	10
B	13.5	11
A	13.7	12
B	14.6	13
A	14.8	14
A	15.0	15
B	15.5	16
A	16.1	17
A	17.3	18

R_1 is the sum of the ranks in group B, i.e.

$$R_1 = 1.5 + 3.5 + 6.5 + 9 + 10 + 11 + 13 + 16 = 70.5$$

R_2 is the sum of the ranks in group A, i.e.

$$R_2 = 100.5$$

$$U_1 = (10)(8) + \frac{8(8+1)}{2} - 70.5 = 45.5$$

$$U_2 = (10)(8) + \frac{10(10+1)}{2} - 100.5 = 34.5$$

Notice that $U_1 + U_2 = n_1 n_2$

$U = \min(U_1, U_2) = 34.5$

One now consults tables of critical values of U (see Appendix, Table 4). The table gives the maximum value U can take before accepting H_1; by convention n_2 is the largest sample. At $\alpha = 0.05$ for a two-tailed test and $n_1 = 8$ and $n_2 = 10$, a value of 17 is found. Since this value is reached and in fact exceeded, the null hypothesis is accepted and one concludes that no difference between the two groups can be shown.

4.2 The Wilcoxon matched-pairs signed-ranks test or the Wilcoxon T-test (related samples)

Example. Let us take the same example as that given in Sect. 3.2.2.

Sample	x_R	x_T	d_i	Rank	Signed rank
1	114	116	−2	1	−1
2	49	42	+7	7.5	+7.5
3	100	95	+5	4	+4
4	20	10	+10	9.5	+9.5
5	90	94	−4	2.5	−2.5
6	106	100	+6	5.5	+5.5
7	100	96	+4	2.5	+2.5
8	95	102	−7	7.5	−7.5
9	160	150	+10	9.5	+9.5
10	110	104	+6	5.5	+5.5

The d_i values are ranked first without regard to sign starting with the smallest value (in the example, d_1 with absolute difference 2). Then the same sign is given as to the corresponding difference. For ties, the same rule as for the Mann–Whitney test is applied.

Some d_i values may be 0. There is some controversy about what to do in this case. Siegel [6] counsels dropping such values from the analysis, while Marascuilo and McSweeney [7] propose giving a rank equal to $(p+1)/2$, p being the number of zero differences, with half of the zero differences receiving positive rank and half negative. The second procedure is the more conservative because it favours retention rather than rejection of the null hypothesis.

The null hypothesis is that methods R and T are equivalent. If H_0 were true, it would be expected that the sum of all ranks for positive differences ($T+$) would be close to the sum for negative differences ($T-$).

$T = \min(T+, T-)$

The smaller the value of T, the larger the significance of the difference.

The critical values of T as a function of n and α are given in Table 5 of the Appendix. In the example, $T+ = 44.0$ and $T- = 11.0$ so $T = 11.0$. If the calculated T value is equal to or smaller than the critical value, the null hypothesis is rejected. For $\alpha = 0.05$, $n = 10$ and for a two-sided test, the critical value of $T = 8$. The null hypothesis is thus accepted and one concludes that no difference between the two methods can be shown.

5. Comparison of two methods by least-squares fitting

When the results obtained for a number of samples with the test procedure are plotted against those obtained with the reference procedure, a straight regression line should be obtained. In the absence of error, this line should have a slope, b, of exactly unity and an intercept on the ordinate, a, of zero and all points should fall on the line. In this book, we have adopted the convention that x values relate to concentrations of a sample and y values to signals used to derive these concentrations. In comparing two procedures by least-squares techniques, we should therefore use the symbols x_1 and x_2. For ease of notation, we shall, however, in this section represent the concentration obtained with one of these procedures by y and the other by x. In practice, one calculates y as a function of x (the x values being the results of the method with highest precision) using the classical least-squares procedure (Chap. 5).

Let us now consider the effects of different kinds of error (see Fig. 5). The presence of random errors in the test method leads to a scatter of the points around the least-squares line and a slight deviation of the calculated slope and intercept from unity and zero, respectively. The random error can be estimated from the calculation of the standard deviation in the y direction, $s_{y/x}$ (also called the standard deviation of the estimate of y on x).

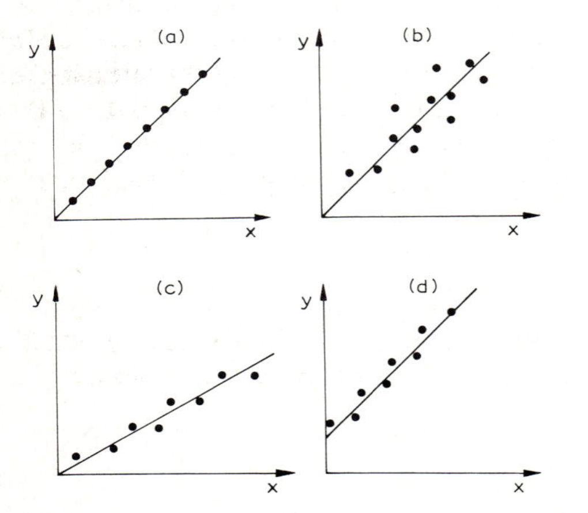

Fig. 5. Use of the regression method in the determination of systematic errors. (a) Ideal behaviour; (b) accurate method with low precision; (c) effect of proportional error; (d) effect of constant error.

A proportional systematic error leads to a change in b so that the difference between b and unity gives an estimate of the proportional error. A constant systematic error shows up in a value of the intercept different from zero. The study of the regression therefore leads to estimates of the three types of error (random, proportional and constant), which enables one to conclude that least-squares analysis is potentially the most useful statistical technique for the comparison of two methods. The least-squares method is a very general technique which enables one to fit data to a theoretical function. One can investigate whether this function does really describe the experimental observations by carrying out a goodness-of-fit test. In the present case, the equation is (see Chap. 5)

$$y = x \tag{2}$$

where y is the result of the test method and x the result of the reference method. Equation (2) is a particular case of

$$y = \alpha + \beta x$$

In fact, both methods are subject to random errors and one can thus calculate two regression equations, i.e. y as a function of x and x as a function of y, supposing, in the former case, that x is free of error and, in the latter case, that y is free of error. One could therefore use a procedure which consists of minimizing the distances of each point (x_i, y_i) from the straight line in the direction perpendicular to the line (orthogonal regression) instead of, classically, in the x or y direction (see Chap. 5) [8]. To be statistically completely correct, one should employ the orthogonal regression because it takes into account errors in x and y. However, the usual least-squares regression is often employed in practice.

If the experimental estimate, a, for α is close enough to zero and the estimate, b, for β is close enough to unity, it will be concluded that eqn. (2) is true and that there are no systematic errors. This calculation requires two steps

(1) the determination of a and b from the experimental data (see Chap. 5); and

(2) a test to investigate whether a and b differ significantly from zero and unity, respectively, or, to put it another way, "does the line $y = x$ fit the data?". To do this, one must carry out an analysis of variance (see Chap. 4) [9] or apply t-tests.

To test whether b is significantly different from $\beta = 1$, one calculates

$$t = \frac{b - \beta}{\sqrt{1 - r^2}} \sqrt{n - 2}$$

which has a Student distribution with $n - 2$ degrees of freedom.

To test whether a is significantly different from $\alpha = 0$, one calculates

$$t = \frac{a - \alpha}{s_a^2}$$

which has a Student distribution with $n - 2$ degrees of freedom. s_a^2 can be calculated according to the formula given in Chap. 5.

References p. 57

6. Recovery experiments

Proportional systematic errors are caused by the fact that the calibration line obtained with standards does not have the same slope as the functional relationship between the measurement result and the concentration in the sample, or, in other words, that the sensitivity (see Chap. 7) is different for standards and sample.

Consider, for example, neutron-activation analysis. In this technique, the concentration of element a in the unknown, $x_{a,\mathrm{u}}$, is estimated by comparing the radioactivity, $A_{a,\mathrm{u}}$, with the activity $A_{a,\mathrm{s}}$ of a standard with known concentration $x_{a,\mathrm{s}}$ of a by using the relationship

$$\frac{A_{a,\mathrm{u}}}{x_{a,\mathrm{u}}} = \frac{A_{a,\mathrm{s}}}{x_{a,\mathrm{s}}}$$

These ratios are, in fact, the sensitivities in the samples and for the standards. The calculation procedure implies that they do not depend on the composition of the matrix, but analytical chemists know that often this is not so. In neutron-activation analysis, it is possible, for example, that a strongly neutron-absorbing isotope is present in the sample. The activity obtained per gram of substance, u, will then be smaller, i.e. the ratio $A_{a,\mathrm{u}}/x_{a,\mathrm{u}}$ is smaller than the ratio $A_{a,\mathrm{s}}/x_{a,\mathrm{s}}$ and a proportional systematic error is obtained. When such a difficulty is suspected, analytical chemists estimate the content of the unknown by the standard addition method, which requires the determination of a calibration line in the particular sample. It can also serve to evaluate the occurrence of proportional systematic errors, and such an approach is then called a recovery experiment. In its simplest form, it consists of the addition of a known amount of the analyte of which the concentrations before and after the addition are determined. The difference

$$x_\mathrm{d} = x_\mathrm{after} - x_\mathrm{before}$$

should ideally be identical with the known amount added, Δx. Owing to the presence of random errors, this does not occur generally. If the standard deviation at both levels of concentration is known, one can test whether or not the difference between Δx and x_d is significant.

One can also carry out several additions of known but different concentrations in such a way as to arrive at the determination of the slope of a calibration line in the sample (Fig. 6).

This procedure can be exploited in several ways.

(a) One can compare the slopes of the regression line obtained in the recovery experiments and of the calibration line obtained with pure standards. These slopes are estimates of a true slope and one should therefore carry out a test to decide whether or not the slopes differ significantly.

(b) A second, but more indirect way, is to compare the results obtained from standard additions with those obtained by using the direct determination. The standard addition result is equal to the value measured without addition divided by the recovery slope. If one uses the measurement values determined from the

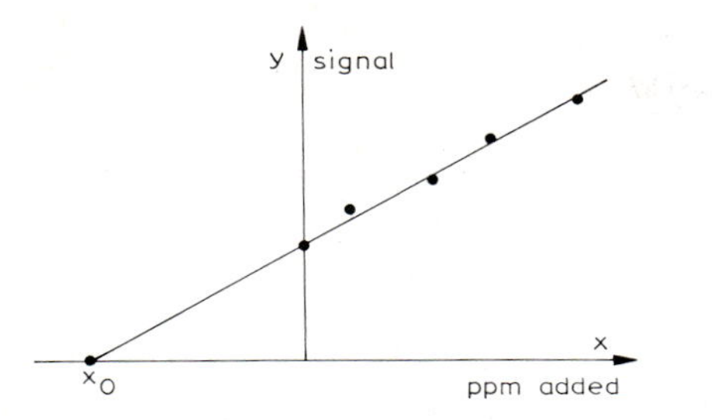

Fig. 6. A standard addition experiment.

regression lines $y = a + bx$ instead of the actual measurement results, this is given by a/b. As the intercept on the abscissa, x_0, (see Fig. 6) is equal to $-a/b$, one can determine the standard addition result graphically by measuring this intercept. One can also calculate the standard deviation on a and b [9] and therefore on the result obtained by the standard addition method.

7. Comparison of the precision of different methods (*F*-test)

It is common practice to compare the precision of two or more procedures by carrying out multi-replicate analyses with each of the procedures. This results in standard deviations, which are compared in order to select the most reproducible procedure. It is not always realized that, as standard deviations obtained from measurements are estimates, they are subject to sampling errors. Estimated standard deviations are subject to a distribution, the standard deviation of which is

$$\sigma_s = \sigma/\sqrt{2n}$$

where n is the number of measurements. Therefore the fact that procedures 1 and 2 yield results such that $s_1^2 > s_2^2$ does not automatically mean that procedure 2 is more precise. The significance of the difference in variances must be tested.

Let us suppose that one carries out n_1 replicate measurements by using procedure 1 and n_2 replicate measurements by using procedure 2, all on the same sample. One asks whether $\sigma_1^2 \neq \sigma_2^2$. If the null hypothesis is true, then the estimates s_1^2 and s_2^2 do not differ very much and their ratio should not differ much from unity. In fact, one uses the ratio of the variances

$$F = \frac{s_1^2}{s_2^2}$$

This ratio is distributed around unity and its mathematical properties are discussed in Chap. 2. It is conventional to calculate the F ratio by dividing the largest variance by the smallest in order to obtain a value equal to or larger than unity. If one sets a significance level of, for example, 5%, one has to compare the calculated F-value with

$F^{0.05}_{(n_1-1),(n_2-1)}$ (1-sided test) or with

$F^{0.025}_{(n_1-1),(n_2-1)}$ (2-sided test)

from a double entry one-tailed F table. If the calculated F-value is smaller than the F-value from the table, one cannot conclude that the procedures are significantly different in precision or, in other words, one concludes that there is only 5% probability that the procedures are different in precision.

8. The χ^2-test

The χ^2-test is used to compare experimental results with expected or theoretical results. The data used in the test are enumerative; they result from counts of occurrences. They are sorted into classes. The most frequent application of the χ^2-test in the context of this book, is the goodness of fit test which compares an observed distribution with a theoretical distribution.

One calculates the test statistic

$$\chi^2 = \sum_{i=1}^{k} \frac{(O_i - E_i)^2}{E_i} \tag{3}$$

where O_i is the observed (absolute) frequency in class i and E_i the expected frequency on the basis of the assumption of the distribution. $\Sigma O_i = \Sigma E_i = n$, the total number of observations.

The calculated χ^2-value is compared with the critical χ^2-value for $k-1$ degrees of freedom and a significance level α, and can be found in tables (Appendix, Table 6).

Example. Thin layer chromatography of 200 compounds was performed. The results are grouped in 10 classes and the class width is 0.1 R_f units. The observed frequencies are

$f_1 = 17$; $f_2 = 22$; $f_3 = 25$; $f_4 = 16$; $f_5 = 15$;

$f_6 = 21$; $f_7 = 12$; $f_8 = 23$; $f_9 = 29$; $f_{10} = 20$

The question is asked whether or not the observed frequencies correspond to a rectangular distribution.

The null hypothesis H_0 = the distribution is rectangular or $f_i = 20$. The alternative hypothesis $H_1 = f_i \neq 20$. Using eqn. (3), χ^2 is calculated as follows.

$$\chi^2 = \frac{(17-20)^2}{20} + \frac{(22-20)^2}{20} + \frac{(25-20)^2}{20} + \frac{(16-20)^2}{20} + \frac{(15-20)^2}{20}$$

$$+ \frac{(21-20)^2}{20} + \frac{(12-20)^2}{20} + \frac{(23-20)^2}{20} + \frac{(29-20)^2}{20} + \frac{(20-20)^2}{20}$$

$$\chi^2 = \tfrac{1}{20}(9+4+25+16+25+1+64+9+81) = 11.70$$

A 5% significance level is chosen. The value of χ^2 at 9 degrees of freedom and $\alpha = 0.05$ equals 16.919. The null hypothesis is accepted which means that the distribution is not significantly different from a rectangular distribution.

References

1 W.H. Beyer, Handbook of Tables for Probability and Statistics, CRC Press, FL, 1981.
2 R.N. Barnett and W.J. Youden, Am. J. Clin. Pathol., 54 (1970) 454.
3 M. Thompson and R.J. Howarth, Analyst, 101 (1976) 690.
4 E.K. Harris and D.L. De Mets, Clin. Chem., 18 (1972) 605.
5 H.F. Martin, B.J. Gudzinowicz and H. Fanger, Normal Values in Clinical Chemistry: A Guide to Statistical Analysis of Laboratory Data, Marcel Dekker, New York, 1975.
6 S. Siegel, Nonparametric Statistics for the Behavioral Sciences, McGraw-Hill, New York, 1956.
7 L.A. Marascuilo and M. McSweeney, Nonparametric and Distribution-free Methods for the Social Sciences. Brooks-Cole, Monterey, CA, 1977.
8 J. Mandell, The Statistical Analysis of Experimental Data, Interscience, New York, 1964.
9 B.E. Cooper, Statistics for Experimentalists, Pergamon Press, Oxford, 1975.
10 W.J. Youden, Statistical Methods for Chemists, Wiley, New York and Chapman and Hall, London, 1951.

Recommended reading

R. Johnson, Elementary Statistics, Duxbury Press, Boston, 1984.

M.J. Cardone, Detection and determination of error in analytical methodology, Parts I and II, J. Assoc. Off. Anal. Chem., 66 (1983) 1257; 66 (1983) 1283.

M. Thompson, Regression methods in the comparison of accuracy, Analyst, 107 (1982) 1169.

G.E. Box, W.S. Hunter and J.S. Hunter, Statistics for Experimenters, Wiley, New York, 1978.

J.O. Westgard and M.R. Hunt, Use and interpretation of common statistical tests in method comparison studies, Clin. Chem., 19 (1973) 49.

P.J.M. Wakkers, H.B.A. Hellendoorn, G.J. Op de Weegh and W. Heerspink, Applications of statistics in clinical chemistry. A critical evaluation of regression lines, Clin. Chim. Acta, 64 (1975) 173.

G.T. Wu, S.L. Twomey and R.E. Thiers, Statistical evaluation of method-comparison data, Clin. Chem., 21 (1975) 315.

G.S. Cembrowski, J.O. Westgard, W.J. Conover and E.C. Toren, Jr., Statistical analysis of method comparison data: testing normality, Am. J. Clin. Pathol., 72 (1979) 21.

R.K. Skogerboe and S.R. Koirtyohann, Accuracy in trace analysis: sampling, sample handling and analysis, Natl. Bur. Stand. U.S. Spec. Publ., 422 (1976) 199.

Software

D.L. Massart, M.P. Derde, Y. Michotte and L. Kaufman, BALANCE, Elsevier Scientific Software, Amsterdam, 1984.

Chapter 4

Evaluation of Sources of Variation in Data. Analysis of Variance

1. Introduction. Definitions

In Chap. 3, the *t*-test was used to decide whether two methods yield significantly different results. The profusion of analytical methods is such that often more than two procedures have to be compared; also, the results of more than two laboratories or analysts could be compared, etc. Such multiple comparisons are possible with the analysis of variance (ANOVA) technique.

The basic problem to which the analysis of variance is applied is the determination of which part of the variation in a population is due to systematic reasons (called factors) and which is due to chance. Scheffe [1] defines ANOVA as a "statistical technique for analysing measurements that depend on several kinds of effects operating simultaneously to decide which kinds of effects are important and to estimate the effects" .

In the comparison of several procedures, each may be subject to systematic error; the procedures then constitute a (controlled) factor. Moreover, the results of the analytical determinations are subject to random errors. The analysis of variance compares both causes of variation with the purpose of deciding whether or not the controlled factor has a significant effect.

An example relates to the work of official or public analysts and concerns the statistical analysis of a collaborative test of a procedure considered for official adoption. In such a test, several laboratories are asked to analyse a number of samples with the same procedure using a predetermined number of replicate determinations. The analysis serves to distinguish between sources of variation between laboratories, between samples, and between replicates. Therefore, there are two controlled factors (laboratories and samples), the variance between replicates being considered to be the effect of chance. An ANOVA with n factors is called an n-way layout, and therefore this example constitutes of two-way layout. It may be that one of the samples consists of a fine powder while another consists of a more granular material that has to be ground in order to obtain smaller particles before the analysis. A laboratory that has efficient grinding equipment may obtain accurate results for both kinds of sample, while another laboratory with inadequate equipment may produce more or less correct results for the first sample but systematically low results for the coarser material. In other words, the effect of the laboratory is not the same for all samples; this is called a sample–laboratory interaction and may have to be taken into account in the ANOVA. ANOVA can be used to detect such an interaction. This is a typical example of the use of ANOVA in optimization studies. If, in the example given, significant sample–laboratory interaction is detected, this should lead to a better specification of the procedure or standardization of the

References p. 73

equipment and therefore to smaller overall errors. Of course, as Youden pointed out, interlaboratory studies are a waste of everyone's time unless the initiating laboratory has taken care to remove as many assignable causes of variation as possible. That is, the method should be "abused" in the initiating laboratory to discover its weakness. When these weaknesses have been corrected, then it should be sent out for study by other laboratories. Determination of the ruggedness of the method (see Chap. 6) may be useful.

Another application of ANOVA is the development of an analytical procedure where one wants to investigate the effect of some factors, parts of the procedure, on the results and also the effect of interactions of these factors. Take, as an example, the case of a colorimetric analysis in which one investigates the effect on the absorbance of the concentration of the reagent A, of the reagent B and of the temperature at which the reaction is allowed to occur. This will be done by performing the experiments at some chosen levels of the factors followed by a 3-way ANOVA. One then concludes whether or not a factor or a combination of factors (i.e. interaction) significantly influences the response. Once one knows which factors are important, one can optimise their value. This is discussed to a much larger extent in Chaps. 16 and 17.

2. The one-way layout

Let us consider two similar examples, the first of which is the comparison of p procedures. To avoid consideration of between-laboratory errors, the procedures are assumed to be carried out by one laboratory. If this were not the case, one would have to carry out an ANOVA with two controlled factors (and perhaps interaction), namely the laboratories and the procedures.

The second example is the comparison of p laboratories. Here again, to avoid between-procedure errors, the laboratories are assumed to carry out the same procedure.

The object of ANOVA in these examples is to compare systematic errors with the random error obtained for the replicates, i.e. the precision of the procedures (in the second example, the precision of the laboratories). It is important to note that we have written "the precision of the procedures" and not "the precisions". This choice illustrates one of the important suppositions behind ANOVA: each of the procedures (laboratories) is considered to have the same precision. It is well known that, in practice, this is improbable. However, ANOVA remains valid when the values of σ do not differ too greatly. Nevertheless, it should be remembered that ANOVA could lead to erroneous conclusions if methods with widely differing precisions are compared.

One-way ANOVA can be illustrated by the following numerical example. Seven laboratories are asked to analyse aflatoxin M1 in milk using the same analytical method. A portion of a well-homogenised milk sample is given to each of the 7 laboratories and each asked to perform 5 independent determinations. The aim of the study is to investigate if there is an effect of the laboratory (between-laboratory

TABLE 1

One-way ANOVA. The determination of aflatoxin M in 7 laboratories

	Laboratory							
	a	b	c	d	e	f	g	
	1.6	4.6	1.2	1.5	6.0	6.2	3.3	
	2.9	2.8	1.9	2.7	3.9	3.8	3.8	
	3.5	3.0	2.9	3.4	4.3	5.5	5.5	
	1.8	4.5	1.1	2.0	5.8	4.2	4.9	
	2.2	3.1	2.9	3.4	4.0	5.3	4.5	
Sum	12.0	18.0	10.0	13.0	24.0	25.0	22.0	Grand mean
Mean	2.4	3.6	2.0	2.6	4.8	5.0	4.4	3.54

precision) on the results. A significant effect could mean that the procedure is not sufficiently well specified. The results (ppb) obtained are given in Table 1.

Let the laboratories be called $j = 1, 2, \ldots, q$ and let there be $i = 1, 2, \ldots, n_j$ determinations for one laboratory. x_{ij} is the ith result from the jth laboratory. The number of determinations is considered here to be the same for all laboratories ($n_1 = n_2 = \ldots = n_q = n_j$). This is, of course, not always the case and certainly not necessary to perform the ANOVA. To introduce statistical notation, a one-way layout table is given in Table 2.

$$\bar{x}_{\cdot j} = \frac{1}{n_j} \sum_{i=1}^{n_j} x_{ij}$$

is the mean of the results obtained in the jth laboratory and

$$\bar{x}_{\cdot\cdot} = \frac{1}{n} \sum_{j=1}^{q} \sum_{i=1}^{n_j} x_{ij}$$

is the grand mean with $n = \sum_{j=1}^{q} n_j$. It estimates the overall mean, μ, that would be obtained for all determinations in all laboratories.

The one-way layout problem is a comparison of the means of a variable measured in several samples. In the example considered here, the question to be

TABLE 2

One-way layout table

Laboratory 1	Laboratory 2 ...	Laboratory j
x_{11}	x_{12}	x_{1j}
x_{21}	x_{22}	x_{2j}
x_{31}	x_{32}	x_{3j}
$\vdots$	$\vdots$	$\vdots$
$x_{n_1,1}$	$x_{n_2,2}$	$x_{n_j,j}$
$\bar{x}_{\cdot 1}$	$\bar{x}_{\cdot 2}$	$\bar{x}_{\cdot j}$

References p. 73

asked is "do the q laboratories yield the same results?". The total variation among the x_{ij} values has two possible causes

(1) the fact that a number of different laboratories are asked to determine x_{ij}s. This source of variation is controlled; and

(2) random effects which cause the precision (or imprecision) of a procedure in the laboratories. This source of variation is uncontrolled.

Each x_{ij} result can be written as the sum of a constant μ (the overall mean), μ_j (a term which measures the effect of the jth laboratory and is called the laboratory bias), and an error term e_{ij} called the residual error or residual. The linear (or additive) model

$$x_{ij} = \mu + \mu_j + e_{ij}$$

can be written for a one-way layout.

The following assumptions are made.

(1) x_{ij} is normally distributed with $\mu + \mu_j$ and σ^2 as the parameters of the distribution

$$x_{ij} : N\left(\mu + \mu_j, \sigma^2\right)$$

$\mu + \mu_j$ is estimated by $\bar{x}_{.j}$, the mean obtained in the jth laboratory; μ is estimated by $\bar{x}_{..}$, the grand mean.

(2) The residuals e_{ij} are normally distributed with mean 0 and variance σ^2. All e_{ij} values are independent. σ is the within-laboratory precision.

(3) $\sum_{j=1}^{q} n_j\mu_j = 0$ for a model with fixed effects and for a model with random effects the μ_j have mean 0 and variance σ_m^2.

The difference between the two models can be explained as follows. Suppose one wants to investigate whether a given number of well-chosen laboratories all yield the same results. The mean values obtained in these laboratories are considered as fixed and the model defined in this way is called a fixed-effect model. When, however, the number of laboratories to be compared is very large, a sample of laboratories is drawn in a random way. The effects, μ_j, of the laboratories are random variables from a distribution with mean 0 and variance σ_m^2. This is called a random-effect model.

In summary, the main basic assumptions in ANOVA are

(1) normality of x and

(2) homogeneity of variances: a test for checking this assumption is given in Sect. 5.

It has to be stressed, however, that ANOVA is a very robust statistical technique, which means that it is insensitive to moderate deviations of the assumptions.

The ANOVA procedure now consists in calculating a number of sums of squares. The total sum of squares of the deviation of all measurements with respect to the general mean, called corrected sums of squares, is given by

$$\mathrm{SS_T} = \sum_{j=1}^{n_j} \sum_{i=1}^{q} (x_{ij} - \bar{x}_{..})^2$$

This sum of squares can be broken down in the following way.

$$x_{ij} - \bar{x}_{..} \equiv (x_{ij} - \bar{x}_{.j}) + (\bar{x}_{.j} - \bar{x}_{..})$$

Squaring yields

$$(x_{ij} - \bar{x}_{..})^2 = (x_{ij} - \bar{x}_{.j})^2 + (\bar{x}_{.j} - \bar{x}_{..})^2 + 2(x_{ij} - \bar{x}_{.j})(\bar{x}_{.j} - \bar{x}_{..})$$

Summation over i and j gives

$$\sum_i \sum_j (x_{ij} - \bar{x}_{..})^2 = \sum_i \sum_j (x_{ij} - \bar{x}_{.j})^2 + \sum_j n_j (\bar{x}_{.j} - \bar{x}_{..})^2 + 2\sum_i (x_{ij} - \bar{x}_{.j}) \sum_j (\bar{x}_{.j} - \bar{x}_{..})$$

The last term of this equation is 0 since the sum of the deviations with respect to the mean always equals zero.

The total sum of squares, SS_T, can thus be written as a sum of two sums of squares

$$SS_T = \sum_i \sum_j (x_{ij} - \bar{x}_{..})^2 = \underbrace{\sum_i \sum_j (x_{ij} - \bar{x}_{.j})^2}_{SS_R} + \underbrace{\sum_j n_j (\bar{x}_{.j} - \bar{x}_{..})^2}_{SS_{LAB}}$$

Let us now analyse both sums of squares

(1) SS_R

SS_R measures the deviations between the observations x_{ij} of a group and the mean of the group. In other words, it is a measure of the variance within the groups. In fact, SS_R divided by the corresponding degrees of freedom, $n - q$, is an unbiased estimate of the variance σ^2, the within-laboratory precision. Again, this only applies when the underlying hypothesis is correct, namely that all laboratories work with the same precision.

(2) SS_{LAB}

SS_{LAB} measures the deviations between laboratories. It has been shown that, for the fixed effect model

$$\frac{SS_{LAB}}{q-1} \text{ estimates } \sigma^2 + \frac{\sum_j n_j \mu_j^2}{q-1}$$

and hence the null-hypothesis is that all μ_j values are equal to zero. If the null hypothesis is accepted, and thus the μ_js are zero, one notes that $SS_{LAB}/(q-1)$ is also an unbiased estimator of σ^2.

It has been shown that, for the random effect model

$$\frac{SS_{LAB}}{q-1} \text{ estimates } \sigma^2 + \sigma_e^2 \frac{\left[n - \left(\sum n_j^2/n\right)\right]}{q-1}$$

The null hypothesis is $\sigma_e^2 = 0$.

Here, again, if the null hypothesis is accepted, i.e. all laboratories yield no different results, $SS_{LAB}/(q-1)$ and $SS_R/(n-q)$ are both unbiased estimators of σ^2.

In both cases (random and fixed models) the null hypotheses are verified with an F-test by calculating

$$F = \frac{SS_{LAB}/(q-1)}{SS_R/(n-q)}$$

and comparing it with the tabulated value of the F-distribution with $q-1$ and $n-q$ degrees of freedom at the chosen significance level.

In most cases, SS_R will be smaller than SS_{LAB}. Indeed, SS_R estimates the within-laboratory error and SS_{LAB} the inter-laboratory error, which is usually larger. In this case, F is larger than 1 and the F table can be used without problems. When F is smaller than 1, the deviation between laboratories is certainly not significant. When a critical value of F is not obtained, one considers that the contribution of the controlled factor to the total variance is not significant. In other words, if procedures are compared, one can conclude with a certain level of confidence that they all yield statistically the same results and, at least if one of the methods is a reference method, all of the procedures are accurate. If laboratories are compared, a not-significant analysis of variance means that all laboratories yield statistically the same results, that there is no laboratory bias, and that the analytical procedure is thus well specified.

The above information can be summarized in the ANOVA table (Table 3).

Since SS_R is the basis against which effects are measured, one should try to make SS_R as small as possible, i.e. do the determinations as precisely as possible. But this, of course, is something analytical chemists should always do.

Let us now return to the numerical example given earlier. To simplify calculations, SS_T and SS_{LAB} can be written as

$$SS_T = \sum_i \sum_j x_{ij}^2 - n\bar{x}_{..}^2$$

TABLE 3

One-way analysis of variance. ANOVA table

Source of variation	Sums of squares (SS)	Degrees of freedom	Mean square (MS)	F-value
Between groups	SS_{LAB}	$q-1$	$\frac{SS_{LAB}}{q-1}$	
Within groups (= residual)	SS_R	$n-q$	$\frac{SS_R}{n-q} = s_R^2$	$F = \frac{SS_{LAB}/q-1}{SS_R/n-q}$
Total	SS_T	$n-1$	$\frac{SS_T}{n-1} = s_T^2$	

TABLE 4

One-way analysis of variance. ANOVA table

Source of variation	Sums of squares (SS)	Degrees of freedom	Mean square (MS)	F
Between laboratories	45.09	6	7.51	9.33
Within laboratories (= residual)	22.54	28	0.81	$\rightarrow P < 0.001$
Total	67.63	34		

and

$$SS_{LAB} = n_j \sum_j \bar{x}_{.j}^2 - n\bar{x}_{..}^2 = \sum_j \frac{\left(\sum_i x_{ij}\right)^2}{n_j} - n\bar{x}_{..}^2$$

The term $n\bar{x}_{..}^2$ is called C, the correction of the mean.

SS_R is then calculated as $SS_R = SS_T - SS_{LAB}$

$$SS_T = 1.6^2 + 2.9^2 + \ldots + 4.5^2 - C$$

$$C = 35 \times 3.54^2 = 438.61$$

$$SS_T = 67.63$$

$$SS_{LAB} = \frac{12^2}{5} + \frac{18^2}{5} + \frac{10^2}{5} + \frac{13^2}{5} + \frac{24^2}{5} + \frac{25^2}{5} + \frac{22^2}{5} - C$$

$$= 45.09$$

$$SS_R = 67.63 - 45.09 = 22.54$$

The ANOVA table (Table 4) is constructed.

The calculated F-value is compared with the tabulated value

$$F_{6,28}^{0.05} = 2.45$$

As 2.45 is smaller then 9.33, the test is highly significant and it is thus concluded that there is an important laboratory effect, which means that one or more laboratories show a bias.

3. A posteriori tests

If the null hypothesis in the analysis of variance is rejected, at least one of the laboratories is at variance with the others. However, the question of which laboratories differ is not answered. A look at the patterns of means obtained by the laboratories will usually tell us which laboratory(ies) are outliers. However, in some cases, one might want to apply statistical reasoning to find those laboratories that differ significantly from the others. One could solve the problem by performing pair-wise comparisons of the group means using t-tests. However, this is not

References p. 73

recommended as the performed tests are mutually dependent (this means that each sample is used several times), resulting in a considerable increase of α, the type 1 error. For example, considering 5 samples from the same population, the probability of some pair showing a significant difference at the 5% level is 40%. Therefore, a number of a posteriori contrast tests were developed which compare all possible pairs of group means without increasing the type 1 error. The groups (in our example, the laboratories) are divided into homogeneous subsets where the difference in the means of any two groups in a subset is not significant at the chosen significance level. One of the most widely used of these tests is the Student–Newman–Keuls (SNK) test. Other similar tests are the Tukey, the Scheffe and the Duncan tests [2].

4. The two-way layout

Let us consider the problem discussed by Amenta [3]. His work was concerned with quality control in a clinical laboratory and involved the analysis of 50 samples from the same pool, two per day at different places in the run of the routine determinations carried out during that day for 25 consecutive days. The following questions were asked.

(1) Is there a significant contribution to the total variance from day-to-day variations?

(2) Is there a significant effect due to the position in the run?

As there are two controlled factors (the position and the day), the technique is called a two-way ANOVA. In this instance, the linear model is

$$x_{ij} = \mu + \mu_j + \nu_i + e_{ij}$$

Here, again, each result x_{ij} can be written as the sum of a constant μ, the general mean, plus μ_j, which measures the effect of the one factor, plus ν_i, which measures the effect of the other factor, plus the residual term e_{ij}. Of course, the same assumptions as for the one-way ANOVA were made.

When the days and the positions have no significant effect, one can conclude that

$$\sigma_T^2 = \sigma_R^2$$

and σ_R^2 is therefore the experimental error that occurs in the absence of other effects. The analysis of variance is based here on the comparison of the terms σ_{DAY}^2 and σ_{POS}^2 with σ_R^2.

The two-way layout table is given in Table 5.

$$\bar{x}_{i\cdot} = \sum_{j=1}^{25} x_{ij}/25$$

$$\bar{x}_{\cdot j} = \sum_{i=1}^{2} x_{ij}/2$$

TABLE 5

Two-way layout table for the clinical laboratory problem

Position $i=1, p\ (p=2)$	Day $j=1,\ldots,q\ (q=25)$					Means for positions
	1	2	3	4 ...	25	
1	$x_{1,1}$	$x_{1,2}$	$x_{1,3}$	$x_{1,4}$	$x_{1,25}$	$\bar{x}_{1\cdot}$
2	$x_{2,1}$	$x_{2,2}$	$x_{2,3}$	$x_{2,4}$	$x_{2,25}$	$\bar{x}_{2\cdot}$
Means for days	$\bar{x}_{\cdot 1}$	$\bar{x}_{\cdot 2}$	$\bar{x}_{\cdot 3}$	$\bar{x}_{\cdot 4}$	$\bar{x}_{\cdot 25}$	Grand mean $\bar{x}_{\cdot\cdot}$

The total sum of squares, SS_T, can be broken down in the following way.

$SS_T = SS_{DAY} + SS_{POS} + SS_R$

or

$$SS_T = \sum_j \sum_i (x_{ij} - \bar{x}_{\cdot\cdot})^2$$

with $n-1$ degrees of freedom.

$$= p \sum_{j=1}^{q} (\bar{x}_{\cdot j} - \bar{x}_{\cdot\cdot})^2$$

SS_{DAY} with $q-1$ degrees of freedom; it measures that part of the total variation in the data, caused by the fact that the samples are analysed on successive days.

$$+ q \sum_{i=1}^{p} (\bar{x}_{i\cdot} - \bar{x}_{\cdot\cdot})^2$$

SS_{POS} with $p-1$ degrees of freedom; it measures the variation in the data caused by the two positions in a run.

$$+ \sum_i \sum_j (x_{ij} - \bar{x}_{i\cdot} - \bar{x}_{\cdot j} + \bar{x}_{\cdot\cdot})^2$$

SS_R with $(p-1)(q-1)$ degrees of freedom; it measures the residual variance; i.e. that part of the total variation which cannot be explained by a controlled factor; it estimates the experimental error.

To simplify calculations, SS_T can be written as

$$SS_T = \sum_i \sum_j x_{ij}^2 - C \text{ (see also Sect. 2)}$$

$$SS_{DAY} = p \sum_j \bar{x}_{\cdot j}^2 - C = p \sum_j \left(\frac{\sum_i x_{ij}}{p} \right)^2 - C$$

$$= \sum_j \frac{\left(\sum_i x_{ij} \right)^2}{p} - C$$

References p. 73

$$\mathrm{SS}_{\mathrm{POS}} = q\sum_i \bar{x}_{i\cdot}^2 - C = \sum_i \frac{\left(\sum_j x_{ij}\right)^2}{q} - C$$

$$\mathrm{SS}_{\mathrm{R}} = \mathrm{SS}_{\mathrm{T}} - \mathrm{SS}_{\mathrm{DAY}} - \mathrm{SS}_{\mathrm{POS}}$$

By dividing the sums of squares by their corresponding degrees of freedom, one obtains the corresponding variances

$$s_{\mathrm{DAY}}^2 = \frac{\mathrm{SS}_{\mathrm{DAY}}}{24}$$

$$s_{\mathrm{POS}}^2 = \frac{\mathrm{SS}_{\mathrm{POS}}}{1}$$

$$s_{\mathrm{R}}^2 = \frac{\mathrm{SS}_{\mathrm{R}}}{(2-1)(25-1)} = \frac{\mathrm{SS}_{\mathrm{R}}}{24}$$

The F ratios

$$F_{\mathrm{DAY}} = \frac{s_{\mathrm{DAY}}^2}{s_{\mathrm{R}}^2}$$

and

$$F_{\mathrm{POS}} = \frac{s_{\mathrm{POS}}^2}{s_{\mathrm{R}}^2}$$

are calculated and compared with the tabulated values. A worked example, using the data of Table 6 is given.

$$C = n\bar{x}_{\cdot\cdot}^2 = 50 \times \left(\frac{6893^2}{50}\right) = 950268.98$$

$$\mathrm{SS}_{\mathrm{T}} = \sum_i \sum_j x_{ij}^2 - C = 138^2 + 137^2 + \ldots + 139^2 - C = 172.02$$

$$\mathrm{SS}_{\mathrm{DAY}} = \sum_j \left(\sum_i x_{ij}\right)^2 / p - C = \tfrac{1}{2}(278^2 + 274^2 + \ldots + 277^2) - C$$

$$= 128.52$$

$$\mathrm{SS}_{\mathrm{POS}} = \sum_i \left(\sum_j x_{ij}\right)^2 / q - C = \tfrac{1}{25}(3436^2 + 3457^2) - C = 8.82$$

$$\mathrm{SS}_{\mathrm{R}} = \mathrm{SS}_{\mathrm{T}} - \mathrm{SS}_{\mathrm{DAY}} - \mathrm{SS}_{\mathrm{POS}} = 34.68$$

The data are summarized in Table 7.

TABLE 6

Two-way layout table for the clinical laboratory problem

	Day												
	1	2	3	4	5	6	7	8	9	10	11	12	13
Position 1	138	137	137	136	137	136	140	139	137	135	132	136	138
Position 2	140	137	136	139	138	137	139	138	139	135	136	137	142
Sum for days	278	274	273	275	275	273	279	277	276	270	268	273	280

	14	15	16	17	18	19	20	21	22	23	24	25	Sum for positions
Position 1	141	137	136	137	138	138	138	136	140	139	140	138	3436
Position 2	139	137	139	135	138	138	138	139	140	141	141	139	3457
Sum for days	280	274	275	272	276	276	276	275	280	280	281	277	Grand sum = 6893

TABLE 7

Two-way analysis of variance. ANOVA table

Source	Sums of squares	Degrees of freedom	Mean square	F
Days	128.52	24	5.36	3.72
Position	8.82	1	8.82	6.12
Residual	34.68	24	1.44	
Total	172.02	49	3.51	

H_0: (1) σ^2_{DAY} is zero
(2) σ^2_{POS} is zero
If both hypotheses are exact, than $\sigma^2_T = \sigma^2_R$

At the 5% confidence level, both null hypotheses are rejected

$F^{0.05}_{24,24} = 1.98$ and $F^{0.05}_{1,24} = 4.26$

which means that there is a significant contribution to the total variance due to variation between days and between positions.

5. Bartlett's test for the comparison of more than two variances

The largest section of this chapter is devoted to the comparison of more than two means through ANOVA. One of the assumptions is that the compared groups (laboratories, procedures) have statistically equal variances. Another reason to compare variances in the scope of the investigation and selection of an analytical procedure, is that, when laboratories are compared, those with significantly lower precision than the others can be discarded or, if procedures are compared, the most imprecise one(s) can be withdrawn.

The parameter proposed by Bartlett is

$$M = \sum_{i=1}^{p} (n_i - 1) \ln s^2 - \sum_{i=1}^{p} (n_i - 1) \ln s_i^2$$

where s_i^2 is the variance of procedure i with n determinations and $n_i - 1$ degrees of freedom and

$$s^2 = \frac{\sum_{i=1}^{p} (n_i - 1) s_i^2}{\sum_{i=1}^{p} (n_i - 1)}$$

If the null hypothesis is true (the variances are equal), M follows a χ^2-distribution with $p - 1$ degrees of freedom.

The example introduced in Sect. 2 (Table 1) can be worked out. The data needed to perform Bartlett's test are summarized in Table 8.

TABLE 8

Data for the comparison of more than two variances

Laboratory (i)	Mean of 5 determinations ($\bar{x}_i$)	Variance (s_i^2)	Degrees of freedom ($n_i - 1$)	$\log_e s_i^2$
a	2.4	0.7906	4	−0.2350
b	3.6	0.8746	4	−0.1340
c	2.0	0.8775	4	−0.1307
d	2.6	0.8456	4	−0.1677
e	4.8	1.0173	4	0.0172
f	5.0	0.9823	4	−0.0179
g	4.4	0.8718	4	−0.1372

$$s^2 = \frac{\sum_{i=1}^{p} (n_i - 1) s_i^2}{\sum_{i=1}^{p} (n_i - 1)} = \frac{1}{28}(4 \times 0.7906 + \ldots + 4 \times 0.8718)$$

$$= 0.8942$$

$$\ln s^2 = -0.1118$$

$$\sum_{i=1}^{p} (n_i - 1) \ln s^2 = 28 \times (-0.1118) = -3.1298$$

$$\sum_{i=1}^{p} (n_i - 1) \ln s_i^2 = 4 \times (-0.2350) + \ldots + 4 \times (-0.1372)$$

$$= -3.2212$$

$$M = -3.1298 - (-3.2212) = 0.0914$$

The null hypothesis can be formulated as H_0 = all variances σ_i^2 are equal. A significance level $\alpha = 0.05$ is chosen.

The calculated M-value is compared will the value of the χ^2-distribution at 6 ($= p - 1$) degrees of freedom.

$\chi_6^2 = 12.592$

The null hypothesis is accepted.

6. Applications

Some applications of the use of ANOVA have already been discussed in preceding sections.

The application of the ANOVA technique that is encountered most frequently in analytical chemistry is the breakdown of a total precision into its components such as between-days and within-days, between-laboratories and within-laboratories, etc. If these components are known, one can decide which component is the weak link and should be improved first. ANOVA is also used to decide whether a certain factor has a meaningful effect on the results. When the factor is the choice of the

References p. 73

procedure, i.e. when ANOVA is used to compare procedures, this means that one determines whether the procedures give the same result. If they do, then one concludes that all of the procedures have the same degree of accuracy. A warning is necessary here. If the procedures to be compared contain common steps, such as the same preliminary separation step, and if this common step introduces a systematic error, one will not be able to detect this. ANOVA does not permit one to observe whether $\mu = \mu_0$ (see Chap. 2).

One of the more important applications of ANOVA in the context of this book is its use as a preliminary step in the experimental optimization of procedures. As will be explained in Chaps. 16–18, the selection of the factors that have an influence on the optimization criterion is an important part of such an optimization.

For this application, it is recommended that one should take into account the fact that the variables may depend on each other. Therefore, factorial analysis (Chap. 17) is indicated.

The problem of how to determine the between-laboratory and within-laboratory components of the precision from collaborative experiments is important in analytical chemistry. Very often, the between-laboratory component is the larger, indicating that some of the parameters of the procedure should be controlled more strictly. A collaborative testing programme for the Association of Official Analytical Chemists typically involves distributing 3–5 sufficiently different samples to 8–15 sufficiently proficient laboratories. Each laboratory is asked to report two replicate determinations on each sample. The statistical analysis involves the following steps.

(1) Reduction of the results reported to equal numbers of replicates. Some laboratories report more than two replicates. The ANOVA calculations can be carried out on unequal numbers of replicates, but in this instance the procedure becomes more complex. Therefore, the AOAC recommends that some of the results are eliminated. This must be done randomly, using a random numbers table, for example.

(2) Elimination of laboratories reporting systematically high or low results, The AOAC recommends that this should be done using a simple non-parametric ranking test as described by Youden and Steiner [4].

(3) Elimination of outlying results using Dixon's test [5].

(4) Examination of the homogeneity of the variance. It was assumed that the variances are the same in the laboratories. If there is a statistically significant difference between the variances from the laboratories, this usually means that one laboratory has been working with much lower precision than the others and this laboratory should be eliminated. In the same way, the residual errors should be normally distributed with constant mean.

(5) The ANOVA procedure, in this instance a two-way lay-out (samples, laboratories) with interaction between samples and laboratories.

(6) Calculation of what Youden and Steiner call reproducibility and repeatability (for definitions, see Chap. 2, Sect. 1.4).

An analytical application of a special kind of layout for ANOVA, called a nested design, was proposed by Wernimont [6]. He studied a procedure for an acetyl

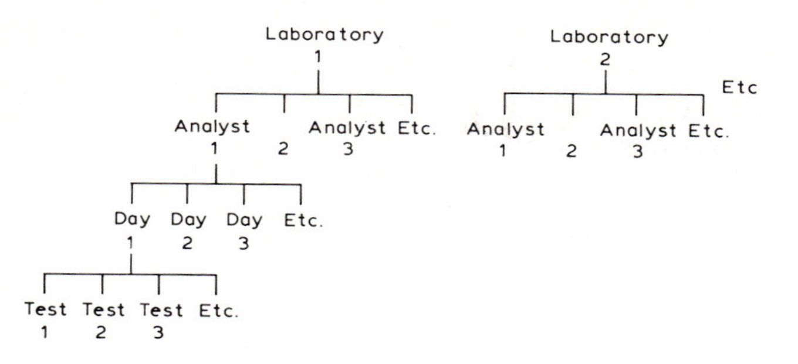

Fig. 1. Nested sampling design. (Reprinted from ref. 6 by permission of The American Chemical Society.)

determination and compared several sources of variation by having two analysts in each of eight laboratories perform two replicate tests on each of three days. The usual ANOVA procedure would have consisted of having the same two analysts perform two replicate tests on three days in the eight laboratories. Clearly, there is little sense in organizing an experiment that would have required two analytical chemists to run from one laboratory to the other and therefore the nested design was preferred. One says that the levels of a factor (analysts) are nested within the levels of another factor (laboratories) if every level of the first factor appears with only one level of the second factor in the observations. Each analyst appears only in one of the eight laboratories. The nested design used by Wernimont is represented in Fig. 1.

Wernimont's conclusion was that the laboratories are responsible for the largest source of variation. The mathematics of nested designs are not discussed in this book, but have been discussed, for example, by Scheffe [1].

One can use layouts of a higher order than the two-way layout. The equations in Sect. 4 can be generalized without much difficulty. The theory was given comprehensively in ref. 1.

References

1 H. Scheffe, The Analysis of Variance, Wiley, New York, 1959.
2 B.J. Winer, Statistical Principles in Experimental Design. International Student Edition, McGraw-Hill, London, 1970.
3 J.S.A. Amenta, Am. J. Clin. Pathol., 49 (1968) 842.
4 W.J. Youden and E.H. Steiner, Statistical Manual of the Association of Official Analytical Chemists, The Association of Official Analytical Chemists, Washington, DC, 1975.
5 W.J. Dixon, Biometrics, 9 (1953) 4.
6 G. Wernimont, Anal. Chem., 23 (1951) 1572.

Recommended books on ANOVA

B.E. Cooper, Statistics for Experimentalists, Pergamon Press, New York, 1969.
B.J. Winer, Statistical Principles in Experimental Design, International Student Edition, McGraw-Hill, London, 1970.

Recommended reading

G.T. Wernimont, in W. Spendley (Ed.), Use of Statistics to Develop and Evaluate Analytical Methods, The Association of Official Analytical Chemists, Arlington, VA, 1985.

C.R. Hicks, Fundamental Concepts in the Design of Experiments, Holt, Rinehart and Winston, New York, 3rd edn., 1982.

M.G. Natrella, Design and analysis of experiments, in J.M. Juran (Ed.), Quality Control Handbook, McGraw-Hill, New York, 1974.

A.L. Wilson The performance characteristics of analytical methods, Talanta, 17 (1970) 31.

NCCLS Proposed Standard PSEP-3, Protocol for Establishing Performance Claims for Clinical Chemical Methods, Replication Experiment, Instrument Evaluation Subcommittee of the Evaluation Protocols Area Committee, National Committee for Clinical Laboratory Standards, Villanova, PA, 1979.

Chapter 5

Calibration

1. Introduction

Calibration is one of the most important steps in chemical analysis. A good precision and accuracy can only be obtained when a good calibration procedure is used. Except in a very few cases (e.g. gravimetry), the concentration of a sample cannot be measured directly, but is determined using another physical measuring quantity, y. The condition for doing this is that an unambiguous empirical or theoretical relationship can be shown between this quantity and the concentration, x. Only the calibration function $y = g(x)$ is directly useful and yields by inversion the analytical calculation function.

The calibration function can be obtained by fitting an adequate mathematical model through the experimental data. The most convenient calibration function is linear, goes through the origin and is applicable over a wide dynamic range. In practice, however, many deviations from this ideal calibration line may occur. Well known, for example, is the curvature towards the x axis of the calibration line in the upper concentration range in spectroscopic methods.

For the majority of analytical techniques, the analyst uses the calibration equation

$$y = a + bx$$

He visually checks the linearity of the calibration graph and restricts the working range to the linear part of the curve. This working range is usually one decade. When working over a wider concentration range, deviation from linearity becomes more probable and must, of course, be checked and/or a new calibration function must be calculated which is valid over the entire range. This is the case for techniques such as ICP and GC–MS, which can be used over a wide concentration range.

In this chapter, we only consider calibration problems for which standards and samples have the same composition and are measured with the same precision. Only linear calibration graphs will be extensively discussed. Curvilinear and non-linear regression are discussed in Chap. 13.

In calibration, univariate regression is applied, which means that all observations are dependent upon a single variable x.

2. The method of least squares

Least-squares regression analysis is used to describe the relationship between signal and concentration. All models describing the relationship between y and x

References p. 92

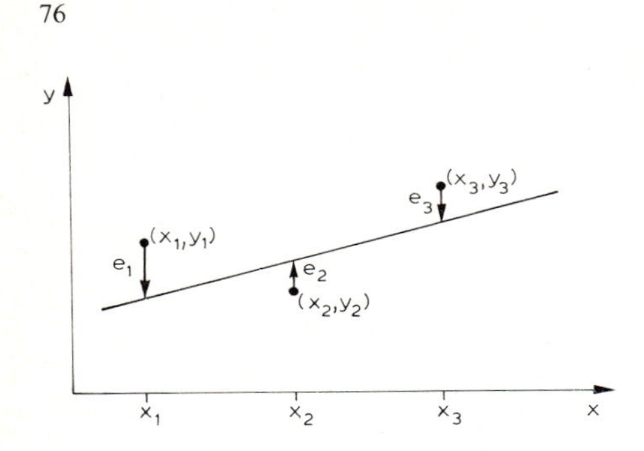

Fig. 1. The fitting of a straight line. The method of least squares.

can be represented by the general function $y = f(x, a, b_1, \ldots, b_m)$ where $a, b_1, \ldots, b_m$ are the parameters of the function.

We adopt the convention that the x values relate to the controlled or independent variable (e.g. the concentration of a standard) and the y values to the dependent variable (the response measurements). This means that the x values have no error. On the condition that errors made in preparing the standards are significantly smaller than the measuring errors (which is usually the case in analytical practice), this assumption is realistic in calibration problems. The values of the unknown parameters $a, b_1, \ldots, b_m$ must be estimated in such a way that the model fits the experimental data points (x_i, y_i) as well as possible.

The true relationship between x and y is considered to be given by a straight line. The relationship between each observation pair (x_i, y_i) can then be represented as

$$y_i = \alpha + \beta x_i + e_i$$

The signal y_i is composed of a deterministic component predicted by the linear model and a random component, e_i (Fig. 1).

One must now find the estimates a and b of the true values α and β. This is done by calculating values of a and b for which Σe_i^2 is minimal.

The components e_i represent the differences between the observed y_i values and the $\hat{y}_i$ values predicted by the model; the e_i are called the residuals.

$$e_i = y_i - \hat{y}_i$$

$$\hat{y}_i = a + bx_i$$

and therefore

$$e_i = y_i - a - bx_i$$

The least squares estimates of α and β are found by minimizing R, the sum of the squared residuals.

$$R = \sum e_i^2 = (y_i - a - bx_i)^2$$

This is done by setting the partial derivatives of R with respect to a and b equal to zero

$$\frac{\partial R}{\partial a} = \sum_i 2(y_i - a - bx_i)(-1) = 0$$

$$\frac{\partial R}{\partial b} = \sum_i 2(y_i - a - bx_i)(-x_i) = 0$$

These are the so-called normal equations. After elaboration, one obtains the expressions

$$b = \frac{\sum_i (x_i - \bar{x})(y_i - \bar{y})}{\sum_i (x_i - \bar{x})^2} = \frac{n\sum_i x_i y_i - \sum_i x_i \sum_i y_i}{n\sum_i x_i^2 - \left(\sum_i x_i\right)^2} \tag{1}$$

$$a = \bar{y} - b\bar{x} = \frac{\sum_i y_i \sum_i x_i^2 - \sum_i x_i \sum_i x_i y_i}{n\sum_i x_i^2 - \left(\sum_i x_i\right)^2} \tag{2}$$

A number of assumptions about the e_i are made.

(1) The e_i are normally distributed random variables with mean 0 and variance σ^2.

$e_i : N(0, \sigma^2)$

(2) The e_i are independent.

(3) All e_i have equal variances σ^2. This means that the variance σ^2 is constant over the entire dynamic concentration range; it is independent of the concentration (x). This property is called homoscedasticity.

The expected value of y_i is thus $a + bx_i$ and the variance of y_i is σ^2. Figure 2 illustrates the assumptions concerning e_i.

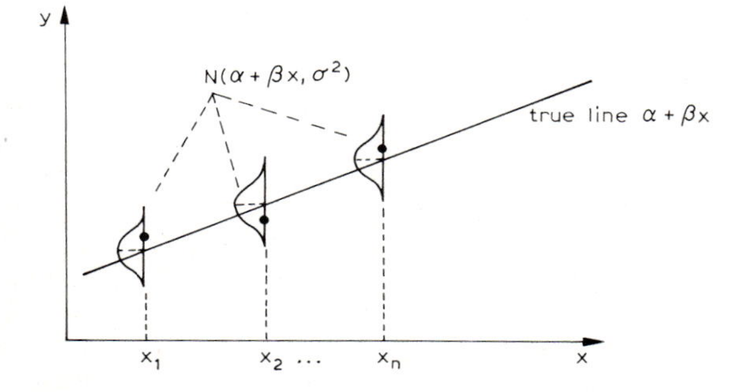

Fig. 2. The method of least squares. Assumptions concerning the residuals.

References p. 92

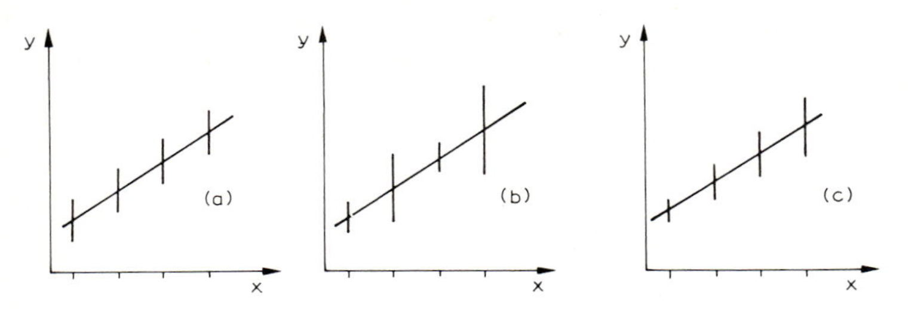

Fig. 3. Illustration of the properties homoscedasticity and heteroscedasticity. (a) Homoscedasticity; (b) heteroscedasticity; (c) heteroscedasticity with constant relative standard deviation.

Heteroscedasticity occurs when the variance of y_i depends upon x; a peculiar case of heteroscedasticity, important in analytical chemistry, is a constant relative standard deviation. An illustration is given in Fig. 3.

The procedure of regression can be summarized as

(1) selection of a model. In this chapter, a linear model is chosen. Non-linear models will be discussed in Chap. 13;

(2) estimation of the parameters of the model by the least-squares method;

(3) validation of the model (see Sects. 3 and 4); and

(4) calculation of confidence limits for the parameters, for the regression function, and for the analytical result (see Sect. 6).

3. Analysis of residuals

Examination of the residuals can provide answers to two types of question. On the one hand, lack of fit to the linear model can be detected and on the other, one can see whether the assumptions about the model are correct.

3.1 Graphical methods

A visual inspection of the residual plots can yield valuable information about the model and its assumptions. However, some skill and judgment are necessary. The first kind of plot that can be obtained is the overall plot which gives the frequency distribution of the residuals.

Consider, as an example, a regression analysis which provides 15 residuals with values of -5, 3, 0, 1, -2, 1, 0, -1, 3, 1, -2, 6, -1, -3, -1 (Fig. 4).

Visual inspection of the diagram allows the normality of the error distribution to be checked. However, if the number of observations is small, no really valid information can be obtained in this way. Another graphical method of checking normality is plotting the residuals on probability paper. The points should fall on a straight line (see also Chap. 3).

Residuals can also be plotted against the estimated y_i, values or against the values x_i of the independent variable. Generally, one obtains one of the patterns shown in Fig. 5.

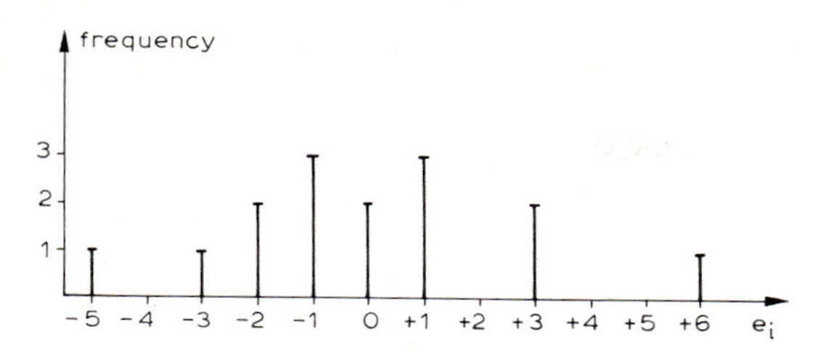

Fig. 4. Frequency distribution of residuals.

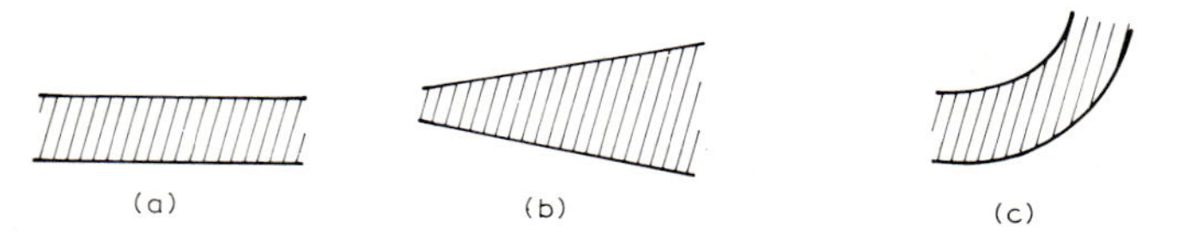

Fig. 5. Examples of patterns of residuals.

The plot (a) is obtained when a constant error term is present and when the correct model is chosen. The plot (b) is obtained when the condition of homoscedasticity is not fulfilled. In this case weighted least-squares or some transformation of the data should be performed (see Sect. 5). The plot (c) suggests a departure from the model.

One can check the following items on a plot of residuals.

(a) The shape of the plot.

(b) The number of positive residuals should be approximately equal to the number of negative residuals.

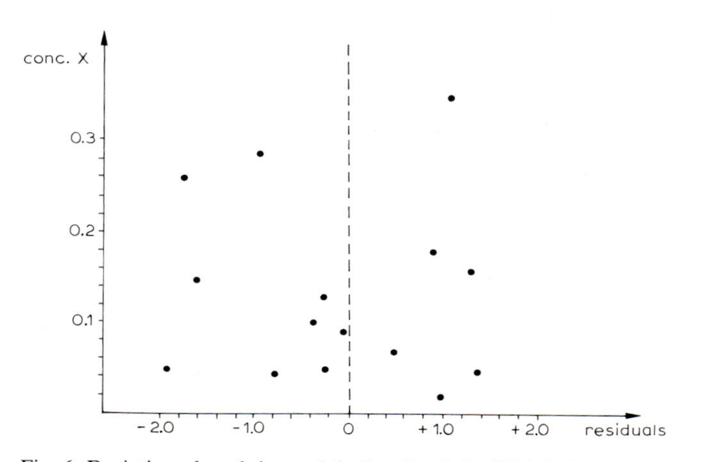

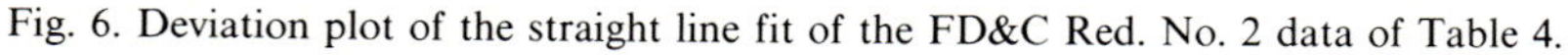
Fig. 6. Deviation plot of the straight line fit of the FD&C Red. No. 2 data of Table 4.

(c) The signs of the residuals should be distributed at random between plus and minus; a random sequence should be obtained.

(d) Some obvious outliers can be detected: they are much greater than the rest of the residuals.

Figure 6 shows the deviation plot of the straight line fit of the data given in Sect. 6, Table 4.

3.2 Statistical methods

There are statistical tests to check the normality of a distribution and tests to compare variances have been mentioned in Chaps. 3 and 4. One should always keep in mind that these tests are very insensitive when the number of replicates is small. It can happen that non-uniformity of variance might not be detected due to a lack of replicates.

4. Analysis of variance

The sum of squares

$$SS_T = \sum_i \sum_j (y_{ij} - \bar{y})^2$$

represents the variation of the y values about the mean $\bar{y}$ value. Part of the variation can be ascribed to the regression line and part to the fact that the observations do not all lie on the regression line. SS_T may be broken down in two sums of squares

(1) the "SS due to regression" (SS_{REG}) is that part of SS_T accounted for by the fitted regression line and

(2) the "SS about regression" or the residual SS (SS_R) contains the remaining variability about the regression line.

The residual SS can again be broken down into two components, viz.

(1) a component due to the variability within each group of replicate measurements. This is called SS_{PE}, the sum of squares due to purely experimental uncertainty or pure error SS and

(2) a component due to the variability of group averages about the regression line. This is called SS_{LOF}, the sum of squares due to lack of fit. Figure 7 gives a schematic representation.

The analysis of variance is performed in the following way. At each x_i ($i = 1, \ldots, k$) there are n_i observations

$$y_{ij} \; (j = 1, \ldots, n_i)$$

The total sum of squares $SS_T = \sum_i \sum_j (y_{ij} - \bar{y})^2$ can be broken down into 3 components

$$y_{ij} - \bar{y} = (y_{ij} - \bar{y}_i) + (\bar{y}_i - \hat{y}_i) + (\hat{y}_i - \bar{y})$$

where $\bar{y}_i$ is the mean value of the replicates y_{ij} at concentration x_i and $\hat{y}_i$ is the value of y at x_i estimated by the calibration function $y = a + bx$.

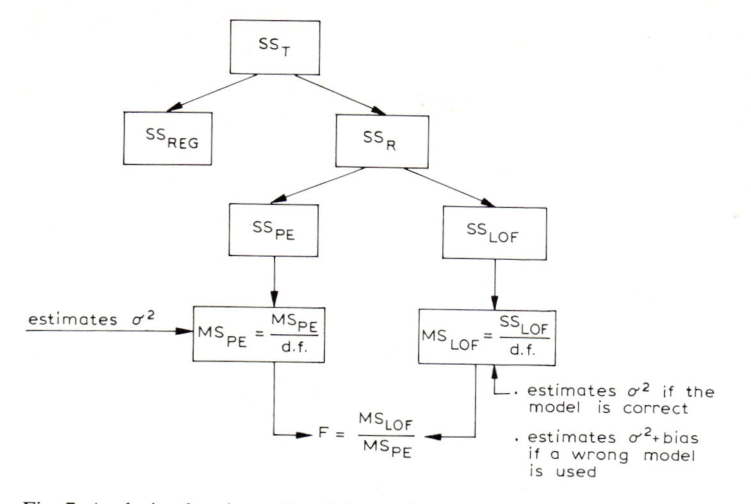

Fig. 7. Analysis of variance. Breakdown of sums of squares.

Squaring and summation over i and j yields

$$\sum_i \sum_j (y_{ij} - \bar{y})^2 = \underset{\mathbf{1}}{\sum_i \sum_j (y_{ij} - \bar{y}_i)^2} + \underset{\mathbf{2}}{\sum_i n_i (\bar{y}_i - \hat{y}_i)^2} + \underset{\mathbf{3}}{\sum_i n_i (\hat{y}_i - \bar{y})^2}$$

The corresponding mean square of component **1** is a pure estimator of σ^2, the variance of pure error. Component **3** measures the variance due to regression and

TABLE 1

Breakdown of sums of squares. ANOVA table

Source of variation	SS	Degrees of freedom	MS	F
Due to regression	$SS_{REG} = \sum_i n_i (\hat{y}_i - \bar{y})^2$	1	MS_{REG}	$\frac{MS_{REG}}{MS_R}$
Variation of group means about the line = lack of fit	$SS_{LOF} = \sum_i n_i (\bar{y}_i - \hat{y}_i)^2$	$k-2$	MS_{LOF}	$\frac{MS_{LOF}}{MS_{PE}}$
Within groups = pure error	$SS_{PE} = \sum_i \sum_j (y_{ij} - \bar{y}_i)^2$	$\sum_i n_i - k$	MS_{PE}	
Total	$SS_T = \sum_i \sum_j (y_{ij} - \bar{y})^2$	$\sum_i n_i - 1$		

component **2** measures the variation of the group means $\bar{y}_i$ about the line. The ANOVA table is given in Table 1.

The linearity test is performed by calculating $F = \mathrm{MS}_{\mathrm{LOF}}/\mathrm{MS}_{\mathrm{PE}}$.

Example. The results of a calibration experiment are given in Table 2.

TABLE 2

Results of a calibration experiment

x_i	1	2	3	5	10
y_{ij}	16.6	21.8	31.0	49.3	82.4
	13.7	22.2	30.3	51.4	85.9
	14.8	21.8	32.3	51.6	85.6
	14.3	21.8	31.2	51.6	84.0
$\sum_j y_{ij}$	59.4	87.6	124.8	203.9	337.9

$$n = \sum_i^k n_i = 20$$

$k = 5$

Calculation of the total sum of squares, corrected for the mean gives

$$\mathrm{SS_T} = \sum_i \sum_j (y_{ij} - \bar{y})^2 = \sum_i \sum_j y_{ij}^2 - \frac{\left(\sum_i \sum_j y_{ij}\right)^2}{n}$$

$$= 16.6^2 + 13.7^2 + \ldots + 84.0^2 - \frac{813.60^2}{20} = 12553.41$$

Calculation of the sum of squares due to pure error gives

$$\mathrm{SS_{PE}} = \sum_i \sum_j (y_{ij} - \bar{y}_i)^2 = \sum_i \sum_j y_{ij}^2 - \sum_i \left[\frac{\left(\sum_j y_{ij}\right)^2}{n_j}\right]$$

$$= 45650.66 - \left(\frac{59.4^2}{4} + \frac{87.6^2}{4} + \frac{124.8^2}{4} + \frac{203.9^2}{4} + \frac{337.9^2}{4}\right) = 18.46$$

Calculation of the sum of squares due to regression gives

$$\mathrm{SS_{REG}} = \sum_i n_i(\hat{y}_i - \bar{y})^2 = b^2 \sum_i n_i(x_i - \bar{x})^2$$

with

$$b = \frac{\sum_i n_i(x_i - \bar{x})(\bar{y}_i - \bar{y})}{\sum_i n_i(x_i - \bar{x})^2}$$

$$\mathrm{SS_{REG}} = \frac{\left(\sum_i n_i(x_i - \bar{x})(\bar{y}_i - \bar{y})\right)^2}{\sum_i n_i(x_i - \bar{x})^2}$$

It is, of course, not necessary to calculate SS_{REG} since, in a calibration experiment, there is always a significant regression. The computation of b, needed to calculate SS_{LOF}, will, however, be given and one notes that the calculation of SS_{REG} from b is straightforward.

$$\sum_i n_i(x_i-\bar{x})^2 = \sum_i n_i x_i^2 - \frac{\left(\sum_i n_i x_i\right)^2}{n}$$

$$= 4\times1^2+4\times2^2+4\times3^2+4\times5^2+4\times10^2 - \frac{(4\times1+4\times2+4\times3+4\times5+4\times10)^2}{20}$$

$$= 203.20$$

$$\sum_i n_i(x_i-\bar{x})(\bar{y}_i-\bar{y}) = \sum_i\left(x_i\sum_j y_{ij}\right) - \frac{\left(\sum_i\sum_j y_{ij}\right)\left(\sum_i n_i x_i\right)}{n}$$

$$=1\times59.4+2\times87.6+3\times124.8+5\times203.9+10\times337.9-\frac{813.60\times84}{20}$$

$$=1590.38$$

$$b=\frac{1590.38}{203.20}=7.83$$

$$SS_{REG}=\frac{1590.38^2}{203.20}=12447.38$$

$$SS_{LOF}=\sum_i n_i(\bar{y}_i-\hat{y}_i)^2$$

$$=\sum_i n_i[\bar{y}_i-\bar{y}-b(x_i-\bar{x})]^2$$

$$=\left[\sum_i\frac{\left(\sum_i y_{ij}\right)^2}{n_i}-\frac{\left(\sum_i\sum_j y_{ij}\right)^2}{n}\right]-\frac{\left[\sum_i n_i(x_i-\bar{x})(\bar{y}_i-\bar{y})\right]^2}{\sum_i n_i(x_i-\bar{x})^2}$$

$$=45632.20-33097.25-12447.38$$
$$=87.57$$

SS_{LOF} also can be calculated as

$$SS_{LOF}=SS_T-SS_{REG}-SS_{PE}$$
$$=12553.41-12447.38-18.46$$
$$=87.57$$

TABLE 3

ANOVA table for the example

Source of variation	SS	Degrees of freedom	MS	F
Due to regression	$SS_{REG}=12447.38$	1	12447.38	2113.16
Variation of group means about the line	$SS_{LOF}=87.57$	3	29.19	23.73
Within-groups	$SS_{PE}=18.46$	15	1.23	
Total	$SS_T=12553.41$	19		

References p. 92

The results are summarized in Table 3.

$F_{3,15}^{0.05} = 3.29$

The lack of fit term is highly significant which means that the model chosen is not an adequate description of the true relationship between y and x. In fact, one should have used a non-linear calibration curve (see Chap. 13).

5. Heteroscedasticity

For many analytical procedures, including methods based on counting measurements and photometric analyses, the condition of uniform variance along the calibration curve is not fulfilled.

When the response measurement is a count, one can, on a theoretical basis, expect the variances to be non-uniform (for a Poisson distribution $\mu = \sigma^2$), even in the presence of random errors such as errors of pipetting, volume, reaction time, temperature, etc.

According to Garden et al. [1] many analytical data show non-uniform variance. They analysed the random errors which are obtained by repeatedly measuring the same sample. These random errors are caused by noise which may be due to several factors. As examples of noise sources, the authors cite: shot noise arising from photomultiplier detectors proportional to the square root of the signal; noise due to fluctuations in the light source proportional to the signal; noise from the instrument electronics which can exhibit constant variance or variance which is a function of the signal; and flame noise in atomic spectroscopy proportional to the concentration of the analyte.

Many analytical chemists, if they happen to check homoscedasticity at all, use too few replicates and therefore cannot prove heteroscedasticity. It is clear that data from one calibration experiment cannot yield the information necessary to check basic assumptions. Past experience of similar experiments can, however, be used.

When the analyst comes to the conclusion that his data exhibit non-uniform variance, he cannot use the simple least-squares procedure to calculate the calibration curve without decreasing accuracy.

The problem can be solved in two ways.

(1) Perform a transformation of the variables in such a way that homoscedasticity is obtained (Sect. 5.2).

(2) Use a weighted least-squares procedure by applying at each point weighting factors inversely proportional to the variance in that point.

5.1 Weighting

5.1.1 Weighted least squares

The idea behind the weighted least-squares method is to attach the most importance to the data that are measured with the greatest precision.

An obvious weighting factor, therefore, is simply inversely proportional to the variance of the responses at each x_i ($w_i = 1/s_{y_i}^2$).

The weighted least-squares procedure consists in minimizing the weighted residuals. Each of the points (x_i, y_i) are weighted according to the inverse of the variance in that point

$$R = \sum_i w_i (y_i - a - bx_i)^2$$

$$b = \frac{\sum_i w_i (x_i - \bar{x})(y_i - \bar{y})}{\sum_i w_i (x_i - \bar{x})^2}$$

$$= \frac{\sum_i w_i \sum_i w_i x_i y_i - \sum_i w_i x_i \sum_i w_i y_i}{\sum_i w_i \sum_i w_i x_i^2 - \left(\sum_i w_i x_i\right)^2} \tag{3}$$

$$a = \bar{y} - b\bar{x}$$

$$= \frac{\sum_i w_i x_i^2 \sum_i w_i y_i - \sum_i w_i x_i \sum_i w_i x_i y_i}{\sum_i w_i \sum_i w_i x_i^2 - \left(\sum_i w_i x_i\right)^2} \tag{4}$$

$$\bar{x} = \frac{\sum_i w_i x_i}{\sum_i w_i}$$

$$\bar{y} = \frac{\sum_i w_i y_i}{\sum_i w_i}$$

The greater the departure from homoscedasticity, the greater is the benefit to be expected from using a weighted least-squares procedure. It is the task of the analyst to decide whether or not he needs to improve his method by using more sophisticated calibration procedures. His choice should depend on the accuracy he would like to obtain but also on considerations of cost, additional effort and computer facilities. In fact, a cost–benefit analysis (Chap. 10) is needed.

An example of the use of weighted least squares is given in Sect. 6.

5.1.2 Determination of weighting factors

An important problem is to make a good estimate of the weighting factors and thus of $s_{y_i}^2$, the response variances at each point i.

In most cases, knowledge of variance must be gained experimentally. One of the most evident ways of achieving this is by measuring a sufficient number of replicates at each concentration x_i. This, of course, implies a great number of

References p. 92

experiments. In the absence of a sufficient number of replicates, a functional relationship between variance and concentration can be assumed. Indeed, one can set up a variance function from which the variances $s^2_{y_i}$ can be estimated. The variance function is calculated by fitting an equation through the data points $(s^2_{y_i}, x_i)$ or $(s^2_{y_i}, \bar{y}_i)$. It has been shown, for example, that in radio immunoassay $s^2_{y_i}$ is nearly linearly related to $\bar{y}_i$. In order to obtain a better estimate of $\sigma^2_{y_i}$, one can, after testing for homogeneity, pool the results for standards and unknowns; also, the results over assays and in local regions of the concentration range can be pooled. In this way, an estimate of $\sigma^2_{y_i}$ can be calculated with a very large number of degrees of freedom.

5.2 Transformation to homoscedasticity

When the condition of uniform variance is not fulfilled, some transformation of the data can be performed depending on the variance function, i.e. the variance of the y_i values, $s^2_{y_i}$, as a function of x_i or $\bar{y}_i$.

It can be shown that if $s^2_{y_i}$ is proportional to $\bar{y}_i$, the transformation $z = \sqrt{y}$ should be carried out and if $s^2_{y_i}$ is proportional to $\bar{y}_i^2$, one should use the transformation $z = \log y$. An application of transformation is given by Agterdenbosch in ICP analysis who found a constant relative standard deviation in the higher concentration range and therefore performed a log transformation to achieve homoscedasticity [2].

One notices that, after transformation of the y values, either by calculating the square root or the logarithm, a linear calibration graph becomes non-linear. To avoid this, one should also transform the x values and use the equations

$$\log y = a + b \log x$$

or

$$\sqrt{y} = a + b\sqrt{x}$$

when the calibration graph is linear. Log–log transformation leads to a straight line only when the intercept is zero or near to zero, which is usually true for calibration.

6. Confidence intervals

The precision of the estimated sample concentration depends on both the measurement error of the sample and the confidence interval of the calibration curve at the value of the sample concentration which in turn is related to the uncertainty of the estimates a and b.

From the expressions for a and b (Sect. 2), one can calculate the variances of the parameters a, b and of the estimation $\hat{y}$ of the "true" y (see also Chap. 2, Sect. 2.5).

$$\hat{y} = \bar{y} + b(x_i - \bar{x})$$

As $\bar{y}$ and b are uncorrelated, $s_{\hat{y}}^2$ can be calculated as

$$s_{\hat{y}}^2 = s_{\bar{y}}^2 + (x_i - \bar{x})^2 s_b^2$$

The least-squares parameter b can also be calculated as

$$b = \frac{\sum_i (x_i - \bar{x}) y_i}{\sum_i (x_i - \bar{x})^2}$$

The x_i are constants so

$$s_b^2 = \frac{s_{y/x}^2}{\sum_i (x_i - \bar{x})^2} \tag{5}$$

and $s_{\hat{y}}^2$ becomes

$$s_{\hat{y}}^2 = \frac{s_{y/x}^2}{n} + (x_i - \bar{x})^2 \frac{s_{y/x}^2}{\sum_i (x_i - \bar{x})^2}$$

or

$$s_{\hat{y}}^2 = s_{y/x}^2 \left(\frac{1}{n} + \frac{(x_i - \bar{x})^2}{\sum_i (x_i - \bar{x})^2} \right) \tag{6}$$

One also can calculate

$$s_a^2 = s_{y/x}^2 \left(\frac{1}{n} + \frac{\bar{x}^2}{\sum_i (x_i - \bar{x})^2} \right) \tag{7}$$

The 95% confidence intervals can be determined.

β: $b \pm t_{n-2}^{0.05} s_b$

α: $a \pm t_{n-2}^{0.05} s_a$

y: $\hat{y} \pm t_{n-2}^{0.05} s_{\hat{y}}$

$$s_{y/x}^2 = \frac{\sum_i (y_i - \hat{y}_i)^2}{n-2}$$

$s_{y/x}^2$ is the variance of the y values with respect to the straight line; there are $n-2$ degrees of freedom, since 2 degrees of freedom were used for determining the parameters of the straight line. Notice that $\sum (y_i - \hat{y}_i)^2$ is SS_R, the SS about regression given in Fig. 7, Sect. 4.

$s^2_{y/x}$ can be calculated using the formula

$$(n-2)s^2_{y/x} = \sum y_i^2 - \frac{\left(\sum y_i\right)^2}{n} - \frac{\left[\sum x_i y_i - \left(\sum x_i \sum y_i/n\right)\right]^2}{\sum x_i^2 - \left[\left(\sum x_i\right)^2/n\right]}$$

or

$$(n-2)s^2_{y/x} = \sum y_i^2 - \frac{\left(\sum y_i\right)^2}{n} - b\left(\sum x_i y_i - \frac{\left(\sum x_i\right)\left(\sum y_i\right)}{n}\right) \tag{8}$$

Using eqn. (6), the 95% confidence limits for the true mean value of y at a given value of x, x_0 can be written as

$$a + bx_0 \pm t^{0.05}_{n-2}\left[\left(\frac{1}{n} + \frac{(x_0-\bar{x})^2}{\sum_i (x_i-\bar{x})^2}\right)s^2_{y/x}\right]^{1/2} \tag{9}$$

This is the confidence interval of the true regression line, sometimes called the Working–Hotelling region. The curves are hyperbolae. The confidence band around the calibration curve depends on the experimental design of the calibration points over the calibration range. The most important terms affecting the width of the confidence band are $(x_0-\bar{x})^2$ and $\Sigma(x_i-\bar{x})^2$. One sees from eqn. (6) that the variance of $\hat{y}$, the predicted value of y at x_0, is a minimum when $x_0=\bar{x}$ and increases as x_0 moves away from $\bar{x}$. This means that the greater the distance between x_0 and $\bar{x}$, the larger is the error one may expect to make when predicting the mean value of y at x from the regression line. The term $\Sigma(x_i-\bar{x})^2$ depends on the distribution of the x_i with respect to $\bar{x}$; the further the extreme points are situated along the x axis, the smaller is the width of the band and also the more precise the estimation of β [eqn. (5)]. Theoretically, the smallest error is thus found when taking an equal number of low and high x values at both ends of the calibration range ("two clouds" type of distribution) (Fig. 8).

However, in that case no evidence is obtained that the straight line describes the true relationship, since no indication for a lack of fit is obtained. In analytical work,

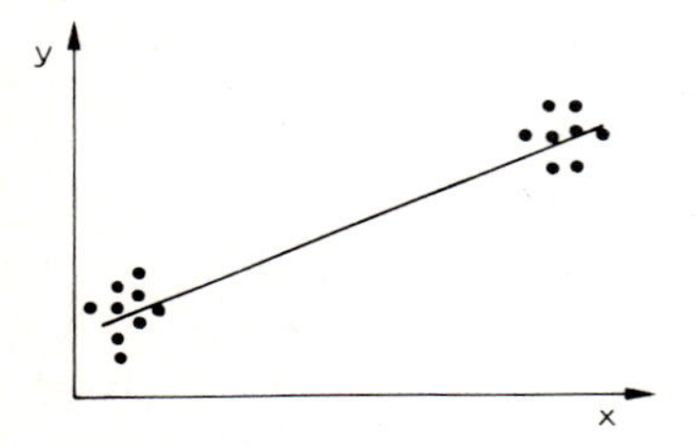

Fig. 8. Calibration line with the theoretically smallest error.

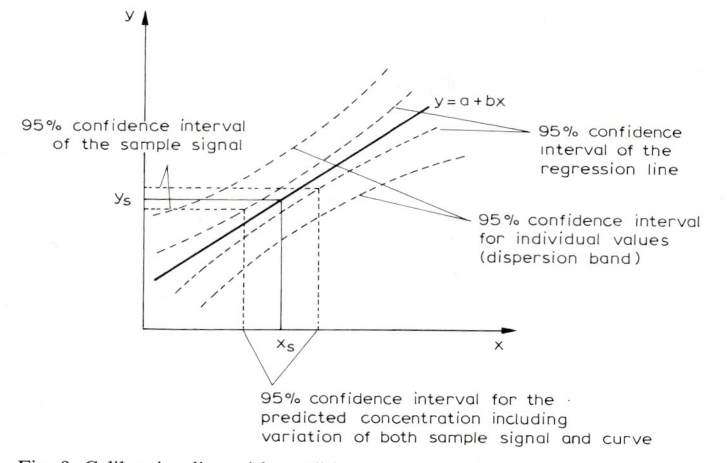

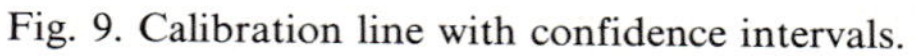
Fig. 9. Calibration line with confidence intervals.

the correctness of the linear model is not always a priori evident and therefore the calibration points are usually evenly spaced over the calibration range.

For a future individual observation $y_0 = a + bx_0$ to be recorded at x_0, one obtains

$$y_0 \pm t_{n-2}^{0.05}\left[\left(1 + \frac{1}{n} + \frac{(x_0 - \bar{x})^2}{\sum_i (x_i - \bar{x})^2}\right) s_{y/x}^2\right]^{1/2} \tag{10}$$

This confidence interval is sometimes called the dispersion band.

TABLE 4

Calibration data for FD&C Red. No. 2 (naftionic acid) by HPLC [3]

Concentration, x_i	Peak area, y_i	Predicted area, $\hat{y}_i$	$e_i = y_i - \hat{y}_i$	$w = 1/s_{y_i}^2$
0.18	26.666	25.723	0.943	0.1406
0.35	50.651	49.482	1.169	0.0390
0.055	9.628	8.254	1.374	1.0788
0.022	4.634	3.642	0.992	4.6568
0.29	40.206	41.097	−0.891	0.0619
0.15	21.368	21.531	−0.163	0.2190
0.044	5.948	6.716	−0.768	2.8266
0.028	4.245	4.480	−0.235	5.5494
0.044	4.786	6.716	−1.930	4.3657
0.073	11.321	10.769	0.552	0.7802
0.13	18.456	18.736	−0.280	0.2936
0.088	12.865	12.866	0.001	0.6042
0.26	35.186	36.904	−1.718	0.0808
0.16	24.245	22.928	1.317	0.1701
0.10	14.175	14.543	−0.368	0.4977

References p. 92

Knowing the confidence band around the calibration line, one can now calculate the confidence band around the predicted concentration x_s of the unknown sample.

$$x_s \pm t_{n-2}^{0.05} \left[\left(\frac{1}{N} + \frac{1}{n} + \frac{(y_s - \bar{y})^2}{b^2 \sum_i (x_i - \bar{x})^2} \right) \frac{s_{y/x}^2}{b^2} \right]^{1/2} \tag{11}$$

where N is the number of replicate y_s measurements for the same sample with unknown concentration x_s. The confidence band around the predicted concentration can also be estimated graphically (Fig. 9).

Example.

Using the equations of Sect. 2, one calculates the regression equation $y = 0.567 + 139.758\ x$. From the data of Table 4, the following terms are calculated

$$\sum x_i = 1.9740 \qquad \bar{x} = 0.1316$$

$$\sum y_i = 284.3810 \qquad \bar{y} = 18.9587$$

$$\sum x_i^2 = 0.4028$$

$$\sum y_i^2 = 8201.3353$$

$$\sum x_i y_i = 57.4184$$

$$\sum (x_i - \bar{x})^2 = \sum x_i^2 - \frac{\left(\sum x_i\right)^2}{n}$$

$$= 0.4028 - \frac{(1.9740)^2}{15} = 0.1431$$

From eqn. (8)

$$s_{y/x}^2 = \frac{1}{13} \left[8201.3353 - \frac{(284.3810)^2}{15} - 139.758 \left(57.4184 - \frac{1.9740 \times 284.3810}{15} \right) \right]$$

$$= 1.1946$$

The 95% confidence intervals can now be calculated

$$\alpha: 0.567 \pm 2.160 \left[\left(\frac{1}{15} + \frac{(0.1316)^2}{0.1431} \right) 1.1946 \right]^{1/2}$$

2.160 is the tabulated t value for $\alpha = 0.05$ and 13 degrees of freedom

$$\alpha: 0.567 \pm 1.023$$

$$\beta: 139.758 \pm 2.160 \left[\left(\frac{1}{0.1431} \right) 1.1946 \right]^{1/2}$$

$$\beta: 139.758 \pm 6.241$$

The confidence interval of the regression line

$$y: 0.567 + 139.758 x_0 \pm 2.160 \left[\left(\frac{1}{15} + \frac{(x_0 - 0.1316)^2}{0.1431} \right) 1.1946 \right]^{1/2}$$

For $x_0 = 0.15$, the 95% confidence interval for the true mean value of y, y_0, is 21.5307 ± 0.6203.

Let us consider as an example that an unknown sample yields the result $y_s = 20.000$. Using the regression equation in reverse, one can calculate the concentration x_s as 0.1390. Using eqn. (11), one can calculate the 95% confidence interval of the true concentration of the sample.

$$x_s = \frac{y_s - a}{b} \pm 2.16 \left[\left(1 + \frac{1}{15} + \frac{(20.000 - 18.9587)^2}{139.758^2 \times 0.1431} \right) \frac{1.1946^2}{139.758^2} \right]^{1/2}$$

$$= 0.1390 \pm 0.0191$$

This means that one may be 95% confident that the true concentration lies between 0.1199 and 0.1581.

The data of Table 4 can also be used to illustrate the weighted least-squares method. Suppose that all the y_i are measured with a constant relative standard deviation of 10%. The w_i are calculated as $1/s_{y_i}^2$ and are given in Table 4.

In order to apply eqns. (3) and (4), one first computes the terms

$\sum w_i = 21.3644$ $\quad\sum w_i x_i = 0.9699$

$\sum w_i y_i = 142.1695$ $\quad\sum w_i x_i y_i = 10.2925$

$\sum w_i x_i^2 = 0.07185$

The regression equation then becomes

$$y = 0.39 + 138.09x$$

7. Some further developments

When a non-linear, i.e. a non-straight line, calibration graph is obtained, a number of different approaches are possible.

(1) Restrict the analyses to the linear working range of the calibration curve by making appropriate dilutions of the samples.

(2) Perform a transformation of the variables in order to obtain linearization, e.g. the function $y = ax^b$, encountered in emission spectrometry, can be linearized by a logarithmic transformation, i.e.

$$\log y = a' + b \log x$$

(3) Construct a calibration graph by fitting a polynomial function through the calibration data. This can be done by fitting polynomials of successively higher degree until an F-test on the residuals indicates that introduction of additional terms does not reduce the residual variance significantly (see also Chap. 13).

(4) Perform a segmentation of the calibration curve. A method, proposed by Schwartz [4] is the use of linear segments obtained by connecting two adjacent points. The result, x_s, is calculated from y_s, the response of the sample, by linear interpolation. This method has several disadvantages and should only be used when no computer facilities are available.

Mitchell et al. [5] also proposed a method based on a segmentation of the calibration curve. The principle of what the authors call the multiple curve procedure is that, for each sample, a number of regression equations (1st-, 2nd- or 3rd-order) are calculated using the data of 3 or more contiguous standards enclosing the sample. The equation yielding the smallest confidence interval is chosen to calculate the unknown concentration. The advantage of this method is that the confidence band of the calibration curve will be the narrowest in the neighbourhood of the sample and that, by restricting the working range, the condition of homoscedasticity is more likely to be fulfilled.

References p. 92

(5) Use cubic spline functions to describe the calibration function mathematically. This technique can be used when a large dynamic range is used and/or if the relationship between signal and concentration is complicated. In the chemical literature, this technique was used for spectrochemical analysis using a photographic plate as a detector [6] and also in RIA analyses [7].

A cubic spline function is an interpolating function for which all intervals between two adjacent points are covered by 3rd-degree polynomials, generally one different from the other, whereby the general condition is satisfied that the overall function is continuous with continuous first and second derivatives over the whole range of data points.

References

1 J.S. Garden, D.G. Mitchell and W.N. Mills, Anal. Chem., 52 (1980) 2310.
2 J. Agterdenbos, Anal. Chim. Acta, 108 (1979) 315.
3 J.S. Hunter, J. Assoc. Off. Anal. Chem., 64 (1981) 574.
4 L.M. Schwartz, Anal. Chem., 49 (1977) 2062.
5 D.G. Mitchell, W.N. Mills and J.S. Garden, Anal. Chem., 49 (1977) 1656.
6 P.F. Frigieri and F.B. Rossi, Anal. Chem., 51 (1979) 54.
7 T.G.R. Rawlins and T. Yrjonen, Int. Lab., (Nov., Dec.) (1979).

Recommended reading

N. Draper and H. Smith, Applied Regression Analysis, Wiley, New York, 2nd edn., 1981.
D.A. Kurtz (Ed.), Trace Residue Analysis: Chemometric Estimations of Sampling, Amount and Error, American Chemical Society, Washington, DC, 1985.
L.M. Schwartz, Nonlinear calibration curves, Anal. Chem., 48 (1976) 2287.
L.M. Schwartz, Nonlinear calibration, Anal. Chem., 49 (1977) 2062.
L.M. Schwartz, Statistical uncertainties of analyses by calibration of counting measurements, Anal. Chem., 50 (1978) 980.
L.M. Schwartz, Calibration curves with nonuniform variance, Anal. Chem., 51 (1979) 723.
D.G. Mitchell and J.S. Garden, Measuring and maximizing precision in analyses based on use of calibration graphs, Talanta, 29 (1982) 921.
K.J. Ellis and R.G. Duggleby, What happens when data are fitted to the wrong equation?, Biochem. J., 171 (1978) 513.
J. Mandel and L.G. Nanni, in S.L. Horn (Ed.), Measurement Evaluation: Quality Assurance Practices for Health Laboratories, American Public Health Association, 1978, p. 209.

Chapter 6

Reliability and Drift

The reliability of a test or method can be defined as its ability to maintain accuracy and precision and there are several ways of studying it. One technique is to follow the method over a long period of time to reach a conclusion about the reliability a posteriori, i.e. from experience. This is the technique adopted in modern routine laboratories where it is one aspect of quality control. The a priori approach is to try and predict the reliability, which can be partly achieved by the determination of the "ruggedness" of a test. We shall discuss both aspects in the following sections. The notion of reliability is related to the notion of drift, which can be defined as a systematic trend in the results as a function of time. Drift has been found to occur in many instances, particularly in automatic apparatus in which many determinations per hour are carried out. It should not be concluded that automatic methods are more prone to show drift than manual methods, but rather that the larger series of determinations carried out with the former methods allows easier detection of drift.

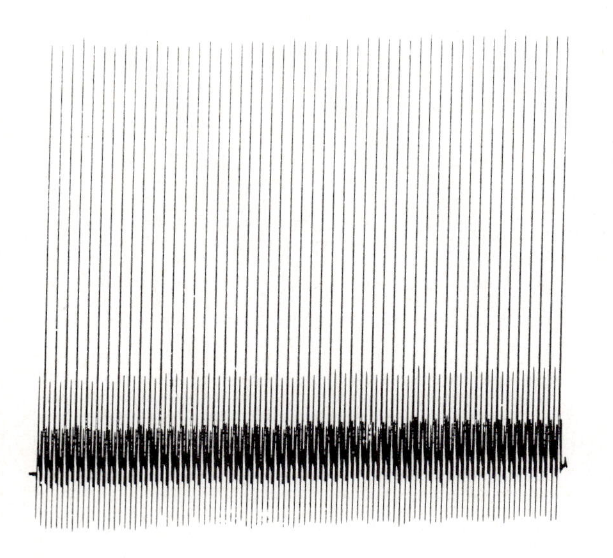

Fig. 1. Sequence of 50 injections of a particular sample in an FIA system.

References p. 105

Most automatic methods carried out as routine are of the continuous type. For that reason, some emphasis will be placed in this chapter on continuous methods. However, the quality control techniques discussed can also be used for routine methods of the non-continuous type.

The sequence of 50 injections of a particular sample in an FIA system (Fig. 1) demonstrates the main sources of disturbance of a continuous analysis, viz.

(1) noise superimposed on both the baseline and the signal,

(2) drift on the baseline, and

(3) drift of the sensitivity coefficient (see Chap. 7).

Usually, a continuous method consists of (1) the measurement of a calibration line relating concentration to signal; (2) the measurement of a certain sequence of samples; and (3) the measurement of control (reference) samples.

Depending on the values obtained for the control samples, various conclusions may follow: (1) the system is still operating under the defined conditions and samples can be measured or (2) the measurement falls outside the "expected" range and recalibration, correction of the results or a checking of the equipment is necessary.

1. The a posteriori approach. Trend detection

1.1 Control chart methods for detection of drift

In clinical chemistry, quality control is defined by the IFCC as "the study of those sources of variation which are the responsibility of the laboratory, and of the procedures used to recognize and minimize them, including all sources (such as random variation and bias) which arise within the laboratory between the receipt of the specimen and the dispatch of the report" [1].

Many methods for trend detection in quality control are currently in use, among which the most common are the control chart methods. In clinical chemistry, they are also called Levey–Jennings chart methods and in industrial analytical chemistry, Shewhart charts. Control samples (samples with known content) are analysed every day or with each run or batch and their values are plotted on a chart as depicted in Fig. 2. The solid line depicts the mean value and the broken lines limits of $\pm 2s$ and $\pm 3s$. These limits have to be determined before starting the quality control scheme. The $2s$ limit is usually called the warning limit and the $3s$ limit the action limit. The laboratory under control follows rules such as: "If one point falls outside the action limits or two consecutive points outside the warning limits but within the action limits, the results are accepted but the procedure is nevertheless investigated", etc.

Clearly, the emphasis is on the detection of time-dependent systematic errors, i.e. errors that influence the accuracy. It is also possible, however, to evaluate a trend in the precision, which necessitates the analysis of replicates. Control chart methods are relevant to our purpose only in that they permit one to observe whether a method remains acceptable or not; they do not allow a quantitative measure of the

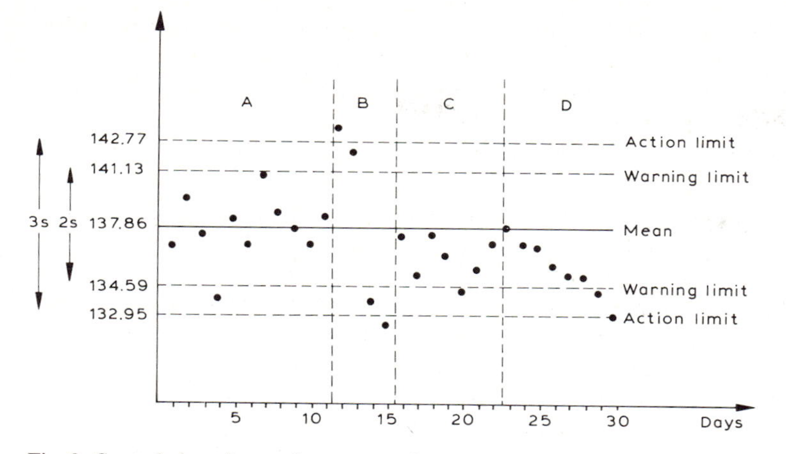

Fig. 2. Control chart for a reference sample.

reliability. As the use of control charts in clinical and analytical laboratories is extremely widespread, we shall discuss them in more detail and give an example.

In a preliminary investigation, the analyst measures the magnitude of the variation he can expect for the determination of a component using a peculiar analytical method in his laboratory. The factors to be controlled are not only the analytical method but also the laboratory equipment, the chemicals, the skill of analysts, etc. In an initial period, where the method is considered stable and ready for routine use, several determinations of a reference specimen will be carried out daily during a certain period of time. The reference specimen used can be commercially available (e.g. lyophilized sera) or prepared in the laboratory from pooled sera. For statistical reasons, at least 20 measurements are required. These data are then used to establish "acceptable limits of errors". An appropriately designed ANOVA scheme allows estimates of the within-day and between-day variability to be calculated and immediately indicate whether day-to-day factors are important. An example of such a design was given in Chap. 4, Sect. 4 [2]. This design consisted of the analysis of 50 samples from the same pool at a rate of two a day from different places in the run of the routine determinations carried out during that day for 25 consecutive days. Using the MS data of the ANOVA table (Table 7 of Chap. 4), one can construct control charts. It is obvious that, when there is a significant between-day variability, the between-day standard deviation should be used to calculate the limits. Amenta proposed the use of data from the ANOVA table as follows. One first calculates the mean of each pair of data. From these 25 values, mean and standard deviation are calculated. It can, however, be shown that this standard deviation can also be calculated as $\sqrt{s^2_{\mathrm{DAY}}/2}$ where s^2_{DAY} is the between-day variance (Chap. 4, Table 7). The $2s$ confidence limits are then calculated as

$$\bar{x} \pm 2 \times \sqrt{s^2_{\mathrm{DAY}}/2} = 137.86 \pm 2 \times 1.64$$

This yields 134.59–141.13 as warning limits and 132.95–142.77 as action limits (Fig. 2).

One also can calculate the standard deviation of the difference (the range) between the two results obtained daily.

$$s_{\text{range}} = \sqrt{2s_{\text{R}}^2} = 1.70$$

where s_{R}^2 is the residual variance. If there is no drift, the average range should be 0. In the same way as above, limits for the range can be calculated as $0 \pm 2s_{\text{range}}$ and $0 \pm 3s_{\text{range}}$ and used to construct a control chart of the ranges. The confidence limits for a single analysis are $\bar{x} \pm 2s_{\text{T}}$ where s_{T} is the total standard deviation. This yields 137.86 ± 3.75 as warning limits for a single analysis.

Figure 2 illustrates an example of a control chart where daily values of the control specimen are plotted. In situation A, the method is under control. In situation B, the method is out of control as the limits are exceeded. In situation C, a trend below the mean is observed: 7 subsequent measurements yield values lower than the mean. It can be calculated that the probability of obtaining a trend of 7 subsequent values in a series of 20 yielding higher or lower values than the mean is less than 5%. When values on several consecutive days distribute themselves on one side of the mean value line, but remain at a constant level, the trend is called a shift. Situation D also illustrates a trend: a progressive falling trend is observed. Trends can be caused by gradual deterioration of a reagent, changes in standard, problems with the equipment, etc. It should be noticed that there is no generally accepted way of calculating the limits. Some laboratories determine their own limits in different ways, other laboratories use the values of the firm which supplied the control serum. In any event, whatever the limits used, it is important for the laboratory to know exactly the meaning of the s used to establish the limits. The laboratories should also know which sources of variation are relevant. By proper design and with application of ANOVA techniques, the different components of the total variation in the data may be estimated.

1.2 The Cusum technique

For a series of control measurements $x_0, x_1, \ldots, x_t$, one determines the cumulative sum of differences between the observed value and the previously determined mean value, $\bar{x}$

$$C_1 = x_1 - \bar{x} \tag{1}$$

$$C_2 = C_1 + (x_2 - \bar{x}) \tag{2}$$

$$C_3 = C_2 + (x_3 - \bar{x}) \tag{3}$$

These values are displayed on a chart such as that shown in Fig. 3. If the deviations from $\bar{x}$ are random, then the C values oscillate around the line at zero, at least, if the mean $\bar{x}$ is an accurate estimate of the true mean value. If not, they will veer away from this line.

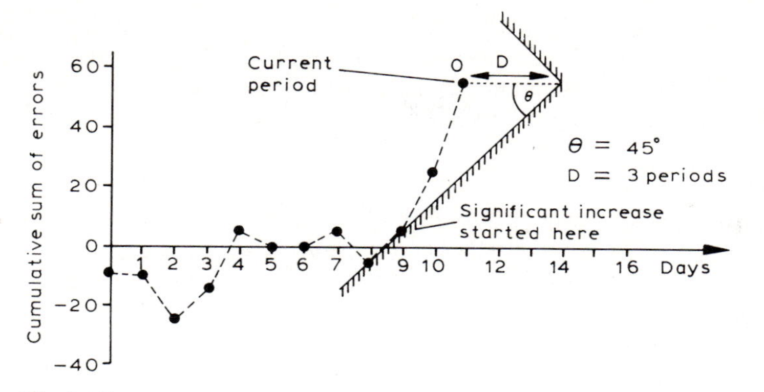

Fig. 3. Illustration of the Cusum method (adapted from ref. 3).

The interpretation of the results obtained is not as straightforward as in the control chart methods. In particular, it is not evident from the Cusum results how one should decide when a trend is significant and when it is not. The most general means of coming to such a decision appears to be the use of a V mask. This is illustrated in Fig. 3.

When one wishes to evaluate a possible trend at time t using C_t, one places the mask so that point O coincides with C_t. If the Cusum line cuts one of the limits of the mask, then the trend is considered to be significant. The difficulty resides in the choice of the angle θ and the distance D, which is somewhat intuitive. Some rules for choosing these parameters were given in ref. 4 and the selection of both the dimensions of the V mask and the process of placing it on the plot have been computerised [5].

1.3 Trigg's monitoring technique

To determine whether or not there is a drift, observations are made at regular times. Such a series of observations is called a time series (see Chap. 14) and the analysis of time series is a classical statistical problem used, for example, in the evaluation of economic trends.

The main difficulty in the analysis of time series as applied here is to separate the long-term effects from irregular, random effects and one of the methods used to do this is the application of moving averages. For a series of control measurements $x_1, x_2, \ldots, x_t$, one defines

$$\frac{x_1 + x_2 + \ldots x_n}{n}, \frac{x_2 + x_3 + \ldots x_{n+1}}{n}, \frac{x_3 + x_4 + \ldots x_{n+2}}{n}, \ldots$$

as the moving averages of order n. Moving averages have the effect of reducing the random variations, thereby smoothing the time series. To avoid too large effects from the extreme values, one often uses weighted moving averages, which are obtained by giving larger weights to the central values than to the extreme values. It appears that these simple methods have not been used (or at least have been used only infrequently) for quality control in analytical chemistry. They, and also the Savitzky–Golay smoothing techniques, are used more frequently in chromatography and spectrometry techniques. They will be discussed in the context of signal analysis in Chap. 15.

A technique, which has been proposed for quality control purposes by Cembrowski et al. [6], is called the Trigg's monitoring technique.

In Trigg's method, one calculates an exponentially weighted average, C_t, of the observations

$$C_t = \alpha x_t + (1-\alpha)C_{t-1} \tag{4}$$

In the same way

$$C_{t-1} = \alpha x_{t-1} + (1-\alpha)C_{t-2} \tag{5}$$

C is also given by

$$C_t = \alpha x_t + \alpha(1-\alpha)x_{t-1} + (1-\alpha)^2 C_{t-2} \tag{6}$$

Equation (4) is equivalent to

$$C_t = \sum_{n=0}^{t-1} \alpha(1-\alpha)^n x_{t-n} + (1-\alpha)^t x_0 \tag{7}$$

In eqn. (7), α is a constant which is generally chosen to be 0.1 or 0.2. If $\alpha = 0.2$, the average time, t, is calculated with a weight of 0.2 for the current observation, $0.2 \times (1-0.2) = 0.16$ for the last but one and 0.128, 0.102, etc., for the preceding ones. The number of observations that should be included depends on α. For $\alpha = 0.2$, the number of observations to be taken into account may be limited to 9. In fact, C_t is a moving average for which the weights of the observations in the calculation decrease with time. The moving average C_t is considered as a predictor for the next observation, x_{t+1}. For the current observation, x_t, the prediction error e_t is given by

$$e_t = x_t - C_{t-1} \tag{8}$$

The smoothed forecast error $\bar{e}_t$ is then calculated according to the same principle as in eqn. (4), i.e.

$$\bar{e}_t = \alpha e_t + (1-\alpha)\bar{e}_{t-1} \tag{9}$$

When $\bar{e}_t$ changes continuously in one direction as a function of time, a changing trend in the results is indicated.

Instead of observing directly the trend in $\bar{e}_t$, one compares $\bar{e}_t$ with the mean absolute deviation, MAD_t, by means of Trigg's tracking signal, T_t

MAD = α latest absolute error + $(1-\alpha)$ previous MAD

or

$$\text{MAD}_t = \alpha|e_t| + (1-\alpha)\text{MAD}_{t-1} \tag{10}$$

$$T_t = \frac{\bar{e}_t}{\text{MAD}_t} \tag{11}$$

TABLE 1

Tracking signal values [6]

Confidence level (%)	Tracking signal	
	$\alpha = 0.1$	$\alpha = 0.2$
70	0.24	0.33
80	0.29	0.40
85	0.32	0.45
90	0.35	0.50
95	0.42	0.58
96	0.43	0.60
97	0.45	0.62
98	0.48	0.66
99	0.53	0.71
100	1.00	1.00

The tracking signal oscillates between +1 and −1 and the more different it is from zero, the more significant is the trend. For example, for $\alpha = 0.2$, a value of $T_t = 0.4$ indicates an 80% confidence level, i.e. that there is an 80% probability that a significant change has taken place (see Table 1). The tracking signal can therefore be used as a criterion to describe drift.

In Table 2, a worked example is given. To be able to apply eqns. (4), (9) and (10), initial values of C_{t-1}, $\bar{e}_{t-1}$ and MAD_{t-1} must be determined. The smoothed forecast error is initially set to zero and the initial mean absolute deviation, MAD, is set to [6]

$$\mathrm{MAD}_{t-1} = \sqrt{\left(\frac{2}{\pi}\right)}\, s$$

The standard deviation, s, was calculated in preceding experiments and was found to be 10.027. C_{t-1}, the immediate past exponentially weighted average, is initially set to the average value of previous determinations.

In the example given, the values of C_{t-1}, $\bar{e}_{t-1}$ and MAD_{t-1} at time zero are 50, 0 and 8, respectively. α is chosen to be 0.2. All calculations to be effected in order to obtain the tracking signal are described in Table 2.

Since up to $t = 8$ the value of T_t does not exceed 0.50, which, according to Table 1, corresponds to a 90% confidence level, no statistically significant change occurs up to that time. At $t = 9$, however, the tracking signal reaches 0.60, which corresponds to a 95% confidence level. At $t = 10$, $T_t = 0.68$, which means that, at this time, an increase has taken place with a 98% level of confidence. In the literature, a level of 97% is considered high enough to infer that a significant change has occurred. This method has also been implemented on a microcomputer [7].

1.4 Other statistical methods

Various other statistical techniques have been applied to test whether or not there is a drift in the results, for example, the chi-square test. This test, which is discussed

TABLE 2

Example of calculation of Trigg's tracking signal

	$t=0$	1	2	3	4	5	6	7	8	9	10
	$x_t=40$	52	36	55	47	61	57	49	60	65	64
C_{t-1}	50	48	48.8	46.24	47.99	47.79	50.43	51.74	51.19	52.95	55.36
$e_t = x_t - C_{t-1}$	−10	4	−12.8	8.76	−0.99	13.21	6.57	−2.74	8.81	12.05	8.64
αx_t	8	10.4	7.2	11	9.4	12.2	11.4	9.8	12	13	12.8
$(1-\alpha)(C_{t-1})$	40	38.4	39.04	36.99	38.39	38.23	40.34	41.39	40.95	42.36	44.29
$C_t = \alpha x_t + (1-\alpha)C_{t-1}$	48	48.8	46.24	47.99	47.79	50.43	51.74	51.19	52.95	55.36	57.09
αe_t	−2	0.8	−2.56	1.75	−0.2	2.64	1.31	−0.55	1.76	2.41	1.73
$(1-\alpha)\bar{e}_{t-1}$	0	−1.6	−0.64	−2.56	−0.65	−0.68	1.57	2.30	1.40	2.53	3.95
$\bar{e}_t = \alpha e_t + (1-\alpha)\bar{e}_{t-1}$	−2	−0.8	−3.2	−0.81	−0.85	1.96	2.88	1.75	3.16	4.94	5.68
$\alpha\|e_t\|$	2	0.8	2.56	1.75	0.2	2.64	1.31	0.55	1.76	2.41	1.73
$(1-\alpha)\mathrm{MAD}_{t-1}$	6.4	6.72	6.02	6.86	6.89	5.67	6.65	6.37	5.54	5.84	6.60
$\mathrm{MAD}_t = \alpha\|e_t\| + (1-\alpha)\mathrm{MAD}_{t-1}$	8.4	7.52	8.58	8.61	7.09	8.31	7.96	6.92	7.30	8.25	8.33
$T_t = \dfrac{\bar{e}_t}{\mathrm{MAD}_t}$	−0.24	−0.11	−0.37	−0.09	−0.12	0.24	0.36	0.25	0.43	0.60	0.68

in more detail in Chap. 3, is used to discriminate between different distributions of data. When a significantly different distribution compared with former distributions is found, this may indicate a problem and the cause of this problem must then be investigated. Laboratory error is only one possible source of such changes. Other possible causes are not within the scope of analytical chemistry; they include changes of population, medical treatment, diet, etc.

Another way of detecting changes of distribution is by means of the non-parametric one-sample runs test. One uses the results of, for example, 20 control determinations and determines the median, the observations with a lower result being denoted by a minus sign and those with a higher result by a plus sign. A sequence of identical signs is called a run. In the following example, the median equals 8 [(7 + 9)/2] and there are 9 runs.

Sample	1	2	3	4	5	6	7	8	9	10	11	12	13	14	15	16	17	18	19	20
Result	5	9	9	10	7	3	7	10	9	7	9	9	4	9	10	9	4	6	2	3
Sign	−	+	+	+	−	−	−	+	+	−	+	+	−	+	+	+	−	−	−	−
Run	—	——	——	——	——	——	——	——	——	—	——	——	—	——	——	——	——	——	——	——

Statistical tables displaying the distribution of the number of runs in samples of size (n_1, n_2) being respectively the number of positive and negative results, indicate that the probability of finding 9 or fewer runs is 0.242. The critical value for rejecting at the 0.05 probability level the hypothesis that there is no drift is 6 runs.

2. The a priori approach. Ruggedness of a method

The reason for a decrease of reliability of a method with time is that it is sensitive to minor changes in procedure, such as variations in concentrations of reagents, heating rates, etc. One can, of course, try new methods and see how they behave over a long period of time in order to test their reliability (see Sect. 1). However, it is preferable to have an idea of the reliability to be expected. This can be obtained by measuring the sensitivity of the method to small variations.

It is clear that insensitivity to small changes in procedure is an important asset for an analytical method. Therefore, this property, which has been called "ruggedness", can be considered as an evaluation criterion. An insufficiently "rugged" method is also subject to large laboratory biases. As we have already stated, it is unfortunate, but well known, that methods proposed in the literature do not always yield the expected good results. Laboratory bias is estimated (see Chaps. 3 and 4) by collaborative research programs using the two-sample method or analysis of variance. These collaborative programs require important organizational efforts, so that it is out of the question to subject all promising methods in an early stage of development to such programs. Here again, an a priori approach, permitting a prediction of the laboratory bias to be expected, would be useful. If can be concluded that a measure of the ruggedness gives an idea of the day-to-day or between-laboratory variations to be expected. This can be done using methods developed first by Plackett and Burman [8] and introduced in analytical chemistry

TABLE 3

Partial factorial experiment for seven factors

Experiment	Factors							Measurement
	A	B	C	D	E	F	G	
1	+	+	+	+	+	+	+	y_1
2	+	+	−	+	−	−	−	y_2
3	+	−	+	−	+	−	−	y_3
4	+	−	−	−	−	+	+	y_4
5	−	+	+	−	−	+	−	y_5
6	−	+	−	−	+	−	+	y_6
7	−	−	+	+	−	−	+	y_7
8	−	−	−	+	+	+	−	y_8

to test official methods by Youden and Steiner [9]. To study the effect of minor and inevitable variations, one could carry out a factorial experiment (see Chap. 17). In this instance, one could use a two-level experiment, one of the levels being that given in the proposed procedure and the other a level which deviates from the former to an extent that can be reasonably conceived to occur in practice. Denoted by plus and minus signs and following the practice introduced by Plackett and Burman, these levels are respectively called the nominal and extreme values. In this type of investigation, one would like to include a large number of parameters and therefore introduce a large number of factors in the factorial experiment. If this number is n, then the number of experiments to be carried out in a complete design is 2^n. As the number of factors is often larger than 5, it is clear that complete factorial experiments are often impractical for this purpose. Several designs have been proposed to obtain an estimate in a much smaller number of experiments. To understand this, let us first consider the design of Table 3 for seven factors using eight experiments.

This means that eight experiments are carried out, each yielding a result, $y_1, \ldots, y_8$. The third experiment, for example, is carried out in such a way that factors A, C and E take their nominal values while the others are at the extreme level. Note that Table 3 is constructed in such a way that each factor occurs four times at the nominal and four times at the extreme level. To determine the effect of changing factor A from the nominal + level to the extreme − level, one compares the mean value of the results obtained at both levels. In this instance, this means carrying out the operation

$$D_A = \frac{y_1 + y_2 + y_3 + y_4}{4} - \frac{y_5 + y_6 + y_7 + y_8}{4}$$

For factor C, the following calculation should be carried out

$$D_C = \frac{y_1 + y_3 + y_5 + y_7}{4} - \frac{y_2 + y_4 + y_6 + y_8}{4}$$

One observes that, in doing this, one divides the experiments into two groups for each factor. In one of these groups, the factor being investigated is at the + level. In each group, all the other factors are twice at the − level and twice at the + level. When carrying out the comparison of averages, the effects of all of the other factors cancel out. In fact, this is completely true only when there is no interaction (see Chaps. 16 and 17) but, as the variations introduced in the factor levels are small, this will be of relatively little importance.

Obtaining the differences $D_A, \ldots, D_G$ is not sufficient in itself and one must determine whether these differences are significantly greater than the experimental error determined by carrying out replicate measurements at the nominal level. These replicate measurements do not involve extra work as they have probably been carried out already by the author of the method in order to measure the repeatability of the proposed procedure. If this is characterized by a standard deviation, s, then when there is no significant factor, the standard deviation on the mean of four measurements is $s/\sqrt{4}$ and the standard deviation, s_D, on the difference between two averages

$$s_D = \frac{\sqrt{2s^2}}{\sqrt{4}} = \frac{s}{\sqrt{2}}$$

(see Chap. 3). The expected mean of the D distribution being zero (again, when there is no significant factor), one can consider that a factor is significant when D is larger than $2s_D = \sqrt{2}\,s$. When, in actual experimentation, a significant factor is noted, steps should be taken to eliminate it or, as this is often impossible, the procedure should state clearly the limits between which the parameter may be allowed to vary.

This design is very elegant but unfortunately it is impossible to propose an analogous device for, for example, six factors with seven experiments. The solution to this difficulty is to continue carrying out the above design in which one of the variables is now a dummy one. As Youden and Steiner stated, one should "associate

TABLE 4

Partial factorial experiment for six factors

Experiment	Factors						Measurement
	A	B	C	D	E	F	
1	+	+	+	+	+	+	y_1
2	+	+	−	+	−	−	y_2
3	+	−	+	−	+	−	y_3
4	+	−	−	−	−	+	y_4
5	−	+	+	−	−	+	y_5
6	−	+	−	−	+	−	y_6
7	−	−	+	+	−	−	y_7
8	−	−	−	+	+	+	y_8

References p. 105

TABLE 5

Partial factorial design for three factors

Experiment	Factors			Measurement
	A	B	C	
1	+	+	+	y_1
2	−	+	−	y_2
3	+	−	−	y_3
4	−	−	+	y_4

with factor G some meaningless operation such as solemnly picking up the beaker, looking at it intently and setting it down again". This means, in fact, that one uses the design of Table 4.

In Table 5, a design is given for the three-factor situation.

So-called cyclical designs with the same properties have been proposed for higher numbers of factors by Plackett and Burman. Consider, for example, the 11-factor, 12-experiment design. The design is obtained from the line

+ + − + + + − − − + −

and describes experiment 1. Experiments 2–11 are obtained by writing down all cyclical permutations of this line and the last, experiment 12, contains only minus signs. The complete design is given by Table 6.

In the same way as for the 6-factor example, arrangements for 8–10 factors can be derived from the 11-factor design and for 12–14 factors from a 15-factor design, using dummy factors. These methods have been called partial factorial experiments. As discussed later, they make it possible to investigate only main factors and are less useful for optimization purposes when interaction occurs.

TABLE 6

Partial factorial design for eleven factors

Experiment	Factors											Measurement
	A	B	C	D	E	F	G	H	I	J	K	
1	+	+	−	+	+	+	−	−	−	+	−	y_1
2	−	+	+	−	+	+	+	−	−	−	+	y_2
3	+	−	+	+	−	+	+	+	−	−	−	y_3
4	−	+	−	+	+	−	+	+	+	−	−	y_4
5	−	−	+	−	+	+	−	+	+	+	−	y_5
6	−	−	−	+	−	+	+	−	+	+	+	y_6
7	+	−	−	−	+	−	+	+	−	+	+	y_7
8	+	+	−	−	−	+	−	+	+	−	+	y_8
9	+	+	+	−	−	−	+	−	+	+	−	y_9
10	−	+	+	+	−	−	−	+	−	+	+	y_{10}
11	+	−	+	+	+	−	−	−	+	−	+	y_{11}
12	−	−	−	−	−	−	−	−	−	−	−	y_{12}

TABLE 7

Partial factorial design for three factors

Experiment	Factors			Measurement
	A	B	C	
1	+	+	+	$y_1 = 0.200$
2	−	+	−	$y_2 = 0.218$
3	+	−	−	$y_3 = 0.240$
4	−	−	+	$y_4 = 0.206$

$$D_A = \frac{y_1 + y_3}{2} - \frac{y_2 + y_4}{2} = 0.008$$

$$D_B = \frac{y_1 + y_2}{2} - \frac{y_3 + y_4}{2} = -0.014$$

$$D_C = \frac{y_1 + y_4}{2} - \frac{y_2 + y_3}{2} = -0.026$$

To illustrate calculations, a simple example of a partial factorial design for 3 factors is given. The factors investigated for, for example, a colorimetric method could be reaction time, reaction temperature, pH, concentration of reagents, etc. The data are given in Table 7.

Factor A:	pH	+ level 8;	− level 8.5
Factor B:	temperature	+ level 20 °C;	− level 22 °C
Factor C:	concentration	+ level 0.10 M;	− level 0.12 M

The standard deviation obtained for replicate measurements at the nominal level is 0.010. When $|D|$ is larger than $\sqrt{2}\,s$, the factor contributes significantly to the result. $\sqrt{2}\,s = 0.014$. One concludes that factor C should be controlled more strictly. In practice, this means that factor C is liable to introduce errors when the method is used. Therefore, the procedure should specify more strictly the level of concentration to be used, e.g. 0.10 M instead of 0.1 M, with perhaps the warning that the concentration level is a critical factor.

A nice example of the Plackett and Burman design for the evaluation of the interference of 7 elements on the AAS determination of Pb, Cd and Ni was given by Feinberg and Ducauze [10].

References

1 J. Buttner, R. Borth, J.H. Boutwell, P.M.G. Broughton and R.C. Bowyer, Clin. Chim. Acta, 98 (1979) 129F.
2 J.S.A. Amenta, Am. J. Clin. Pathol., 49 (1968) 842.
3 C.D. Lewis, Med. Biol. Eng., 9 (1971) 315.
4 H.M. Taylor, Technometrics, 10 (1968) 479.
5 C.D. Lewis, Qual. Assur., 6 (1980) 3.
6 G.S. Cembrowski, J.O. Westgard, A.A. Eggert and E.C. Toren, Jr., Clin. Chem., 21 (1975) 1396.
7 T.F. Hartley and T.W. Huber, Trends Anal. Chem., 3 (1984) 251.
8 R.L. Plackett and J.P. Burman, Biometrika, 23 (1946) 305.

9 W.J. Youden and E.H. Steiner, Statistical Manual of the A.O.A.C., Association of Official Analytical Chemists, Washington, DC, 1975.
10 M. Feinberg and C. Ducauze, Analusis, 8 (1980) 185.

Recommended reading

J.O. Westgard and T. Groth, Design and evaluation of statistical control procedures: applications of a computer "quality control simulator" program, Clin. Chem., 27 (1981) 1536.

G. Kateman and F.W. Pijpers, Quality Control in Analytical Chemistry, Wiley, New York, 1981.

J.O. Westgard, P.L. Barry, M.R. Hunt and T. Groth, A multi-rule Shewart chart for quality control in clinical chemistry, Clin. Chem., 27 (1981) 493.

R.W.H. Edwards, Internal analytical quality control using the cusum chart and truncated V-mask procedure, Ann. Clin. Biochem., 17 (1980) 205.

Chapter 7

Sensitivity and Limit of Detection

1. Introduction

Many problems in analytical chemistry are problems of detecting and determining elements or compounds in small amounts of sample (microanalysis), of determining very low concentrations or small amounts in larger samples (trace analysis) or even of determining low concentrations in small samples. Progress in analytical chemistry might well be measured by the shift of the detection limit towards lower values.

Comparing analytical procedures according to their limits of detection is not easy. In many papers describing analytical procedures, no detection limits are given and, to the analyst facing the problem of choosing a procedure from several possibilities, this omission is very disappointing. Even more disappointing is the lack of uniformity in describing performances with respect to the smallest amounts or concentrations that can be detected or determined with certainty.

Often, a procedure is said to be very "sensitive" when the limit of detection is low and the "limit of detection" and "sensitivity" are often considered to be synonymous. However, sensitivity is defined as the slope of the curve obtained when the results of the measurements are plotted against the amounts that are determined. In analytical chemistry, sensitivity defined in this way is equal to the slope of the analytical calibration curve and throughout this book, this definition of sensitivity will be used. The lower limit of detection is to be understood as the limit below which detection is not possible with a specified degree of certainty.

The IUPAC definition [1] states that "the limit of detection, expressed as the concentration, c_L, or the quantity, q_L, is derived from the smallest measure, x_L, that can be detected with reasonable certainty for a given analytical procedure". The definition, given in ref. 2 states that "the limit of detection is the lowest concentration of an analyte that an analytical process can reliably detect". To clarify terms such as "reasonable certainty", or "reliably detect", one has to take into account that random measurement errors are associated with an analytical method. Although these definitions clarify the meaning of the term, they certainly are not sufficient when the detection limit is to be used as a performance characteristic of an analytical procedure. It appears that quantification of this characteristic gives rise to considerable confusion, as has been clearly demonstrated by Currie [3]. Figure 1 gives several values of limits of detection for a specific radioactivity measurement process. These values were calculated using different definitions. The differences can be partly ascribed to differences in formulating the problem (when is a component detected and with what certainty?).

References p. 114

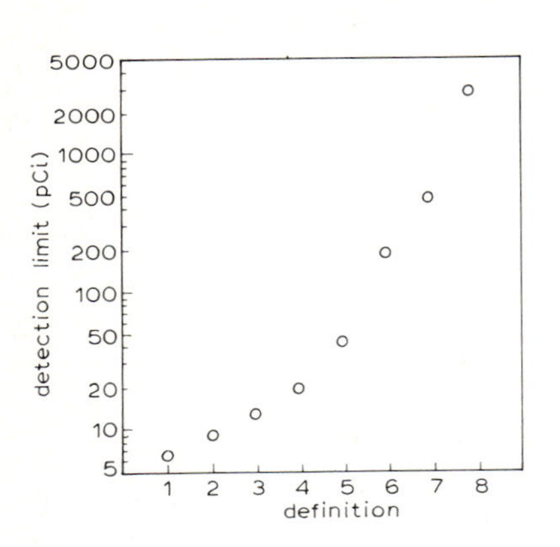

Fig. 1. "Ordered" detection limits calculated from literature definitions. The detection limit for a specific radioactivity measurement process is plotted in ascending order according to commonly used alternative definitions [3]. Reprinted with the permission of The American Chemical Society.

Uncertainties in the lower limits to the detection of elements and compounds arise because of the presence of uncertainties (errors or noise) in the measured analytical result. Therefore, a definition and quantification of detection limits must be based upon statistics.

The confusion about the term was also expressed in a report by Long and Winefordner [4] who stated that "the limit of detection is the lowest concentration level that can be determined to be statistically different from an analytical blank". Significant problems have been encountered in expressing these values because of the various approaches to the term "statistically different". The discussion of the detection limit and related quantities in this chapter is based on papers by Kaiser [5], Currie [3], Boumans [6], and Long and Winefordner [4].

2. Sensitivity and the analytical calibration function

The sensitivity of a procedure designed for a quantitative analysis can be defined as the slope of the analytical calibration function $y = \mathrm{f}(x)$. This calibration function relates the result, y, of the measuring process (output, analytical signal) to the concentration or amount, x, of the component to be determined. The output can be a meter reading, an electric current or voltage, a weight, etc. The sensitivity, S, can be written as the differential quotient

$$S = \frac{\mathrm{d}y}{\mathrm{d}x} \tag{1}$$

For linear relationships between x and y and in the absence of a blank, the sensitivity is simply the ratio between y and x. Figure 2 illustrates the concept of sensitivity. The calibration lines in Fig. 2 were obtained by plotting the absorbance against the iron concentration. It appears that the determination of iron with

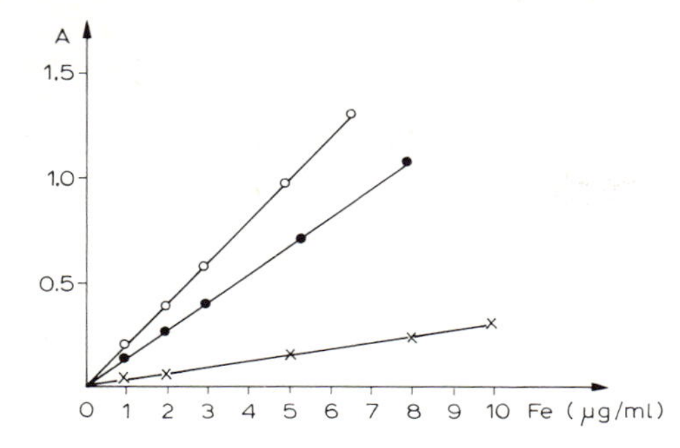

Fig. 2. Calibration lines for the photometric determination of iron (definition of sensitivity), ○, With 2-pyridine aldoxime, $dA/dc = 0.18$; ●, with *o*-phenanthroline, $dA/dc = 0.14$; ×, with 2,6-pyridinedicarboxylic acid, $dA/dc = 0.028$. Thickness of cell = 1 cm [7].

2-pyridine aldoxime has a greater sensitivity than the procedure with *o*-phenanthroline and 2,6-pyridinedicarboxylic acid.

For purposes of characterizing analytical procedures, the sensitivity is of limited importance. The sensitivity of an analytical procedure can often be changed. For example, connecting an amplifier to the output of an instrument can easily bring the output from the millivolts to the volts range and, according to the definition, the sensitivity is then increased by a factor of 1000. Similarly, the sensitivity of a photometric determination can be increased by increasing the optical path length.

Sensitivities are seldom constant over large concentration ranges and sensitivities are therefore meaningful only when concentrations or concentration ranges are specified. Here again, sensitivities can be easily manipulated; a wide variety of linearizing devices are available. This does not mean, of course, that any calibration graph is acceptable to the analytical chemist and one must at least be cautious about non-linear calibration graphs. In some instances, theoretical considerations can lead to a fully justified linearization of the calibration graph and the use of logarithms in spectrophotometry (Lambert–Beer laws) and potentiometry (Nernst) is well known in this respect. In other instances, non-linear graphs are due to saturation effects, sometimes even resulting in a change of slope from positive to negative.

The range over which the sensitivity can be considered to be constant has, of course, lower and upper boundaries. Often, the lower boundary will be the detection limit (as defined in Sect. 1) and the concentration where the sensitivity begins to change (going from lower to higher concentrations) can be regarded as the upper boundary. In general, such a change will be gradual and the upper boundary cannot be specified unless a specification is given of what is to be considered to be a straight calibration graph. A definition of the upper boundary is the concentration where the response differs by a certain percentage (for instance, 3%) from the response that might be expected from the sensitivity near the detection boundary. As far as we know, no generally accepted definition of the upper limit and thus of the linear range for characterizing a procedure has been proposed. The concept of

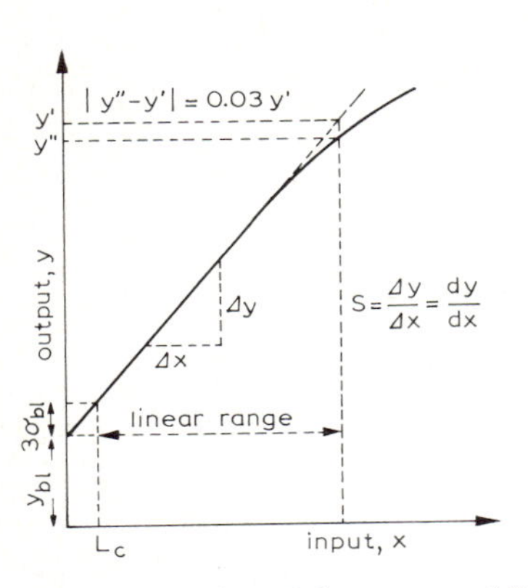

Fig. 3. Illustration of the concept of linear range.

the linear range is illustrated in Fig. 3. The linear range is usually expressed as the number of decades between the lower and upper limit.

3. Decision limit

An analytical chemist has to accept that random errors are unavoidable and therefore also that there are limits to the detection (and thus to the determination) of elements and compounds. He intuitively may feel that it makes no sense to try to detect or determine amounts that are smaller than the random errors inherent to the procedure used. In fact, a rough estimate of the detection limit could be made by taking the value of the standard deviation (in units of concentration or amount). However, this rough picture needs some refinement.

The concentration or amount of the component to be determined, x, can be calculated from the measurement, y, by making use of the calibration function, $y = f(x)$. The discussion will be given in terms of signals. Usually y, the measured signal, is regarded as the difference between two measurements, i.e. a measurement of the unknown sample, y_u, and a measurement of the blank y_{bl}. The problem can thus be formulated in two essentially identical ways: it can be questioned whether y_u differs significantly from y_{bl} or whether $y_u - y_{bl}$ differs significantly from zero. The answer to this question can be obtained by means of statistics. However, some assumptions have to be made about the distribution of errors. The case of a normal distribution of the reading of the blank is represented in Fig. 4. The standard deviation is denoted by σ_{bl} and the true value of the blank by μ_{bl}.

It is clear that the probability of measuring signals $y_{bl} > L_c$ will be

$$\alpha = \int_{L_c}^{\infty} p(y_{bl})\, d y_{bl}$$

where $p(y_{bl})$ represents the distribution function of y_{bl}. If signals larger than the decision limit, L_c, are interpreted as "component present", then a fraction, α, of the measurements of the blank will be misinterpreted.

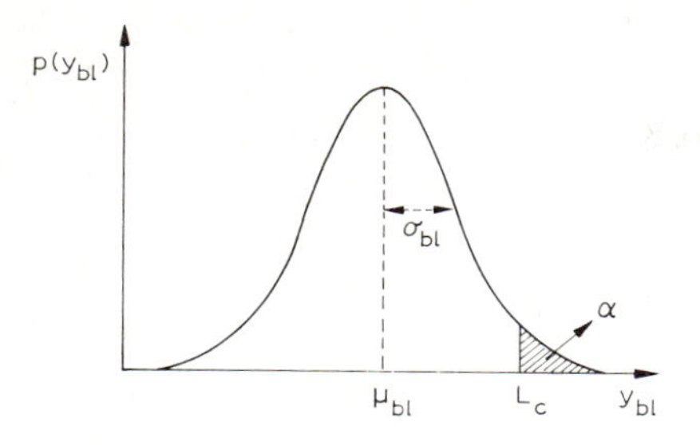

Fig. 4. Normal distribution of y_{bl}.

The critical value, L_c, thus depends on both the standard deviation of the distribution and the risk one is willing to take of making a wrong decision. The decision limit can be expressed in terms of signals by

$$L_c = \mu_{bl} + k_c\sigma_{bl} \quad (2)$$

Introducing a value of $k_c = 3$ leads to a probability $1 - \alpha = 99.87\%$ if y_{bl} is normally distributed. One thus runs a risk of $\alpha = 0.13\%$ that a true blank would be interpreted as having the component present. This error is called the α error (see also Chap. 3).

L_c is thus the lowest meaningful signal (and through the calibration constant, the lowest concentration) that the analytical procedure will ever yield. For $k_c = 3$, L_c is equal to Kaiser's detection limit and also to IUPAC's detection limit.

The decision limit, L_c, cannot, in principle, be used by itself as a quality criterion for the analytical procedure. Indeed, if the limit of decision is defined in this way, the probability of not detecting the analyte when it is present with a concentration yielding a signal L_c is 50%, which is unacceptable. This is illustrated by Fig. 5 where the two probability distribution functions of y_{bl} and y_u, the signal of the unknown sample, overlap. Thus, Fig. 5 represents the situation of a large number of repeated measurements on a sample with a concentration corresponding to the decision limit L_c.

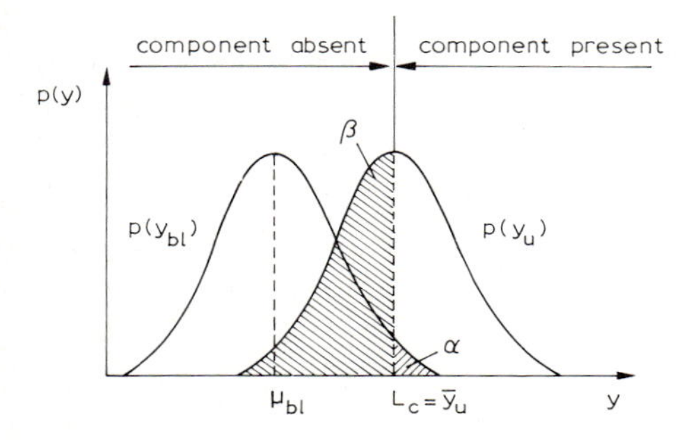

Fig. 5. Illustration of decision limit.

References p. 114

The standard deviations σ_{bl} and σ_u are considered to be equal (which is usually the case for small concentrations). Signals larger than L_c would be interpreted as "component present". However, a fraction, β, of the measurements on a sample with a content L_c of the component to be detected will yield signals smaller than L_c. β is given by

$$\beta = \int_{-\infty}^{L_c} p(y_u)\, dy_u$$

From Fig. 5, it is apparent that $\beta = 0.5$ and the statement about the absence of the component is very unreliable. To express this differently: the error of the first type (deciding that the component is present when it is not) is small (α), whereas the error of the second type (deciding that the component is absent when it is present) is large (β) (see Chap. 3). Signals larger than L_c can be interpreted as the detection of the component with great certainty, whereas signals smaller than L_c can be interpreted as the absence of the component with poor certainty ($< 50\%$).

4. Detection limit

The a posteriori decision about the presence of a component from a measured signal has resulted in a definition of the decision limit as given above. In order to characterize an analytical procedure, it is necessary to define a level, L_d, specifying the detection capabilities of the analytical procedure. This level, the detection limit, should correspond to a concentration that, with great probability, will yield signals that can be distinguished from the signals obtained from the blank. This, of course, corresponds to reducing the error of the second type and thus of reducing β. In other words, the limit of detection should be defined in such a way that α and β are well balanced. In Fig. 6, a situation is represented where $\alpha = \beta$. The mean of y_u, $\bar{y}_u$, can be used for defining a detection limit, L_d.

$$L_d = \mu_{bl} + k_d\sigma_{bl} = L_c + k'_d\sigma_{bl} \tag{3}$$

Here, again, the standard deviations of the distributions $p(y_{bl})$ and $p(y_u)$ have been

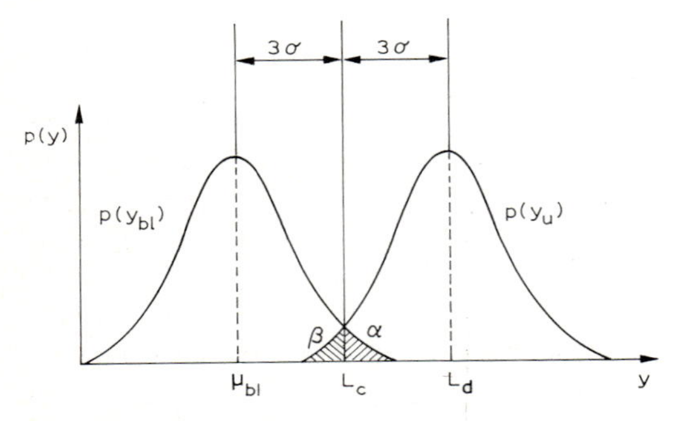

Fig. 6. Illustration of detection limit.

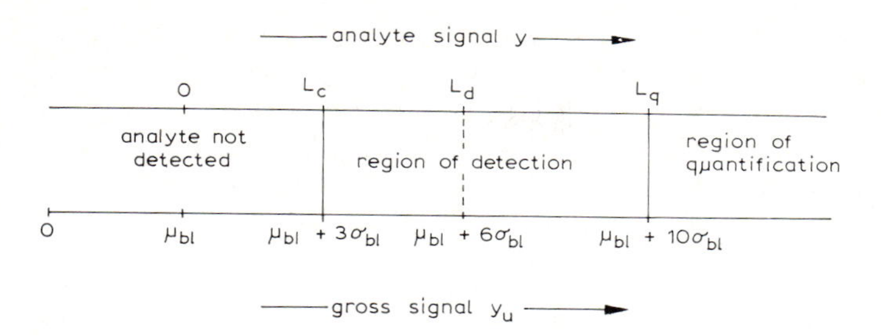

Fig. 7. Illustration of decision limit, detection limit, and quantification limit. Adapted from ref. 2.

assumed to be identical. If a concentration is equal to the detection limit, for which $k_d = 6$ or $k'_d = 3$, it can be detected with 99.87% certainty. Smaller concentrations can be detected with less confidence.

5. Determination limit

A determination or quantification limit can be defined as the limit at which a given procedure will be sufficiently precise to yield a satisfactory quantitative estimate of the unknown concentration. In other words, the limit of determination is the concentration that can be determined with a fixed maximum relative standard deviation. Such a limit, L_q, can be defined in terms of μ_{bl} and σ_{bl}, again assuming that the standard deviations for blank and unknown are identical.

$$L_q = \mu_{bl} + k_q \sigma_{bl} \tag{4}$$

k_q is 20 if the maximum allowed relative standard deviation is 5% or 10 if it is 10%. The relative standard deviation obtained from measurements at the level L_q, is thus $1/k_q$. This means that the relative standard deviation of the "quantitative" measurement at the decision level L_c is 33.33% and at L_d 16.67%.

The definitions of the decision limit, the limit of detection and the limit of quantification are illustrated in Fig. 7.

6. Discussion

One realizes from the literature that there is some confusion about the way detection limits are defined, determined, and interpreted. Until there is consensus about definitions, analytical chemists should be clear about the way in which limits have been calculated. Definitions have been formulated here in terms of the limiting mean of the blank, μ_{bl}, and the standard deviation of the blank. In practice, only a limited number of experiments will be available for the estimation of these quantities. Therefore, we must take into account that the calculated standard deviation is only an estimate, so that the constants k_c and k_d in eqns. (2) and (3) should be replaced by t values for $(n-1)$ degrees of freedom derived from Student's t-distri-

References p. 114

bution. Surely, the detection limit does not change by orders of magnitude if a reasonable number of measurements of the blank have been made. If the estimates are used for calculation of the limits of decision and detection, an uncertainty is introduced and deciding whether a measurement of the unknown differs significantly from the blank should therefore be carried out with the *t*-test. The detection limit is easily calculated from the standard deviation and its reliability can be taken into account when the number of degrees of freedom is known. It should be noted that, in general, it is not permissible to calculate detection limits from standard deviations obtained from measurements at concentration levels much higher than the detection limit, or at least the analyst has to be aware of the pitfalls in doing so. Standard deviations are usually a function of concentration (Chaps. 2 and 5).

Detection limits should be regarded as characteristics of well-described analytical procedures. It makes no sense, for instance, to specify the detection limit for titrations in general. A change in conditions will lead to a change in the procedure and possibly to a change in the limits. A range of limits can, however, be given for a number of techniques.

The nature of the procedure (usually the measurement) will lead to a formulation of the detection limit either in terms of amounts of the compound to be detected or in terms of concentrations. With specified amounts of sample, concentrations can easily be converted into amounts and vice versa. However, it is essential to specify the units when quoting values for the performance characteristics.

The discussion about detection limits was given here in terms of signals. Signals are converted into concentrations making use of the calibration function. As the measurements of the slope and the intercept are also subject to errors, these errors can be taken into account to calculate detection limits.

Long and Winefordner [4] have discussed methods to determine the detection limit which take into account errors in the measurement of the analytical sensitivity.

References

1 IUPAC, Spectrochim. Acta, 33 B (1978) 242.
2 G.H. Morrison, Anal. Chem., 52 (1980) 2241A.
3 L.A. Currie, Anal. Chem., 40 (1968) 586.
4 G.L. Long and J.D. Winefordner, Anal. Chem., 55 (1983) 712A.
5 H. Kaiser, Anal. Chem., 42 (1970) 26A.
6 P.W.J.H. Boumans, Spectrochim. Acta, 33B (1978) 625.
7 H. Specker, Angew. Chem. Int. Ed. Engl., 7 (1968) 252.

Recommended reading

S. Ebel and K. Kamm, Statistische Definition der Bestimmungsgrenze, Fresenius Anal. Chem., 316 (1983) 382.

Chapter 8

Selectivity and Specificity

1. Interferences and matrix effects

One of the simpler analytical problems is the quantification of one element or compound in a sample. Therefore a property y is measured to derive the concentration or amount x. In many cases, the quantification of x through the measurement of y is disturbed because the property y also depends on factors other than x, e.g. temperature, amount of sample and reagents, instrumental factors, etc. Apart from inherent (random) fluctuations, the relationship between x and y is deterministic (see Chap. 12 for a definition of this term). Thus, the analytical calibration function, $y = f(x)$, should be regarded as a characteristic for the analytical procedure.

As a matter of fact, analytical chemists would wish analytical quantification to be as simple as that. Daily practice, however, proves the contrary. Analytical calibration functions are usually influenced by the presence of other compounds in the sample, generally denoted as the matrix of the sample. For instance, the relationship $y = f(x)$ found for the determination of calcium in a "synthetic" solution will not necessarily be valid for the determination of calcium in a real sample such as sea water. In the terminology of the preceding chapter, the sensitivity of the analytical sensors (e.g. the absorbance at a given wavelength) may be influenced by the presence of other compounds. This, of course, constitutes an important complication and much work has been done to cope with this difficulty. One solution is to describe carefully the sample matrix when describing a procedure. In many instances, however, this is not feasible and a solution has to be found in devising suitable calibration methods or in developing better analytical procedures.

Generally speaking, every sample is a multicomponent mixture, which consists of two distinctive parts: (1) the analytes, which are the compounds to be determined quantitatively and (2) the matrix, which is the set of all compounds that may influence the measurements, but for which quantification is not wanted. The basis for the quantification is the analytical signal, which is provided by the analytical sensor. An example of an analytical method with one sensor is the measurement of the voltage of an ion selective electrode in a sample solution. A UV–vis spectrometer can be considered as a multi-sensor device, though it is equipped with only one detector. A measurement at a given wavelength for the quantifying of the analytes in the sample can be considered as a measurement with a given sensor.

The determination of an analyte may be disturbed in three ways.

(1) The matrix and/or other analytes influence the sensitivity of the sensor for the analyte to be determined. For example, in X-ray fluorescence, the peak height usually depends on the content of the corresponding elements and on the constitu-

References p. 126

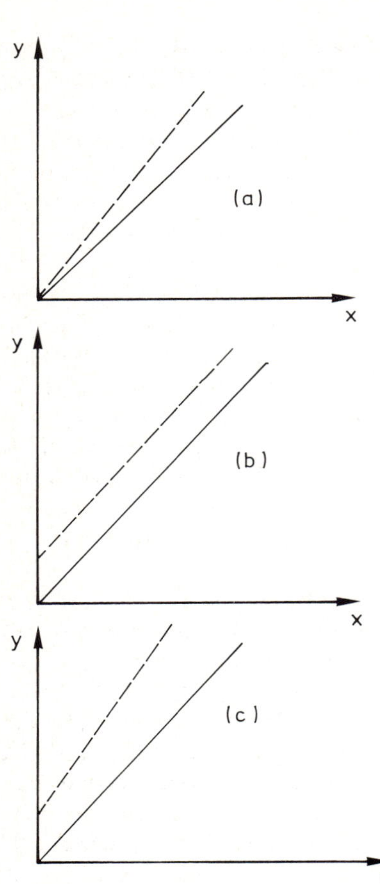

Fig. 1. Possible disturbances to the determination of an analyte. (a) Matrix effect; (b) interference; (c) combined interference and matrix effect.

tion of the matrix [Fig. 1(a)]. In analytical chemistry, this effect is generally known as a matrix effect.

(2) Some elements present in the matrix contribute to the analytical response of the sensor, but without influencing the sensitivity of the sensor for the analyte itself. Gas chromatographic detectors and UV–vis spectrometers are typical examples of such sensors. When two compounds of a mixture elute together, the detector response (e.g. a flame ionization detector) is the sum of the responses generated by both compounds. In UV–vis spectrometry, many compounds absorb radiation of the same wavelength. When several of such compounds are present in the sample, the measured absorbance is the sum of the absorbances of all separate compounds. The effect is called interference and is shown schematically in Fig. 1(b).

(3) The most complex situation is encountered when matrix effects and interferences occur simultaneously [Fig. 1(c)].

Depending on the complexity of the effects influencing the analytical response, special calibration measures must be taken to obtain correct analytical results. Interferences require a multicomponent analysis approach, which forces the analyst to determine more compounds than he is really interested in. Matrix effects can be handled by the addition of standards of the analyte to the sample (standard addition method). Samples with a combined interference and matrix effect often require a solution along chemical paths, e.g. separations or a search for proper masking agents. Very recently, a generalized standard addition method (GSAM) has been developed [1], which proved to be an appropriate calibration procedure in some particular instances of a joint presence of an interferent and matrix effect (e.g. ICP). The application of multivariate statistics such as partial least squares (PLS) [2] may also solve problems of a combined interference and matrix effect provided that calibration standards of known constitution of the analytes are available.

2. Qualitative definition of specificity and selectivity

Interferences and matrix effects are of great concern to every analytical chemist. Both determine the complexity of the calibration procedure needed to obtain accurate results. The smaller the probability that interferences or matrix effects will disturb a given analytical procedure, the better its quality. Therefore, it is not surprising that analytical chemists have tried hard to express the "amount" of interference and/or matrix effects of novel analytical methods in terms such as "very selective" or "very specific" and so on. At present, however, there does not seem to be much uniformity in the analytical literature when describing selectivity, specificity, interferences and matrix effects. IUPAC [3] recommends that an interfering substance for an analytical procedure be defined as one that causes a predeterminate systematic error in the analytical result.

An example may help to clarify the meaning of the terms specificity and selectivity. If a reagent forms a coloured complex with one analyte only, the reagent is called specific for that particular analyte. In the terminology introduced so far, the analytical sensor is sensitive for only one analyte in the sample. If the reagent, however, forms coloured complexes with many constituents in the sample, but with a distinct colour for each constituent, the procedure of the complexing reaction might be called selective. Expressed in the above terminology, a selective measuring system consists of sensors which are specific for one of the analytes.

When considering the problem of selectivity (and of specificity) in more detail, the analyst will discover that a distinction between non-selective and selective is quite artificial. Selectivity and specificity depend on the set of analytes considered and therefore on the analytical problem at hand. An AAS method may be fully specific for the determination of Ca in samples where Al and PO_4^{3-} are absent. On the other hand, the presence of Al and PO_4^{3-} suppresses the sensitivity for Ca. As has been stressed by Belcher [4] and Betteridge [5], it is necessary to avoid the use of

References p. 126

qualitative terms such as "highly selective" and therefore a more quantitative approach is imperative. An inherent condition for a useful quantitative measure for selectivity and specificity is that it provides some quantitative information about the expected quality of the analytical result. Selective procedures are better than non-selective procedures because, on the average, the quality of the analytical result is better. AAS procedures require less attention for interferences and matrix effects than a UV–vis spectrometric procedure. Thus, the probability for systematic errors in AAS is smaller than in UV–vis spectrometry, which is considered as "better".

3. Quantitative definition of selectivity and specificity

3.1 Propagation of error

In principle, a multicomponent analysis with very specific sensors and one with less specific sensors should yield equal analytical answers when a proper multivariate calibration is performed and when error-free responses are obtained. Because the measured value deviates from the true value, due to the uncertainty of the measurement, this is not true. The propagation or amplification of the measurement error into the analytical result is much stronger for non-selective than for selective procedures: thus, when one takes the presence of interfering substances into account, a systematic error is avoided but an extra amplification of the measurement error may be the consequence. This is demonstrated in the following example.

Example. Suppose a system is analysed consisting of 2 analytes in concentration $x_1 = 1$ and $x_2 = 2$. The measurements are first carried out with two specific sensors, which respond only to the presence of one of the analytes (e.g. with a sensitivity of 10). Then the error-free responses at the sensors are

sensor 1: $y_1 = 10x_1 + 0x_2 = 10$

sensor 2: $y_2 = 0x_1 + 10x_2 = 20$

x_1 and x_2 are calculated as $x_1 = 10/10 = 1$ and $x_2 = 20/10 = 2$ which is the correct answer.

When measuring with two non-specific sensors (e.g. with a respective sensitivity for component 1 equal to 4 and 6 and for component 2 equal to 3 and 7), the error-free responses are

$y_1 = 4x_1 + 3x_2 = 10$

$y_2 = 6x_1 + 7x_2 = 20$

and here, again, one obtains $x_1 = 1$ and $x_2 = 2$. Thus, both procedures, the selective and the non-selective, yield the same unbiased answer.

However, when the responses y_1 and y_2 are affected by a measurement error, the situation becomes quite different. Suppose that, in both cases, the values $y_1 = 9$ (-10%) and $y_2 = 22$ $(+10\%)$ have been measured. The estimated concentrations for both systems are then

system 1: $x_1 = 0.9(-10\%)$; $x_2 = 2.2(+10\%)$

system 2: $x_1 = -0.3(-130\%)$; $x_2 = 3.4(+50\%)$

No error amplification occurs in the fully selective procedure, while the non-specific method exhibits a very strong error amplification.

The selectivity of both systems is fully determined by the matrix of the calibration factors of the sensors. This matrix is called the **K**-matrix [1]. It can be shown that the values of the elements in the **K**-matrix, which define the selectivity of the method, determine the error amplification not only in the case of the multicomponent analysis but in the standard addition method and generalized standard addition method (GSAM) as well. Therefore, a definition of selectivity (and specificity) should be based on a numerical evaluation of the **K**-matrix.

3.2 Quantitative definition

An idealized model for the analytical calibration function is

$$y_i = k_{i1}x_1 + k_{i2}x_2 + \ldots + k_{im}x_m + e_i$$

The concentration (or quantity) of component i, x_i, can be derived from the response y_i provided that concentrations of all the other components present are known together with the sensitivities of the sensor k_i for these components. If the entire composition is to be determined, a set of n responses $y_1, y_2, \ldots, y_i, \ldots, y_n$ must be measured at the sensors $s_1, s_2, \ldots, s_i, \ldots, s_n$. n must be equal to or greater than m for the problem to be solved.

Thus, the analytical calibration function becomes

$$\begin{aligned} y_1 &= k_{11}x_1 + k_{12}x_2 + \ldots + k_{1m}x_m + e_1 \\ y_2 &= k_{21}x_1 + k_{22}x_2 + \ldots + k_{2m}x_m + e_2 \\ &\vdots \\ y_n &= k_{n1}x_1 + k_{n2}x_2 + \ldots + k_{nm}x_m + e_n \end{aligned} \tag{1}$$

The analytical procedure then usually consists of two steps.

(a) The calibration step in which the values of the **K**-matrix are quantified. The most general calibration procedure for that purpose is the generalized standard addition method (GSAM) [1].

(b) The determination step where the concentrations $(x_1, \ldots, x_m)$ are calculated from the responses $(y_1, \ldots, y_n)$ obtained for the sample. For this purpose, the k values obtained in the calibration step are substituted in eqn. (1).

The mathematical model of a multicomponent analysis as given by eqn. (1) is the simplest possible such model. However, it will demonstrate the possibilities and limitations of expressing selectivity and specificity in an effective way. Equation (1) (with $n = m$) represents a selective method if each measurement depends on only one component in the sample, e.g. a gas chromatographic determination of m components where the concentrations are derived from the areas of m well-resolved peaks. Of all $n \cdot m$ coefficients, only m sensitivity coefficients retain a value. The lower the resolution of the chromatographic system, the more signals (or peaks) of the other compounds will overlap with the signal of the analyte and therefore the less selective the procedure becomes.

Full selectivities (and specificities) are rare for analytical procedures. A quantitative expression for both parameters has been formulated by Kaiser [6] who intro-

References p. 126

duced a selectivity parameter based on the elements of a square **K**-matrix containing all sensitivity coefficients of a determined system (number of sensors, n, is equal to the number of analytes, m).

The **K**-matrix of the analytical system given in eqn. (1) is

$$\mathbf{K} = \begin{bmatrix} k_{11} & k_{12} & \dots & k_{1m} \\ k_{21} & k_{22} & \dots & k_{2m} \\ \cdot & \cdot & \dots & \cdot \\ k_{n1} & k_{n2} & \dots & k_{nm} \end{bmatrix}$$

The selectivity of the **K**-matrix was defined by Kaiser as

$\xi = \text{Min } \xi_i$

where

$$\xi_i = \frac{|k_{ii}|}{\sum_{j=1}^{m} |k_{ij}| - |k_{ii}|} - 1 \qquad (2)$$

As an example, the selectivity ξ is calculated for the chlorine–bromine system, measured at the wavenumbers 22000 and 24000 cm^{-1} (Table 1), giving the **K**-matrix.

$$\mathbf{K} = \begin{bmatrix} 4.5 & 168 \\ 8.4 & 211 \end{bmatrix}$$

For $i = 1$

$$\xi_1 = \frac{|k_{11}|}{|k_{11}| + |k_{12}| - |k_{11}|} - 1 = \frac{4.5}{4.5 + 168 - 4.5} - 1 = -0.973$$

For $i = 2$

$$\xi_2 = \frac{|k_{22}|}{|k_{21}| + |k_{22}| - |k_{22}|} - 1 = \frac{211}{8.4 + 211 - 211} - 1 = 24.1$$

Min $\xi = -0.973$

Thus in eqn. (2), for each i equation, the sum of the (absolute values of the) sensitivity coefficients k_{ij} with $i \neq j$ ($\Sigma |k_{ij}| - |k_{ii}|$) is determined. If this sum is small in comparison with k_{ii}, ξ is large. Full selectivity corresponds to a value of infinity.

From the example it follows that the sensitivities of the compounds at the first wavelength ($\nu = 22000$ cm^{-1}) are the reason for the overall low selectivity of the total system. In fact, this first equation in the system is the weakest part of the procedure. Or, in general, the row of the **K**-matrix which yields the smallest value is the weakest part of the procedure and according to Kaiser [6], this minimum value determines the selectivity of the whole procedure.

Substitution of the wavenumber 22000 cm^{-1} by 30000 cm^{-1} gives the selectivity as

$$\frac{100}{100 + 4.7 - 100} - 1 = 20.3$$

which lifts the selectivity of the total system to 20.3.

A drawback of the definition of Kaiser is that the selectivity parameter is not directly related to the quality of the analytical result. Recalling the examples with

$$\mathbf{K} = \begin{bmatrix} 10 & 0 \\ 0 & 10 \end{bmatrix}$$

and

$$\mathbf{K} = \begin{bmatrix} 4.5 & 168 \\ 8.4 & 211 \end{bmatrix}$$

the selectivity coefficients $\xi = \infty$ and $\xi = -0.973$ are found. A procedure with a low value of ξ is to be considered as a poor procedure, but the term "poor" is still not quantified.

Another approach is to calculate what is called the condition of the **K**-matrix. The m analytical results of an m-component analysis and their corresponding absolute errors can be represented by the vectors

$$\mathbf{x} = \begin{bmatrix} x_1 \\ x_2 \\ \cdot \\ x_m \end{bmatrix}$$

and

$$\Delta\mathbf{x} = \begin{bmatrix} \Delta x_1 \\ \Delta x_2 \\ \cdot \\ \Delta x_m \end{bmatrix}$$

Vectors and matrices can be compared by their magnitude, called the norm. The norm of the error vector is a measure of the magnitude of the absolute error of a set of m analytical results. One of the definitions of the norm of a vector is based on the geometrical representation of vectors in the m-dimensional space, namely the length of a vector, where

$$\text{the norm of } \mathbf{x} = \|\mathbf{x}\| = \left[\sum_{i=1}^{m} (x_i)^2 \right]^{1/2}$$

The definition of the norm of a matrix **K** is somewhat more complex.

When the matrix **K** is square, the norm of **K**, $\|\mathbf{K}\|$, equals λ_1, which is the largest eigenvalue of **K**. It can be derived that the norm of the inverse matrix $\mathbf{K}^{-1}$, $\|\mathbf{K}^{-1}\|$, is the reciprocal value of the smallest eigenvalue of **K**, i.e. $1/\lambda_m$. The norm of a non-square matrix is defined as the square root of the largest eigenvalue of the positive definite matrix $\mathbf{K}' \cdot \mathbf{K}$. Besides the norm of **x**, $\|\mathbf{x}\|$, we can also define the norm of the vector of absolute errors $\|\Delta\mathbf{x}\|$. Consequently, the relative error in the analytical result in an m-component analysis may be expressed as $\|\Delta\mathbf{x}\|/\|\mathbf{x}\|$.

References p. 126

An additional source of uncertainty in the analytical result is that in the measured sensitivities, which can be expressed as an error matrix $\Delta\mathbf{K}$, where

$$\Delta\mathbf{K} = \begin{bmatrix} \Delta k_{11} & \Delta k_{12} & \cdots & \Delta k_{1m} \\ \Delta k_{21} & \Delta k_{22} & \cdots & \Delta k_{2m} \\ \cdot & \cdot & \cdots & \\ \Delta k_{n1} & \Delta k_{n2} & \cdots & \Delta k_{nm} \end{bmatrix}$$

where an element Δk_{ij} of the matrix $\mathbf{K}$ represents the absolute error in the measured selectivity k_{ij}.

An important relationship exists between the relative error in the analytical result and the relative error in the measurements [1,7], namely

$$\frac{\|\Delta\mathbf{x}\|}{\|\mathbf{x}\|} \leqslant \|\mathbf{K}\| \cdot \|\mathbf{K}^{-1}\| \left\{ \frac{\|\Delta y\|}{\|\mathbf{y}\|} + \frac{\|\Delta\mathbf{K}\|}{\|\mathbf{K}\|} \right\} \tag{3}$$

where $\|\Delta\mathbf{x}\| / \|\mathbf{x}\|$ gives the relative error in the measurement.

Because $\|\mathbf{K}\| \cdot \|\mathbf{K}^{-1}\|$ is the definition of the condition, COND(**K**), of matrix **K**, eqn. (3) can be written

$$\frac{\|\Delta\mathbf{x}\|}{\|\mathbf{x}\|} \leqslant \mathrm{COND}(\mathbf{K}) \left\{ \frac{\|\Delta\mathbf{y}\|}{\|\mathbf{y}\|} + \frac{\|\Delta\mathbf{K}\|}{\|\mathbf{K}\|} \right\} \tag{4}$$

Because, in most cases, the relative error in the k values is much smaller than the relative error in the measurements, eqn. (4) becomes

$$\frac{\|\Delta\mathbf{x}\|}{\|\mathbf{x}\|} \leqslant \mathrm{COND}(\mathbf{K}) \frac{\|\Delta\mathbf{y}\|}{\|\mathbf{y}\|} \tag{5}$$

From eqn. (5), the important conclusion follows that the condition of the **K**-matrix is a direct measure for the amplification of the relative measurement error into the relative error of the analytical result. Thus COND(**K**) relates the relative magnitude of the different sensitivity coefficients in the **K**-matrix, with the amplification of the measurement error into the analytical result.

Recalling the 2-component example given in Sect. 3 where

$$\mathbf{K} = \begin{bmatrix} 10 & 0 \\ 0 & 10 \end{bmatrix}$$

we calculate the eigenvalues

$$\begin{vmatrix} 10-\lambda & 0 \\ 0 & 10-\lambda \end{vmatrix} = 0;\ (10-\lambda)^2 = 0;\ \lambda_1 = \lambda_2 = 10$$

$$\mathrm{COND}(\mathbf{K}) = \lambda_1/\lambda_2 = 1$$

This means that no error amplification is expected or that the relative error in the analytical result should be equal to the relative error in the measurements.

Verification indeed shows that for

$$\frac{\|\Delta y\|}{\|\mathbf{y}\|} = \frac{\left[(10-9)^2 + (22-10)^2\right]^{1/2}}{(10^2+20^2)^{1/2}} = 0.1$$

a relative error $\| \Delta \mathbf{x} \| / \| \mathbf{x} \|$ is found which is equal to

$$\frac{\left[(1-0.9)^2+(2.2-2)^2\right]^{1/2}}{(1^2+2^2)^{1/2}}=0.1$$

For the other system in the same example with $\mathbf{K}=\begin{bmatrix} 4 & 3 \\ 6 & 7 \end{bmatrix}$

$$(4-\lambda)\cdot(7-\lambda)-18=0$$

and

$$\lambda_1=10$$

$$\lambda_2=1$$

Thus, COND($\mathbf{K}$) = 10.

A relative error

$$\| \Delta \mathbf{x} \| / \| \mathbf{x} \| = \left[(1+0.3)^2+(2.2-3.4)^2\right]^{1/2}/5^{1/2}=0.8$$

is found for $\| \Delta \mathbf{y} \| / \| \mathbf{y} \| = 0.1$, i.e. the error has been amplified with a factor 8, while from COND($\mathbf{K}$) = 10, a maximal error amplification up to a factor of 10 was expected.

Although by expressing a set of $n \cdot m$ sensitivity coefficients into one selectivity parameter, such as COND($\mathbf{K}$), much information concerning the procedure has been lost, some essential information has been preserved. Moreover, the influence of the replacement of a moderately selective sensor by a more selective one can be forecast.

The error propagation formula given in eqn. (5) gives the maximal amplification of the measurement error into the analytical result. $\| \Delta \mathbf{y} \|$ in eqn. (5) represents the difference between the measured and true value, irrespective of the type of error. In the case where systematic errors are absent and only random errors with known distribution influence the measurements, one can determine the probability of measuring a value with a given deviation from the true value. This probability is expressed by a probability distribution function characterized by its standard deviation σ_y. In Chap. 13, an expression will be derived for the amplification of the uncertainty of the measurements into the uncertainty of the concentration, provided the measurements contain no bias.

4. Choosing the optimal set of wavenumbers

For a multicomponent analysis, one can use an overdetermined system or keep the number of measurements equal to the minimum ($n = m$) required for the determination.

The question arises of which set of wavenumbers is to be preferred. The search for such an optimal set requires the use of an optimization criterion. Two suitable criteria to be optimized are precision and sensitivity. For a one-component analysis, an optimization for sensitivity leads to the choice of the wavenumber where the absorption peak has a maximum. This also results in a minimal relative error.

According to Kaiser [6], it is possible to define the sensitivity of a multicomponent procedure as the absolute value of the determinant of the sensitivity matrix **K**. This is possible only if the number of measurements is equal to the number of compounds ($n = m$). Then

$$\mathbf{K} = \begin{bmatrix} k_{11} & k_{12} & \cdots & k_{1n} \\ k_{21} & k_{22} & \cdots & k_{2n} \\ \cdot & \cdot & \cdots & \cdot \\ \cdot & \cdot & \cdots & \cdot \\ k_{n1} & k_{n2} & \cdots & k_{nn} \end{bmatrix}$$

A maximum sensitivity corresponds to a determinant with large diagonal elements and small off-diagonal elements. In fact, a high sensitivity implies a high selectivity (Sect. 3.2), which means that a highly sensitive procedure is a procedure in which each measurement is largely dependent on the concentration of only one of the components. The sensitivity is therefore a parameter that can be used for comparing different sets of wavenumbers. However, as will be shown below, a maximum sensitivity as defined by Kaiser does not necessarily correspond with the best relative precision.

In order to illustrate the principle of using the sensitivity as an optimization criterion, we choose a chlorine–bromine 2-component system, with the absorptivities shown in Table 1. From the six wavenumbers, it is possible to choose several pairs, to be exact $C_2^6 = 15$ pairs. For each of these pairs, it is possible to calculate the sensitivity, i.e. the absolute value of the determinant of the absorptivities. The values of these determinants are given in Table 2. The combination 24×10^3 and 30×10^3 cm^{-1} has the largest sensitivity. This corresponds with the respective absorption maxima of bromine and chlorine.

Although the optimization procedure is theoretically relatively simple, in practice its application requires a large number of calculations. With p wavenumbers from which a set of n ($n \geqslant m$) is to be chosen, $C_p^n = p!/(p-n)!n!$ systems have to be compared. For a relatively simple situation of $p = 30$ and $n = 6$, the number of

TABLE 1

Absorptivities of Cl_2 and Br_2 in chloroform

Wavenumber ($\times 10^3$ cm^{-1})	Absorptivities	
	a_{Cl_2}	a_{Br_2}
22	4.5	168
24	8.4	211
26	20	158
28	56	30
30	100	4.7
32	71	5.3

TABLE 2

Sensitivities for the chlorine–bromine system in chloroform with combinations of two wavenumbers

	Wavenumber ($\times 10^3$ cm^{-1})					
	22	24	26	28	30	32
22	0	460	1650	9300	16800	12000
24		0	2890	11600	21050	15000
26			0	8250	15705	11000
28				0	2740	1830
30					0	189
32						0

determinants to be calculated is 593 775. Therefore, a straightforward procedure as described here is not the most appropriate.

For the optimization of overdetermined systems for sensitivity, Junker and Bergman [8] defined the sensitivity of overdetermined systems as the root of the determinant of the product of the calibration matrix and its transpose, i.e.

$$\text{sensitivity} = (|\mathbf{K}'\mathbf{K}|)^{1/2}$$

Junker and Bergman [8] developed an optimization algorithm which calculates the n ($n \geqslant m$) best wavenumbers out of a set of p wavenumbers and which is based on the maximization of the determinant and thus optimizes sensitivity. They quoted a reduction in computer time of about 1000-fold in comparison with the straightforward manner. Because the optimization procedure is terminated at any number, n, of wavenumbers that one wishes to retain, it is clear that the true optimum is not necessarily found at the preselected value of n, but at some other unknown value of n instead.

Instead of selecting a set of wavelengths with the best sensitivity, one can select the set of wavelengths with the smallest error amplification. According to eqn. (5), the maximal relative precision is obtained for a minimal condition of the matrix **K**. Therefore the Cond(**K**) is a suitable criterion for the optimization for precision. The values of the Cond(**K**) of all 15 pairs of wavenumbers are listed in Table 3. The

TABLE 3

Conditions for the chlorine–bromine system in chloroform with combinations of two wavenumbers

	Wavenumber ($\times 10^3$ cm^{-1})					
	22	24	26	28	30	32
22		157	20	3.1	1.7	2.4
24			24.1	3.9	2.1	3.0
26				3.2	1.6	2.3
28					4.9	4.7
30						7.7

References p. 126

smallest error amplification (factor 1.6), and thus the best precision, is obtained when using the combination 26×10^3 and 30×10^3 cm^{-1}. This result shows that the pair of wavelengths with the best sensitivity does not necessarily give the smallest error amplification.

In the previous paragraph the condition of an overdetermined system (more wavelengths than unknown concentrations) has been defined as

$$\text{Cond}(\mathbf{K}) = \frac{\lambda_1}{\lambda_m}$$

For a non-square matrix **K**, λ_1 is the square root of the largest eigenvalue of the square matrix $\mathbf{K}' \cdot \mathbf{K}$ and λ_m is the square root of the smallest eigenvalue of the square matrix $\mathbf{K}' \cdot \mathbf{K}$. Thus, overdetermined systems may also be optimized for the best relative precision by choosing the set of wavelengths with the smallest Cond(**K**). Fast algorithms for the optimization of precision by minimizing the Cond(**K**) have not been reported.

References

1 C. Jochum, P. Jochum and B.R. Kowalski, Error propagation and optimal performance in multicomponent analysis, Anal. Chem., 53 (1981) 85.
2 M. Sjöstrom, S. Wold, W. Lindberg, J.A. Persson and H. Martens, A multivariate calibration problem in analytical chemistry solved by partial least squares models in latent variables, Anal. Chim. Acta, 150 (1983) 61.
3 C. den Boef and A. Hulanicki, Recommendations for the usage of selective, selectivity and related terms in analytical chemistry, Pure Appl. Chem., 45 (1983) 553.
4 R. Belcher, Sensitivity index, Talanta, 23 (1976) 883.
5 D. Betteridge, Selectivity index, Talanta, 12 (1965) 129.
6 H. Kaiser, On the definition of selectivity, specificity and sensitivity of analytical methods, Z. Anal. Chem., 260 (1972) 252.
7 J. Stoer, Einführung in die Numerische Mathematik, Springer, Berlin, 1972.
8 A. Junker and G. Bergman, Selection, comparison and valuation of optimum working conditions for qualitative multi-component analysis. Part I: Two-dimensional overdetermined systems, sensitivity as optimisation parameter, Z. Anal. Chem., 272 (1974) 267.

Recommended reading

Selectivity as a performance criterion is discussed in

J. Inczedy, Some remarks on the quantitative expression of the selectivity of an analytical procedure, Talanta, 29 (1982) 595.
A.G. Wilson, Performance characteristics of analytical methods. IV, Talanta, 21 (1974) 1109.

Different approaches to the selection of a restricted set of wavelengths are found in

C.W. Brown, P.F. Lynch, R.J. Obremski and D.S. Lavery, Matrix representations and criteria for selecting analytical wavelengths for multicomponent spectroscopic analysis, Anal. Chem., 54 (1982) 1472.
J. Sustek, Method for the choice of optimal analytical positions in spectrophotometric analysis of multicomponent systems, Anal. Chem., 46 (1974) 1676.
P.C. Thijssen, G. Kateman and H.C. Smit, Optimal designs with information theory in least squares problems, Anal. Chim. Acta, 157 (1984) 99.
S. Ebel, E. Glaser, S. Abdulla, U. Steffens and V. Walter, Optimization of wavelengths in spectrophotometric multicomponent analysis, Z. Anal. Chem., 313 (1982) 24.

Chapter 9

Information

1. Introduction

In the analytical chemical literature, qualitative analytical methods are often referred to as "good", "valuable", "excellent", "specific", etc., with no further explanation of these terms. An objective interpretation of such terms is not easy and therefore the resulting choice of method often does not have a completely rational basis. Whereas quantitative analytical methods can be evaluated by using criteria such as precision, accuracy, reliability, limit of detection, selectivity, and others discussed in the preceding chapters, no generally accepted criteria exist for qualitative analysis.

Information theory permits a mathematical evaluation of qualitative methods by calculation of the expected or average amount of information obtained from the analysis.

The aim of an analysis is to reduce the uncertainty with respect to the sample (and therefore the system) to be analysed. It will be appreciated that the reduction of uncertainty is considered to be equivalent to obtaining information. This corresponds with the common use of the terms uncertainty and information. A newspaper offers only news (information) if the reader has not yet been informed about the events through other communication channels. If the *has* been informed, he is (almost) certain about the contents of the newspaper. The same is true of qualitative analysis: the analysis is carried out because there is an uncertainty about the identity of the components in the sample. After the analysis, the state of uncertainty is (hopefully) turned into a state of certainty (or, at least, of less uncertainty); in other words, the analysis has yielded a certain amount of information.

Information theory is related to classical probability theory. For a large number of possible identities, the probability of each will, in general, be small. Following this reasoning, we can arrive at an expression for the information obtained from an analysis. Assume a simple model of the analytical problem in which each of the possible identities has the same probability before analysis (in fact, in practice some identities are more likely to be found than others). Before the experiment, the uncertainty can be expressed in terms of the number of possible identities, n_0, each having a probability $p(x_0) = 1/n_0$. Due to experiment i, the number of possible identities is reduced to n_i with probabilities $p(x_i) = 1/n_i$. The information I_i obtained from the ith experiment is defined by

$$I_i = \log_2(n_0/n_i) \qquad (1)$$

where $\log_2$ is the logarithm to the base 2.

References p. 135

This expression can be replaced by

$$I_i = \log_2[p(x_i)/p(x_0)] \tag{2}$$

I_i is expressed in bits (binary digits) and is called the specific information.

A numerical example will illustrate the concepts introduced so far. Let us assume that, in a qualitative analysis, it is known that the sample to be analysed is one of 100 possible substances and that the measurement yields a signal corresponding to 10 possible identities. If the qualitative analysis technique were, for instance, thin layer chromatography (TLC), the signal would be a particular R_f value. In that case, one would know, for instance, that finding $R_f = 0.15$ means that only 10 identities remain possible. Then, the application of eqns. (1) or (2) leads to the specific information $I_i = \log_2(100/10)$ or $I_i = \log_2(0.1/0.01) = 3.32$ bits.

The information obtained in this way depends on the outcome of the experiment and different outcomes may yield different quantities of specific information. Suppose that, in the TLC experiment, 10 substances have identical R_f values (or R_f values that cannot be distinguished) and that the other 90 substances have, for instance, R_f values of zero. Then in 10% of the experiments the information obtained will be 3.32 bits, whereas in 90% an information of only $I = \log_2(100/90) = 0.15$ bits is obtained. The average of the specific informations or information content of such a TLC procedure is then $I = (0.1 \times 3.32) + (0.9 \times 0.15) = 0.47$ bits (assuming that all 100 substances are to be found with the same probability).

In symbol form, the equation can be written as

$$I = \frac{n_1}{n_0} \log_2\left(\frac{n_0}{n_1}\right) + \frac{n_2}{n_0} \log_2\left(\frac{n_0}{n_2}\right) \tag{3}$$

with values of n_0, n_1 and n_2 of 100, 10 and 90, respectively. It is customary to write n_0 in the denominator. After generalization, this yields

$$I = \sum_i - \frac{n_i}{n_0} \log_2\left(\frac{n_i}{n_0}\right) = \sum_i p(x_i) I_i \tag{4}$$

where I is the information content of the procedure, n_0 the number of possible identities before the experiment (with equal probabilities) and n_i the number of possible identities after interpretation of the experiment with signal i. I_i is the specific information obtained from the experiment with result y_i and $p(x_i)$ is the probability of measuring a signal y_i. It should be stressed here that the value of the information content is only of interest when used in a relative way, i.e. as a means to compare the performance of one qualitative procedure with another.

A more general model can now be introduced. Readers who are not acquainted with the Bayes' theorem should omit the rest of this section until they have read Chap. 26 and go directly to Sect. 2 of the present chapter. This model will represent a set of possible identities before the experiment $(x_1, x_2, \ldots, x_j, \ldots, x_n)$, each having a probability $p(x_i)$. The uncertainty before the experiment, H, can be expressed by means of the Shannon equation

$$H = \sum_{j=1}^{n} -p(x_j) \log_2 p(x_j) \tag{5}$$

The uncertainty is also called entropy because of its analogy with the entropy expression as used in thermodynamics. Similarly, after the experiment with result y_i

$$H_i = \sum_{j=1}^{n} -p(x_j/y_i) \log p(x_j/y_i) \tag{6}$$

where $p(x_j/y_i)$ is the (conditional) probability (also called Bayes' probability, see Chap. 26) of identity x_j, provided that the experiment has yielded a signal i ($i = 1, \ldots, m$). The uncertainty, H_i, remaining after a signal i was obtained (also called the entropy) depends, of course, on the signal measured. The difference $H - H_i$ is equal to the specific information. In order to arrive at an equation for the information content, we have to subtract the weighted average H_i from H, which leads to the expression

$$I = H - \sum_{i=1}^{m} p(y_i) H_i \tag{7}$$

where $p(y_i)$ is the probability of measuring a signal i. By making use of Shannon's uncertainty equation [1], eqn. (7) can be written as

$$I = \sum_{j=1}^{n} -p(x_j) \log_2 p(x_j) - \sum_{i=1}^{m} p(y_i) \sum_{j=1}^{n} -p(x_j/y_i) \log_2 p(x_j/y_i) \tag{8}$$

Calculation of the information content generally requires a knowledge of the following probabilities.

(a) The probabilities of the identities of the unknown substance before analysis, $p(x_j)$. The first term on the right-hand side of eqn. (8) represents what is known about the analytical problem in a formal way, or the "pre-information". The analytical problem in terms of the probabilities $p(x_j)$ is essential for calculating the information content. An infinite number of possible identities each having a very small probability (approaching zero) represents a situation without pre-information. The uncertainty is infinitely large and solving the analytical problem requires an infinite amount of information.

(b) The probabilities of the possible signals, $p(y_i)$. These probabilities depend on the relationship between the identities and the signals (tables of melting points, R_f values, spectra, etc.) and also on the substances expected to be identified, $p(x_j)$. If an identity is not likely to be found, the corresponding signal is not likely to be measured. It should be noted that, in replicate experiments, one identity can lead to different signals because of the presence of experimental errors.

(c) The probabilities of the identities when the signal is known, $p(x_j/y_i)$. In fact, these probabilities are the result of the interpretation of the measured signals in terms of possible identities. To this end, the following "interpretation" relationship can be used.

$$p(x_j/y_i) = \frac{p(x_j)\cdot p(y_i/x_j)}{\sum_j p(x_j)\cdot p(y_i/x_j)} \tag{9}$$

This relationship shows that the probabilities for the identities after analysis can be calculated from the pre-information, $p(x_j)$, and the relationships between the identities and the signals, $p(y_i/x_j)$. Equation (9) is found in the literature as Bayes' theorem (Chap. 26). It should be observed that one particular signal can correspond with more than one identity.

2. An application to thin layer chromatography

In TLC, the signal that permits the identification of an unknown substance is an R_f value. If we assume that substances of which the R_f values differ by 0.05 can be distinguished, the complete range of R_f values can be divided into 20 groups (0–0.05, 0.06–0.10, ...). Such a simplified model leads to a situation where substances with R_f values of, for instance, 0.05 and 0.06 are considered to be separated, which clearly is not true. However, the model allows an easy calculation of approximate values of the information content. Furthermore, at least in this application, it is not important to distinguish exactly which substances are separated and which are not, as the purpose is rather to see how well the substances are spread out over the plate.

Each of the 20 groups of R_f values can then be considered as a possible signal ($y_1, y_2, \ldots, y_{20}$) and there is a distinct probability [$p(y_1), p(y_2), \ldots, p(y_{20})$] that an unknown substance will have an R_f value within the limits of one of the groups. Let us consider a TLC procedure that is used to identify a substance belonging to a set of n_0 substances; n_1 substances fall into group 1, n_2 into group 2, etc. If all substances have the same a priori probability to be the unknown compound, eqn. (4) can be used for calculating the information content. To understand further the meaning of the information content, let us investigate some extreme conditions.

(a) All substances fall into the same group n_i. In this instance $n_i/n_0 = 1$ and thus $I = I_i = 0$. As all of the substances yield the same R_f value, the experiment does not indicate anything to the observer. No information is obtained because there is no uncertainty as to which event (signal, R_f value) will occur: whatever the unknown substance, the result will always be the same.

(b) All n_0 substances ($n_0 < 20$) fall into different groups. The information content is maximal as each substance yields a different R_f value. The information content, from eqn. (4) with all $n_i = 1$, is now equal to

$$I = -n_0 \frac{1}{n_0} \log_2\left(\frac{1}{n_0}\right) = \log_2 n_0 \tag{10}$$

TABLE 1

hR_F values of DDT and related compounds and information content of the proposed separation [2]

Solvent system	p,p′-DDT	o,p′-DDT	p,p′-DDE	o,p′-DDE	p,p′-DDD	DDA	DDMU	DBP	Kel-thane	DPE	DBH	BPE	DDM	I
1	25	35	41	36	10	0	32	2	0	27	0	0	35	2.28
2	48	55	61	65	28	0	55	8	2	44	0	0	54	2.47
3	69	72	75	72	52	0	67	24	3	63	0	0	67	2.35
4	76	76	77	75	64	0	74	45	8	72	2	4	75	2.50
5	67	69	72	69	52	0	70	69	7	66	1	2	71	1.98
6	67	70	75	68	51	0	72	45	10	66	2	4	70	2.87
7	70	70	75	70	63	0	74	61	23	66	6	13	69	2.78
8	78	79	83	79	77	6	83	75	53	77	27	39	80	2.56
9	69	71	76	69	60	0	73	56	19	67	4	9	71	3.03
10	35	44	49	45	16	0	48	4	0	35	0	0	42	2.62
11	58	63	65	62	42	0	62	18	3	55	0	0	61	2.41
12	60	64	71	64	43	0	69	50	8	60	2	5	67	2.71
13	63	66	71	64	48	0	68	55	16	62	4	9	67	2.97
14	73	74	77	72	65	0	75	68	48	77	24	36	76	2.57
15	83	84	85	83	78	5	83	81	58	82	46	55	83	1.47
16	84	84	84	83	80	0	84	82	75	84	71	74	83	1.70
17	87	88	89	87	81	20	87	83	64	86	59	63	86	1.47
18	59	68	73	69	39	8	67	34	34	60	33	34	67	2.31
19	92	94	96	93	83	32	92	82	72	91	61	70	93	2.35
20	78	80	80	77	69	17	78	67	43	74	24	35	78	2.50
21	82	82	82	80	81	0	81	81	81	81	80	81	82	0.39
22	80	80	80	77	79	5	78	78	79	79	79	79	80	0.39
23	35	45	50	45	13	0	40	2	0	36	0	0	39	2.35
24	40	51	52	48	16	0	44	3	0	42	0	0	42	2.28
25	71	74	77	72	59	0	74	64	21	70	5	13	72	2.72
26	77	77	78	77	72	30	78	72	56	78	43	47	78	2.03
27	27	40	43	42	10	0	35	2	0	32	0	0	35	2.08
28	70	75	77	75	51	2	74	20	4	69	4	4	75	2.04
29	65	69	76	68	48	0	72	57	7	64	2	5	70	2.93
30	85	85	85	84	84	4	80	83	84	84	83	84	84	0.77
31	54	67	74	69	22	0	61	6	6	54	6	6	62	2.62
32	100	100	100	100	94	35	100	94	92	100	83	92	100	1.57
33	93	94	96	94	90	7	96	93	75	94	67	70	93	1.88

TABLE 2

Example of computation of the information content

The data are those for system 13 in Table 1.

R_f groups	Substances falling in the group	n_i/n_0	$\log_2(n_i/n_0)$	$(-n_i/n_0)\log_2(n_i/n_0)$
0 -0.05	DDA, DBH	2/13	−2.697	0.415
0.06–0.10	BPE	1/13	−3.715	0.286
0.16–0.20	Kelthane	1/13	−3.715	0.286
0.46–0.50	p,p'-DDD	1/13	−3.715	0.286
0.51–0.55	DBP	1/13	−3.715	0.286
0.61–0.65	DPE, p,p'-DDT, o,p'-DDE	3/13	−2.114	0.488
0.66–0.70	o,p'-DDT, DMU, DDM	3/13	−2.114	0.488
0.71–0.75	p,p'-DDE	1/13	−3.715	0.286
				$I = 2.82$

It can be shown that this is indeed the maximum value which can be obtained. It is equal to the information necessary to obtain an unambiguous, complete identification of each substance. When $n_0 > 20$, the maximum value that can be attained for I is obtained when the substances are spread as evenly as possible over the 20 signal categories.

From these extreme conditions, it follows that, in order to obtain a maximum information content, the TLC system should cause an equal spread of the R_f values over the entire range. The results for an application [2] concerning DDT and related compounds are summarized in Table 1. The computation of I for system 13 is given as an example in Table 2.

The best separations are obtained with solvents 9, 13, and 29 and further investigations should be aimed at optimizing small changes in the best three solvents (for instance, by applying one of the techniques described in Chaps. 16–19).

3. The information content of combined procedures

The object of a qualitative analysis is to obtain an amount of information permitting unambiguous identification. This, in practice, is often not possible with a single test and therefore experiments have to be combined. For example, in toxicological analysis of basic drugs, one will combine techniques such as UV and IR spectrometry, TLC and GLC or one will use two (or more) TLC procedures, etc. in order to obtain the necessary amount of information. Hence, the next question is how to calculate the information content of two or more methods.

When two TLC systems are combined, one can consider the combination of two R_f values which fall in the range 0.00–0.05 as one event (signal y_{11}), an R_f value of 0.00–0.05 for system 1 and 0.05–0.10 for system 2 as a signal y_{12}, etc. As before, one can define a probability $p(y_{ij})$ for signal y_{ij} so that $n_{ij}/n_0 = p(y_{ij})$. In the

TABLE 3

R_f values of eight substances in three different solvents

Substance	Solvent I	Solvent II	Solvent III
A	0.20	0.20	0.20
B	0.20	0.40	0.20
C	0.40	0.20	0.20
D	0.40	0.40	0.20
E	0.60	0.20	0.40
F	0.60	0.40	0.40
G	0.80	0.20	0.40
H	0.80	0.40	0.40
Information content (bit)	2	1	1

general case of system 1 containing m_1 classes and system 2 containing m_2 classes, eqn. (4) can be converted into

$$I = \sum_{i=1}^{m_1} \sum_{j=1}^{m_2} \frac{n_{ij}}{n_0} \log_2\left(\frac{n_{ij}}{n_0}\right) \tag{10}$$

At first sight, one might assume that I is the sum of the information content of the systems 1 and 2

$$I = I(1) + I(2) \tag{11}$$

This is true only if the information yielded by systems 1 and 2 is not correlated, i.e. if no part of the information is redundant. This can be understood more easily by considering a simple example represented by the R_f values for eight substances in three different solvents (Table 3).

With solvent I, one obtains 2 bits of information, while 3 bits are necessary for the complete identification of each possible substance. Solvents II and III each yield 1 bit of information. First running a plate with solvent I and then with solvent II does, indeed, permit complete identification: 3 bits are obtained with this combination. Although solvents I and III have clearly different R_f values, the combination of I and III does not yield any more information than that obtained with solvent I. The information content of a procedure in which both solvents are used is still 2 bits. Both of these cases are extreme and the combination of two TLC procedures, and in general of any two procedures, will lead to an amount of information less than that which would be obtained by adding the information content of both procedures but at least equal to the information content of a single procedure. In practice, it is improbable that two chromatographic systems would yield completely "uncorrelated information" and even when combining methods such as chromatography and spectrophotometry some correlation should be expected.

This is due to correlations between the physical quantities (signals) from which the identities of the unknown compounds (or concentrations in quantitative analysis) are derived. As a result of these correlations, the measurement of two (or more)

References p. 135

physical quantities yields partly the same information (also called mutual information).

For instance, both melting and boiling points usually increase with increasing molecular weight. When a high melting point has been observed, the boiling point is also expected to be high. If the correlation between melting point and boiling point were perfect, it would make no sense to determine both quantities for identification purposes. However, the correlation is not perfect because melting and boiling points are not determined solely by the size of the molecule but are also governed by factors such as its polarity. From this crude physical description, it is clear that the measurement of the boiling point will yield an additional information even if the melting point is known. However, this additional amount of information is smaller than that obtained in the case of an unknown melting point.

The most important conclusions from this section are that the highest information content for individual systems is obtained when the substances are distributed evenly over the different classes and that, for combinations of methods, the "correlated information" should be kept as low as possible.

Neither conclusion is surprising. Analytical chemists know that a TLC separation is better when the substances are divided over the complete R_f range and they also understand that two TLC systems in combination should not be too similar. Information theory allows one to formalize this intuitive knowledge and to quantify it, so that an optimal method can be devised.

To calculate the amount of information obtained from a combination of procedures, one takes into account all possible combinations of (two) signals. In fact, each combination is considered as one (composite) signal. The probability for each combination is introduced into Shannon's equation and thus the influence of the correlation upon the information content is implicitly taken into account.

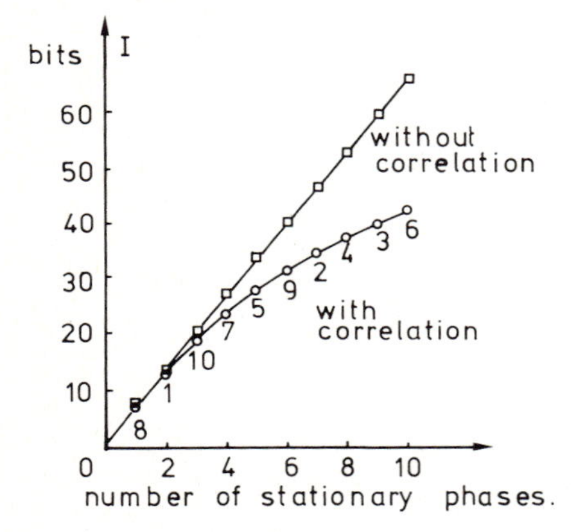

Fig. 1. Amount of information as a function of the number of stationary phases [4]. Reprinted with the permission of the American Chemical Society.

Mathematically more sophisticated methods also exist. They fall outside the scope of the present book but a result can be given. Information theory can, for instance, be used as a tool for optimizing combinations of analytical methods [3,4]. For instance, in gas chromatography, many different liquid stationary phases have been described (several hundred). Quite clearly, one does not want to keep in stock so many stationary phases because many of these have the same separation characteristics or, to put it in the terminology used in this chapter, yield "correlated" information. Figure 1 [4] gives the amount of information which can be extracted by using gas chromatography for a particular set of compounds.The best stationary phase, i.e. the one with the highest information content, yields 7 bits for this particular group of compounds. In fact, in this particular instance, many of the stationary phases yield about the same quantity of information so that a combination of 5 stationary phases should yield 35 bits. However, as the figure shows, only 28 bits are obtained. The 7 bits of information lost are due to correlation. The mathematics used to obtain this result are quite complex and of minor importance for our purpose. Of greater importance, however, are the rules for qualitative analysis which can be derived from them. These rules are as follows.

(1) Optimal combined methods for qualitative analysis make use of individually good systems.

(2) Individually good systems are characterized by high spread of the analytical signals (for instance, many different R_f values) and low errors (substances with small difference in R_f value can be discriminated).

(3) Optimal combinations also require that the individual systems should yield uncorrelated information, meaning that very dissimilar systems should be combined. This explains the power of combinations of methods with very different principles, such as GC/MS (combination of a chromatographic method with a spectrochemical one).

Information theory has not been applied much in analytical chemistry. It is mainly of philosophical importance. Practical applications include the development of a strategic approach to HPLC and optimal retrieval from MS data banks.

References

1 E. Shannon and W. Weaver, The Mathematical Theory of Information, University of Illinois Press, Urbana, IL, 1949.

2 D.L. Massart, The use of information theory in thin layer chromatography, J. Chromatogr., 79 (1973) 157.

3 A. Eskes, P.F. Dupuis, A. Dijkstra, H. De Clercq and D.L. Massart, The application of information theory and numerical taxonomy to the selection of GLC stationary phases, Anal. Chem., 47 (1975) 2168.

4 F. Dupuis and A. Dijkstra, Application of information theory to analytical chemistry. Identification by retrieval of gas chromatographic retention indices, Anal. Chem., 47 (1975) 379.

Recommended reading

K. Eckschlager and V. Stepanek, Analytical Measurement and Information: Advances in the Information Theoretic Approach to Chemical Analysis, Research Studies Press, Letchworth, Hertfordshire, 1985.

K. Eckschlager and V. Stepanek, Information Theory as Applied to Chemical Analysis, Wiley, New York, 1979.

J.C.W.G. Bink and H.A. Van't Klooster, Classification of organic compounds by infrared spectroscopy with pattern recognition and information theory, Anal. Chim. Acta, 150 (1980) 53.

P. Cleij and A. Dijkstra, Information theory applied to qualitative analysis, Fresenius Z. Anal. Chem., 298 (1979) 97.

S.L. Grotch, Matching of mass spectra when peak height is encoded to one bit, Anal. Chem., 42 (1970) 1214.

T.A.H.M. Janse and G. Kateman, Enhancement of performance of analytical laboratories. A theoretical approach to planning, Anal. Chim. Acta, 150 (1983) 219.

S.P. Perone and C.L. Ham, Measurement and control of information content in electrochemical experiments, J. Res. Natl. Bur. Stand., 90 (1985) 531.

D.R. Scott, Determination of chemical classes from mass spectra of toxic organic compounds by SIMCA pattern recognition and information theory, Anal. Chem., 58 (1986) 881.

G. Van Marlen, A. Dijkstra and H.A. Van't Klooster, Calculation of the information content of retrieval procedures applied to mass spectral data bases, Anal. Chim. Acta, 112 (1979) 233.

G. Van Marlen and A. Dijkstra, Information theory applied to selection of peaks for retrieval of mass spectra, Anal. Chem., 48 (1976) 595.

Chapter 10

Costs

1. The cost of a determination

The cost of an analysis depends on many factors. To a large extent, it depends on the nature of the analytical procedure, but possibly as much on the organization of the analytical laboratory. Analysis involves the use of labour, instruments, chemicals, and energy and all of these factors can be expressed in financial terms. Some of the costs can be directly related to an apparatus or a determination (direct costs). In addition, there are the so-called overhead costs (administration, maintenance of buildings, etc.). In this section, we will discuss only direct costs.

Direct costs are divided into fixed and variable costs. Fixed costs do not depend on the number of determinations carried out and consist, for instance, of

(a) the price of the apparatus,

(b) the cost of service contract, and

(c) the reagents, such as calibration samples, that have to be used only once for each series of determinations, independently of the number of determinations in the series. The following equation for fixed costs per day has been given in the literature [1].

$$K_{\mathrm{f}} = \frac{L[1 + (S/100)T_1]}{T_2 T_3} + \frac{G}{T_4} + R_{\mathrm{f}} \qquad (1)$$

where K_{f} is the fixed costs per day for a series of determinations, L the price of the apparatus, S the service cost per year as a percentage of L, T_1 is T_2 minus the guarantee period in years, T_2 is the expected number of years that the apparatus can be used, T_3 the number of work days per year, G the cost of the glassware, T_4 the expected number of days that the glassware can be used, and R_{f} the fixed costs of reagents and other materials per series of analyses per day.

Variable costs depend on the number of determinations carried out in the period of time considered. They consist essentially of labour costs and cost of materials consumed (for instance reagents and disposable vials).

The variable costs per daily series of analyses can be expressed as

$$K_{\mathrm{v}} = R_{\mathrm{v}} n + P t_n \qquad (2)$$

where t_n is the time required for one series of n analyses (hours), P the cost of labour per hour, and R_{v} the cost of reagents and other materials per sample.

The total cost per day is, of course

$$K_{\mathrm{t}} = K_{\mathrm{f}} + K_{\mathrm{v}} \qquad (3)$$

This leads to figures relating total cost and cost per determination to the number of determinations carried out (see Figs. 1 and 2).

References p. 146

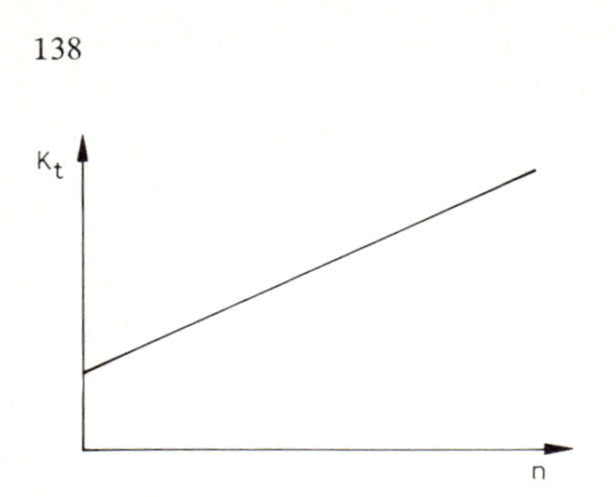

Fig. 1. Total cost, K_t, for the analysis of a series of n samples as a function of the number of determinations.

These considerations permit one to select the most economical apparatus, method or procedure as a function of given work load. A figure such as Fig. 1 can be drawn for each procedure. The total cost increases linearly with the number of samples analyzed but with different characteristics as shown in Fig. 3. This compares a manual method (A) to an automated one (B). A is characterized by a low cut-off on the price axis (low investments) and a steep gradient (mainly due to the labour component), while B has a high cut-off and a rather small gradient. The two lines therefore cross. The point where they cross is called the critical series length, n_c. Before this point is reached, the manual method is to be preferred and when the number of samples increases beyond it, one should use the automatic apparatus.

Two examples are given in Figs. 4 and 5. The first concerns the determination of inorganic nitrogen in fertilizers and the other the enzymatic determination of glucose.

The following general conclusions can be drawn.

(1) The cost per analysis is almost independent of the number of analyses when the procedure requires much labour and cheap instruments (or no instruments). This situation usually applies to "classical" analyses (manual titration, gravimetric analysis, etc.) such as in the nitrogen determination and to the classical distillation procedure. The cost of chemicals is usually negligible when applying such procedures.

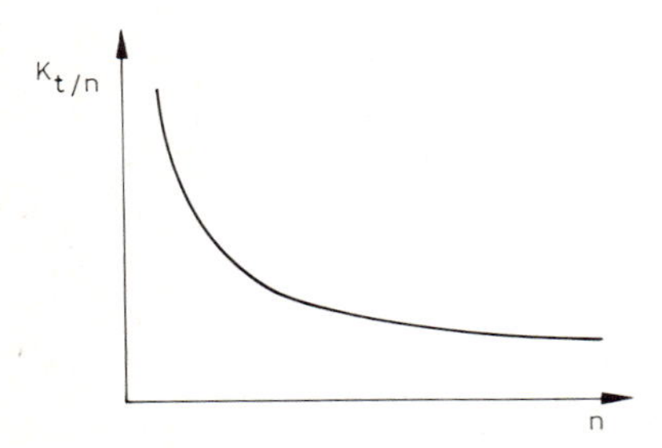

Fig. 2. The cost per determination, K_t/n, as a function of the number of determinations.

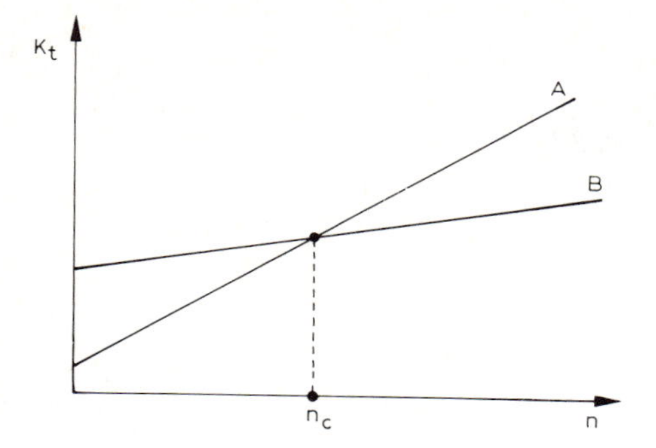

Fig. 3. The total cost, K_t, of a series of n determinations for the analysis of a certain constituent using methods (or apparatus) A and B as a function of the number of determinations, n_c, is the critical series length.

(2) For fully automated equipment, the cost per analysis is inversely proportional to the number of samples analyzed.

(3) The actual situation is usually intermediate between (1) and (2). Expressed as an equation, the total cost, K_t, per analysis for a series of n analyses per unit time

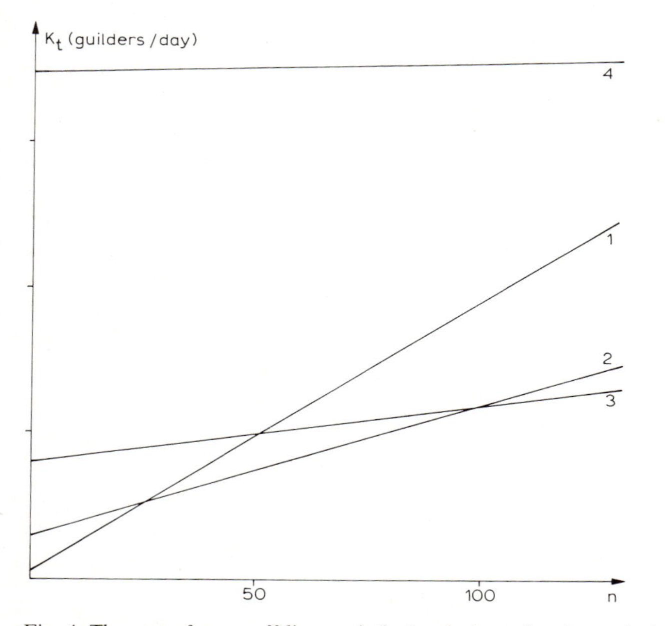

Fig. 4. The cost of some off-line analytical techniques for the analysis of inorganic nitrogen. 1, Classical distillation; 2, specific electrode; 3, Technicon autoanalyzer; 4, activation analysis. Adapted and simplified from ref. 2.

References p. 146

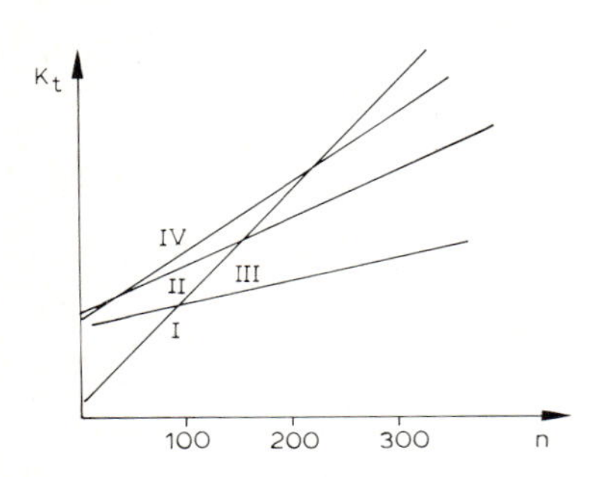

Fig. 5. The total cost, K_t, as a function of the series length, n, for the enzymatic determination of glucose with two different centrifugal analyzers (II–IV) and two manual methods (I). Lines III and IV were obtained with the same apparatus but different reagents. The best solutions are, evidently, the manual for series below about 100 (the exact value of n_c is 93) and III for longer series. Adapted from ref. 1.

can be approximated as

$$K_t = \frac{a}{n} + b \tag{4}$$

where a and b are the fixed and variable costs.

Since the cost per determination decreases asymptotically (Fig. 2), one may state that, in general, for rather large series the price is constant.

(4) From the above, it is clear that the method to be preferred will depend on the laboratory. In general, simple non-instrumental and non-automated methods are most economical when only a few determinations are to be made. Instrumentation and automation should be considered, and are justified, only for a large series of analyses. Of course, it is possible that an advanced and costly instrumentation and automation scheme may be attractive for other reasons, for instance in order to achieve better reproducibility.

It should be noted that the approach given here is only an approximation: only those parameters which can be known more or less accurately before an apparatus is bought are included. When the apparatus is in use and its performance is better known, additional parameters can be included to evaluate cost. An important parameter, for instance, is the reliability of an instrument. What is the main time between failures and the time needed to effect repairs, i.e. the main failure time? If, moreover, the measuring system is well known, for instance if one knows the distribution of the times between the arrival of two samples, still better evaluations are possible. This is discussed in some detail in the chapter on operations research (Chap. 25, in the section on queueing).

A cost–benefit analysis is then needed to determine whether the better reproducibility justifies the extra investment.

Costs of analytical determinations are, of course, also determined by the way a laboratory is organized. In Chap. 25, operations research methods are used to optimize laboratories, for instance to select the optimal routing of samples, the determination of priorities, etc.

2. The cost and the benefit of analytical information

2.1 Cost–benefit analysis. Opportunity costs

Once it has been decided that an analysis must be carried out, the computations in Sect. 1 can give guide lines as to which procedure should be selected. However, one can also ask whether an analysis is necessary and also how many determinations should be carried out and how thorough the analysis should be. In other words, how much analytical information does one want. In fact, one should ask the question why one produces analytical data. A somewhat cynical colleague [3] offered the three possible answers:

(a) because everyone does,

(b) because my instrument produces them (or as another cynical colleague put it: a lot of mediocre science is done just to keep big instruments running), and

(c) because we think that data are carriers of relevant information.

One should add a fourth answer, namely

(d) because we think the value of the information is worth more than the cost of obtaining it

and hope that this is the correct answer.

This means that, in evaluating analytical procedures or programs, one should place them in an economical context and compare the costs with the benefit derived from the information obtained. In this section, we will discuss some cost-related principles from economics relevant to this problem.

Cost–benefit analysis is a conceptual framework originally designed for comparing or evaluating government investments. It can, however, be extended to other investments. The term investment here means the use of human resources (labour, capital) and, in this sense, carrying out or developing an analysis is an investment.

In cost–benefit analysis, one lists all costs and benefits and tries to compare them. This involves quantifying them on a common money basis. This, of course, is extremely difficult. Consider, for example, a clinical analysis. Its cost is fairly simple to evaluate and express in money terms. The benefit (if there is any) is a correct diagnosis and possibly the saving of human life. But how does one evaluate this? Studies have been performed where the human life is equated to the expected economic production of an individual, but clearly this is not a complete answer. Moreover, the term benefit can mean different things depending on the point of view of the person who measures the benefit. The medical doctor who has to carry out a diagnosis and the public health authorities of a country have different criteria when they evaluate the benefit of a certain clinical chemical test. Nevertheless, economists consider it good practice to list benefits and costs and to try and quantify them.

Somewhat easier to achieve is a cost-effectiveness analysis. This is a closely related technique, but easier to undertake because one avoids the principal difficulty in cost–benefit analysis, namely the evaluation in money terms of benefits and costs by determining the financial implication to achieve a certain result which can be

References p. 146

quantified according to some non-monetary scale. While it is very difficult to evaluate, in money terms, the benefit of a program designed to analyse mutagenic mycotoxins in foodstuffs, it is much easier to evaluate the cost of setting up a project designed to analyse the mycotoxin content of the diet of the average U.S.A. inhabitant with the object of measuring, within 20%, for instance, the exposure of the population.

Another economic concept of fundamental importance in the evaluation of an investment or project is what is called the opportunity cost. Pearce [4] defines it as the "value of the foregone alternative actions". Opportunity cost can only arise in a world where the resources available to meet wants are limited so that all wants cannot be satisfied. Economists explain this by first considering an isolated individual, often called Robinson Crusoe. Robinson Crusoe is not confronted with money economics but still the opportunity cost applies: if Robinson Crusoe spends a few days picking strawberries, he cannot spend the same time harvesting grain and the cost of strawberries may be thought of as the foregone amount of grain or even as the amount of time he could otherwise have spent sunbathing. It is not difficult to think of analogies in the analytical laboratory. If an analytical chemist in an academic laboratory decides to use his time and that of his staff comparing two methods of mycotoxin analysis, he will not be able to use the same resources to measure mycotoxins in peanuts, one of its principal sources.

2.2 The law of marginal diminishing utility and related laws

The concepts of cost–benefit analysis and opportunity costs may be used to investigate an analytical project. The concept of marginal utility can be used to evaluate the extent of a project. By extent, we then mean a characteristic of the project which can be quantified such as information, precision, sensitivity. While it is difficult to do this in the strict economic sense, i.e. on a money basis, it is very instructive to use it at least conceptually. It is our opinion that, in many instances, some reflection on these laws would save a lot of time and money for unnecessary analyses or superfluous quality of analysis.

The term "marginal" will be used quite often in the rest of this chapter. It is defined by Pearce [4] as follows. "A marginal unit is the extra unit of something." The marginal cost, for instance, is the extra cost of producing an extra unit. The marginal cost of the 100th item produced is the total cost of producing all 100 items minus the cost for the first 99.

The law of diminishing marginal utility is due to Gossen, one of the important economists of the Austrian school, which ruled economic thinking for a long time. Gossen concluded that the marginal utility of a product diminishes when the fulfilment of a need increases. Another economist of the same school illustrated this with a simple example. In a tropical forest lives a colonist who has harvested five sacks of grain. He needs the first sack to remain alive, the second one to be sufficiently nourished, the third one to keep chickens, the fourth to be able to make brandy and the fifth to befriend the parrots. The law of diminishing marginal utility is given by Fig. 6.

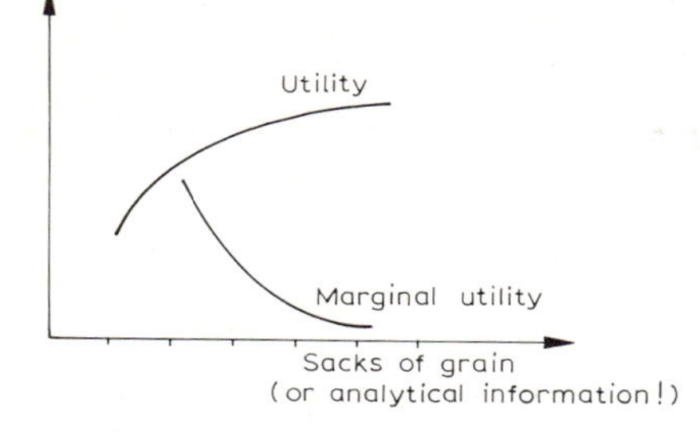

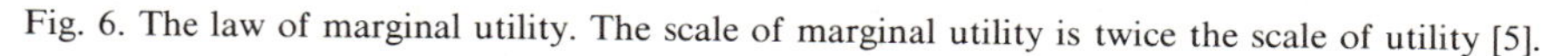

Fig. 6. The law of marginal utility. The scale of marginal utility is twice the scale of utility [5].

It is not difficult to translate this into analytical terms. If one analyzes mercury in fish and can only arrive at the conclusion that the concentration is less than 0.5 ppm, this will enable one to decide that the fish may or may not be sold. If one can determine that it is 0.4 ppm, this will satisfy a government agency and get one's laboratory certified. A correct result of 0.42 ppm allows the method to be published and a result of 0.423 ppm permits one to boast to ones colleagues about ones skill as an analyst.

Other laws in which the word "marginal" is used and which also relate to analytical chemistry are the law of marginal returns and the law of marginal costs. The following example is taken from a handbook on economics. When Robinson works on his land, the production obtained depends on the amount of work he puts into it. At the beginning, the additional production per day (the marginal return) may grow but soon this gain per additional day of work decreases. In the same way,

TABLE 1

Production and cost under the assumption of a fixed cost of 1000 monetary units (MU) and constant variable cost/day [5]

Number of working days	Total production (kg)	Marginal production (kg)	Total variable cost (MU)	Total cost (MU)	Marginal cost (MU)	Marginal cost/unit (MU)
1	37		150	1150		
2	88	51	300	1300	150	2.94
3	147	59	450	1450	150	2.54
4	216	69	600	1600	150	2.17
5	290	74	750	1750	150	2.03
6	366	76	900	1900	150	1.97
7	441	75	1050	2050	150	2.00
8	512	71	1200	2200	150	2.11
9	567	55	1350	2350	150	2.73
10	610	43	1500	2500	150	3.49
11	627	17	1650	2650	150	8.82
12	636	9	1800	2800	150	16.67
13	637	1	1950	2950	150	150.00

References p. 146

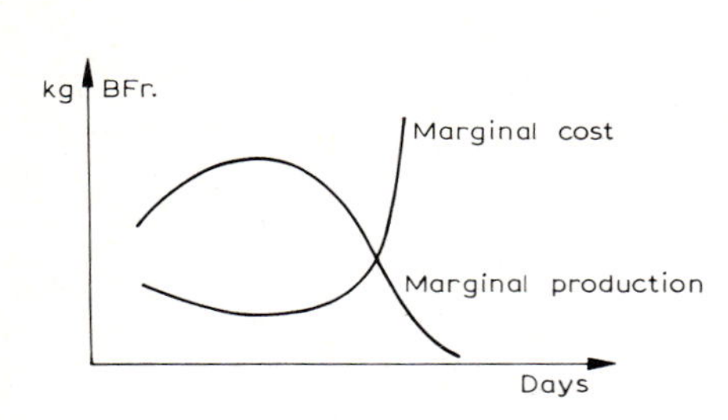

Fig. 7. Marginal production (kg) and cost (BFr.) as a function of the investment (in days) as derived from Table 1 [5].

the marginal costs per unit of product may initially decrease but after a time they increase. This is shown with the example of Table 1 and Fig. 7.

Let us look now for examples where the marginal laws apply in analytical chemistry. A very simple example of the marginal laws can be found in the precision with which the concentration in a sample is determined. By carrying out n replicate measurements, the cost of the analysis can increase by a factor n when the cost is determined mostly by personnel costs while the precision with which the real concentration is estimated improves with a factor of $\sqrt{n}$ (because the standard deviation on the mean decreases by the same factor). Analytical chemistry produces information. In Chap. 9, information was defined as a reduction of uncertainty. It is clear that this reduction is more important when the analytical result is known to be more precise. Information theory can be used to derive an inverse relationship between information and precision. One obtains Fig. 8, which is an illustration of the law of marginal returns.

Going one step further in the evaluation process, one may wonder whether the improvement of precision, which we considered as the marginal return, is indeed necessary, i.e. we can question its utility. Indeed, since we are trying to put analytical chemistry into an economic perspective, we should relate the utility to the economic objective.

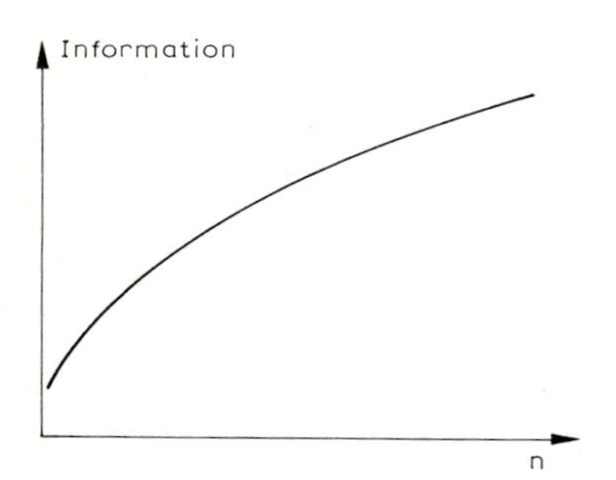

Fig. 8. Information, $I \approx \log_2(1/s)$, as a function of the number of samples, n. This is an illustration of the law of marginal returns [5].

TABLE 2

Data from Acland and Lipton [6] transformed to illustrate the law of marginal utility

σ_A/σ_N	P	I	U
0.1	0.99	3.32	99
0.2	0.99	2.32	99
0.3	0.98	1.73	98
0.4	0.97	1.32	97
0.5	0.95	1.00	95
0.6	0.94	0.73	94
0.8	0.90	0.32	90
1.0	0.86	0.00	86

σ_A = analytical standard deviation.
σ_N = standard deviation of normal range.
P = probability of recognizing a normal patient as such as given by Ackland and Lipton.
I = information which is proportional to the value given in this column. This value is $\log_2(\sigma_A/\sigma_N)$
U = specificity (in %).

One area of analytical chemistry in which a relationship between the quality of an analytical method and its utility can be made is clinical chemistry. The fewer analytical errors are made, the better the medical doctor will be able to use the result and arrive at a correct diagnosis. Acland and Lipton [6] investigated the necessary precision for a clinical test. Let us assume that the utility of a clinical test is directly proportional to the percentage of "normal" patients it recognizes as such. This is what clinical chemists call the specificity of a method (see Chap. 26). To answer Ackland and Lipton's question, one must determine the probability of finding a normal person outside the normal range due to insufficient precision. Table 2 gives the probability and the resulting specificity. One can therefore produce the graph of Fig. 9, which is a good illustration of the law of diminishing marginal utility.

Another example from process control will be given in Chap. 27, Sect. 3.

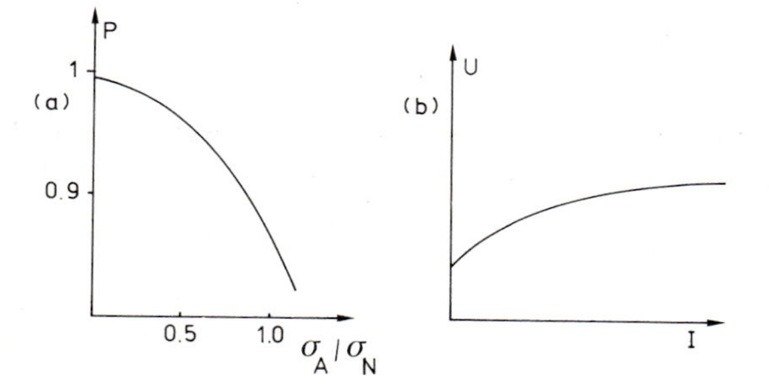

Fig. 9. (a) The probability, P, of recognizing a normal patient as a function of the precision of the determination, σ_A, compared with the biological variation, σ_N. (b) The specificity, U, as a function of information, I [5].

All these examples relate to the precison and/or accuracy of analysis. In fact, it can be shown that the "laws" are useful in decision making in analytical chemistry in general and that very recent developments in analytical chemistry, such as the application of information theory, time series, and pattern recognition, have to do with determining the functions describing the marginal laws. An example from the domain of pattern recognition will show this.

The main purpose of applying pattern recognition in analytical chemistry (see Chap. 23) is to classify a sample into one or other category on the basis of a pattern of analytical results obtained for the sample. This pattern consists of the results of different analytical determinations. For instance, one can try to classify patients into one of three categories (euthyroid, hypothyroid or hyperthyroid) by the use of five analytical results (total serum thyroxine, T4; total serum tri-iodothyronine, T3; T3 resin uptake, RT3U;serum thyroid-stimulating hormone, TSH; and increase of TSH after injection of TSH-stimulating hormone, ΔTSH). This example is discussed further in Chap. 20, Sects. 7 and 8. One of the questions which can be asked is what the effect of selecting a smaller number of tests is on the quality of the diagnosis. So-called feature selection techniques permit the determination of the combination of m tests out of n possibilities which results in an optimal classification, i.e. diagnostic efficiency can be achieved. Table 2 of Chap. 20 contains a measure of the diagnostic efficiency called classification success. If we plot this against the number of tests, one again obtains the law of marginal returns, at least for the EU–HYPER data. (The EU–HYPO data illustrate that, in some cases, the utility even decreases when the number of tests grows.)

Clearly, the "marginal laws" are quite generally applicable in analytical chemistry and, very often, the analysts' problem is to determine which cost/utility ratio is the optimal one for his purpose. This, in fact, is a cost–benefit or cost–efficiency analysis.

References

1 R. Haeckel, Rationalisierung des Medizinischen Laboratoriums, G-I-T Verlag Ernst Giebeler, Darmstadt, 1976.
2 F.A. Leemans, Anal. Chem., 43 (1971) 36A.
3 O. Christie, in a lecture before the Analytical Division of the Royal Chemical Society.
4 D.W. Pearce, Dictionary of Modern Economics, Macmillan, New York, 1981.
5 D.L. Massart, Trends Anal. Chem., 1 (1982) 348.
6 J.D. Ackland and S. Lipton, J. Clin. Pathol., 24 (1971) 369.

Recommended reading

Articles that describe cost calculations in the laboratory can be found in the clinical chemistry literature. Some examples are given below. References to articles on cost–benefit studies are given in Chap. 26

J.F. Leijten, F. Van der Geer, M.N.M. Scholten and H.M.J. Goldschmidt, The costing of tests in a laboratory for clinical chemistry and haematology, Ann. Clin. Biochem., 21 (1984) 109.

A.F. Krieg, M. Israel, R. Fink and L.K. Shearer, An approach to cost analysis of clinical laboratory services, Am. J. Clin. Pathol., 69 (1978) 525.
J. Stilwell, Costs of a clinical chemistry laboratory, J. Clin. Pathol., 34 (1981) 589.

Chapter 11

The Time Constant

Continuous procedures are designed to measure (changes in) the composition of a continuous product stream. An important aspect of a continuous analyzer is its ability to monitor a fluctuating concentration of the analytes. The response of the instrument has to be fast with respect to the fluctuation rate of the signal to be measured, otherwise the amplitude of the fluctuations as shown by the analyzer will be smaller than the true value. The response of an instrument is expressed by its response function. To quantify the response function of a continuous analyzer, one measures the response of the analyzer on a well-defined disturbance at the input (Fig. 1). Often used disturbances are an impulse (or concentration spike), a step (or a concentration jump), or a periodic function.

As an example to illustrate the response of a continuous procedure, consider a flow cell as used in many instruments which operate continuously. In some instances, such a flow cell has the properties of an ideal diluter with good mixing characteristics. The sample which enters the cell is mixed with its entire contents. A sudden (stepwise) change of the concentration in the sample stream from 0 to c_∞ causes a more gradual change from concentration 0 to c_∞ in the sample cell. If the volume of the flow cell is V ml and the flow equals v ml s^{-1}, the differential equation describing the concentration $c(t)$ in the flow cell as a function of time is

$$\frac{\mathrm{d}c(t)}{\mathrm{d}t} = \frac{v}{V}[c_\infty - c(t)]$$

Solution of this equation gives

$$\ln[c(t) - c_\infty] + a = -\frac{vt}{V}$$

Substituting $c = 0$ for $t = 0$ gives

$$a = -\ln(-c_\infty)$$

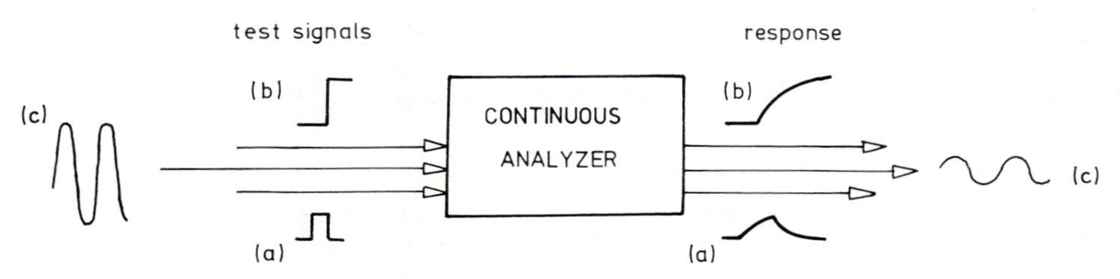

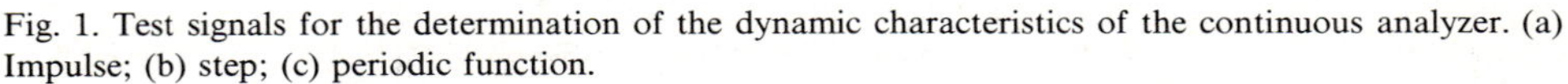

Fig. 1. Test signals for the determination of the dynamic characteristics of the continuous analyzer. (a) Impulse; (b) step; (c) periodic function.

References p. 156

The concentration profile as a function of time is thus

$$c(t) = c_\infty[1 - \exp(-vt/V)] \tag{1}$$

The shape of the concentration profile or response function is completely defined by the value v/V. From eqn. (1), it is seen that V/v has the dimension of time. V/v is called the time constant of the cell

$$T_x = \frac{V}{v}$$

Substitution of T_x in eqn. (1) gives

$$c(t) = c_\infty[1 - \exp(-t/T_x)] \tag{2}$$

Assuming a linear relationship between the absorbance and the concentration $c(t)$, where $A(t) = Sc(t)$, the measured response in absorbance units is

$$A(t) = Sc_\infty[1 - \exp(-t/T_x)] = A(\infty)[1 - \exp(-t/T_x)] \tag{3}$$

where S is the sensitivity as defined in Chap. 7.

A 5 ml flow cell with a sample flow of 0.5 ml s^{-1}, for instance, is a system with a time constant $T_x = 5/0.5 = 10$ s. In Fig. 2, response functions are plotted for $T_x = 1$, 10, and 100 s. From these plots, it is seen that the time required to reach 95% of the equilibrium value depends on the time constant and is, in fact, a measure of the response function of an analyzer. Based on this property, IUPAC [1] recommends that the response of an ion selective electrode (ISE) is defined by the time, t_α, required for the ISE potentials to reach α% of the equilibrium potential after a step change in sample concentration (activity), where $\alpha = 90\%$. The values 50, 95, and 99

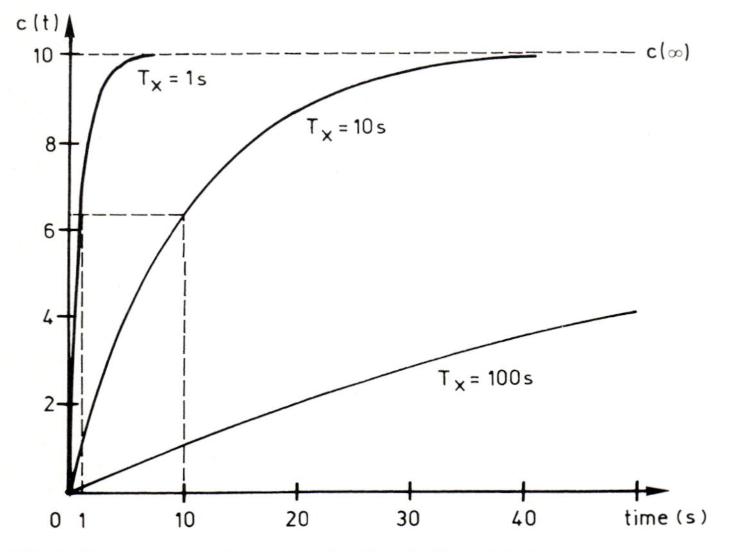

Fig. 2. Step response of systems with $T_x = 1$, 10, and 100 s.

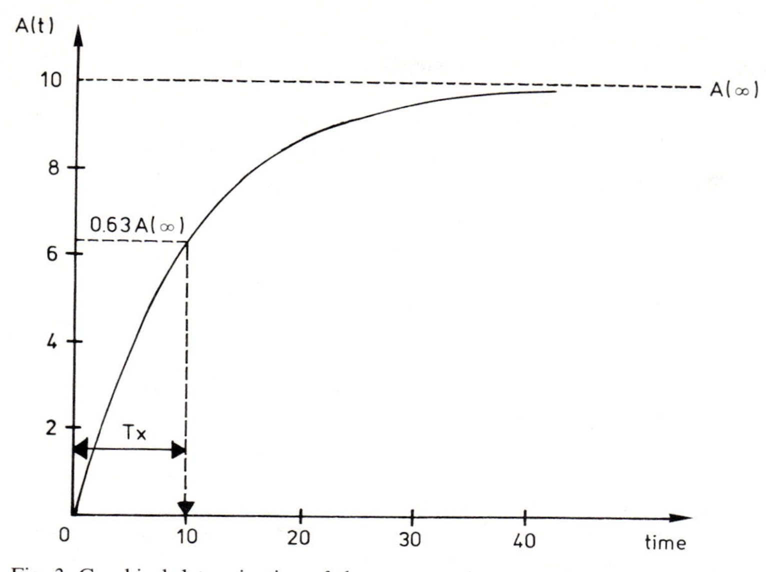

Fig. 3. Graphical determination of the response time or time constant, T_x, from the step response of a first-order system.

have also been proposed for α. If the response of an ISE electrode can be described with one time constant (as it can for most fast electrodes), t_α is related to T_x by

$$\frac{\alpha}{100}=\frac{E(\alpha)}{E(\infty)}=1-\exp(-t_\alpha/T_x)$$

$$\frac{t_\alpha}{T_x}=-\ln[(1-(\alpha/100)]$$

The time constant, T_x, of a continuous analyzer can be determined from its step response in several ways, e.g.

(a) a non-linear regression of eqns. (2) or (3) through the measured absorbances, $A(t)$, as a function of time, t. Non-linear regression is discussed in Chap. 13.

(b) a graphical determination of the time required to reach 63% of the final response (Fig. 3). For $t=T_x$

$$A(T_x)=A(\infty)[1-\exp(-1)]=0.63A(\infty)$$

Consequently, the time needed to reach a signal which is 63% of the steady-state value equals the time constant.

(c) For $t=1$ (expressed in the same units as the time constant)

$$A(1)=A(\infty)[1-\exp(-1/T_x)]$$

or

$$T_x=\frac{-1}{\ln\{1-[A(1)/A(\infty)]\}}$$

TABLE 1

Step reponse in a 5 ml flow cell on a concentration jump in a sample stream with a flow of 0.5 ml s^{-1}

t	0	2	4	6	8	10	12	14
$A(t)$	0	1.8	3.3	4.5	5.5	6.3	7.0	7.7

t	16	18	20	25	30	35	40	∞
$A(t)$	8.0	8.3	8.8	9.2	9.5	9.7	9.8	10.0

An advantage of the non-linear regression method is that no steady-state value, $A(\infty)$, is needed. $A(\infty)$ and T_x are both estimated by a non-linear regression. An example is given below.

Figure 3 shows the step response of a 5 ml flow cell on a concentration jump in a sample stream with a flow of 0.5 ml s^{-1}. The values are given in Table 1. Non-linear regression gives (Chap. 13) $T_x = 10.0$ and $A(\infty) = 10.02$. From $A(1)$, a value $T_x = 9.8$ is obtained. The latter estimator is less precise because it considers only two points of the response function, $A(1)$ and $A(40)$.

A characteristic of a first-order process is that the time constant, T_x, fully defines the response on a stepwise change. From eqn. (3) and Fig. 3, it is clear that a time $5T_x$ is required before the value of $A(t)$ is within 1% of its final value. Substituting $c(t) = 0.99c(\infty)$ in eqn. (2) gives $0.99 = 1 - \exp(-t/T_x)$ or $t = 4.6T_x$. This fact must be taken into account when operating continuous analyzers. Many continuous flow analyzers are in use in analytical and chemical laboratories for the analysis of discrete samples. These samples are usually arranged in a sample stream in order to make it possible to use continuous flow analyzers. In practice, this is realized by a stream of samples which are interchanged with streams of a blank (washing fluid). The result is a series of stepwise changes in concentration from 0 to $c(1)$, back from $c(1)$ to 0, from 0 to $c(2)$ etc. with a time interval, t, between the steps. This time interval between successive sample plugs should be long enough to reach the equilibrium levels for $c(1)$, back to the blank, $c(2)$ etc. Since it takes a time $5T_x$ before a virtually constant response is obtained for the sample and the same time

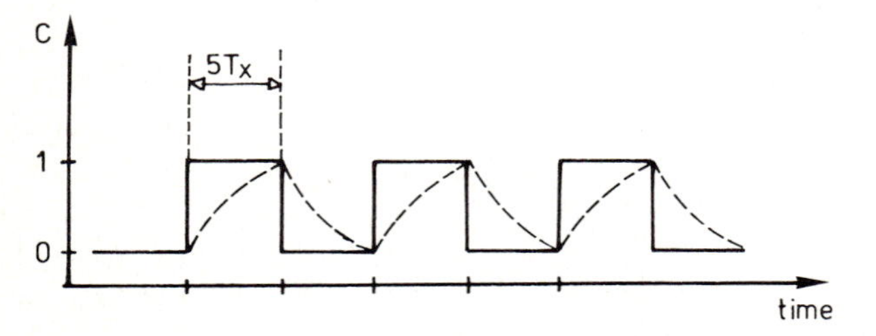

Fig. 4. Signals measured on a sequence of blank ($c = 0$) and a sample ($c = 1$) in a continuous analyzer. ———, Concentration profile at the input of the analyzer (separation $= 5T_x$); - - - - - -, concentration profile at the output (detector) of the analyzer.

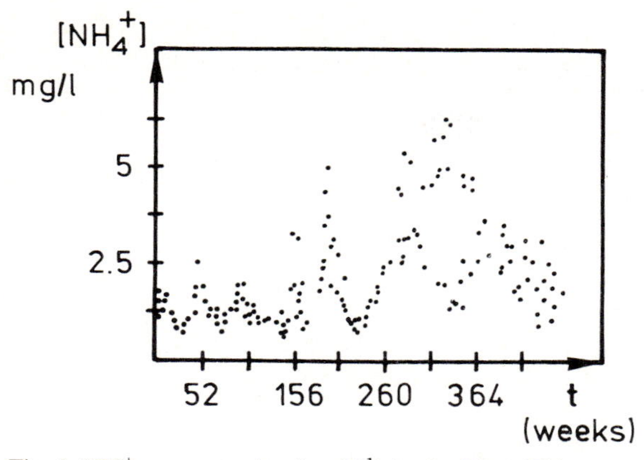

Fig. 5. NH_4^+ concentration (mg l^{-1}) in the River Rhine (1975–1976). The data were taken from ref. 2.

before a steady blank level is measured again (Fig. 4), the time between the samples should be at least $10T_x$.

One has also to take this effect into account when monitoring time varying concentrations with a continuous analyzer (e.g. an ion selective electrode).

In the same way that the response of an analyzer is characterized by its time constant, $(T_x)_a$, the rate of the fluctuations of a product stream (e.g. of a chemical plant or the concentration of some species in a river) is also characterized by a time constant $(T_x)_p$. An example is given in Fig. 5. Obviously one cannot apply a step response on a river and has therefore to apply other methods to determine the time constants of such systems. These methods are discussed in Chap. 14.

The extent to which an analyzer with time constant $(T_x)_a$ may be suitable to follow a process with a time constant $(T_x)_p$ is indicated by the ratio of the observed variance of the concentration pattern, s_p^2, to the true variance, σ_p^2, which is given by

$$\frac{s_p^2}{\sigma_p^2} = \frac{(T_x)_p}{(T_x)_p + (T_x)_a} \qquad (4)$$

It is clear that, for "fast analyzers", $(T_x)_a \ll (T_x)_p$, the observed variations are equal to the true variations. For slow analyzers, the amplitude of the fluctuations are reduced and even disappear for large values of $(T_x)_a$. As a rule of thumb, one may safely state that, if one is interested in a reliable estimate of the standard deviation, the time constant of the continuous analyzer should be smaller by a factor 10–100 than the time constant of the process. The time constant of the NH_4^+ fluctuations in the river Rhine is about 68 days. Therefore, one may conclude that the time constant of analyzers for the monitoring of the NH_4^+ concentration should be less than 1 day to 1 week. In this particular case, any continuous analyzer will be satisfactory as far as the time constant is concerned.

Besides continuous flow analysis (CFA) and flow injection analysis (FIA), every continuous detector can be considered as a continuous part of an analytical system.

References p. 156

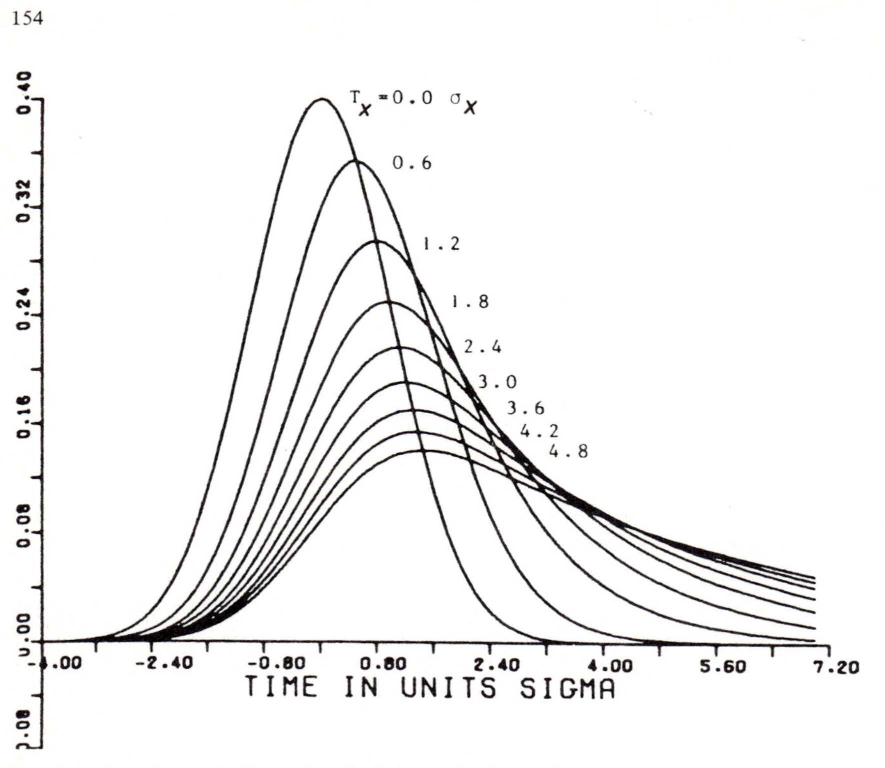

Fig. 6. A Gaussian peak detected with detectors having various time constants.

Therefore response characteristics, T_x, or the response function of such detectors determine in general to what extent the shape of the observed signal will correspond with the actual shape.

It is obvious that when increasing the frequency of periodically changing concentrations, a point will be reached when the instrument cannot follow the changes. Figure 6 illustrates how a Gaussian-shaped curve is deformed at the output of a system with various time constants. The overall effect is a signal broadening and loss of resolution. Figure 6 shows the shapes found for a Gaussian peak with a standard deviation σ_x when recorded with detectors with various time constants. Peak distortion becomes visible for T_x/σ_x ratios exceeding 0.1. Hence the time constant of the detector should certainly be less than 1/30th of the half-height peak width (about 3σ) in order to avoid deformation. Infrared spectroscopists know very well the phenomenon that the resolution in IR spectra is very sensitive to the scan speed due to a slow response of the detector. Measurement of the response time of the instrument under actual operating conditions is necessary to derive the proper scan speed. A typical response time of an infrared spectrometer is 2 s. Requiring that the ratio between T_x and the half-height peak width is less than 1/30, one calculates the proper scan speed as follows. Assuming a peak width of 40 cm^{-1},

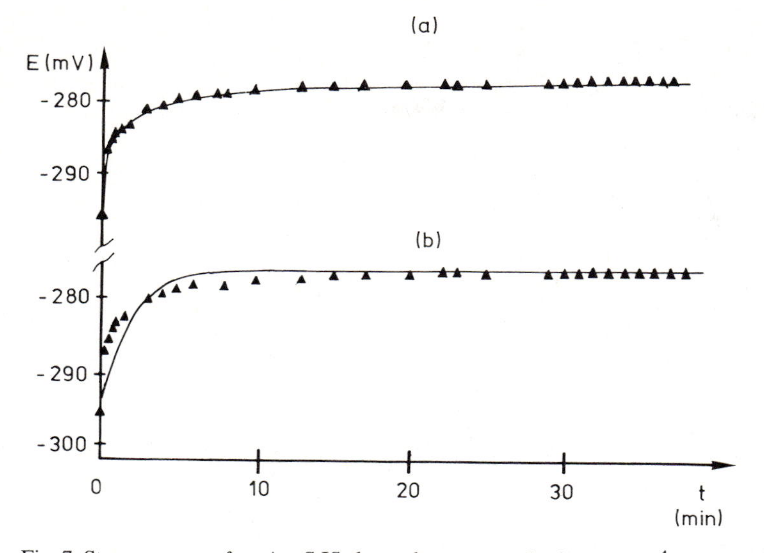

Fig. 7. Step response of an Ag_2S IS electrode on a transfer from a 10^{-4} M to a 10^{-5} M $AgNO_3$ solution. (Adapted from ref. 3.) (a) Fit with a sum of two exponential functions ($T_x = 0.2$ and 4 min); (b) fit with one exponential function ($T_x = 0.6$ min).

$T_x/40$ should be less than $1/30$. For the calculation of this ratio, it is necessary to convert T_x into cm^{-1} according to T_x (in cm^{-1}) $= T_x$ (in s) $\times$ the scan speed. Thus, T_x (in s) $\times v/40 < 1/30$. This gives $v < 40/60 = 0.6$ cm^{-1} s^{-1}. Equally, the shape of a gas chromatographic elution profile as measured by the detector and shown by the recorder depends on many factors. One of these factors is the time constant of the detector (and the recorder) used for monitoring the effluent of the column. A too-slow response time in chromatography causes skewed and broader peaks than the true peaks. Although such a distortion does not affect the peak area, peak distortion should be avoided as much as possible for the accurate determination of the retention times and the automatic processing of chromatograms with (partially) overlapping peaks. Avoidance is better than a cure, even if mathematical procedures are available to restore the true peak shape or simply to sharpen the peaks (see Chap. 15).

So far, it has been assumed that the step response of a continuous analyzer has a shape as shown in Fig. 2. This shape is characterized by one parameter, the time constant. Figure 7 shows the response of an Ag_2S IS electrode on a transfer from a 10^{-4} M to a 10^{-5} M $AgNO_3$ solution. The best fit of this response with eqn. (2) gives a poor result. The reason is that the dynamic behaviour is no longer first order and therefore cannot be adequately described by one time constant. In this particular case, a best fit was obtained with the function

$$E(t) = 296 + 10 \exp[1 - (t/0.2)] + 9 \exp[1 - (t/4)]$$

This expression can be considered as the sum of two first-order responses, respectively, with $T_x = 0.2$ and 4 min.

In general

$$E(t) = E(0) + K(1) \exp[1 - (t/T_1)] + K(2) \exp[1 - (t/T_2)] \quad (5)$$

It can be shown that the expression (5) is found for a system which is composed of two virtually parallel first-order systems. Such a system is second order. Some slow electrodes may sometimes require the inclusion of a third time constant, which indicates a third-order response.

A drawback of the determination of T_x from the step response is the necessity to collect data over a sufficiently large range of the response function in order to obtain a reliable estimate of $A(\infty)$ and thus T_x. For ISE with slow response times (several minutes), this is often a difficult matter. An alternative method is to analyze the characteristics of the baseline noise of the continuous analyzer. This can be done in the frequency domain by a direct Fourier analysis (see Chap. 15) or in the time domain by an autocorrelation analysis (Chap. 14). The frequency characteristics may also be observed directly by feeding a periodically fluctuating concentration to the system. This method, which is very useful when characterizing electronic systems is, however, less feasible when characterizing analytical systems.

References

1 IUPAC Inf. Bull., (1978) 69.
2 Kwaliteitsonderzoek in de Rijkswateren, Quarterly Reports, 1975 and 1976, RIZA, The Netherlands.
3 A. Shatkay, Transient potentials in ion-specific electrodes, Anal. Chem., 48 (1976) 1039.

Recommended reading

L.A. Zadeh and C.A. Desoer, Linear System Theory, McGraw-Hill, New York, 1963.
A. Papoulis, Probability, Random Variables and Stochastic Processes, McGraw-Hill, New York, 1965.

Chapter 12

Signals and Data

An important aspect of chemical analysis is the collection and subsequent treatment of analytical data. Very often, a large number of data is collected, which has to be reduced to a few numbers. For instance, one has to derive the peak positions, peak heights or areas from the original thousand or more data points in a spectrum, chromatogram or other signal.

In some instances, the quality of the measured data is not sufficient to allow a direct derivation of the analytical result. Bad quality of data may be caused by a poor signal-to-noise ratio or by distortions introduced by the instrument. For example, peaks may be distorted by a poor resolution of the monochromator or because of a poor response time of the detector. As a consequence, the quality of the data has to be improved. Basically, there are two approaches: data enhancement and data restoration. Data enhancement is a process which improves the quality of the data without using information on the degradation phenomenon. Methods for improving the signal-to-noise ratio fall in this category. Methods which use knowledge of the degradation phenomenon are called restoration methods. Data restoration is oriented toward the modelling of the degradation and applying the inverse process in order to recover the true or undistorted signal (a spectrum).

1. Classification of signals and data

1.1 Deterministic signals

Deterministic signals are completely defined by one or more independent variables or factors ($x_1, x_2, \ldots, x_n$). A deterministic signal remains invariant when remeasured over equal ranges of the factors. True deterministic signals are very rare because unknown and uncontrollable factors usually influence the signal as well. The effect of these factors is called noise. A noise-free spectrum of the absorbance as a function of the wavelength would be a deterministic signal. In that case, the absorbance is fully defined by the wavelength, analyte concentration, and the instrument setting. A second (noise-free) scan of the spectrum should exactly reproduce the readings. In other words, once the signal has been measured, one can exactly forecast the signal under unchanged conditions. In some instances, a deterministic signal can be represented by a mathematical function or model. The linear relationship between concentration and absorbance in spectrometry is forecast by the Lambert–Beer law. After having measured a number of points of a calibration curve, one can assume an equation or model which describes that curve and calculate its parameters. Very often, models for calibration curves can only be

Reference p. 163

based upon some convenient mathematical equation with no regard for any underlying chemical or physical mechanism. They are called empirical models. In the case of noise-free measurements, the model exactly forecasts the value to be measured for any other known concentration.

1.2 Stochastic signals

In contrast to a deterministic signal, a stochastic signal is influenced by a number of uncontrolled factors. A noisy baseline (Fig. 1) is an example of a stochastic signal. A new scan of a noisy baseline of a spectrometer operated under identical conditions is different from previous scans.

Although the exact measurements are not predictable, some information on the signal is available. The most probable value of the baseline is the average value. Excessively low and high values are much less probable. These probabilities are expressed by a probability distribution function such as a Gaussian distribution. A stochastic signal is thus described by the parameters of a probability distribution function, giving the probability to measure a value at a given level.

Recalling the example of the noisy baseline, one can find another characteristic of a stochastic signal, namely the speed of variation. When the values of the baseline

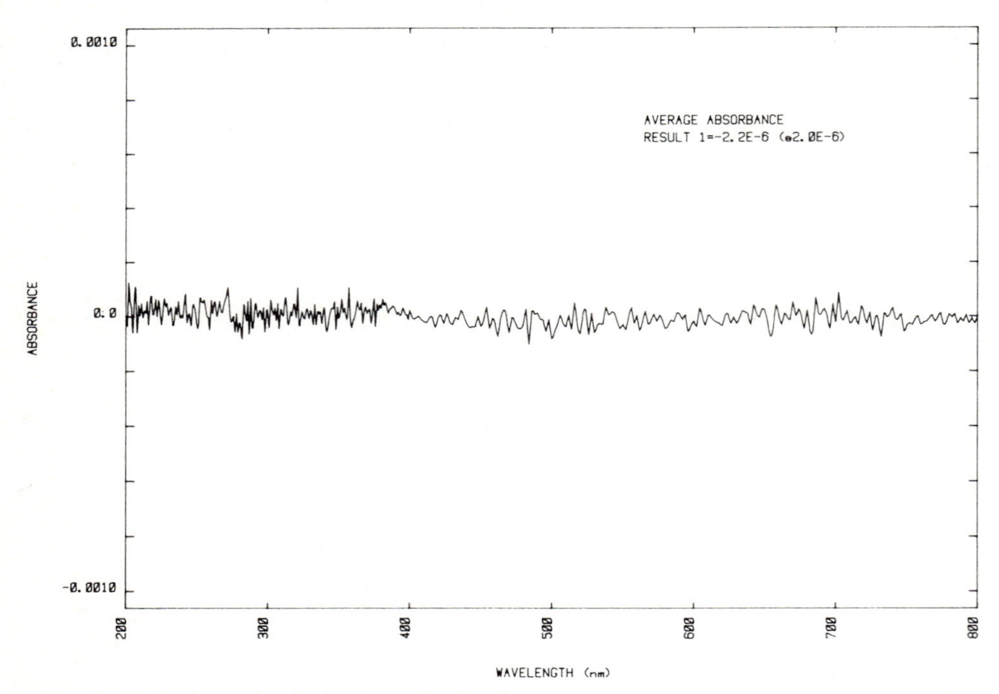

Fig. 1. Example of a stochastic signal: a noisy baseline.

are slowly changing with time, it is reasonable to expect that a high value of the baseline will be followed by another relatively high value (high in the sense of greater than the mean value). In other words, there is some degree of correlation between the observations. This aspect is discussed in Chap. 14.

1.3 Signals with a deterministic and a stochastic part

In practice, analytical signals are a combination of a deterministic and a stochastic part. The deterministic part is usually considered to be the true signal, e.g. a spectrum, a chromatogram etc., while the noise is the stochastic part. The contribution of the deterministic part in analytical signals is usually larger than the contribution of the stochastic part. Another example is the temperature of a river (Fig. 2), which has a strong annual periodicity. After subtraction of the deterministic part (in this case a sine wave with a period of 365 days) a stochastic time series is left. How to model and to separate deterministic and stochastic contributions is discussed in Chap. 14.

1.4 Univariate and multivariate data and signals

Univariate signals are measured as a function of one controlled variable: time in chromatography, wavelength in spectrometry, volume in titrimetry, m/e in mass spectroscopy, concentration in calibration, voltage and current in electrochemistry. In many instances, the qualitative information is obtained from the position along the x axis. The quantitative information is usually present along the axis of the dependent variable, e.g. peak height or area in chromatography.

Multivariate signals and data form a so-called data table. In HPLC with a multiwavelength detector, for example, many chromatograms are measured simultaneously (Fig. 3).

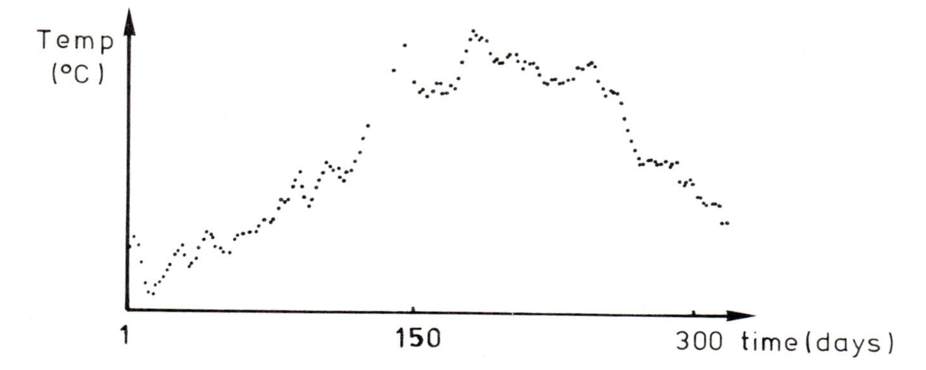

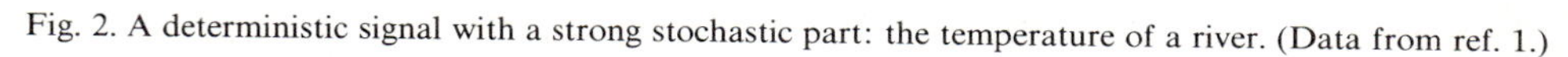
Fig. 2. A deterministic signal with a strong stochastic part: the temperature of a river. (Data from ref. 1.)

Reference p. 163

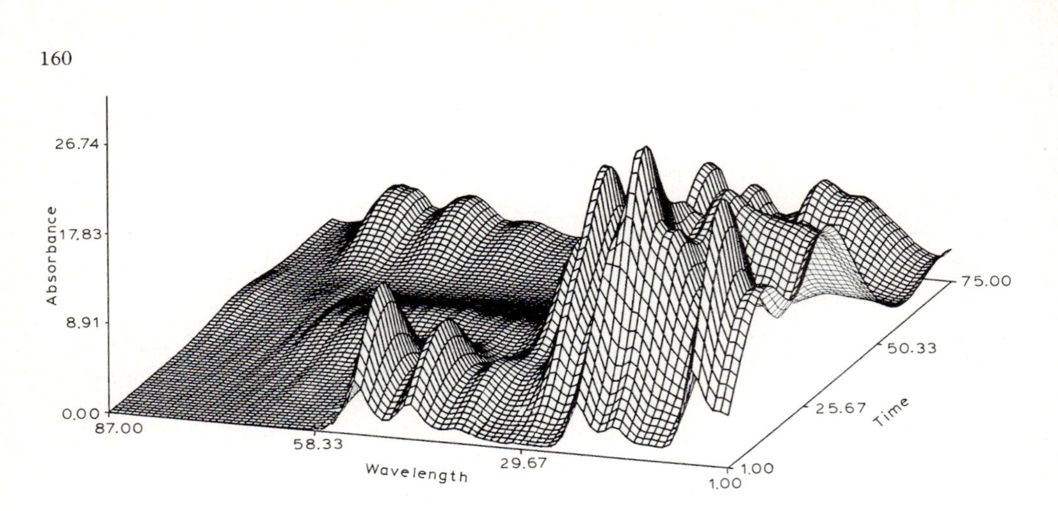

Fig. 3. Multivariate signals of absorbances measured with an HPLC/diode array detector.

A large class of multivariate methods are the so-called hyphenated methods, which consist of two univariate methods in series. For instance, the combination chromatography/mass spectrometry produces a data table with the axes time and m/e. Recently, HPLC/UV–vis combinations have been introduced, which produce a data table with the axes time and wavelength (Fig. 3). These are examples of a multivariate data table.

A data table is also obtained when a number of features, or variables, are measured on a number of objects. For instance, the concentrations of a number of heavy metals in rain water which have been collected at a number of sites. The concentrations of these metals at one site are a row of the data table. The length of the row (or number of columns in the data table) is the number of heavy metals considered. The number of rows is the number of sites. These sites are usually called

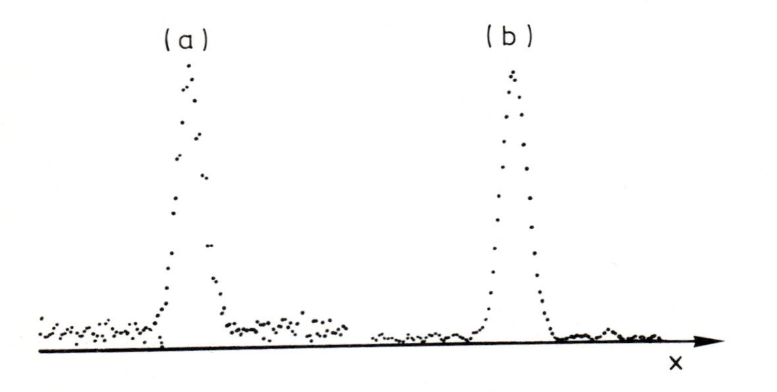

Fig. 4. Example of signal enhancement. (a) Measured signal; (b) enhanced signal.

the objects of the data table. In some instances, several classes or categories of objects can be distinguished. For example, some sites where rain water is collected are situated in industrial areas and others in rural areas.

2. Purpose of signal and data processing

2.1 Signal enhancement by reduction of the noise level

The principal objective of signal enhancement is to treat the data in such a way that the result is more suitable for a specific application than the original signal (Fig. 4). Signal enhancement can be achieved by treating the data directly in the domain in which they have been measured. Such methods are indicated by the term "smoothing". Signals are sometimes transformed to another domain for further processing, whereafter the processed signal is re-transformed to the original domain of the measurements. A well-known transformation is the Fourier transformation. Signal enhancement can also be achieved in the Fourier domain. Such a method is called "filtering". These methods are discussed in Chap. 15.

2.2 Signal restoration

Resolution of peaks or signals is a major problem in spectrometry and chromatography. Although many instrumental improvements, such as better stationary phases in chromatography or better monochromators in spectrometry, have led to a major improvement of resolution, there remain situations where resolution needs to be mathematically improved to solve the analytical problem at hand. Two approaches are possible: if the source of the signal degradation is known and can be described in mathematical terms, one can attempt to restore the signal by applying the inverse process (Fig. 5). This is called restoration or deconvolution (Chap. 15). When the broadening factors are not known, restoration techniques cannot be applied and have to be replaced by "pseudo-deconvolution" techniques such as peak sharpening by the subtraction of the second derivative from a peak-shaped signal.

2.3 Characterization and modelling of the stochastic and deterministic parts of the signal

2.3.1 Univariate data

The observability of chemical compositions depends on the quality of the analytical signal. One indicator of quality is the signal-to-noise ratio. Another indicator is resolution. The determination of the properties of signals and noise is important as it allows the proper measures for improvement to be taken, either instrumentally or mathematically. An important property in that respect is the frequency characteristic of signal and noise. The mathematical methods to determine the frequency characteristics of signals and noise can also be used for the characterization of a stochastic process, e.g. the pH fluctuations in a river or the

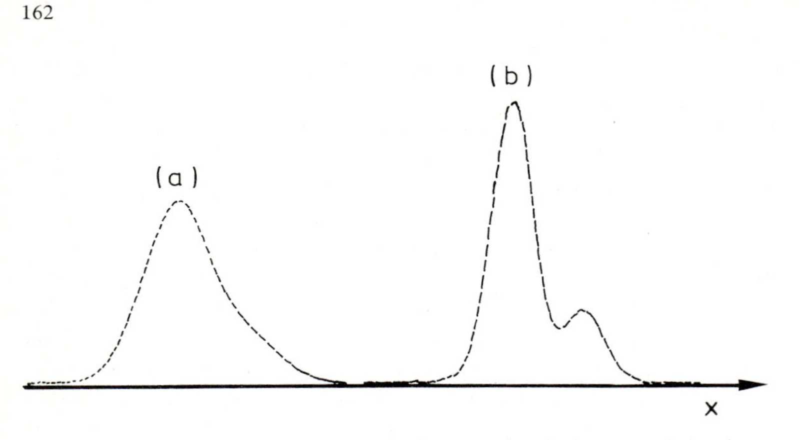

Fig. 5. Example of signal restoration. (a) Measured signal; (b) restored signal.

product constitution at the outlet of a chemical reactor. In Chap. 14, a time series model is introduced as a tool for process modelling, i.e. the characterization of the stochastic part and the deterministic part, e.g. a periodicity or trend. Such a model can also be applied under the difficult circumstances in which the deterministic part is weaker than the stochastic part.

Often, a lack of resolution of deterministic signals is a major concern. For instance, many situations occur in chromatography where the substances cannot be completely separated. In that case, a mathematical decomposition or curve fitting of the peaks may be necessary to obtain the requested quantitative information. The heart of curve fitting is a mathematical function or model which describes the shape of the underlying signals or peaks. A composite profile is then modelled as a sum of mathematical functions of single peaks. The parameters of this model are then adjusted in order to describe adequately the deterministic part in the signal. Parameters often encountered are the peak position, peak height and half-height width. The mathematical method for adjustment is based on regression. This is discussed in Chap. 13.

2.3.2 Multivariate data

By analogy with univariate data, multivariate data, which represent measurements (absorbances in HPLC/UV–vis) or analytical results (concentration), contain a stochastic part (noise) and a deterministic part. These are characterized by the application of multivariate statistics.

One aspect of multivariate statistics, namely correlation, is discussed in Chap. 14. By correlation analysis, one tries to detect a relationship between one column (or row) in the data table and another column (or row). For example, a column of the data table may represent the sodium concentration in rain water collected at all sampled sites and another column represents all values for the chloride concentration. By a correlation analysis, one can determine whether on the average high Na concentrations correspond with high Cl concentrations or vice versa. Other aspects of multivariate statistics are discussed in Chaps. 20–23.

2.4 Data reduction

The purpose of data reduction is the replacement of the large amount of measurements by a few characteristic numbers in which all relevant information has been preserved. A large diversity of methods is available depending on the type of data measured and the type of information wanted. In many instances, data reduction is obtained by fitting a model through the data points. The model is further used to describe the data instead of the data themselves. Fitting models is one of the principal mathematical procedures used by the analytical chemist. In calibration, the model is usually a straight line; in multicomponent analysis, the model is a system of linear equations; in optimization, the model is a polynomial with several independent variables, etc. The basic mathematical procedure for fitting models is regression, which is discussed in Chap. 13.

Reference

1 Kwaliteitsonderzoek in the Rijkswateren, Quarterly Report, 1982, RIZA, The Netherlands.

Chapter 13

Regression Methods

1. Introduction

Two types of model can be distinguished for describing the relationship between one or more controlled variables or factors, x, and a dependent variable or response, y: models which are linear with respect to the parameters and others which are not. A class of linear models consists of polynomials with the mathematical form

$$y = \beta_0 + \beta_1 \cdot x \ldots \beta_m \cdot x^m = \sum_{i=0}^{m} \beta_i \cdot x^i \tag{1}$$

for $m = 0$ y is a constant;

for $m = 1$ the linear model $y = \beta_0 + \beta_1 x$ of a straight line is obtained (first-order model). Most calibration curves which relate the measured response to the concentration are straight lines with a slope β_1 and an intercept β_0.

for $m > 2$ an mth-order polynomial describes the relationship between y and x. To avoid confusion, it is important to note that linear models do not necessarily describe straight lines. Polynomials also belong to the group of linear models. A common property of the linear models is that they are linear (i.e. first order) in all parameters $\beta_0, \ldots, \beta_m$. Moreover, in addition to polynomials, any model of the form

$$y = \beta_0 + \beta_1(\text{any operator on } x) + \beta_2(\text{any operator on } x) + \ldots$$

where the operator on x can be, for example, $1/x$, $\ln x$, $\log x$, $\sin x$, etc., is linear in the parameters. For example

$$y = \beta_0 + \frac{\beta_1}{x} + \beta_2 \log x$$

For models which are non-linear with respect to the parameters, various types of relationship are possible. Frequently encountered models are

the exponential function

$$y = \beta_1 \exp(\beta_2 \cdot x)$$

[e.g. a first-order step response function (Chap. 11)]

the Gaussian function

$$y = \beta_1 \exp - [\beta_2(x - \beta_3)]^2$$

(e.g. the shape of a GLC peak)

the Lorenz function

$$y = \frac{\beta_1}{1 + \beta_2(x - \beta_3)^2}$$

(e.g. the shape of an NMR peak)

All these models, linear and non-linear, can be represented by the relationship $y = f(x, \beta_0, \ldots, \beta_m)$, where $(\beta_0, \ldots, \beta_m)$ are the parameters of the model.

Because of experimental uncertainty, each observation, i, is given by

$$y_i = f(x_i, \beta_0, \ldots, \beta_m) + e_i$$

The function $f(x_i, \beta_0, \ldots, \beta_m)$ is the deterministic part, while e_i is the random or stochastic part.

This chapter discusses how to model the deterministic part of a signal in the presence of a stochastic part. Mathematically, this means that we want to estimate the values of the unknown parameters in the model by fitting the function as well as possible through the observation pairs (x_i, y_i). Of course, the expression "as well as possible" needs translation into a mathematical "goodness of fit" criterion. That criterion is readily found when we consider the following general relationship between the observation pairs (x_i, y_i)

$$y_i = f(x_i, \beta_0, \ldots, \beta_m) + e_i \qquad i = 1, 2, \ldots, n \tag{2}$$

where n is the number of observations, e_i represents the difference between the value predicted by the model $f(x_i, \beta_0, \ldots, \beta_m)$ and the observation y_i (sometimes called the residual).

It is assumed that the x values are under control (no error) and do not contribute to the difference between the observation pairs and the model. For most of the analytical applications, this is a valid assumption. In many instances in chemical analysis, the controlled variable, x, is the concentration of a calibration sample, while the dependent variable, y, is the analytical response. As a rule, the precision of analytical response is considerably poorer than the obtainable precision in making standard solutions. If the errors in x are not negligible, however, the distance between model and observation can be measured in a direction perpendicular to the line (Chap. 5). In this case, the application of the conventional regression methods, which are described below, produces less reliable results.

The term e_i represents the difference between the observation and the model. The estimation of the parameters consists in finding values for the parameters so that the distance between the observations and the model is minimal. Mathematically, this means minimize Σe_i^2. This is called a criterion of goodness of fit and is expressed mathematically as

$$\text{Minimize} \sum_{i=1}^{n} e_i^2$$

Regression consists of the estimation of the parameters of the model, a test of the validity of the model and the calculation of the confidence limits of the parameters (see also Chap. 5).

2. Univariate regression

With univariate regression, we mean that all observations (or responses) are dependent upon a single variable, x, without restrictions on the complexity of that relationship. For example, the response

$$y = \beta_0 + \beta_1 x + \beta_2 x + \beta_3 \ln x$$

depends only upon the value of one controlled variable x.

There are two types of univariate model: either the response function is linear with respect to all parameters or it is not linear in one or more parameters. In the first instance, the parameters are estimated by linear regression and in the latter, by non-linear regression.

2.1 Linear regression

2.1.1 Parameter estimation

The least squares solution of fitting a straight line through a set of data was discussed in detail in Chap. 5. In this chapter, a more general formulation of regression will be given using matrix notation.

Let us suppose that one wants to fit an mth-order polynomial with $p = m + 1$ parameters ($\beta_0, \ldots, \beta_m$) through n observations ($y_1, \ldots, y_n$) obtained at n values of the independent variable ($x_1, \ldots, x_n$). A fit of a polynomial with p parameters requires the solution of a system of p equations with p unknowns. The forms of these equations and their solutions are rather complex. Therefore, there is a need for a compact and clear notation applicable to any linear regression problem. This kind of solution is obtained by using matrix algebra. An additional advantage of the matrix notation is the easy translation of the expressions into computer code.

For an mth-order polynomial, eqn. (1) becomes

$$\begin{aligned} y_1 &= \beta_0 + \beta_1 \cdot x_1 + \beta_2 \cdot x_1^2 + \ldots + \beta_m \cdot x_1^m + e_1 \\ y_2 &= \beta_0 + \beta_1 \cdot x_2 + \beta_2 \cdot x_2^2 + \ldots + \beta_m \cdot x_2^m + e_2 \\ &\vdots \\ y_n &= \beta_0 + \beta_1 \cdot x_n + \beta_2 \cdot x_n^2 + \ldots + \beta_m \cdot x_n^m + e_n \end{aligned} \quad (3)$$

The following vectors and matrices are introduced.

(1) Measurement vector containing all observations.

$$\mathbf{y} = \begin{bmatrix} y_1 \\ y_2 \\ \cdot \\ y_n \end{bmatrix}$$

(2) Residual vector containing the differences between observations and model.

$$\mathbf{e} = \begin{bmatrix} e_1 \\ e_2 \\ \cdot \\ e_n \end{bmatrix}$$

(3) Parameter vector.

$$\boldsymbol{\beta} = \begin{bmatrix} \beta_0 \\ \beta_1 \\ \cdot \\ \beta_m \end{bmatrix}$$

(4) Independent variable matrix.

$$\mathbf{X} = \begin{bmatrix} 1 & x_1 & x_1^2 & x_1^3 \ldots & x_1^m \\ 1 & x_2 & x_2^2 & x_2^3 \ldots & x_2^m \\ & & & \ldots & \\ 1 & x_n & x_n^2 & x_n^3 \ldots & x_n^m \end{bmatrix} \tag{4}$$

Equation (3) can then be rewritten as

$$\underset{|n \times 1|}{\mathbf{y}} = \underset{|n \times p|}{\mathbf{X}} \cdot \underset{|p \times 1|}{\boldsymbol{\beta}} + \underset{|n \times 1|}{\mathbf{e}} \tag{5}$$

The least squares criterion, Min Σe_i^2, in matrix notation means that the square length $|\mathbf{e}|$ of the vector $\mathbf{e}$ should be minimal or

$$(e_1\ e_2 \ldots e_n) \begin{bmatrix} e_1 \\ e_2 \\ \vdots \\ e_n \end{bmatrix} = \mathbf{e}' \cdot \mathbf{e} = |\mathbf{e}| \text{ is minimal} \tag{6}$$

From eqn. (5), it follows that

$$\mathbf{e} = \mathbf{y} - \mathbf{X} \cdot \boldsymbol{\beta}$$

and

$$\mathbf{e}' = (\mathbf{y} - \mathbf{X} \cdot \boldsymbol{\beta})'$$

Thus

$$|\mathbf{e}| = \mathbf{e}' \cdot \mathbf{e} = (\mathbf{y} - \mathbf{X} \cdot \boldsymbol{\beta})'(\mathbf{y} - \mathbf{X} \cdot \boldsymbol{\beta})$$

should be minimized by adjusting the parameter vector, **b**, or by setting the partial derivatives equal to zero. This operation provides the normal equations

$$(\mathbf{X}' \cdot \mathbf{X}) \cdot \mathbf{b} = \mathbf{X}' \cdot \mathbf{y} \tag{7}$$

with the solution

$$\mathbf{b} = (\mathbf{X}' \cdot \mathbf{X})^{-1} \cdot \mathbf{X}' \cdot \mathbf{y} \tag{8}$$

b is called the least-squares estimator of the true value $\boldsymbol{\beta}$. It follows from eqn. (8) that a least-squares estimate is only found when the inverse of the square matrix $(\mathbf{X}' \cdot \mathbf{X})$ exists.

Once the estimates **b** are obtained from a fit of an mth-order polynomial through a set of n observations **y** at the points $x_1, \ldots, x_n$, one can predict an observation $\hat{y}_0$ at a specific unmeasured value of $x = x_0$. For this, one substitutes the value x_0 in the model

$$\hat{y}_0 = b_0 + b_1 \cdot x_0 + \ldots + b_m \cdot x_0^m$$

Or, in matrix notation

$$\hat{y}_0 = (b_0 \; b_1 \ldots b_m) \begin{bmatrix} 1 \\ x_0 \\ \vdots \\ x_0^m \end{bmatrix} = \mathbf{b}' \cdot \mathbf{x}_0$$

The predicted value at x_0 is then

$$\hat{y}_0 = \mathbf{b}' \cdot \mathbf{x}_0 \tag{9}$$

In matrix algebra, the expression $(\mathbf{X}' \cdot \mathbf{X})^{-1} \cdot \mathbf{X}'$ encountered in eqn. (8) is very important. It is called the generalized (or Moore–Penrose) inverse of $\mathbf{X}$ ($\mathbf{X}^{-1}$ is the inverse of matrix $\mathbf{X}$). In analytical chemistry, this term will play an important role in the evaluation of the design of analytical experiments (Chap. 17).

Obviously, a special case of eqn. (8) provides the algebraic solution of the first-order linear regression, which was derived in Chap. 5. In this case, $m = 1$ and one obtains

$$\mathbf{b} = \begin{bmatrix} b_0 \\ b_1 \end{bmatrix}$$

$$\mathbf{y} = \begin{bmatrix} y_1 \\ y_2 \\ \cdot \\ y_n \end{bmatrix}$$

$$\mathbf{X} = \begin{bmatrix} 1 & x_1 \\ 1 & x_2 \\ \cdot & \cdot \\ 1 & x_n \end{bmatrix}$$

References p. 188

and

$$(\mathbf{X}' \cdot \mathbf{X}) = \begin{bmatrix} 1 & 1 & \dots & 1 \\ x_1 & x_2 & \dots & x_n \end{bmatrix} \begin{bmatrix} 1 & x_1 \\ 1 & x_2 \\ \cdot & \cdot \\ 1 & x_n \end{bmatrix} = \begin{bmatrix} n & \Sigma x_i \\ \Sigma x_i & \Sigma x_i^2 \end{bmatrix}$$

Thus, according to eqn. (7), the normal equations are

$$\begin{bmatrix} n & \Sigma x_i \\ \Sigma x_i & \Sigma x_i^2 \end{bmatrix} \begin{bmatrix} b_0 \\ b_1 \end{bmatrix} = \begin{bmatrix} 1 & 1 & \dots & 1 \\ x_1 & x_2 & \dots & x_n \end{bmatrix} \begin{bmatrix} y_1 \\ \cdot \\ y_n \end{bmatrix}$$

which is equivalent to

$$n \cdot b_0 + b_1 \sum x_i = \Sigma y_i$$

$$b_0 \Sigma x_i + b_1 \Sigma x_i^2 = \Sigma x_i y_i$$

which is also discussed in Chap. 5, Sect. 2.

Although the form of eqn. (8) is quite simple, the operations involved are not. For instance, the inversion of a matrix is a complex operation which is difficult to carry out by hand. An easier way is to use computer algorithms, especially when higher-order models are fitted through the data. These algorithms may be based on a Gauss elimination and LU decomposition [1] (L = lower-triangular matrix and U = upper-triangular matrix) or on a search for a solution by a Gauss–Seidel iteration [1] or Jacobi iteration [1]. An obvious feature of the outlined least-squares method is that all data should be collected first before an estimation of the parameters can be obtained. Besides the required matrix inversion, another drawback of the method is that all calculations must be redone when a single new pair of observations is added to the existing data set. Recursive least-squares estimation methods exist for linear regression which do not suffer from these handicaps and which do not require a matrix inversion [2].

It should also be stressed at this point that eqn. (8) was derived under the assumption that the variance of e, var(e), is constant over the entire dynamic range of the measurements. A direct application of eqn. (8) if var(e) is not constant may lead to incorrect results and weights are necessary.

Deviations, e, between the model and the measurements are less serious for measurements with an inherently large random error than for measurements with a small error. An obvious weighting factor, therefore, is inversely proportional to the variance of the measurements.

Consider the weighted least-squares fit for the general case. The weight matrix of the n residuals is defined as

$$\mathbf{W} = \begin{bmatrix} w_1 & & & 0 \\ & w_2 & & \\ & & \ddots & \\ 0 & & & w_n \end{bmatrix}$$

When the vector of the residuals, **e**, is multiplied by this matrix **W**, one obtains

$$\mathbf{W}\cdot\mathbf{e}=\begin{bmatrix} w_1 & & & 0 \\ & w_2 & & \\ & & \ddots & \\ 0 & & & w_n \end{bmatrix}\begin{bmatrix} e_1 \\ e_2 \\ \vdots \\ e_n \end{bmatrix}=\begin{bmatrix} w_1\cdot e_1 \\ w_2\cdot e_2 \\ \vdots \\ w_n\cdot e_n \end{bmatrix}$$

which means that each residual is multiplied (weighted) with the corresponding diagonal term of **W**.

For weighting with the reciprocal value of the variance, one defines

$$\mathbf{W}=\begin{bmatrix} 1/s_1^2 & & & 0 \\ & 1/s_2^2 & & \\ & & \ddots & \\ 0 & & & 1/s_n^2 \end{bmatrix}$$

where s_i^2 is the variance of observation y_i.

The least-squares estimation of the parameters now proceeds in an analogous way to the unweighted linear regression. The least-squares estimates **b** of $\boldsymbol{\beta}$ are found by minimizing the square of length $|\mathbf{W}\cdot\mathbf{e}|$ of vector $\mathbf{W}\cdot\mathbf{e}$.

$$|\mathbf{W}\cdot\mathbf{e}|=(\mathbf{W}\cdot\mathbf{e})'(\mathbf{W}\cdot\mathbf{e})=(\mathbf{W}\cdot\mathbf{y}-\mathbf{W}\cdot\mathbf{X}\cdot\mathbf{b})'(\mathbf{W}\cdot\mathbf{y}-\mathbf{W}\cdot\mathbf{X}\cdot\mathbf{b})$$

Renaming the vector $\mathbf{W}\cdot\mathbf{y}=\mathbf{z}$ and the matrix $\mathbf{W}\cdot\mathbf{X}=\mathbf{Z}$

$$|\mathbf{W}\cdot\mathbf{e}|=(\mathbf{z}-\mathbf{Z}\cdot\mathbf{b})'(\mathbf{z}-\mathbf{Z}\cdot\mathbf{b})$$

By analogy with eqn. (8), the estimates **b**, giving least-squares weighted residuals, are

$$\mathbf{b}=(\mathbf{Z}'\cdot\mathbf{Z})^{-1}\mathbf{Z}'\cdot\mathbf{z}$$

or

$$\mathbf{b}=((\mathbf{W}\cdot\mathbf{X})'(\mathbf{W}\cdot\mathbf{X}))^{-1}(\mathbf{W}\cdot\mathbf{X})'(\mathbf{W}\cdot\mathbf{y})$$

Because **W** is a diagonal matrix (all off-diagonal elements are zero), the expression for **b** can be somewhat simplified by defining a new diagonal matrix $\mathbf{U}=\mathbf{W}'\cdot\mathbf{W}$. Then

$$\mathbf{b}=(\mathbf{X}'\cdot\mathbf{U}\cdot\mathbf{X})^{-1}(\mathbf{X}'\cdot\mathbf{U}\cdot\mathbf{y})$$

with

$$\mathbf{U}=\begin{bmatrix} w_1 & & & 0 \\ & w_2 & & \\ & & \ddots & \\ 0 & & & w_n \end{bmatrix}\begin{bmatrix} w_1 & & & 0 \\ & w_2 & & \\ & & \ddots & \\ 0 & & & w_n \end{bmatrix}=\begin{bmatrix} w_1^2 & & & 0 \\ & w_2^2 & & \\ & & \ddots & \\ 0 & & & w_n^2 \end{bmatrix}$$

An example of a weighted straight line regression is given in Chap. 5.

References p. 188

2.1.2 Confidence intervals for the parameters

If the uncertainty in one parameter is independent of the others, one can calculate the variances $v(b_0), \ldots, v(b_m)$ of the parameters separately. In linear regression, however, the estimation of one parameter is usually dependent upon the estimate of the other and thus also their uncertainties. The information necessary to determine confidence intervals is contained in the variance–covariance matrix. The diagonal terms of the variance–covariance matrix contain the variances of the parameters. The off-diagonal terms are the covariances between the parameters. Covariance and variance–covariance matrices are discussed in Chap. 14.

For a straight line, the variance–covariance matrix $\mathbf{V}(b)$, is

$$\mathbf{V}(b) = \begin{bmatrix} v(b_0) & \text{cov}(b_0, b_1) \\ \text{cov}(b_1, b_0) & v(b_1) \end{bmatrix}$$

where $v(b_0)$ is the estimated variance associated with b_0, $v(b_1)$ is the estimated variance associated with b_1 and $\text{cov}(b_0, b_1) = \text{cov}(b_1, b_0)$ is the estimated covariance between the two parameters.

For an mth-order polynomial, one finds

$$\mathbf{V}(b) = \begin{bmatrix} v(b_0) & \text{cov}(b_0, b_1) & \ldots & \text{cov}(b_0, b_m) \\ \text{cov}(b_1, b_0) & v(b_1) & \ldots & \text{cov}(b_1, b_m) \\ \cdot & \cdot & \ldots & \cdot \\ \cdot & \cdot & \ldots & \cdot \\ \text{cov}(b_m, b_0) & \cdot & \ldots & v(b_m) \end{bmatrix} \tag{10}$$

The diagonal elements are the variances of the parameters, in the same order as they appear in the model. One can show that the variance–covariance matrix of the model parameters is given by

$$\mathbf{V}(b) = s_{pe}^2(\mathbf{X}'\mathbf{X})^{-1}$$

where s_{pe}^2 is the estimate of the variance σ_{pe}^2 of the pure experimental error, which can be calculated if the observations are replicated (Chap. 5). If the model does not show a lack of fit, then the variance of the residuals

$$s_e^2 = \text{SS}_e/(n-p)$$

is also a valid estimate of σ_{pe}^2. SS_e is the sum of squares of the residuals. Thus

$$\mathbf{V}(b) = s_e^2(\mathbf{X}'\mathbf{X})^{-1} \tag{11}$$

Having specified the variance–covariance matrix of the parameters, we can derive the variance of $\hat{y}_0$ predicted at x_0 by the regression equation (9).

$$\hat{y}_0 = \mathbf{b}' \cdot \mathbf{X}_0$$

The variance of the difference between predicted and true value is

$$\mathbf{V}(\hat{y}_0) = \mathbf{X}_0' \cdot \mathbf{V}(b) \cdot \mathbf{X}_0 = \mathbf{X}_0'(\mathbf{X}' \cdot \mathbf{X})^{-1}\mathbf{X}_0 \cdot s_e^2 \tag{12}$$

where $\mathbf{X}_0$ is the vector

$$\begin{bmatrix} 1 \\ x_0 \\ \cdot \\ x_0^m \end{bmatrix}$$

Note that $\mathbf{X}$ is the matrix of the parameter coefficients [eqn. (4)]. This matrix is defined by the design of the measurements, i.e. the numerical value of $(\mathbf{X}' \cdot \mathbf{X})^{-1}$ and therefore the design of the observations influences the precision of the model parameters obtained (the choice of the values of the controlled variable). Various designs and their influence on the precision of the calculated model parameters are discussed in Chap. 17. It is also noted that eqns. (11) and (12) are valid whether the measurement errors are normally distributed or not. For $m = 2$ (or $p = 3$)

$$\mathbf{X} = \begin{bmatrix} 1 & x_1 & x_1^2 \\ \vdots & \vdots & \vdots \\ 1 & x_n & x_n^2 \end{bmatrix}$$

If one assumes that the measurement errors are normally distributed and are independent, confidence intervals, including the true values, of the individual parameters can be derived from the variances of the parameter estimates

$$b_i \pm t_{n-p}^{0.975} \sqrt{v(b_i)}$$

where $v(b_i)$ is the ith diagonal term of matrix $\mathbf{V}(b)$ [eqn. (10)] and $t_{n-p}^{0.975}$ is the tabulated t value at the 97.5% level of confidence.

It can be shown that, for any value of the controlled variable, x_0, the confidence interval within which the true response will be found with a given probability is

$$\mathbf{b}'\mathbf{X}_0 \pm t_{n-p}^{0.975} s_e \sqrt{\mathbf{X}_0'(\mathbf{X}' \cdot \mathbf{X})^{-1} \cdot \mathbf{X}_0)} \tag{13}$$

Example (data from ref. 3). Table 1 lists the observations which are fitted with a second-order model. The model is

$$y = \beta_0 + \beta_1 x + \beta_2 x^2 + e$$

TABLE 1

Observations to be fitted with a second-order model (from ref. 3)

Controlled variable	y (measured)	y (estimated)	e
−3	30	34.5	−4.5
−2	48	45.0	3
−1	68	63.8	4.2
0	98	91.1	6.9
+1	120	126.8	−6.8
+2	160	171.0	−11.0
+3	232	223.6	8.4

According to eqn. (8)

$\mathbf{b} = (\mathbf{X}' \cdot \mathbf{X})^{-1} \cdot \mathbf{X}' \cdot \mathbf{y}$

where

$$\mathbf{X} = \begin{bmatrix} 1 & x_1 & x_1^2 \\ 1 & x_2 & x_2^2 \\ 1 & x_3 & x_3^2 \\ 1 & x_4 & x_4^2 \\ 1 & x_5 & x_5^2 \\ 1 & x_6 & x_6^2 \\ 1 & x_7 & x_7^2 \end{bmatrix} = \begin{bmatrix} 1 & -3 & 9 \\ 1 & -2 & 4 \\ 1 & -1 & 1 \\ 1 & 0 & 0 \\ 1 & 1 & 1 \\ 1 & 2 & 4 \\ 1 & 3 & 9 \end{bmatrix}$$

$$(\mathbf{X}'\mathbf{X}) = \begin{bmatrix} 1 & 1 & 1 & 1 & 1 & 1 & 1 \\ -3 & -2 & -1 & 0 & 1 & 2 & 3 \\ 9 & 4 & 1 & 0 & 1 & 4 & 9 \end{bmatrix} \begin{bmatrix} 1 & -3 & 9 \\ 1 & -2 & 4 \\ 1 & -1 & 1 \\ 1 & 0 & 0 \\ 1 & 1 & 1 \\ 1 & 2 & 4 \\ 1 & 3 & 9 \end{bmatrix} = \begin{bmatrix} 7 & 0 & 28 \\ 0 & 28 & 0 \\ 28 & 0 & 196 \end{bmatrix}$$

The inverse of this 3×3 matrix is

$$(\mathbf{X}'\mathbf{X})^{-1} = \begin{bmatrix} 0.333333 & 0 & -0.047619 \\ 0 & 0.035714 & 0 \\ -0.047619 & 0 & 0.011905 \end{bmatrix}$$

$$\mathbf{b} = \begin{bmatrix} b_0 \\ b_1 \\ b_2 \end{bmatrix} = (\mathbf{X}'\mathbf{X})^{-1}(\mathbf{X}'\mathbf{y}) = \begin{bmatrix} 91.143 \\ 31.5 \\ 4.214 \end{bmatrix}$$

Thus the fitted model is

$y = 91.143 + 31.5x + 4.214x^2$

The sum of squares of the residuals $\sum_i e_i^2 = 333$ with $n - p = 7 - 3 = 4$ degrees of freedom.

The variance–covariance matrix of the parameters is

$$\mathbf{V}(b) = s_e^2(\mathbf{X}' \cdot \mathbf{X})^{-1} = 333/4 \begin{bmatrix} 0.33 & 0 & -0.048 \\ 0 & 0.036 & 0 \\ -0.048 & 0 & 0.012 \end{bmatrix}$$

$$= \begin{bmatrix} 27.5 & 0 & -4.0 \\ 0 & 3.0 & 0 \\ -4.0 & 0 & 1 \end{bmatrix}$$

The 95% confidence ($t_4^{0.975} = 2.77$) interval for b_0 is

$91.143 \pm 2.77\sqrt{27.47} = 91.143 \pm 14.5$

Equally, the 95% confidence intervals for b_1 and b_2 are, respectively, 31.5 ± 5.5 and 4.2 ± 2.77. The 95% confidence interval for the true response at $x = 0.5$ is

$$2.77\sqrt{83.25[1 \quad 0.5 \quad 0.25] \begin{bmatrix} 0.3333 & 0 & -0.048 \\ 0 & 0.0357 & 0 \\ -0.048 & 0 & 0.012 \end{bmatrix} \begin{bmatrix} 1 \\ 0.5 \\ 0.25 \end{bmatrix}} = 107.94 \pm 13.4$$

A word of caution is indicated here. Although eqn. (11) is the exact expression of the confidence intervals of the individual parameters $b_0, b_1, \ldots, b_m$, it does not give

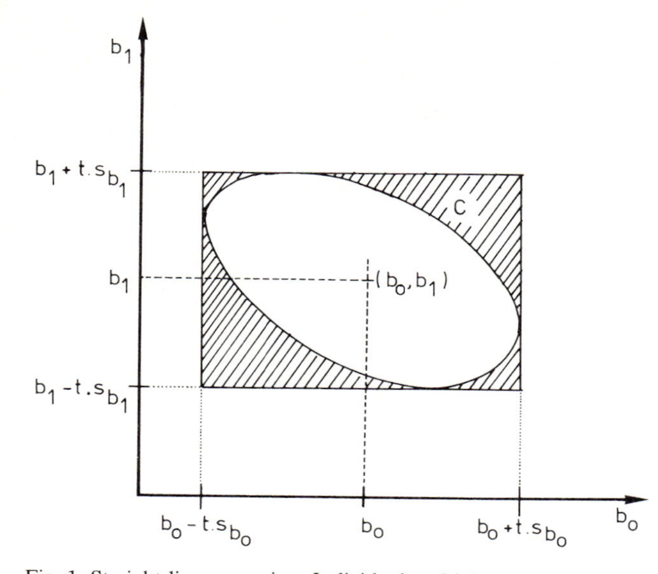

Fig. 1. Straight line regression. Individual and joint confidence intervals of the parameters.

the joint probability region for combinations of the parameters (b_0 and b_1 and ,..., and b_m). This is illustrated in Fig. 1 in which the individual confidence intervals for b_0 and b_1 of the straight line model are plotted along two orthogonal axes. From this plot, one could wrongly conclude that the confidence region of the combinations (b_0, b_1) is given by the rectangle given in Fig. 1 and that, for example, it is the combination indicated by the point C is highly probable. This would only be true if the error found in one parameter is independent of the errors in the other estimates. Because this is not the case, the shape of the joint confidence region is an ellipse (Fig. 1), showing that point C is, in fact, highly improbable. The larger the off-diagonal values are in comparison with the diagonal elements in the variance–covariance matrix in eqn. (10), the more will the shape of the ellipse deviate from the rectangle and the more one should become cautious in making statements on joint probabilities. A discussion of joint confidence intervals can be found in ref. 4.

2.2 Non-linear regression

2.2.1 Parameter estimation

In many cases, linear models are not adequate to fit the (x_i, y_i) measurement pairs obtained in chemical analysis. The potentiometric titration is such an example. Another large class of examples is found in chromatography and spectroscopy (IR, NMR, etc.) where the various compounds or substituents contribute to the signal, e.g. chromatographic peaks, or overlapping peaks in an IR spectrum. The analytical problem frequently turns out to be the determination of the area or the height of overlapping bands in a composite profile. In general, a much larger number of parameters needs to be calculated in overlapping peak systems than in the linear regression problems discussed so far. For example, a Gaussian model for a chromatographic peak

$$f(x) = \beta_1 \exp\left[-(x-\beta_2)^2/\beta_3^2\right]$$

contains 3 parameters, where β_1 is the peak height, β_2 is the peak position, and β_3 is related to the width of the peak. A model for N peaks thus contains $3N$ parameters. The model for a system of N GLC peaks becomes

$$f(x) = \sum_{i=1}^{N} \beta_{1,i} \exp\left[-(x-\beta_{2,i})^2/\beta_{3,i}^2\right]$$

In non-linear regression

$$\sum (y_i - f(x))^2 = \mathbf{e}' \cdot \mathbf{e}$$

is minimized. The set of measurements can no longer be represented as a simple product of the independent variable matrix $\mathbf{X}$ and the parameter vector $\boldsymbol{\beta}$ such as

$$\mathbf{y} = \mathbf{X} \cdot \boldsymbol{\beta} + \mathbf{e}$$

However, if approximate values $(\mathbf{b})_0$ of the parameters are known, then $f(x)$ can be rewritten as a linear function of the adjustments, $\Delta\mathbf{b} = \boldsymbol{\beta} - (\mathbf{b})_0$. These can be used to estimate the parameters according to the least-squares criterion that $\mathbf{e}' \cdot \mathbf{e}$ is minimal.

According to Taylor's expansion theorem, any continuous function $f(x, b)$ can be expanded in polynomial terms of the form

$$f(x, b+\Delta b) = f(x, b) + \frac{\delta f}{\delta b}\Delta b + \frac{\delta^2 f}{2!\,\delta b^2}\Delta b^2 + \frac{\delta^3 f}{3!\,\delta b^3}\Delta b^3 + \ldots$$

For an exponential function with $b = 4$ and $\Delta b = 0.1$, one finds

$$\exp(4.1x) = \exp(4x) + x \cdot \exp(4x)0.1 + \frac{x^2}{2!} \cdot \exp(4x)(0.1)^2 + \frac{x^3}{3!} \cdot \exp(4x)(0.1)^3$$

Substituting $x = 1$ gives

$$\exp(4.1) = 60.34029$$

which is approximated by

$$\exp(4) + 0.1\exp(4) + 0.005\exp(4) + \ldots$$

The expansion with the first three terms approximates the exact function value within 3 significant figures, namely $f(x) = 60.33$.

For $\Delta b = 0.01$, the second and all higher order terms can be neglected, e.g.

$$\exp(4.01) = 55.147$$

which is approximated by

$$\exp(4.00) + \exp(4.00) \cdot (0.01) = 55.15$$

Thus, any function can be made first-order in the parameters by a Taylor expansion by neglecting the second and higher order terms on condition that Δb is relatively small. For a function with several parameters, the expression becomes

$$\begin{aligned} y_i &= \mathrm{f}(x_i, b_1, \ldots, b_m) + e_i \\ &= \mathrm{f}[x_i, (b_1)_0, \ldots, (b_m)_0] \\ &\quad + \left.\frac{\delta \mathrm{f}_i}{\delta b_1}\right|_{\mathbf{b}=(\mathbf{b})_0} \cdot \Delta b_1 + \left.\frac{\delta \mathrm{f}_i}{\delta b_2}\right|_{\mathbf{b}=(\mathbf{b})_0} \cdot \Delta b_2 + \ldots + \left.\frac{\delta \mathrm{f}_i}{\delta b_m}\right|_{\mathbf{b}=(\mathbf{b})_0} \cdot \Delta b_m + e_i \end{aligned} \quad (14)$$

where (x_i, y_i) are the measurement pairs i;

$$(\mathbf{b})_0 = \begin{bmatrix} (b_1)_0 \\ (b_2)_0 \\ \cdot \\ \cdot \\ (b_m)_0 \end{bmatrix}$$

is the vector of the initial values of the parameters;

$$\mathrm{f}[x_i, (b_1)_0, \ldots, (b_m)_0]$$

is the value of the model for $x = x_i$, with the initial values of the parameters substituted in the model;

$$\left.\frac{\delta \mathrm{f}_i}{\delta b_j}\right|_{\mathbf{b}=(\mathbf{b})_0}$$

is the first derivative to b_j of the model in point i, with the initial values of the parameters substituted in the model; and e_i is the residual, or difference between the value of the model and measurement y_i.

Equation (14) is linear in the parameters $\Delta b_1, \Delta b_2, \ldots, \Delta b_m$ and its form is very similar to that of eqn. (1) obtained for linear regression. The only difference is the form of the design matrix **X**, which now contains the first derivatives of the model to all parameters in all x values

$$\begin{bmatrix} \frac{\delta \mathrm{f}_1}{\delta b_1} & \frac{\delta \mathrm{f}_1}{\delta b_2} & \cdots & \frac{\delta \mathrm{f}_1}{\delta b_m} \\ \frac{\delta \mathrm{f}_2}{\delta b_1} & \frac{\delta \mathrm{f}_2}{\delta b_2} & \cdots & \frac{\delta \mathrm{f}_2}{\delta b_m} \\ \cdot & \cdot & & \cdot \\ \frac{\mathrm{df}_n}{\delta b_1} & \frac{\delta \mathrm{f}_n}{\delta b_2} & \cdots & \frac{\delta \mathrm{f}_n}{\delta b_m} \end{bmatrix} = \mathbf{X} = \mathbf{J}$$

where $\partial f_i / \partial b_j$ is the partial derivative of the model to b_j in data point i.

References p. 188

In non-linear regression, this matrix is called the Jacobian matrix and is often denoted as **J**. Let us introduce two vectors: $\Delta\mathbf{y}$ which is the vector of the differences between the measurements **y** and the model outcome with the estimated parameters

$$\Delta\mathbf{y} = \begin{bmatrix} y_1 - \hat{y}_{1,0} \\ \vdots \\ y_n - \hat{y}_{n,0} \end{bmatrix}$$

and $\Delta\mathbf{b}$ which is the vector of the differences between the true and initially estimated parameter values

$$\Delta\mathbf{b} = \begin{bmatrix} b_1 - (b_1)_0 \\ \vdots \\ b_m - (b_m)_0 \end{bmatrix}$$

where $(b_1, \ldots, b_m)$ are the true but unknown values of the parameters.

Equation (14) can be represented in the matrix notation

$$\Delta\mathbf{y} = \mathbf{J} \cdot \Delta\mathbf{b} + \mathbf{e}$$

As demonstrated for linear regression, the least-squares estimate $\Delta\mathbf{b}$ can be obtained by minimizing

$$\mathbf{e}' \cdot \mathbf{e} = (\Delta\mathbf{y} - \mathbf{J} \cdot \Delta\mathbf{b})' \cdot (\Delta\mathbf{y} - \mathbf{J} \cdot \Delta\mathbf{b}) \qquad (15)$$

Minimization is obtained by adjusting the parameter increment vector $\Delta\mathbf{b}$. The partial derivatives of $\mathbf{e}' \cdot \mathbf{e} = |\mathbf{e}|$ with respect to all increments are equated to zero. In a similar way as outlined for the linear problem, the elaboration of eqn. (15) leads to a set of normal equations

$$(\mathbf{J}' \cdot \mathbf{J})\ \Delta\mathbf{b} = \mathbf{J}' \cdot \Delta\mathbf{y} \qquad (16)$$

giving an estimate $\Delta\mathbf{b}$

$$\Delta\mathbf{b} = (\mathbf{J}' \cdot \mathbf{J})^{-1} \cdot \mathbf{J}' \cdot \Delta\mathbf{y} \qquad (17)$$

The estimated parameter values **b** are then

$$\mathbf{b} = (\mathbf{b})_0 + \Delta\mathbf{b}$$

TABLE 2

Dataset for an exponential model

Factor (x)	Response (measured) (y)	$(\hat{y})_0$ (start)	e	$(\hat{y})_1$ (1st iteration)	e
−1	72	53.8	18.2	77.27	−5.27
0	11	12	1.0	11.25	−0.25
+1	2	2.68	0.68	1.64	0.36

Example. The non-linear regression algorithm is demonstrated on the fit of an exponential function $f(x) = b_1 \exp(-b_2 \cdot x)$ through the data given in Table 2. $(b_1)_0 = 12$ and $(b_2)_0 = 1.5$ are considered good guesses to start the calculations. The Jacobian matrix is

$$\begin{bmatrix} \left.\frac{\delta f}{\delta b_1}\right|_{x=-1} & \left.\frac{\delta f}{\delta b_2}\right|_{x=-1} \\ \left.\frac{\delta f}{\delta b_1}\right|_{x=0} & \left.\frac{\delta f}{\delta b_2}\right|_{x=0} \\ \left.\frac{\delta f}{\delta b_1}\right|_{x=1} & \left.\frac{\delta f}{\delta b_2}\right|_{x=1} \end{bmatrix} = \begin{bmatrix} 4.48 & 53.8 \\ 1 & 0 \\ 0.223 & -2.677 \end{bmatrix}$$

As an example, the calculation of the first element of the Jacobian matrix is given in detail.

The model is	$b_1 \cdot \exp(-b_2 \cdot x)$
The derivative $\delta f/\delta b_1$ is	$\exp(-b_2 \cdot x)$
Thus, $\delta f/\delta b_1$ for $x = -1$ is	$\exp[-b_2 \cdot (-1)]$
The current value of b_2 is	$(b_2)_0 = 1.5$
Thus, $\delta f/\delta b_1$ for $x = -1$ is	$\exp(1.5) = 4.48$

The $\mathbf{\Delta y}$ vector is

$$\begin{bmatrix} 72 - 12 \cdot \exp(1.5) \\ 11 - 12 \cdot \exp(0) \\ 2 - 12 \cdot \exp(-1.5) \end{bmatrix} = \begin{bmatrix} 18.2 \\ -1 \\ -0.677 \end{bmatrix}$$

The parameter increment vector $\mathbf{\Delta b}$ can now be calculated.

$$\left[\begin{bmatrix} 4.48 & 1 & 0.223 \\ 53.78 & 0 & -2.677 \end{bmatrix} \begin{bmatrix} 4.48 & 53.78 \\ 1 & 0 \\ 0.223 & -2.67 \end{bmatrix}\right]^{-1} \begin{bmatrix} 4.48 & 1 & 0.223 \\ 53.78 & 0 & -2.677 \end{bmatrix} \begin{bmatrix} 18.2 \\ -1 \\ -0.677 \end{bmatrix}$$

$$\begin{bmatrix} 21.12 & 240.33 \\ 241.33 & 2899.45 \end{bmatrix}^{-1} \begin{bmatrix} 80.4 \\ 980.6 \end{bmatrix}$$

$$\begin{bmatrix} \Delta b_1 \\ \Delta b_2 \end{bmatrix} = \begin{bmatrix} 0.834 & -0.0691 \\ -0.0691 & 0.0061 \end{bmatrix} \begin{bmatrix} 80.38 \\ 980.0 \end{bmatrix} = \begin{bmatrix} -0.746 \\ 0.427 \end{bmatrix}$$

Thus

$b_1 = 12 - 0.746 = 11.25$

$b_2 = 1.5 + 0.427 = 1.93$

The sum of squares of the residuals is reduced from 332.7 found with the initial estimates to 27.96 calculated using the renewed estimates.

The accuracy of the estimates obtained for the parameters depends on the validity of the approximation of the model equation used in a Taylor expansion

References p. 188

neglecting the higher-order terms. This, in turn, depends on the accuracy of the initial guesses of the values of the parameters. One can generally assume that the calculated parameters will be better than the initial guesses. As a rule, the sum of squares of the residuals between model outcome and measurements will not be minimal. This is due to the fact that, in most cases, the initial guesses are not good enough to neglect the higher-order terms in the Taylor expansion. Better estimates are then obtained by repeating the whole procedure, but now starting with the values obtained in the previous step as initial guesses. This process of iteration is continued until the parameter adjustments produce only negligibly small changes of the squared errors. One problem is the speed of convergence of the iteration procedure, which depends on the initial choice of the parameters. If that choice is close to the optimum, only a few cycles of iteration are needed, but if the initial choice is poor, the method will fail. However, there are ways to overcome these difficulties [5].

Recalling the example of Table 2, the substitution of the calculated parameters in the Jacobian matrix restarts the next cycle. Table 3 shows the course of the estimation of the parameters and demonstrates the convergence of the sum of squares of the residuals. The iteration is stopped after the sum of squares of the residuals is found to be within the experimental error.

2.2.2 Validation of the non-linear model

If the model is adequate, the mean square differences between model and observations should be equal to the variance of the random part of the signal at the end of the process of iteration. Unfortunately, models are very often an approximation of the underlying deterministic signal. For example, a Gaussian model, which is a symmetrical function, is frequently used to describe GLC peaks, which are usually slightly asymmetric. The examination of the residual plot between model and observations may provide an indication of the goodness-of-fit over the signal range. In order to conclude that there is a lack of fit, one can examine the residuals by means of an analysis of variance, as is discussed in Chap. 5.

In many instances, a simple plot of the residuals as a function of the controlled variable may already indicate whether the regression appears to be correct as has been demonstrated in Chap. 5. Translation of a significant deviation into model modifications, however, remains a difficult matter. Sometimes, it suffices to include more parameters in the model, but sometimes no better model is found. The parameter values obtained under such conditions depend to a great extent on the problem at hand. A very common feature of a non-linear regression experiment, however, is that the uncertainty in the parameters, or the confidence interval, is defined, not by the random part of the signal alone, but also by the deficiency of the model. For the convenience of the reader, a simple scheme is offered in Table 4 in which the limits of applicability of curve fitting of Gaussian and/or Lorenzian peaks are shown. An important conclusion is that the number of bands must be known in order to obtain reliable results from curve fitting, provided there are no more than three peaks between two inflection points. If the peak shape is described

TABLE 3

Iterative estimation of the parameters of a non-linear function by a least-squares method

Iteration step	b_1	b_2	e
0	12	1.5	332.7
1	11.25	1.927	27.96
2	11.09	1.872	0.104
3	11.09	1.871	0.104

by an approximate model, curve fitting only provides acceptable results when the number of inflection points is twice the number of bands [6].

3. Multivariate linear regression

In the foregoing paragraphs, all observations (or responses) were dependent on one variable only but, of course, the number of variables is not necessarily limited to one. There are many problems where the observation, y, depends on a set of m variables $(x_1, x_2, \ldots, x_m)$. For instance, the absorbance in multicomponent analysis depends on the concentration of all analytes present. This dependency can be described by a mathematical function of the form

$$y = \beta_0 + \beta_1 x_1 + \beta_2 x_2 + \beta_3 x_3$$

or more generally

$$y = \mathrm{f}(x_1, x_2, \ldots, x_m, \beta_0, \beta_1, \ldots, \beta_m) \tag{18}$$

To determine the estimates $(b_0, \ldots, b_m)$ of the unknown parameters $(\beta_0, \ldots, \beta_m)$, n observations are made of the variable y, $(y_1, \ldots, y_n)$ for n different combinations of the m controlled variables $(x_{1,1}, \ldots, x_{m,1}; \ldots; x_{1,n}, \ldots, x_{m,n})$ $(n \geqslant m)$. As mentioned in Sect. 1, every observation (y_i) can be expressed as the sum of the value of the model (f_i) and an uncertainty e_i.

$$y_i = \mathrm{f}(x_{1,i}, x_{2,i}, \ldots, x_{m,i}; \beta_0, \beta_1, \ldots, \beta_m) + e_i \qquad (i = 1, 2, \ldots, n) \tag{19}$$

The problem is now to estimate the parameter set $(\beta_0, \ldots, \beta_m)$ in such a way that the residual vector $|\mathbf{e}| = \mathbf{e}' \cdot \mathbf{e}$ is minimal. In the next sections some examples of the least squares estimation are given.

3.1 Fitting of plane surfaces

If the response (y) is first order in all independent variables (x_i), without the occurrence of cross terms $(x_i \cdot x_j)$, which represent interactions (see Chap. 17), the response surface in the multidimensional space $(x_1, \ldots, x_m)$ is a plane. The general linear function can be expressed as

$$y = \beta_0 + \beta_1 \cdot x_1 + \ldots + \beta_m \cdot x_m \tag{20}$$

References p. 188

TABLE 4

Limits and applicability of curve-fitting (from ref. 6)

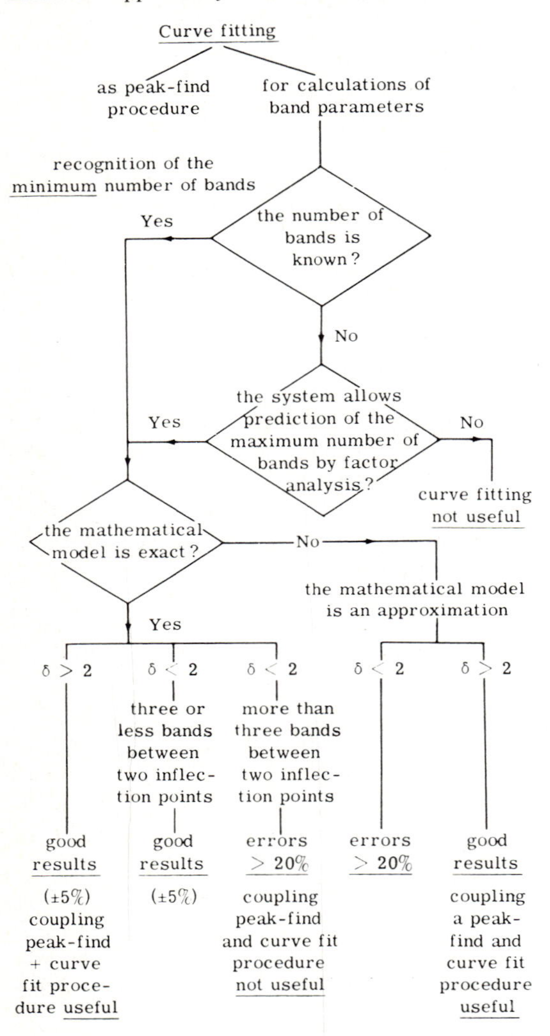

In this particular case of eqn. (20) the number of parameters is $p = m + 1$. Analogous to eqn. (3) the observations can be represented by

$$\begin{aligned} y_1 &= \beta_0 + \beta_1 \cdot x_{1,1} + \beta_2 \cdot x_{2,1} + \ldots + \beta_m \cdot x_{m,1} + e_1 \\ y_2 &= \beta_0 + \beta_1 \cdot x_{1,2} + \beta_2 \cdot x_{2,2} + \ldots + \beta_m \cdot x_{m,2} + e_2 \\ &\cdot \quad \cdot \quad \cdot \quad \cdot \quad \cdot \quad \cdot \\ y_n &= \beta_0 + \beta_1 \cdot x_{1,n} + \beta_2 \cdot x_{2,n} + \ldots + \beta_m \cdot x_{m,n} + e_n \end{aligned} \tag{21}$$

TABLE 5

An orthogonal design of 4 combinations of 3 factors $n = 4,\ m = 3$

Experiment (i)	Values (x)			
	$1,i$	$2,i$	$3,i$	y
1	1	1	−1	−2.1
2	1	−1	1	3.9
3	1	−1	−1	−0.1
4	−1	−1	−1	−1.8

This set of equations can be rewritten in shorthand matrix notation as

$$\mathbf{y} = \mathbf{X} \cdot \boldsymbol{\beta} + \mathbf{e} \tag{22}$$

where $\mathbf{y}$ is the measurement vector, $\mathbf{e}$ the residual vector and $\boldsymbol{\beta}$ the parameter vector, while the independent variable matrix $\mathbf{X}$ is now

$$\mathbf{X} = \begin{bmatrix} 1 & x_{1,1} & x_{2,1} & x_{3,1} & \cdots & x_{m,1} \\ 1 & x_{1,2} & x_{2,2} & x_{3,2} & \cdots & x_{m,2} \\ \cdots & & \cdots & & \cdots & \cdots \\ 1 & x_{1,n} & x_{2,n} & x_{3,n} & \cdots & x_{m,n} \end{bmatrix}$$

Equation (22) is completely identical to eqn. (5). Hence, the least-squares estimates, $\mathbf{b}$, of $\boldsymbol{\beta}$ can be found, in analogy with eqn. (8), from

$$\mathbf{b} = (\mathbf{X}' \cdot \mathbf{X})^{-1} \cdot \mathbf{X}' \cdot \mathbf{y} \tag{23}$$

As an example, we give an orthogonal design with three variables x_1, x_2, and x_3, which is shown in Table 5. The model of the responses is

$$y = \beta_1 \cdot x_1 + \beta_2 \cdot x_2 + \beta_3 \cdot x_3$$

Here

$$\mathbf{X} = \begin{bmatrix} 1 & 1 & -1 \\ 1 & -1 & 1 \\ 1 & -1 & -1 \\ -1 & -1 & -1 \end{bmatrix}$$

and

$$(\mathbf{X}' \cdot \mathbf{X}) = \begin{bmatrix} 4 & 0 & 0 \\ 0 & 4 & 0 \\ 0 & 0 & 4 \end{bmatrix}$$

$$\begin{bmatrix} b_1 \\ b_2 \\ b_3 \end{bmatrix} = \begin{bmatrix} 4 & 0 & 0 \\ 0 & 4 & 0 \\ 0 & 0 & 4 \end{bmatrix}^{-1} \begin{bmatrix} 1 & 1 & 1 & -1 \\ 1 & -1 & -1 & -1 \\ -1 & 1 & -1 & -1 \end{bmatrix} \begin{bmatrix} -2.1 \\ 3.9 \\ -0.1 \\ -1.8 \end{bmatrix}$$

Thus

$$\begin{bmatrix} b_1 \\ b_2 \\ b_3 \end{bmatrix} = \begin{bmatrix} 1/4 & 1/4 & 1/4 & -1/4 \\ 1/4 & -1/4 & -1/4 & -1/4 \\ -1/4 & 1/4 & -1/4 & -1/4 \end{bmatrix} \begin{bmatrix} -2.1 \\ 3.9 \\ -0.1 \\ -1.8 \end{bmatrix}$$

References p. 188

with the result

$b_1 = 0.875$

$b_2 = -1.025$

$b_3 = 1.975$

The residuals are

	i			
	1	2	3	4
$\hat{y}_i$	-2.125	3.875	-0.075	-1.825
e_i	-0.025	-0.025	0.025	-0.025

By analogy with univariate linear regression [eqn. (11)], the variance–covariance matrix of the parameters b is given by

$$\mathbf{V}(b) = (\mathbf{X}' \cdot \mathbf{X})^{-1} s_e^2$$

where $s_e^2 = \mathrm{SS_e}/(n-p)$. Thus, the (95%) confidence limits of b_i are given by

$$b_i \pm t_{n-p}^{0.975} s_e \sqrt{v(b_i)}$$

where p is the number of parameters and n the number of experiments. The predicted response, $\hat{y}_0$, for the combination of the controlled variables $x_{1,0}, x_{2,0}, \ldots, x_{m,0}$ is given by

$$\hat{y}_0 = b_0 + b_1 \cdot x_{1,0} + b_2 \cdot x_{2,0} + \ldots + b_m \cdot x_{m,0}$$

The 95% confidence interval for the true response y_0 at $\mathbf{x}_0$ is given by

$$\hat{y}_0 \pm t_{n-p}^{0.975} s_e \sqrt{\mathbf{x}_0'(\mathbf{X}' \cdot \mathbf{X})^{-1} \cdot \mathbf{x}_0}$$

with

$$\mathbf{x}_0 = \begin{bmatrix} x_{1,0} \\ x_{2,0} \\ \cdot \\ x_{m,0} \end{bmatrix}$$

For the foregoing example, we find

$$s_e^2 = \frac{\mathrm{SS_e}}{(n-p)} = \frac{0.0025}{(4-3)} = 0.0025$$

$$\mathbf{V}(b) = (\mathbf{X}' \cdot \mathbf{X})^{-1} \cdot s_e^2 = \begin{bmatrix} 1/4 & 0 & 0 \\ 0 & 1/4 & 0 \\ 0 & 0 & 1/4 \end{bmatrix} 0.0025$$

$$V(b_1) = 0.0025/4 = 6 \times 10^{-4}$$

$$V(b_2) = 0.0025/4 = 6 \times 10^{-4}$$

$$V(b_3) = 0.0025/4 = 6 \times 10^{-4}$$

The 95% confidence interval for b_1 is

$$1.975 \pm t_1^{0.975}\sqrt{0.0025/4} = 1.975 \pm (12.7)2.5 \times 10^{-2}$$
$$= 1.975 \pm 0.3$$

The 95% confidence interval for the true response y at $x_1 = x_2 = x_3 = 1$ is

$$\hat{y} = 0.875(1) - 1.025(1) + 1.975(1) = 1.825$$

$$1.825 \pm t_1^{0.975}\sqrt{0.0025\begin{bmatrix}1 & 1 & 1\end{bmatrix}\begin{bmatrix}1/4 & 0 & 0\\ 0 & 1/4 & 0\\ 0 & 0 & 1/4\end{bmatrix}\begin{bmatrix}1\\1\\1\end{bmatrix}}$$

$$= 1.825 \pm (12.7)(0.05)(0.86)$$
$$= 1.825 \pm 0.55$$

3.2 Multicomponent analysis by multiple linear regression

The term multicomponent analysis is used for procedures in which several components in a sample are determined simultaneously. For the analysis of an m-component mixture, at least $n = m$ measurements are required provided that linearity (here in the sense of straight line relationships) and additivity of the signals can be assumed. The contribution to the signal of each analyte at a given sensor (e.g. a wavelength in UV–vis spectrometry) is weighted by the sensitivity coefficients, k_i $(i = 1, 2, \ldots, m)$, of each analyte (in spectrometry, k_{ij} is the molar absorptivity of component i at wavelength j). For $m = 2$, for example, the response at sensor 1 is

$$y_1 = k_{11} \cdot c_1 + k_{12} \cdot c_2 + e_1$$

where c_1, c_2 are the unknown concentrations, k_{11}, k_{12} are the respective molar absorptivities, and e_1 is the measurement noise.

The measurement at sensor 2 gives

$$y_2 = k_{21} \cdot c_1 + k_{22} \cdot c_2 + e_2$$

The set of equations has the same form as eqn. (21) without the b_0 term and can be written in the shorthand matrix notation as

$$\mathbf{y} = \mathbf{K} \cdot \mathbf{c} + \mathbf{e} \qquad (24)$$

with

$$\mathbf{K} = \begin{bmatrix} k_{11} & k_{12} \\ k_{21} & k_{22} \end{bmatrix}$$

$$\mathbf{c} = \begin{bmatrix} c_1 \\ c_2 \end{bmatrix}$$

and

$$\mathbf{y} = \begin{bmatrix} y_1 \\ y_2 \end{bmatrix}$$

References p. 188

TABLE 6

Absorptivities of Cl_2 and Br_2 in chloroform (from ref. 7)

Wavenumber ($cm^{-1} \times 10^3$)	Absorptivities		Absorbance of unknown (simulated)
	Cl_2	Br_2	
22	4.5	168	34.10
24	8.4	211	42.95
26	20	158	33.55
28	56	30	11.70
30	100	4.7	11.00
32	71	5.3	7.98

The unknown concentrations are then

$$\mathbf{c} = (\mathbf{K}' \cdot \mathbf{K})^{-1}\mathbf{K}' \cdot \mathbf{y} \tag{25}$$

As no limitations were set to the maximum number of observations in the multiple linear regression problem, the solution is also valid when the responses are measured at more sensors than analytes (a so-called over-determined system).

Example. In order to illustrate the least-squares calculation, we consider a chlorine–bromine system. In Table 6, the absorptivities of Cl_2 and Br_2 in chloroform at six wavenumbers are given [7]. For an optical path length of 1 cm, and concentrations c_1 and c_2 of Cl_2 and Br_2, respectively, the measurements $y_1, y_2, y_3, \ldots, y_6$ are obtained at the wavenumbers (22, 24, 26, 28, 30, 32×10^3 cm^{-1})

$$y_1 = 4.5c_1 + 168c_2 = 34.10$$
$$y_2 = 8.4c_1 + 211c_2 = 42.95$$
$$y_3 = 20c_1 + 158c_2 = 33.55$$
$$y_4 = 56c_1 + 30c_2 = 11.70$$
$$y_5 = 100c_1 + 4.7c_2 = 11.00$$
$$y_6 = 71c_1 + 5.3c_2 = 7.98$$

The concentrations c_1, c_2 are given by eqn. (25), which becomes

$$\begin{bmatrix} c_1 \\ c_2 \end{bmatrix} = \left[\begin{bmatrix} 4.5 & 8.4 & 20 & 56 & 100 & 71 \\ 168 & 211 & 158 & 30 & 4.7 & 5.3 \end{bmatrix} \begin{bmatrix} 4.5 & 168 \\ 8.4 & 211 \\ 20 & 158 \\ 56 & 30 \\ 100 & 4.7 \\ 71 & 5.3 \end{bmatrix} \right]^{-1}$$

$$\times \begin{bmatrix} 4.5 & 8.4 & 20 & 56 & 100 & 71 \\ 168 & 211 & 158 & 30 & 4.7 & 5.3 \end{bmatrix} \begin{bmatrix} 34.10 \\ 42.95 \\ 33.55 \\ 11.70 \\ 11.00 \\ 7.98 \end{bmatrix}$$

This gives

$$\begin{bmatrix} c_1 \\ c_2 \end{bmatrix} = \begin{bmatrix} 18667.81 & 8214.7 \\ 8211.7 & 98659.18 \end{bmatrix}^{-1} \begin{bmatrix} 4.5 & 8.4 & 20 & 56 & 100 & 71 \\ 168 & 211 & 158 & 30 & 4.7 & 5.3 \end{bmatrix} \begin{bmatrix} 34.10 \\ 42.95 \\ 33.55 \\ 11.70 \\ 11.00 \\ 7.98 \end{bmatrix}$$

$$\begin{bmatrix} c_1 \\ c_2 \end{bmatrix} = \begin{bmatrix} 5.56\times10^{-5} & -4.63\times10^{-6} \\ -4.63\times10^{-6} & 1.052\times10^{-5} \end{bmatrix} \begin{bmatrix} 4.5 & 8.4 & 20 & 56 & 100 & 71 \\ 168 & 211 & 158 & 30 & 4.7 & 5.3 \end{bmatrix} \begin{bmatrix} 34.10 \\ 42.95 \\ 33.55 \\ 11.70 \\ 11.00 \\ 7.98 \end{bmatrix}$$

$$\begin{bmatrix} c_1 \\ c_2 \end{bmatrix} = \begin{bmatrix} -0.00053 & -0.00051 & 0.00038 & 0.00298 & 0.00554 & 0.00392 \\ 0.00174 & 0.00218 & 0.00157 & 0.00006 & -0.00041 & -0.00027 \end{bmatrix} \begin{bmatrix} 34.10 \\ 42.95 \\ 33.55 \\ 11.70 \\ 11.00 \\ 7.98 \end{bmatrix}$$

$c_1 = 0.099241$

$c_2 = 0.199843$

In general, it can be expected that the precision of the procedure increases with an increasing number of measurements. To some extent, the effect of using an over-determined system is the same as the effect of repeated measurements on the precision.

By analogy with eqn. (11), the variance of the concentrations is

$$\mathbf{V}(c) = s_e^2(\mathbf{K}'\cdot\mathbf{K})^{-1}$$

In fact, this equation gives the error amplification of the measurement error (s_{pe}^2) to the analytical result, $\mathbf{V}(c)$. As a most important conclusion, it follows that the error propagation depends on the choice of the wavelengths in multicomponent analysis (the **K** matrix).

The residuals are

$$e_i = \hat{y}_i - y_i$$

$$s_e^2 = \frac{\mathrm{SS}_e}{(n-m)}$$

where n is the number of measurements (wavelengths) and m the number of analytes.

$$\mathrm{SS}_e = \sum_{i=1}^{n} (\hat{y}_i - y_i)^2$$

The absorbances predicted by the model, $\hat{y}_i$, and the residuals are tabulated in Table 7.

$s_e^2 = 0.0563/(6-2) = 1.41\times10^{-2}$

$v(c_1) = C_{11}s_e^2$

$v(c_2) = C_{22}s_e^2$

TABLE 7

Predicted absorbances and residuals for the data of Table 5

Wavenumber ($\mathrm{cm}^{-1}\times10^3$)	$\hat{y}$	$\hat{y}-y$	$(\hat{y}-y)^2$
22	34.02	−0.08	6.4×10^{-4}
24	43.01	0.06	3.6×10^{-4}
26	33.57	0.02	4.0×10^{-4}
28	11.59	−0.11	1.2×10^{-2}
30	10.93	−0.07	4.9×10^{-4}
32	8.15	0.17	2.9×10^{-2}
			0.0563 = sum

where C_{11} and C_{22} are the corresponding diagonal elements of the $(\mathbf{K}' \cdot \mathbf{K})^{-1}$ matrix.

$v(c_1) = (5.56 \times 10^{-5})(1.41 \times 10^{-2}) = 7.84 \times 10^{-7}$

$v(c_2) = (1.05 \times 10^{-5})(1.41 \times 10^{-2}) = 1.48 \times 10^{-7}$

The 95% confidence limits of the true concentrations are

$$c_1 \pm t_4^{0.975}\sqrt{v(c_1)} = 0.099241 \pm 2.78\sqrt{7.84 \times 10^{-7}}$$

$$= 0.099 \pm 2.5 \times 10^{-3}$$

$$c_2 \pm t_4^{0.975}\sqrt{v(c_2)} = 0.199843 \pm 2.78\sqrt{1.48 \times 10^{-7}}$$

$$= 0.200 \pm 1.1 \times 10^{-3}$$

References

1 N.W. Johnson, R. Dean Riess, Numerical Analysis, Addison Wesley, Reading, MA, 1982.
2 H.N.J. Poulisse, Multicomponent analysis computations based on Kalman filtering, Anal. Chim. Acta, 112 (1979) 361.
3 S.N. Deming and S.L. Morgan, The use of linear models and matrix least squares in clinical chemistry, Clin. Chem., 25 (1979) 840.
4 N.R. Draper and H. Smith, Applied Regression Analysis, Wiley, New York, 1981.
5 D.W. Marquardt, An algorithm for least-squares estimation of nonlinear parameters, J. Soc. Ind. Appl. Math., 11 (1963) 431.
6 B.G.M. Vandeginste and L. de Galan, Critical evaluation of curve fitting in infrared spectrometry, Anal. Chem., 47 (1975) 2124.
7 Landolt-Bornstein, Zallen Werte und Functionen, Teil 3, Atom und Molekular Physik, Springer Verlag, Berlin, 1951.

Recommended reading

General

C. Daniel and F.S. Wood, Fitting Equations to Data, Wiley, New York, 2nd edn., 1980.

D.A. Kurtz, Trace Residue Analysis — Chemometric Estimations of Sampling, Amount and Error, ACS Symposium Series 284, American Chemical Society, Washington, DC, 1985, Chaps. 8–12.

Calibration graphs

J. Agterdenbos, F.J.M.J. Maessen and J. Balke, Calibration in quantitative analysis. Part I. General considerations, Anal. Chim. Acta, 108 (1979) 315.

J. Agterdenbos, F.J.M.J. Maessen and J. Balke, Calibration in quantitative analysis. Part II. Confidence regions for the sample content in the case of linear calibration relations, Anal. Chim. Acta, 132 (1981) 127.

J.S. Hunter, Calibration and the straight line: Current statistical practices, J. Assoc. Off. Anal. Chem., (1981) 574.

R.G. Krutchkoff, Classical and inverse regression methods of calibration, Technometrics, 9 (1967) 425.

D.G. Mitchell, W.N. Mills, J.S. Garden and M. Zdeb, Multiple curve procedure for improving precision with calibration-curve-based analysis, Anal. Chem., 49 (1977) 1655.

G.R. Phillips and E.M. Eyring, Comparison of conventional and robust regression in analysis of chemical data, Anal. Chem., 55 (1983) 1134.

Non-constant variance techniques

J.S. Garden, D.G. Mitchell and W.N. Mills, Non-constant variance techniques for calibration-curve-based-analysis, Anal. Chem., 52 (1980) 2310.

D.A. Kurz, The use of regression and statistical methods to establish calibration graphs in chromatography, Anal. Chim. Acta, 150 (1983) 105.

Multicomponent analysis

C. Jochum, P. Jochum and B.R. Kowalski, Error propagation and optimal performance in multicomponent analysis, Anal. Chem., 53 (1981) 85.

C.W. Brown, P.F. Lynch, R.J. Obremski and D.S. Lavery, Matrix representations and criteria for selecting analytical wavelengths for multicomponent spectroscopic analysis, Anal. Chem., 54 (1982) 1472.

H.J. Kisner, C.W. Brown and G.I. Kavarnos, Multiple analytical frequencies and standards for the least-squares spectrometric analysis of serum lipids, Anal. Chem., 55 (1983) 1703.

P. Jochum and E.L. Schrott, Deconvolution of multicomponent ultraviolet and visible spectra, Anal. Chim. Acta, 157 (1984) 211.

B. Hoyer and L. Kryger, Application of the generalized standard addition method to correct for the effects of peak overlap and intermetallic compound formation in potentiometric stripping analysis, Anal. Chim. Acta, 167 (1985) 11.

Non-linear regression

M.A. Maris, C.W. Brown and D.S. Lavery, Non-linear multicomponent analysis by infrared spectrophotometry, Anal. Chem., 55 (1983) 1694.

W.F. Maddams, The scope and limitations of curve fitting, Appl. Spectrosc., 34 (1980) 245.

I.C. Copeland, The use of non-linear least squares analysis, J. Chem. Educ., 61 (1984) 778.

G.C. Allen and R.F. McMeeking, Deconvolution of spectra by least-squares fitting, Anal. Chim. Acta, 103 (1978) 73.

T. Brubaker, R. Tracy and C.L. Pomernacki, Linear parameter estimation, Anal. Chem., 50 (1978) 1017A.

Kalman filtering

S.D. Brown, The Kalman filter in analytical chemistry, Anal. Chim. Acta, 181 (1986) 1.

T. Brubaker, Nonlinear parameter estimation, Anal. Chem., 51 (1979) 1385A.

S.C. Rutan, Kalman filtering approaches for solving problems in analytical chemistry, J. Chemomet., 1 (1987) 7.

Partial least squares

H. Martens and T. Naes, Multivariate calibration. I. Concepts and distinctions, Trends Anal. Chem., 3 (1984) 204.

T. Naes and H. Martens, Multivariate calibration II. Chemometric methods, Trends Anal. Chem., 3 (1984) 266.

A. Lorber, L.E. Wangen and B.R. Kowalski, A theoretical foundation for the PLS algorithm, J. Chemomet., 1 (1987) 19.

M. Sjostrom, S. Wold, W. Lindberg, J.A. Persson and H. Martens, A multivariate calibration problem in analytical chemistry solved by partial least squares models in latent variables, Anal. Chim. Acta, 150 (1983) 61.

W. Lindberg, J.A. Persson and S. Wold, Partial least-squares method for spectrofluorimetric analysis of mixtures of humic acid and lignin sulfonate, Anal. Chem., 55 (1983) 643.

W. Lindberg, J. Ohman, S. Wold and H. Martens, Simultaneous determination of five different food proteins by high-performance liquid chromatography and partial least-squares multivariate calibration, Anal. Chim. Acta, 174 (1985) 41.

Chapter 14

Correlation Methods

1. Introduction

In Chap. 13, regression was discussed for the quantitative evaluation of the relationship between one dependent variable, y, and one or more independent variables, x. It was supposed that a given functional relationship exists between the variables y and x, e.g. $y = ax + b$.

In many situations, two (or more) variables are measured on the same object(s) or individual(s). In environmental chemistry, for example, often several environmental variables (concentrations) are simultaneously determined in a sample. For the interpretation of the results, it is desirable to detect a possible relationship between the measured variables. The problem discussed in this chapter is to find a quantitative measure for the relationship between two or more (random) variables.

Consider an example taken from environmental chemistry. Table 1 lists, among others, the Na and Cl concentrations found in samples of rain water collected at

TABLE 1

Rain data [1]

Values are given as ppb of species listed.

Station	SO_4	Cl	Na	Mg	As	Sb	Pb
1	3700	2680	2000	190	1.8	0.93	58
2	3000	180	280	55	0.9	0.08	20
3	2600	1680	1800	160	3.8	0.46	10
4	1500	230	230	50	0.6	0.08	3
5	1300	4090	2300	240	1.0	0.12	15
6	3000	2820	2600	220	8.0	1.04	67
7	1300	820	800	115	1.0	0.14	5
8	1300	770	1100	90	4.0	0.13	8
9	1600	680	550	55	16.0	0.80	12
10	2500	550	540	50	1.8	0.15	7
11	1400	1270	1300	120	1.0	0.20	9
12	4300	1590	1800	195	4.0	0.73	7
13	4300	1590	1800	195	4.0	0.73	7
14	2600	2850	2600	185	47.0	2.08	21
15	2200	590	450	75	5.0	0.46	5
16	3100	770	850	95	7.7	0.28	6
17	4300	2640	2400	270	4.9	0.40	56
18	2700	2180	2100	250	9.0	0.38	10
19	2200	500	900	125	5.5	0.08	11
20	1100	820	1100	125	0.7	0.03	3
21	2200	1360	1700	170	7.0	0.30	23
22	3500	360	360	45	1.5	0.12	3

References p. 212

TABLE 2

Frequency table of (Na, Cl) observation pairs

Na (ppm)	Cl (ppm) 0–0.49	0.5–0.99	1.0–1.49	1.5–1.99	2.0–2.49	2.5–2.99	> 3.0
0 –0.49	3	1					
0.5–0.99	1	4					
1.0–1.49		2	1				
1.5–1.99			1	3		1	
2.0–2.49					1	1	1
2.5–2.99					1	1	
> 3.0							

various sites in the Puget Sound area [1]. The data can be arranged according to the frequency of the occurrence of given combinations of the content of these ions in rain water. Table 2 indicates that various combinations of the values of both variables occur with different probabilities. Combinations lying on the diagonal of the table are more probable than combinations in the lower left and upper right corners of the table. The observation pairs (Na, Cl) constitute a sample of a bivariate frequency distribution, f(x, y), expressing the probability of finding an

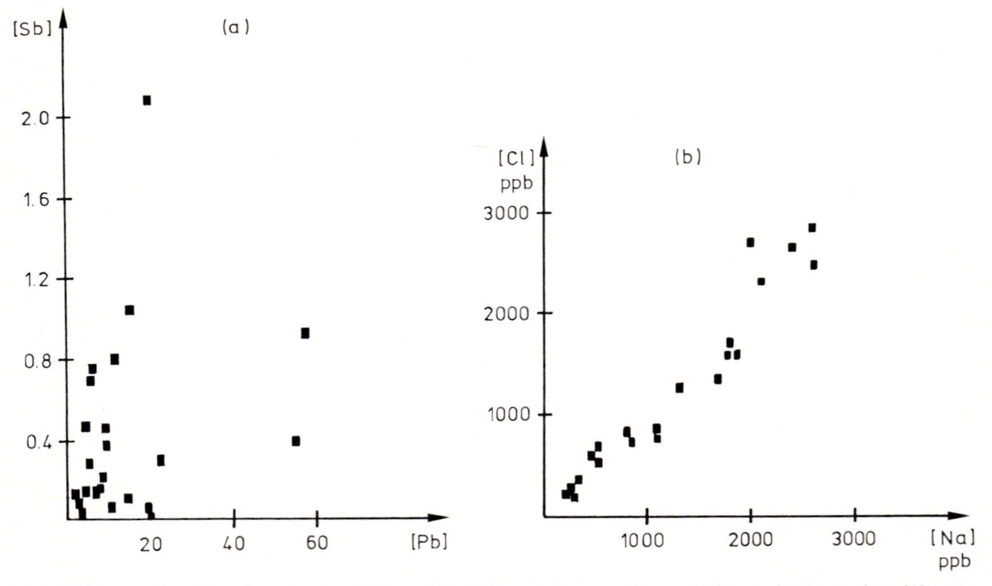

Fig. 1. Scatter plot of two random variables with different degrees of correlation (rain-water data [1]). (a) Sb vs. Pb concentration; (b) Cl vs. Na concentration.

observation pair (x, y). The scatter plot of the data [Fig. 1(b)] provides a graphical representation of this bivariate distribution. An important parameter that characterizes a bivariate distribution is the correlation between both variables. The bivariate normal distribution is discussed in more detail in Chap. 23.

In the situation depicted in Fig. 1(a), the concentration of Pb measured at a given location gives no a priori (i.e. before the actual measurement of the Sb concentration) information on the Sb concentration at the same location. It is said that there is no correlation between the Pb and Sb concentrations. The situation in Fig. 1(b) is clearly different. Once a value for the Na concentration has been obtained, a fairly good estimation of the Cl concentration can be made. There is a distinct relationship between the concentration of Cl and the concentration of Na. This association between two random variables can be expressed by the covariance and the correlation coefficient. They express quantitatively the linear relationship between two variables.

2. Covariance and correlation

Both sets of measurements of Na and Cl, respectively denoted as $x(1), \ldots, x(n)$ and $y(1), \ldots, y(n)$, are samples of size n from two populations characterized by random variables x and y. The covariance of the two variables, $\gamma(x, y)$, is defined as the expectancy that the values (x_i, y_i) measured on the objects $1, \ldots, i, \ldots, n$ deviate in a similar way from the respective means μ_x and μ_y. That is

$$\gamma(x, y) = \frac{1}{n} \sum_{i=1}^{n} (x_i - \mu_x)(y_i - \mu_y) \tag{1}$$

where n is the number of observation pairs (x_i, y_i). In practice, the number of observations is limited and the true population means are not known but only estimates $\bar{x}$ and $\bar{y}$ are available. This gives an estimate of the covariance

$$\mathrm{cov}(x, y) = \frac{1}{n-1} \sum_{i=1}^{n} (x_i - \bar{x})(y_i - \bar{y}) \tag{2}$$

The denominator, $n-1$, is the number of independent observation pairs, i.e. the number of degrees of freedom.

The covariance between the (Na, Cl) concentrations (ppb) listed in Table 1 is

$$\frac{1}{21} \sum_{i=1}^{22} (x_{\mathrm{Na},i} - \bar{x}_{\mathrm{Na}})(y_{\mathrm{Cl},i} - \bar{y}_{\mathrm{Cl}}) = 742656$$

where $\bar{x}_{\mathrm{Na}} = 1343$ ppb and $\bar{y}_{\mathrm{Cl}} = 1390$ ppb.

By elaborating eqn. (2), a more convenient way to calculate the covariance is obtained.

$$\mathrm{cov}(x, y) = \frac{\sum_{i=1}^{n} x_i y_i - \left(\sum_{i=1}^{n} x_i \sum_{i=1}^{n} y_i \right) / n}{n-1}$$

References p. 212

If a strong linear relationship exists between the x and y values, e.g. $y = ax$, then the absolute value of the covariance is high. In that case, one finds that

$$\text{cov}(x, y) = \frac{1}{n-1} \sum_{i=1}^{n} (x_i - \bar{x})(y_i - \bar{y}) = \frac{1}{n-1} \sum_{i=1}^{n} a(x_i - \bar{x})^2 = as_x^2$$

When the correlation between the variables is low, a high (with respect to the mean) value of y may be alternately accompanied by a low or high value of x (with respect to the mean $\bar{x}$). This means that one may find observation pairs: (x_1, y_1) and (x_2, y_2) where $x_1 > \bar{x}$ and $y_1 < \bar{y}$, and $x_2 > \bar{x}$ and $y_2 > \bar{y}$. Consequently, some terms in the summation in eqn. (2) will cancel out, causing a smaller absolute value for the covariance. Theoretically, the covariance lies between $-\infty$ and $+\infty$ and is dependent on the scale chosen. For example, if the Na/Cl data were given in ppm instead of ppb, then $\text{cov}(x, y) = 0.743$ would have been found.

To obtain a parameter independent of the scales in which x and y are measured, the covariance is divided by the product of the standard deviations of x and y, which gives the correlation coefficient.

$$r(x, y) = \frac{\text{cov}(x, y)}{s_x s_y} = \frac{\left[\sum_{i=1}^{n} (x_i - \bar{x})(y_i - \bar{y})\right]/(n-1)}{\sqrt{\left[\sum_{i=1}^{n} (x_i - \bar{x})^2 \cdot \sum_{i=1}^{n} (y_i - \bar{y})^2\right]/(n-1)^2}} \tag{3}$$

The correlation coefficient $r(x, y)$ is more useful to express the relationship between two variables because of its independence of the chosen scales. For the Na/Cl data listed in Table 1, this gives

$$r(x, y) = 0.742/(0.793 \times 1.032) = 0.91$$

The correlation coefficient is bounded between -1 and $+1$. Maximal absolute correlation, $|r(x, y)| = 1$, is found when the value of one variable (y) is perfectly linearly related with the other (x), e.g. $y = ax$. For non-related variables, $r(x, y)$ is close to zero. When the correlation coefficient is not significantly different from zero, the random variables are called uncorrelated. However, this does not necessarily imply there is no relationship between the two variables, but only that there is no linear relationship. For instance, in Fig. 2 there is a strong relationship between x and y, but the relationship is not linear and, as a consequence, the correlation is near zero.

Figure 3 gives an impression of the degree of correlation between data having various correlation coefficients. These figures show clearly that a correlation coefficient of $r = 0.32$ may be interesting but is of no practical importance in forecasting the outcome of one test from the result of another. From an analytical point of view, the knowledge of correlations between analytical results may lead to important decisions about the design of analytical procedures. If one knows that the outcome of two analyses, x and y, are considerably correlated and therefore y can

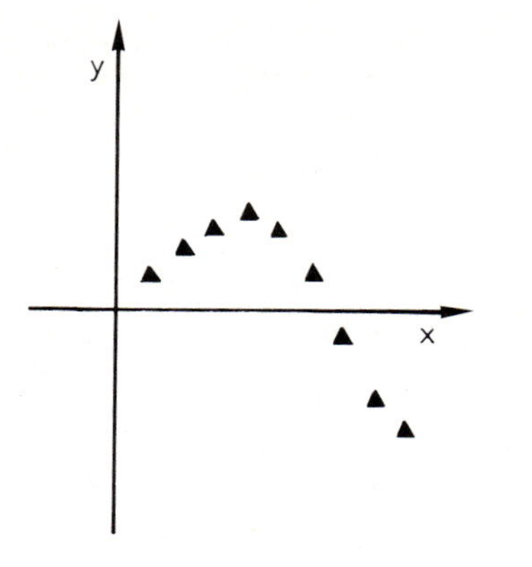

Fig. 2. Dependent variables without correlation.

be satisfactorily forecast from the determination of x [$y = f(x)$], then the actual measurement of y will not deviate appreciably from the estimate already available. In other words, the extra information acquired by the determination of y is limited. On the other hand, when x and y exhibit only moderate correlation, no good estimate of y will be obtained from the determination of x. Therefore the determination of y will considerably augment the amount of information. The relationship between correlation and information is discussed in the chapter on information theory (Chap. 9).

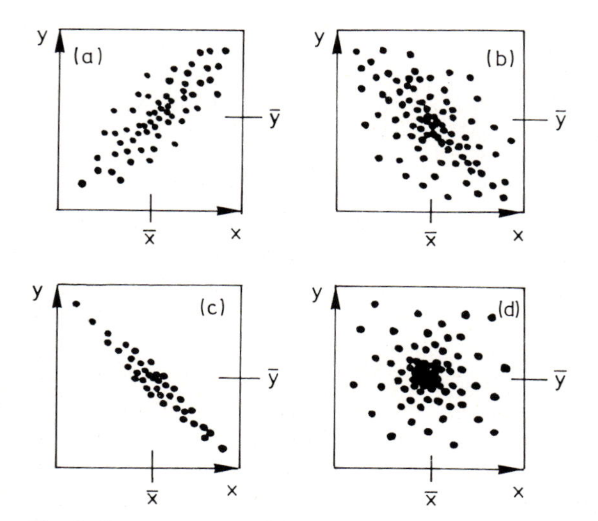

Fig. 3. Plots of random variables with various degrees of correlation. (a) $r = 0.75$; (b) $r = -0.32$; (c) $r = -0.95$; (d) $r = 0$. (Adapted from ref. 2.)

In the situation where a "tight" relationship has been detected between two variables x and y (e.g. $r > 0.9$), as a next step, the value of one random variable, y, can be forecast quantitatively from the value of the other by applying a straight line regression analysis on the data (Chap. 13).

3. Correlation and regression

In Chap. 5, an expression was derived for the regression coefficient of a straight line through a set of n observation pairs (x, y), where y is considered to be dependent upon x.

$$b_1 = \frac{n \sum_{i=1}^{n} x_i y_i - \sum_{i=1}^{n} y_i \sum_{i=1}^{n} x_i}{n \sum_{i=1}^{n} x_i^2 - \left(\sum_{i=1}^{n} x_i \right)^2}$$

One observes that the numerator in the expression for b_1, is equal to $n(n-1)$ times $\text{cov}(x, y)$, while the denominator equals $n(n-1)$ times s_x^2, so that the regression coefficient b_1 becomes

$$b_1 = \frac{n(n-1)\,\text{cov}(x, y)}{n(n-1)s_x^2} = \frac{\text{cov}(x, y)}{s_x^2}$$

Substituting that result in the expression for the correlation coefficient, one finds

$$b_1 = \frac{\text{cov}(x, y)}{s_x^2} = \frac{r(x, y)s_x s_y}{s_x^2} = \frac{r(x, y)s_y}{s_x}$$

Thus, the regression coefficient of a straight line model is related to the correlation coefficient.

Because the relationship $y = b_0 + b_1 x$ also holds for the mean values $\bar{y}$ and $\bar{x}$, one finds that

$$b_0 = \bar{y} - \bar{x} \cdot r(x, y)\frac{s_y}{s_x}$$

As a result, the straight line which fits the data (x_i, y_i) can be expressed in terms of the correlation coefficient, $r(x, y)$ and the standard deviations of the two variables

$$(y_i - \bar{y}) = r(x, y)\frac{s_y}{s_x}(x_i - \bar{x}) \tag{4}$$

The importance of this relationship will become apparent when discussing autocorrelation and autoregression in Sect. 14.4.

$\text{Cov}(x, y)$ and $r(x, y)$ are estimates of the true population parameters $\gamma(x, y)$ and $\rho(x, y)$ and therefore confidence intervals can be given for the calculated estimates (provided some conditions are satisfied). The $100(1-\alpha)\%$ confidence interval of ρ is obtained by considering a new variable $z = 0.5 \ln[(1+r)/(1-r)]$. This variable has an approximate normal distribution with a mean $0.5 \ln[(1+\rho)/(1-\rho)]$ and a standard deviation $\sqrt{(1/(n-3))}$.

For the calculation of the 95% confidence interval of the correlation coefficient, $r = 0.742$ found for the data in Table 1, we proceed in the following manner.

$$z = 0.5 \ln[(1+0.742)/(1-0.742)] = 0.950$$

and the standard deviation of z is

$$\frac{1}{\sqrt{19}} = 0.23$$

The 95% confidence interval for $z = 0.5\ln[(1+\rho)/(1-\rho)]$ is therefore

$$z \pm (1.96 \times 1/\sqrt{19}) = 0.950 \pm (1.96 \times 0.23)$$

or

$$0.499 < z < 1.404$$

Solving $z_1 = 0.499$ and $z_2 = 1.404$ for ρ gives $\rho_1 = 0.87$ and $\rho_2 = 0.47$. So the correlation between the Na and Cl concentration is probably at least as good as $\rho = 0.47$ and could be as high as $\rho = 0.87$.

Values of $z = 0.5\ln[(1+r)/(1-r)]$ are tabulated in most books on statistics.

4. The variance–covariance matrix

In previous sections, covariance and correlation were calculated between two random variables x and y. Of course, it is possible that more than two variables have been measured on the same object. Recalling the example of Na and Cl measurements in rain water at various sites, more components can be monitored, e.g. Mg and Zn (Table 1).

A general representation of the data for a system with n sampling stations and m compounds, is

Station	Compound				
	1	2	3		m
1	x_{11}	x_{12}	x_{13}	...	x_{1m}
2	x_{21}	x_{22}	x_{23}	...	x_{2m}
.	.	.	.	...	.
.	.	.	.	...	.
n	x_{n1}	x_{n2}	x_{n3}	...	x_{nm}
Compound mean	$\bar{x}_1$	$\bar{x}_2$	$\bar{x}_3$		$\bar{x}_m$
Compound variance	s_1^2	s_2^2	s_3^2		s_m^2

where

$$s_j^2 = \frac{1}{n-1} \sum_{i=1}^{n} (x_{ij} - \bar{x}_j)^2$$

and x_{ij} is the concentration of compound j at station i.

The covariance of every possible combination of constituents can be calculated. For example, the covariance of the concentration of Mg and SO_4 over all sampling sites listed in Table 1 is given by

$$\mathrm{cov}(\mathrm{Mg}, \mathrm{SO_4}) = \frac{1}{n-1} \sum_{i=1}^{n} \left[(x_{i,\mathrm{Mg}} - \bar{x}_{\mathrm{Mg}}) \cdot (x_{i,\mathrm{SO_4}} - \bar{x}_{\mathrm{SO_4}}) \right]$$

A convenient notation for these variances and covariances is the so-called variance–covariance matrix or, in short, the covariance matrix of the form

$$\mathbf{C} = \begin{pmatrix} s_1^2 & \mathrm{cov}(1,2) & \cdots & \mathrm{cov}(1,m) \\ \mathrm{cov}(2,1) & s_2^2 & \cdots & \mathrm{cov}(2,m) \\ \cdot & \cdot & \cdots & \cdot \\ \mathrm{cov}(m,1) & \cdot & & s_m^2 \end{pmatrix}$$

From the definition of the covariance given in eqn. (2), it follows that $\mathrm{cov}(x, x) = s_x^2$. Therefore, the diagonal elements of the variance–covariance matrix are the variances of the variables. The matrix **C** is an estimation of the true variance–covariance matrix of the population

$$\Gamma = \begin{pmatrix} \sigma_1^2 & \gamma_{12} & \gamma_{13} & \cdots & \gamma_{1m} \\ \gamma_{21} & \sigma_2^2 & \gamma_{23} & \cdots & \gamma_{2m} \\ \gamma_{31} & \gamma_{32} & & \cdots & \gamma_{3m} \\ \cdot & \cdot & \cdot & \cdots & \cdot \\ \cdot & \cdot & \cdot & \cdots & \cdot \\ \gamma_{m1} & \gamma_{m2} & \gamma_{m3} & \cdots & \sigma_m^2 \end{pmatrix}$$

Because $\mathrm{cov}(j, k) = \mathrm{cov}(k, j)$, the variance–covariance matrix is symmetrical.

If one arranges all compound concentrations, m, at all stations, n, in a $(n \times m)$ matrix

$$\mathbf{X} = \begin{pmatrix} x_{11} & x_{12} & & x_{1m} \\ x_{21} & x_{22} & \cdots & x_{2m} \\ \cdot & \cdot & \cdots & \cdot \\ \cdot & \cdot & \cdots & \cdot \\ x_{n1} & x_{n2} & \cdots & x_{nm} \end{pmatrix}$$

with column means $\bar{x}_1 \quad \bar{x}_2 \quad \cdots \quad \bar{x}_m$
and if one substracts the column means from the elements of that column (thus the columns in the matrix **U** have zero mean) a data matrix **U** is obtained of the form

$$\begin{pmatrix} x_{11}-\bar{x}_1 & x_{12}-\bar{x}_2 & \cdots & x_{1m}-\bar{x}_m \\ x_{21}-\bar{x}_1 & x_{22}-\bar{x}_2 & \cdots & x_{2m}-\bar{x}_m \\ \cdot & \cdot & \cdots & \cdot \\ \cdot & \cdot & & \cdot \\ x_{n1}-\bar{x}_1 & x_{n2}-\bar{x}_2 & \cdots & x_{nm}-\bar{x}_m \end{pmatrix} = \begin{pmatrix} u_{11} & \cdots & u_{1m} \\ u_{21} & \cdots & u_{2m} \\ \cdot & \cdots & \cdot \\ \cdot & \cdots & \cdot \\ u_{n1} & \cdots & u_{nm} \end{pmatrix} = \mathbf{U}$$

With $\bar{u}_1 = \bar{u}_2 = \ldots \bar{u}_m = 0$.

The variance–covariance matrix is readily calculated by a matrix multiplication.

$\mathrm{cov}(\mathbf{X}) = [1/(n-1)]\mathbf{U}' \cdot \mathbf{U}$

since

$$1/(n-1)\begin{pmatrix} u_{11} & u_{21} & \cdots & u_{n1} \\ u_{12} & u_{22} & \cdots & u_{n2} \\ \cdot & \cdot & \cdots & \cdot \\ \cdot & \cdot & \cdots & \cdot \\ u_{1m} & u_{2m} & \cdots & u_{nm} \end{pmatrix}\begin{pmatrix} u_{11} & u_{12} & \cdots & u_{1m} \\ u_{21} & u_{22} & \cdots & u_{2m} \\ \cdot & \cdot & \cdots & \cdot \\ \cdot & \cdot & \cdots & \cdot \\ u_{n1} & u_{n2} & \cdots & u_{nm} \end{pmatrix} =$$

$$1/(n-1)\begin{pmatrix} \sum\limits_{i=1,n} u_{i1}^2 & & & & \\ \sum\limits_{i=1,n} u_{i1}\cdot u_{i2} & \sum\limits_{i=1,n} u_{i2}^2 & & & \\ \sum\limits_{i=1,n} u_{i1}\cdot u_{i3} & \sum\limits_{i=1,n} u_{i2}\cdot u_{i3} & \sum\limits_{i=1,n} u_{i3}^2 & & \\ \cdot & \cdot & \cdot & \cdot\ \cdot & \cdot \\ \cdot & \cdot & \cdot & \cdot\ \cdot & \cdot \\ \cdot & \cdot & \cdot & \cdot\ \cdot & \sum\limits_{i=1,n} u_{in}^2 \end{pmatrix}$$

$$\mathrm{cov}(\mathbf{X}) = \begin{pmatrix} s_1^2 & \mathrm{cov}(1,2) & \cdot & \cdots & \mathrm{cov}(1,m) \\ \mathrm{cov}(2,1) & s_2^2 & \cdot & \cdots & \cdot \\ \mathrm{cov}(3,1) & \cdot & s_3^2 & \cdots & \cdot \\ \cdot & \cdot & \cdot & \cdots & \cdot \\ \mathrm{cov}(m,1) & \cdot & \cdot & \cdots & s_m^2 \end{pmatrix}$$

where $\mathrm{cov}(i, j)$ is the covariance between the columns i and j in matrix **X**.

In many instances, it is useful to replace the covariances in the variance–covariance matrix by the correlation coefficients, giving the correlation matrix. Every term $\mathrm{cov}(i,j)$ in $\mathrm{cov}(\mathbf{X})$ is then divided by the product of the standard deviations $s_i \cdot s_j$. As a result, the diagonal elements of the correlation matrix become unity while the off-diagonal elements have values between -1 and $+1$. The correlation matrix of the rain data (Table 1) is given in Table 3.

$$\mathbf{R} = \begin{pmatrix} 1 & r_{12} & r_{13} & \cdots & r_{1m} \\ r_{21} & 1 & r_{23} & \cdots & r_{2m} \\ r_{31} & r_{32} & 1 & \cdots & r_{3m} \\ \cdot & \cdot & \cdot & & \cdot \\ \cdot & \cdot & \cdot & & \cdot \\ r_{m1} & r_{m2} & r_{m3} & \cdots & 1 \end{pmatrix}$$

An alternative and more direct way to calculate the correlation matrix is obtained by a z-transform of the data matrix before the matrix multiplication. This operation scales the data to zero mean and unity standard deviation.

TABLE 3

Correlation matrix of data in Table 1

Values in the lower left half are correlations > 0.8.

	Na	Mg	Pb	SO_4	As	Sb	Cl
Na	1	0.94	0.52	0.34	0.39	0.57	0.92
Mg	0.94	1	0.54	0.38	0.19	0.36	0.88
Pb			1	0.52	0.10	0.33	0.53
SO_4				1	0.12	0.35	0.26
As					1	0.85	0.30
Sb					0.85	1	0.50
Cl	0.92	0.88					1

One can show that the correlation matrix, **R**, of the original data matrix, **X**, equals $[1/(n-1)]\mathbf{Z}' \cdot \mathbf{Z}$. The correlation matrix of the rain data is given in Table 3.

The rain data can be represented in the m-dimensional space of m constituents, in which each axis represents the concentration of one constituent. For a system with 3 constituents (x, y, z), the concentrations x_i, y_i and z_i at a given location i represent a point in the 3-dimensional space as is shown in Fig. 4. All points in the

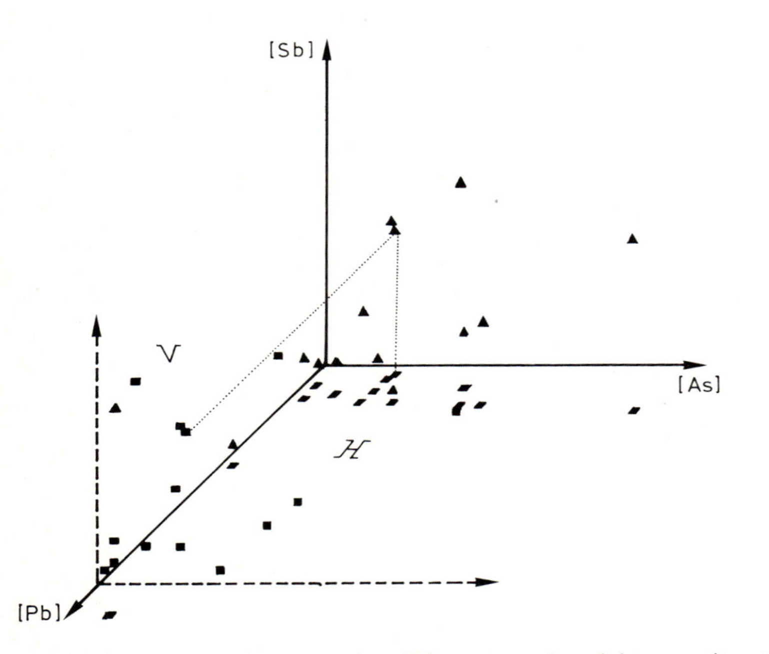

Fig. 4. Three-dimensional representation of the concentration of three constituents, Sb, As, and Pb, in a sample. Data from ref. 1. H is a horizontal plane defined by As–Pb and V a vertical plane defined by Sb–As. ▲, Point in three-dimensional space; ■, projection on V; ▰ , projection on H.

3-D space can be projected on 2-D planes which are parallel to the planes (x, y), (x, z), and (y, z), respectively. Every term r_{ij} in the correlation matrix represents the correlation of the data projected in a plane parallel to the (i, j) plane in the data space. This gives a geometrical interpretation of the correlation matrix. In Fig. 4, all data points have been projected on a plane (V) parallel to the (Sb, As) plane. The correlation coefficient of these projected data is r(Sb, As) listed in the correlation matrix in Table 3.

The variance–covariance matrix is an important concept in analytical chemistry, for example for the calculation of the information obtainable from combined analytical methods. The calculation of the variance–covariance matrix is also a first step in a principal component analysis (Chap. 21) and to solve classification problems (Chap. 23).

5. Time series of random variables

The concentration of a constituent in an object can be monitored as a function of time. Such a series of data measured in time is called a time series, e.g. the pH fluctuation of a river as a function of time (Fig. 5).

Most of the systems sampled by analytical chemists fluctuate in time and can therefore be described by a time series, e.g. various types of technological processes, environmental processes (concentration of pollutants in surface waters), and human processes (concentration of constituents in blood, urine etc). The analytical process itself can often be considered as a time series. In routine chemical analysis, it is common practice to measure reference samples every day or within each run. These measurements also constitute a time series. In Chap. 11, we introduced the concept of time constant, which is a characteristic of the dynamic behaviour of a continuous analyzer. In this section, the time constant will be related to the autocorrelation function and autoregressive modelling of time series. The autocorrelation function and an autoregressive model of a time series provide information necessary to control the statistical fluctuations of analyzers. Also, sampling strategies can be

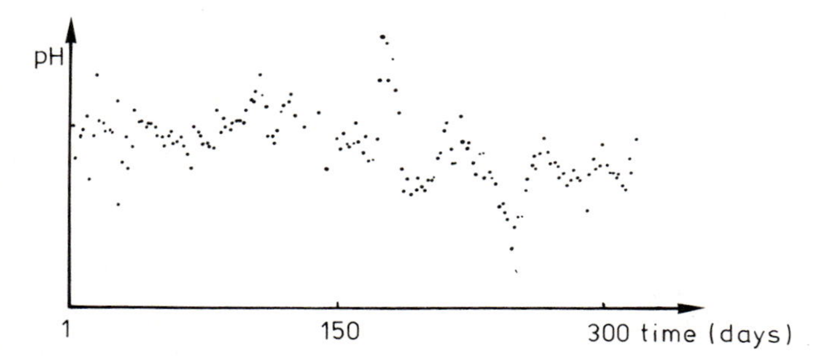

Fig. 5. Example of a time series. The pH of the River Rhine at Lobith during 1982 at intervals of two days. (Adapted from ref. 3.)

References p. 212

designed on the basis of autoregressive models in order to predict the concentration course of a chemical system (Chap. 27). Continuous time series are transformed into discrete time series by sampling. The system state (concentration) is then measured at fixed time intervals.

5.1 Autocovariance and autocorrelation

For the discussion of the concept of autocorrelation, it is useful to define a new independent variable for a time series. When a set of measurements $x(t)$, $x(t+\Delta t), \ldots$ has been collected on a stationary process at equidistant time intervals, Δt, the observation times can be expressed as the number of observation intervals elapsed from the first moment considered, t_0

$$t_1 = t_0 + \Delta t$$

$$t_2 = t_0 + 2\Delta t$$

or, in general

$$t_\tau = t_0 + \tau\Delta t$$

and

$$x(\tau) = x(t_0 + \tau\Delta t)$$

This is illustrated in Fig. 6 where the new variable indicates the number of sampling intervals between two observations.

A very important feature of a time series is the correlation (or covariance) between the observations separated by the same time span, i.e. for $\tau = 1$, the correlation between the observation pairs $x(1)$, $x(2)$; $x(2)$, $x(3)$; $x(3)$, $x(4)$; ...; $x(n-1)$, $x(n)$ and, in general, $x(1)$, $x(1+\tau)$; $x(2)$, $x(2+\tau)$; $x(3)$, $x(3+\tau)$; ...; $x(n-\tau)$, $x(n)$.

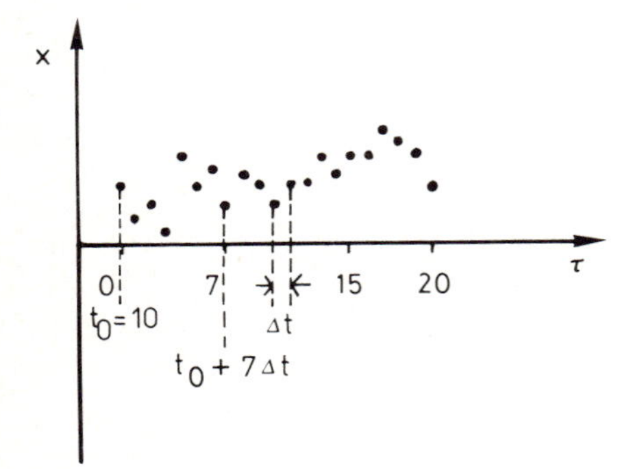

Fig. 6. A discrete time series with the number of intervals as independent variable with $t_0 = 10$ and $\Delta t = 0.1$.

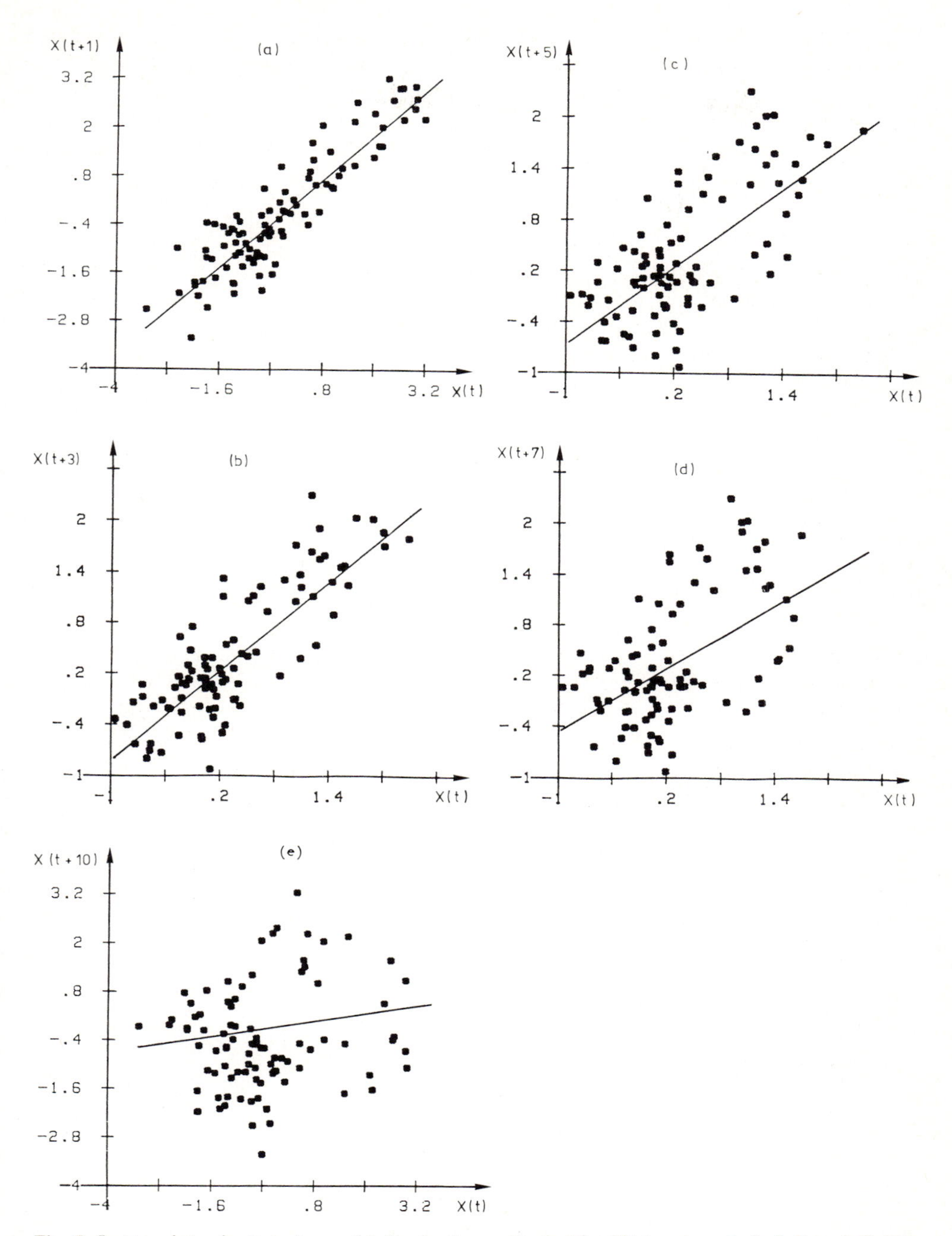

Fig. 7. Scatter plots of $x(t+\tau)$ vs. $x(t)$ for the time series in Fig. 10(a) and $\tau = 1$, 3, 5, 7, and 10. The line is the regression line $x(t+\tau) = b_0 + b_1 x(t)$. (a) $r = 0.904$, $b = 0.908$; (b) $r = 0.812$, $b = 0.831$; (c) $r = 0.697$, $b = 0.771$; (d) $r = 0.591$, $b = 0.611$; (e) $r = 0.171$, $b = 0.159$.

References p. 212

Scatter plots of $x(t+\tau)$ vs. $x(t)$ of the process displayed in Fig. 10(a) are shown in Fig. 7 for all $t = 1, \ldots, n-\tau$ and $\tau = 1, 3, 5, 7$, and 10. In addition, a straight line $x(t+\tau) = b_1 x(t) + b_0$, has been fitted through the data points by the application of linear regression. These regression lines are plotted in the scatter plots. From these plots, the following observations and conclusions follow. With increasing time lag, τ, the data points show less correlation and the regression coefficient approaches zero. Since the regression coefficient of $x(t+\tau)$ vs. $x(t)$ is associated with the correlation coefficient, the expression for the correlation coefficient between $x(t)$ and $x(t+\tau)$ can be derived by substituting x and y in eqn. (3) by $x(t)$ and $x(t+\tau)$. This gives

$$r[x(t),\, x(t+\tau)] = \frac{\sum_{t=1}^{n-\tau} [x(t) - \bar{x}][x(t+\tau) - \bar{x}]}{(n-1-\tau) \cdot s_x(t) \cdot s_x(t+\tau)} \tag{5}$$

where $n-\tau$ is the number of observation pairs $[x(t),\, x(t+\tau)]$.

From the condition that the time series is stationary, it follows that

$$s_x(t) = s_x(t+\tau)$$

Hence

$$r(t,\, t+\tau) = r(\tau) = \frac{\sum_{t=1}^{n-\tau} [x(t) - \bar{x}][x(t+\tau) - \bar{x}]}{(n-1-\tau) s_x^2}$$

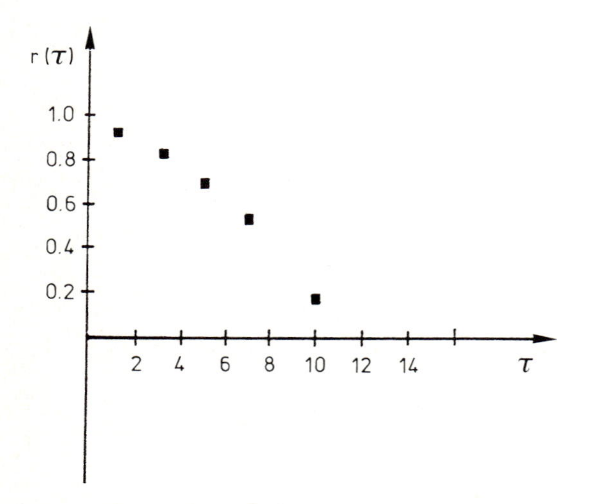

Fig. 8. Autocorrelation function of the time series in Fig. 10(a). The $r(\tau)$ values for $\tau = 1, 3, 5, 7$, and 10 are taken from Fig. 7.

For $\tau = 0$, it is found that

$$r(0) = \frac{\sum_{t=1}^{n} [x(t) - \bar{x}][x(t) - \bar{x}]}{(n-1)s_x^2} = 1$$

As an example, the correlation at $\tau = 1$ is calculated for the residuals listed in Table 1 of Chap. 13. If $r(1)$ is low, evidence is obtained that these residuals are random as they should be for a valid model.

In this example, $n = 7$, $\tau = 1$, and $\bar{x} = 0.03$.

$$r(1) = \frac{\sum_{t=1}^{6} [x(t) - \bar{x}][x(t+1) - \bar{x}]}{5s_x^2} = \frac{1/5(-35.5)}{1/6(332.3)} = -0.128$$

The correlation coefficients given in Fig. 7 can be plotted as a function of the time span τ. Such a plot is called the autocorrelation function (Fig. 8) or autocorrelogram. Usually, an exponentially decreasing autocorrelation function is found.

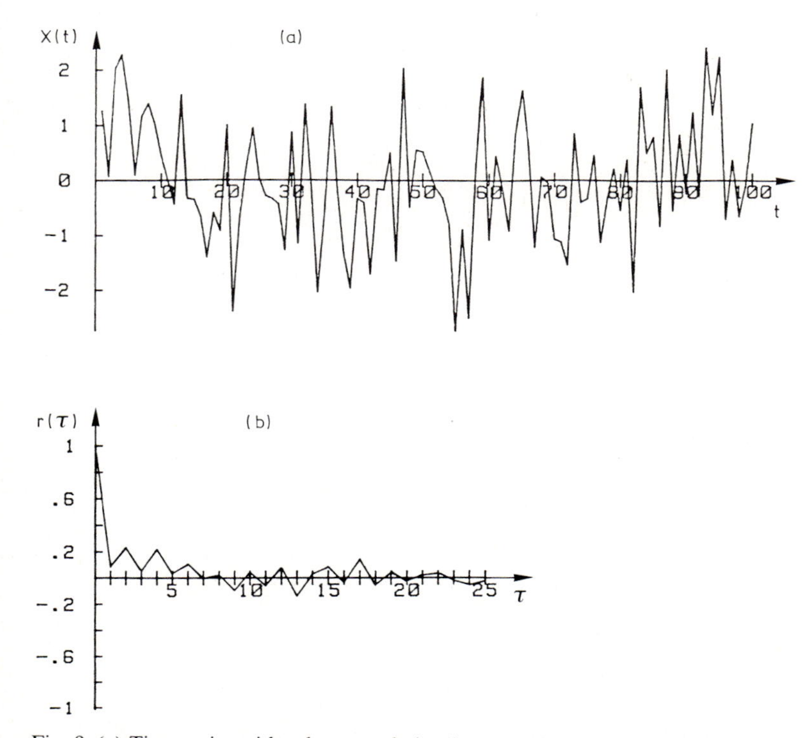

Fig. 9. (a) Time series with a low correlation between the observations. (b) Autocorrelation function of (a).

References p. 212

For computer implementation of eqn. (5), it is advisable to use a modified equation which does not require the prior calculation of the mean value.

$$r(\tau) = \frac{\sum_{t=1}^{n-\tau} x(t) \cdot x(t+\tau) - \sum_{t=1}^{n-\tau} x(t) \sum_{t=1}^{n-\tau} x(t+\tau)/(n-\tau-1)}{(n-1-\tau)s_x} \tag{6}$$

5.2 Autocorrelograms of some characteristic processes

(a) Random series with low and high correlation between the observations.

The autocorrelation functions of the processes in Figs. 9(a) and 10(b) can be approximated by a decreasing exponential function. This function is defined by one parameter, the so-called time constant, T_x, of the time series (Fig. 11). The autocorrelation function then becomes

$$r(\tau) = \exp(-\tau/T_x) \tag{7}$$

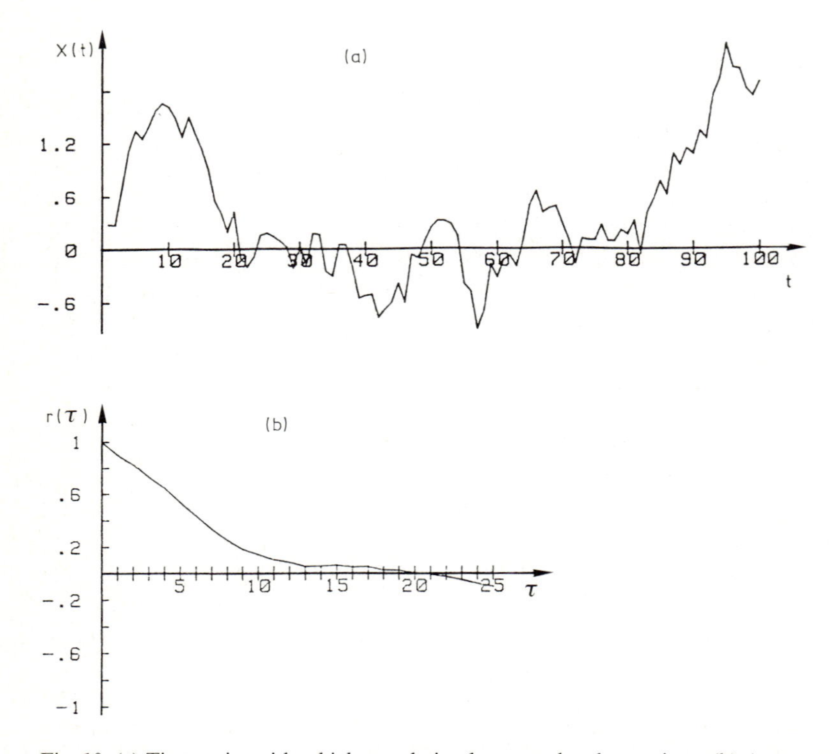

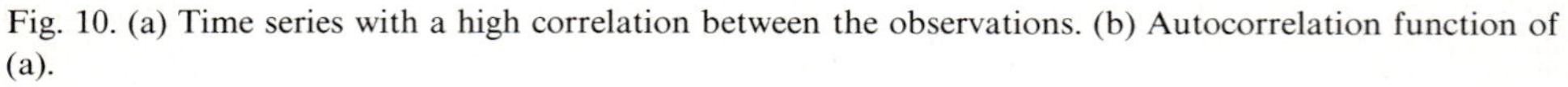
Fig. 10. (a) Time series with a high correlation between the observations. (b) Autocorrelation function of (a).

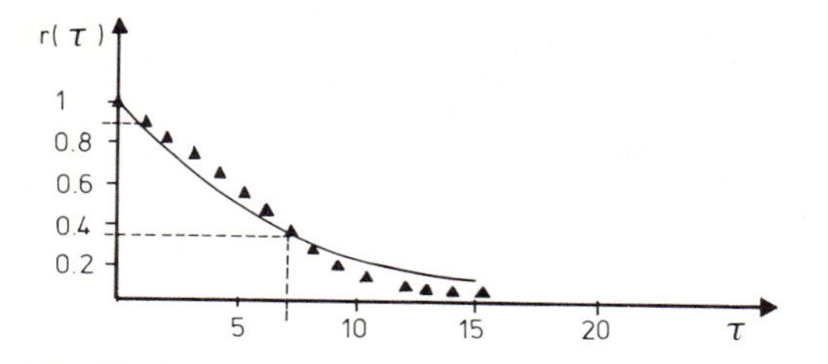

Fig. 11. (a) Autocorrelation of a first-order random process ($T_x = 7$). (b) Best fit of (a) with $\exp(-\tau/T_x)$.

The value of the time constant of a process can be derived from the autocorrelation function in three ways.

(i) from $r(1)$: for $\tau = 1$, eqn. (7) becomes

$$r(1) = \exp(-1/T_x)$$

Thus

$$T_x = -1/\ln[r(1)]$$

(ii) For $\tau = T_x$

$$r(T_x) = \exp(-1) = 0.37$$

The value, $\tau = T_x$, can be found from a plot of the autocorrelation function by determining the τ-value for which $r = 0.37$ (Fig. 11).

(iii) From a fit of the function $\exp(-\tau/T_x)$ through the autocorrelation data by non-linear regression (see Sect. 13.2.2).

In Chap. 11, the time constant of a process was obtained by monitoring the response of the process to a well-defined perturbation (a step or an impulse): one has to disturb the process in order to derive its dynamic characteristics. By an autocorrelation analysis, one determines the time constant from the analysis of the process values themselves. In the case of determining the time constant of a continuous analyzer, one calculates the autocorrelation function of the baseline or by a sudden increase of the input (concentration).

(b) Random series with drift.

The shape of the autocorrelation function can be interpreted in terms of deviations from the stationary behaviour of the time series, for instance the presence of drift in the signal. Another deviation from the stationary behaviour of the time series (periodicity) is discussed under (c). Consequently, autocorrelation analysis is frequently applied to decompose a series into its stochastic and deterministic parts, which are sometimes hard to detect in the plot of the original data.

The steep drop of the autocorrelation function from $\tau = 0$ to $\tau = 1$ in the autocorrelogram [Fig. 12(b)] of the time series in Fig. 12(a) indicates that the time constant of this series is very small (a few times Δt). At higher τ, one would expect the autocorrelation function to fluctuate about the τ axis. However, the autocorrela-

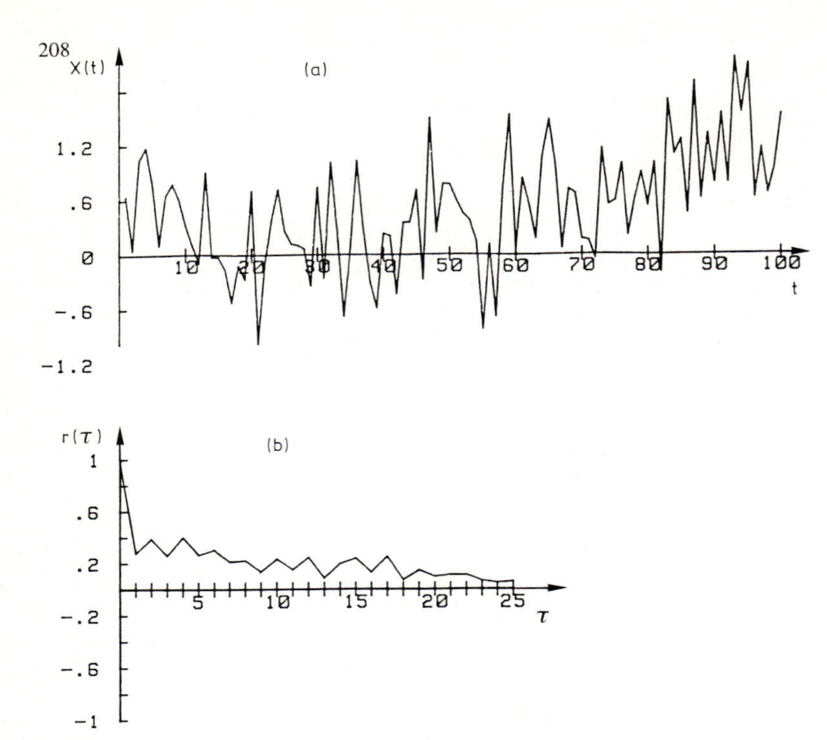

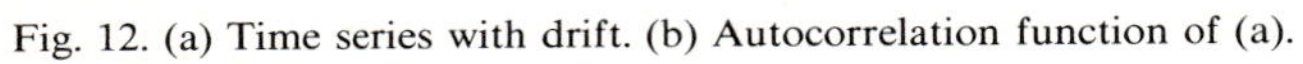

Fig. 12. (a) Time series with drift. (b) Autocorrelation function of (a).

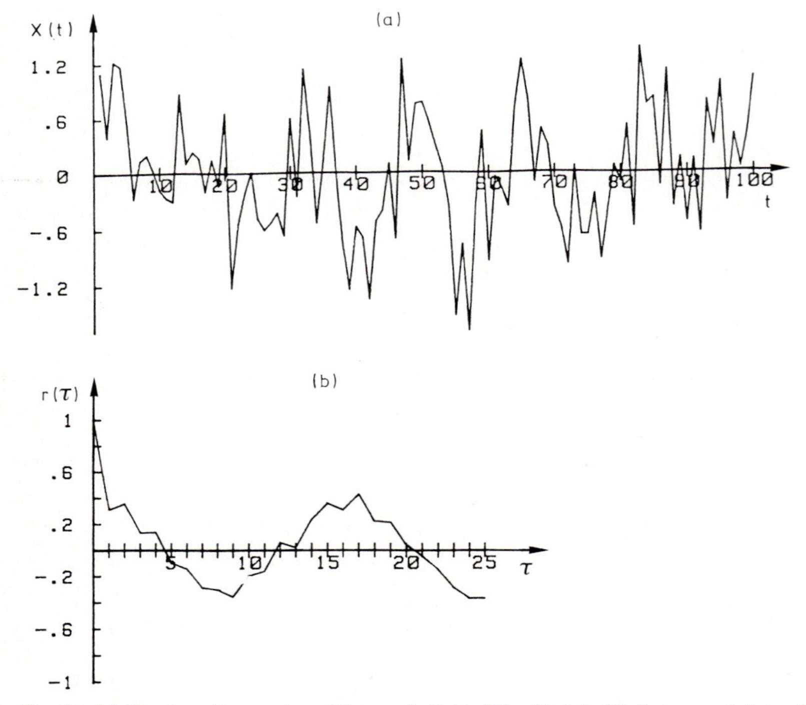

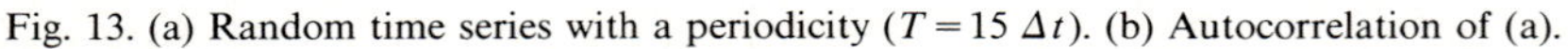

Fig. 13. (a) Random time series with a periodicity ($T = 15\ \Delta t$). (b) Autocorrelation of (a).

tion values all remain positive. This unexpected phenomenon is introduced by a correlation caused by the presence of drift in the series.

(c) Random series with periodicity.

The detection of a periodicity in a time series by an autocorrelation analysis is demonstrated in Fig. 13. Although the observations hardly indicate that the time series exhibits a periodicity, the autocorrelogram, on the contrary, is clearly positive on that point. Autocorrelation analysis enables one to determine the period, T, of the fluctuations. In the example shown in Fig. 13, one can read that the period of the fluctuations is equal to 15 sampling intervals.

5.3 Practical examples

(1) Surveillance of water quality.

Autocorrelation analysis has been applied for the description of concentration fluctuations in the River Rhine with a view to determining an optimal sampling

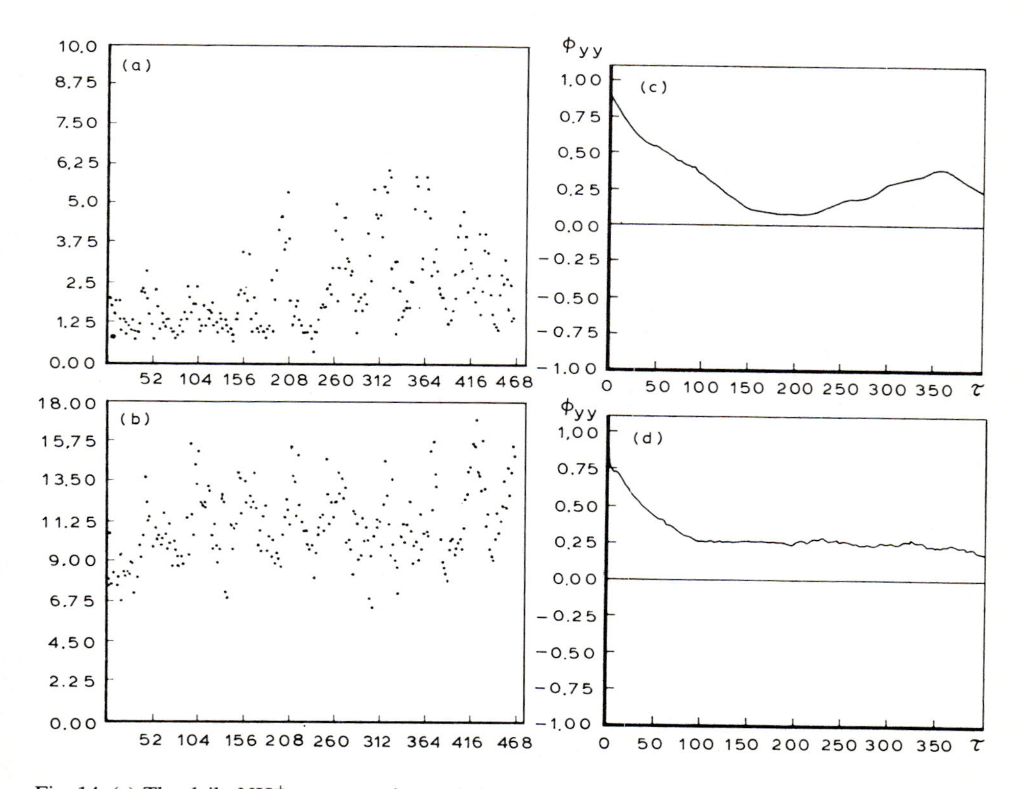

Fig. 14. (a) The daily NH_4^+ concentration and (b) the daily NO_3^- concentration in the River Rhine for the period 1971–1975. (Reprinted from ref. 4.) (c) The autocorrelation function of (a). (d) The autocorrelation function of (b).

strategy for the surveillance of water quality. The autocorrelation functions have been calculated for the concentrations of various constituents (NH_4^+, NO_3^-, NO_2^-, Cl^-, etc) monitored over a period of 5 years. The autocorrelation analysis revealed the characteristics of the underlying "stochastic" and "deterministic" compounds in the fluctuations. As an example, we display in Fig. 14 the autocorrelograms of the NH_4^+ and NO_3^- concentrations in the river Rhine at Bimmen during the period 1971–1975 [4]. Visual inspection leads to the conclusion that both autocorrelograms [Fig. 14(c) and (d)] show a rapid drop of the autocorrelation at $\tau = 1$ day, a positive autocorrelation at high τ values, and a time constant of the stochastic fluctuations of about 100 days. Furthermore, the autocorrelogram of the NH_4^+ concentration exhibits some periodicity. The steep drop at $\tau = 1$ can be ascribed to the noise introduced by experimental error of the analytical determination. The positive values at high τ indicate a trend (the concentration increases over the years). The autocorrelogram for NH_4^+ indicates a periodicity with $T = 365$ days (1 year!) [4]. Examples of correlograms of several processes are also found in ref. 5.

(2) Characterization of detector noise and drift.

Figure 15 shows the autocorrelation function of flame-ionization detector noise. The detector shows no drift because the autocorrelation function stabilizes about zero correlation and the time constant of the noise is smaller than 0.05 s.

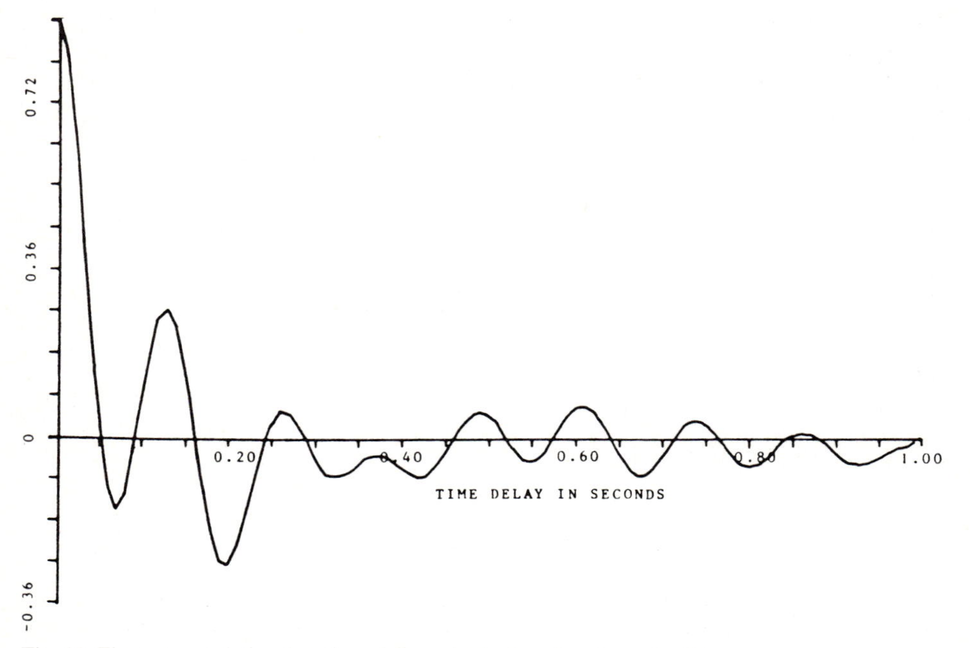

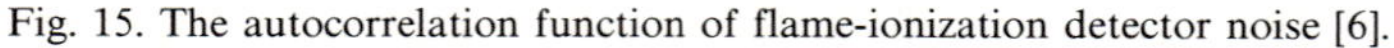

Fig. 15. The autocorrelation function of flame-ionization detector noise [6].

5.4 Autoregression

When discussing Fig. 7, we mentioned that the regression coefficients of the lines fitted through the data are related to the correlation coefficients. In other words, in some way the autocorrelation coefficients are an indication of the linear relationship between $x(t)$ and $x(t+\tau)$ in a time series. Equation (4) relates regression to correlation. This relationship also holds for data in a time series where the identities

$$y_i \equiv x(t+\tau);\; x_i \equiv x(t)$$

$$\bar{y} = \bar{x};\; s_y \equiv s_x;\; r(x, y) = r(\tau)$$

can be introduced. This gives

$$x(t+\tau) - \bar{x} = r(\tau) \cdot [x(t) - \bar{x}] + e(t+\tau) \tag{8}$$

Equation (8) is an important tool for deducting optimal sampling schemes for process control. Indeed, if one knows the autocorrelation coefficient of a process for a time lag τ, then eqn. (8) enables one to forecast quantitatively the system state at a time τ ahead of the measurement at time t (Fig. 16).

This is of importance in estimating the true value from the sample value in process control. Indeed, before any action can be undertaken on a process, information about the sample needs to be translated into information about the sampled system. The sample, however, represents the state of the system as it was at the time of taking the sample. It is clear that when the time span between sampling and availability of the analytical result is too long, the analysis does not give information about the actual state of the system at the time action can be taken. Elaboration of

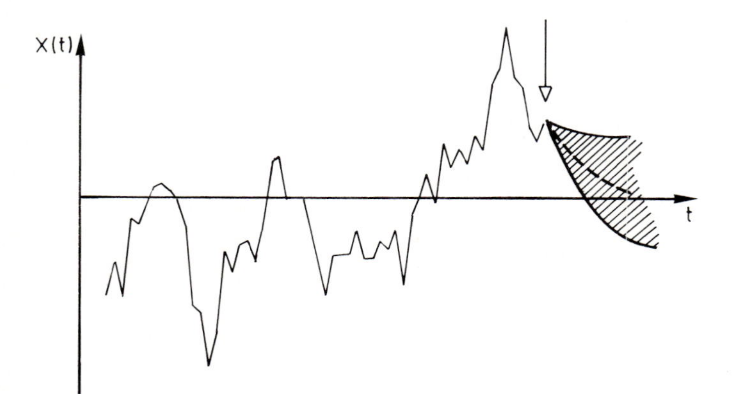

Fig. 16. The autocorrelation function as a predictor for future observations of a first-order time series. ———, Past observations; - - -, exponential prediction function, $\exp(-t/T_x)$. The shaded area is the confidence interval of the predictor.

eqn. (8) for various values of τ gives the relations

$$x(t+1)-\bar{x}=r(1)[x(t)-\bar{x}]+e(t+1)$$

$$x(t+2)-\bar{x}=r(2)[x(t)-\bar{x}]+e(t+2)$$

$$x(t+3)-\bar{x}=r(3)[x(t)-\bar{x}]+e(t+3)$$

$$x(t+\tau)-\bar{x}=r(\tau)[x(t)-\bar{x}]+e(t+\tau)$$

As $r(1)\ldots r(\tau)$ represent the values of the autocorrelation function for $\tau=$ 1, 2, 3..., it follows that the autocorrelation function is a predictor for process states beyond the last observation. One derives that systems with small time constants need to be sampled more frequently than systems with larger time constants in order to obtain the same quality of prediction and that the analysis time itself should be shorter than the time constant of the sampled process, provided one wants to know something about the process between the sampling actions.

The total delay between the sampling action and the availability of the analytical result is usually longer than the analysis time. Delays in the laboratory hold up the analytical result and are influenced by the laboratory organization. The characteristics of the analytical method, sampling strategy, and laboratory organization together define the ability to control the sampled process. These aspects will be discussed in more detail in Chap. 27.

References

1 E.J. Knudsen, D.L. Duewer, G.D. Christian and T.V. Larson, Application of factor analysis to the study of rain chemistry in the Puget Sound, in B.R. Kowalski (Ed.), Chemometrics, Theory and Application, ACS Symposium Series 52, American Chemical Society, Washington, DC, 1977.
2 H. De Jonge, Inleiding tot de Medische Statistiek, Vol. II, Walters Noordhoff, Groningen, 1964.
3 Kwaliteitsonderzoek in de Rijkswateren, Quaterly reports 1982, RIZA, The Netherlands.
4 P.J.W.M. Muskens and W.G.J. Hensegens, Water Res., 11 (1977) 509.
5 G.E.P. Box and G.M. Jenkins, Time Series Analysis (Forecasting and Control), Holden-Day, San Francisco, 1976.
6 H.C. Smit and H.L. Walg, Baseline noise and detection limits in signal-integrating analytical methods, applications to chromatography, Chromatographia, 8 (1975) 311.

Recommended reading

R.G. Brown, Smoothing, Forecasting and Prediction of Discrete Time Series, Prentice Hall, New York, 1962.
K.R. Betty and G. Horlick, Autocorrelation analysis of noisy periodic signals, utilizing a serial analog memory, Anal. Chem., 48 (1976) 1899.
R.B. Lam, D.T. Sparks and T.L. Isenhour, Cross correlation signal/noise enhancement with applications to quantitative gas chromatography/Fourier transform infrared spectrometry, Anal. Chem., 54 (1982) 1927.
C. Liteanu and E. Hopirtean, Statistical approach to the stability of an analytical system based on ion-sensitive membrane electrode using serial correlation Z. Anal. Chem., 288 (1977) 59.
D.E. Sands, Correlation and covariance, J. Chem. Educ., 54 (1977) 90.

P.J.W.M. Muskens and G. Kateman, Sampling of internally correlated lots. The reproducibility of gross samples as a function of sample size, lot size and number of samples. Part I: Theory, Anal. Chim. Acta, 103 (1978) 1.

G. Kateman and P.J.W.M. Muskens, Sampling of internally correlated lots. The reproducibility of gross samples as a function of sample size, lot size and number of samples. Part II: Implications for practical sampling and analysis, Anal. Chim. Acta, 103 (1978) 11.

P.J.W.M. Muskens, The use of autocorrelation techniques for selecting optimal sampling frequency. Application to surveillance of surface water quality, Anal. Chim. Acta, 103 (1978) 445.

Chapter 15

Signal Processing

Analytical signals are measured in many types of dimensions or domains. Typical domains are the time domain (chromatography), the wavelength domain (spectrometry), and the spatial domain (surface analysis). In many analytical applications, data are processed in the domain in which they have been measured. An example is a spectrum where operations such as the calculation of the peak positions or the resolution of overlapping peaks are carried out in the original domain of the data: the wavelength domain. In some situations, however, specific information can be acquired and some operations can be carried out in an easier way by switching to another domain.

Consider the idealized case in which the analytical response varies with time according to a sum of two sine functions [Fig. 1(a)] characterized by their amplitudes, A, and periods, T_1 and T_2. Obviously, the signal contains two frequencies. By switching from the time domain to the frequency domain, i.e. by measuring the signal as a function of frequency instead of time, the signal becomes two single pulses at the frequencies $u_1 = 1/T_1$ and $u_2 = 1/T_2$ [Fig. 1(b)]. The heights of the pulses correspond to the amplitudes of the two sine functions.

In general, one can assume every signal as being composed of a sum of sine and cosine functions, each with a specific frequency ($1/T$) and amplitude. As a result, pulses will be found in the frequency domain at all frequencies coinciding with the periods of the underlying sine or cosine functions present in the signal. Only those periods T are considered which fit a number of times in the measurement time of the signal.

From this simple example of a transform, the preliminary conclusion is that both domains yield specific information.

(1) The time domain gives information about the overall amplitude but no direct information about the frequencies present in the signal.

(2) The frequency domain gives information about the frequencies and their amplitude present in the signal but the information about the overall amplitude is no longer visible.

An operation in which the data in one domain are transformed to another, e.g. time to frequency, is called a transformation or transform. Transformations can be done in two ways, mathematically or by using appropriate hardware. For example, the Fourier transform of an IR spectrum (FTIR) can be measured directly in the frequency domain or may be obtained by applying a mathematical operation on the spectrum recorded in the wavelength domain.

Data transforms are not limited to "time to frequency" (and vice versa) transformations nor are they restricted to one-dimensional data, but data matrices may be

References p. 252

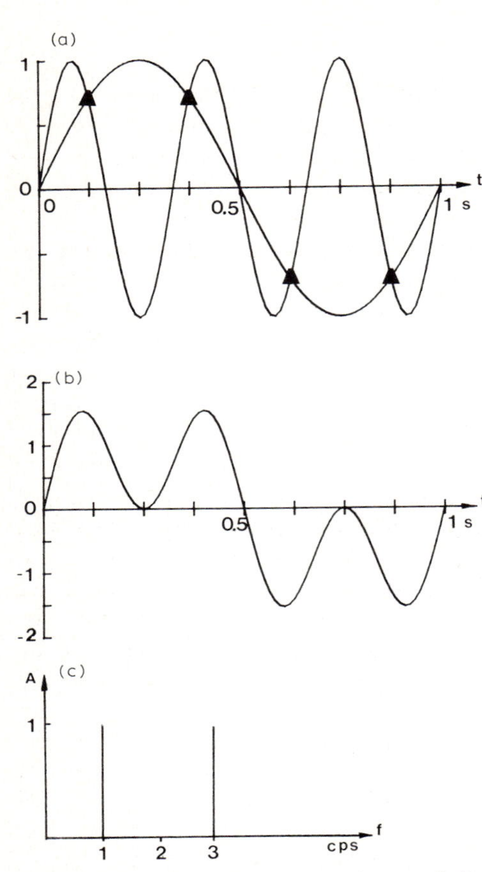

Fig. 1. A signal consisting of a sum of two periodic functions with $T = 1$ s and $T = 1/3$ s (the time axis is considered to be infinite). (a) The two sine functions in the time domain. (b) The sum of the two sine functions. (c) Representation of (b) in the frequency domain. A is the amplitude of the sine functions.

transformed as well. This is of importance when multidimensional data are measured in analytical chemistry, e.g. in surface analysis where the signal is measured as a function of the position on the surface.

Transformations are widespread in signal and data processing for two main reasons: information may be obtained that is hard to assess in the original domain and processing of the data in the other domain may allow some very specific operations on the signal (such as signal-to-noise enhancement). In analytical chemistry, data processing is very often carried out in the sequence: transform to another domain—processing—reverse transform to the original domain.

In this section, a survey will be given of some applications of the Fourier transform (FT) in analytical chemistry.

1. The Fourier transform

Modern signal processing in analytical chemistry is usually performed by computer. Therefore uniformly spaced samples are taken from the continuous signal. A general characteristic is that the data sequence is measured over a finite interval of time or other variable. For example, in gas chromatography, the time range is $(0, t)$, in IR spectrometry the wavenumber range is (4000–650 cm^{-1}).

A continuous signal, $g(x)$, which is discretized in n intervals, is transformed in a function $f(x)$, which can be represented as

$$f(x) = g(x_0 + x\,\Delta x) \qquad \text{for } x = 0, 1, \ldots, n \tag{1}$$

for $x = 0 \quad f(0) = g(x_0)$

for $x = 1 \quad f(1) = g(x_0 + \Delta x)$

for $x = 2 \quad f(2) = g(x_0 + 2\Delta x)$

Such a discretization procedure is schematically shown in Fig. 2(a) and (b).

Since analysts are mainly concerned with the processing of discretized data, attention will be focussed on the Fourier transform of discrete functions of the form of eqn. (1). The continuous formulation of the Fourier transform can be found in many textbooks (see suggested reading).

The discrete Fourier transform of the sequence $f(0), f(1), \ldots, f(n-1)$ of n uniformly spaced samples, taken from a signal measured over a time $(n-1)\,\Delta x$, is defined as

$$F(u) = \frac{1}{n} \sum_{x=0}^{n-1} f(x) \cdot \exp(-j2\pi ux/n) \tag{2}$$

with $j = \sqrt{-1}$ where the values $u = 0, 1, 2, \ldots, n-1$ are discrete samples of the

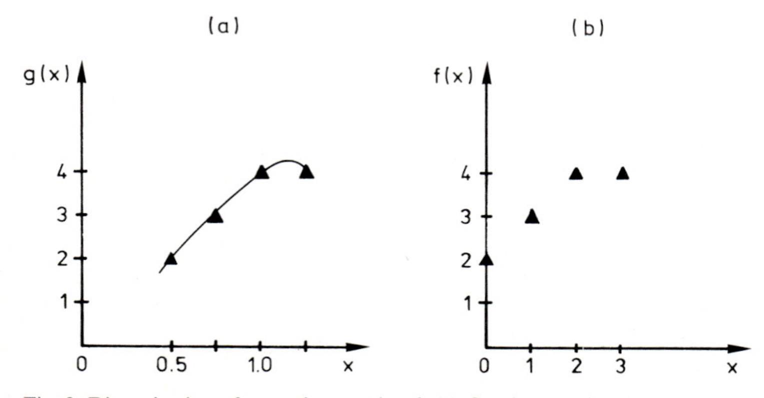

Fig. 2. Discretization of a continuous signal. (a) Continuous signal ($x_0 = 0.5$). (b) Discrete version of (a) ($\Delta x = 0.25$).

References p. 252

continuous transform at values 0, Δu, $2\Delta u, \ldots, (n-1)\Delta u$. A Fourier transform returns a complex number consisting of a real and an imaginary part. The term $f(x) \cdot \exp(-2\pi jux/n)$ in eqn. (2) can also be represented by $f(x)[\cos(2\pi ux/n) - j\sin(2\pi ux/n)]$. The magnitude of a complex number is defined as

$$|F(u)| = \left[R^2(u) + I^2(u)\right]^{1/2}$$

where $R(u)$ and $I(u)$ are the real and imaginary parts of $F(u)$, respectively.

Equation (2) can be written in the more general form

$$F(u) = \sum_{x=0}^{n-1} f(x) \cdot K(u, x) \tag{3}$$

where $K(u, x)$ represents the transform applied on $f(x)$. $K(u, x)$ is called a transform kernel, which equals $\exp(-j2\pi ux/n)$ for the Fourier transform. Various other kernels have been defined, leading to, for example, the so-called Walsh and Hadamard transforms. All these transforms are based on the same general expression, eqn. (3), with a specific definition of the kernel $K(u, x)$ in order to allow some specific operations.

From eqn. (2), it is seen that n discrete data in the time domain are transformed in n discrete data in the Fourier domain. The frequency interval in the Fourier domain, Δu, is related to the measurement time of the data in the time domain and equals $1/(n\Delta x)$ with $n\Delta x$ equal to the total measuring time. As an illustration of eqn. (2), the Fourier transform of the signal shown in Fig. 2(b) is calculated. The first Fourier term ($u = 0$) is always real and represents the component in the signal with a zero frequency. By analogy with the difference between an a.c. and a d.c. voltage in electronics, this component with zero frequency is often called the "DC term". It is equal to the mean value of the data. Thus, for signals with zero mean, the DC term is zero. The other terms in eqn. (2) represent the amplitudes of the sine and cosine functions with a frequency equal to $u/(n\Delta x)$. The imaginary part of the transform represents the sine function, the real part the cosine function. For the example of Fig. 2(b), it is calculated that

$$\begin{aligned}
F(0) &= \tfrac{1}{4} \sum_{x=0}^{3} f(x) \exp(-j2\pi 0x/n) \\
&= \tfrac{1}{4} \sum_{x=0}^{3} f(x) \exp(0) \\
&= \tfrac{1}{4} \sum_{x=0}^{3} f(x) \\
&= \tfrac{1}{4}[f(0) + f(1) + f(2) + f(3)] \\
&= \tfrac{1}{4}(2 + 3 + 4 + 4) \\
&= 3.25
\end{aligned}$$

The other terms are given by

$$F(1) = \tfrac{1}{4} \sum_{x=0}^{3} f(x) \exp(-j2\pi x/4)$$

$$= \tfrac{1}{4} \sum_{x=0}^{3} f(x) \exp(-j\pi x/2)$$

$$= \tfrac{1}{4}(2e^{0} + 3e^{-j\pi/2} + 4e^{-j\pi 2/2} + 4e^{-j\pi 3/2})$$

$$= \tfrac{1}{4}(-2 + j)$$

The real part is thus $-1/2$ and the imaginary part $j/4$.

$$F(2) = \tfrac{1}{4} \sum_{x=0}^{3} f(x) \exp(-j4\pi x/4)$$

$$= \tfrac{1}{4} \sum_{x=0}^{3} f(x) \exp(-j\pi x)$$

$$= \tfrac{1}{4}(2e^{0} + 3e^{-j4\pi/4} + 4e^{-j4\pi 2/4} + 4e^{-j4\pi 3/4})$$

$$= -\tfrac{1}{4}(1 + j \cdot 0)$$

$$F(3) = \tfrac{1}{4} \sum_{x=0}^{3} f(x) \exp(-j6\pi x/4)$$

$$= \tfrac{1}{4} \sum_{x=0}^{3} f(x) \exp(-j3\pi x/2)$$

$$= \tfrac{1}{4}(2e^{0} + 3e^{-j3\pi/2} + 4e^{-j3\pi 2/2} + 4e^{-j3\pi 3/2})$$

$$= -\tfrac{1}{4}(2 + j)$$

The magnitude of $F(u)$, $|F(u)|$, as a function of u is called the Fourier spectrum of $f(x)$. The square of the spectrum

$$|F(u)|^2 = R^2(u) + I^2(u)$$

is usually called the energy, amplitude, or power spectrum of $f(x)$. These quantities are often used instead of working directly with the real and imaginary components. Referring to the example, the Fourier and power spectra are

$$|F(0)| = 3.25 \qquad |F(0)|^2 = 10.56$$

$$|F(1)| = \left[(2/4)^2 + (1/4)^2\right]^{1/2} = 0.559 \qquad |F(1)|^2 = 0.312$$

$$|F(2)| = \left[(1/4)^2 + (0/4)^2\right]^{1/2} = 0.25 \qquad |F(2)|^2 = 0.062$$

$$|F(3)| = \left[(2/4)^2 + (1/4)^2\right]^{1/2} = 0.559 \qquad |F(3)|^2 = 0.312$$

As $\Delta u = 1/(4\,\Delta x) = 1/(4 \times 0.25) = 1$ cps, the frequencies corresponding with these 4 points, $u = 0, 1, 2, 3$ are 0, 1, 2 and 3 cps (cycles per second).

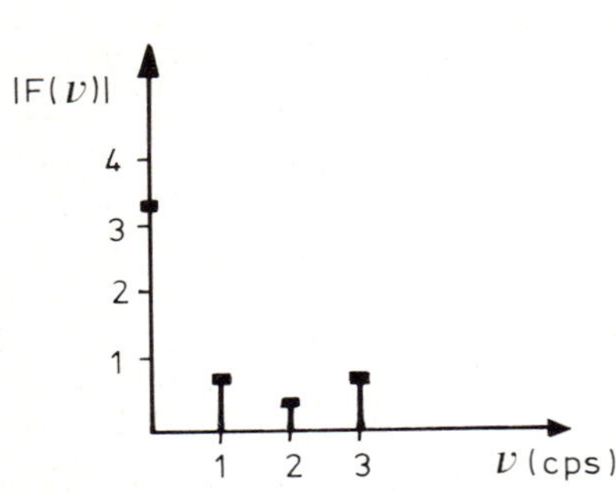

Fig. 3. The Fourier spectrum of the discrete signal in Fig. 2.

In Fig. 1(a), a sine function is plotted with a frequency of 3 cps, corresponding to F(3) in the example, which is sampled every 0.25 s ($=\Delta x$). It is seen that the same points also fit a sine function with a frequency equal to 1 cps. This phenomenon is called "folding". The highest frequency, f_{max}, which cannot be folded for $\Delta x = 0.25$ is 2 cps, which is equal to $1/(2\Delta x)$ and is called the Nyquist frequency. This frequency is found at the data point $u = n/2$. Consequently the maximal frequency (f_{max}) observed in the data of the example is not 3 cps but 2 cps. The points F(3) and F(1) represent the same frequency in the data, as the arguments of their exponents differ by 2π. When one wishes to observe in the signal frequencies higher than $f = 1/(2\Delta x)$ when digitizing, one should increase the sampling frequency (smaller Δx). It is noted here that Δx for continuous signals which are not discretized equals zero and therefore f_{max} is infinite. This means that the amplitudes of all frequencies from zero to infinity can be calculated in steps of 1/(measuring time).

Summarizing, the Fourier transform of a continuous signal measured during an infinite time is also continuous. If the measuring time is equal to T, the Fourier transform becomes discrete (interval $= 1/T$), but the highest observable frequency is unlimited. If a discrete signal is measured during a certain time, the highest observable frequency is limited to the Nyquist frequency.

1.1 The fast Fourier transform (FFT)

The way the FT was calculated in the example is quite satisfactory as long as n remains relatively small. Examination of eqn. (2) reveals that for each of the n values of u, $n-1$ additions are required of n multiplications of complex numbers. The number of complex multiplications and additions, therefore, is proportional to n^2. Even for the fastest computer system, this operation involves a large amount of calculation time. For that reason, so-called fast Fourier transform (FFT) algorithms have been developed. They are available in many software packages (see Borman's review of commercial software, details of which are given at the end of this chapter). The number of operations in FFT is proportional to $n \log_2(n)$ so that the FFT algorithm permits considerable saving of computer time. For example, the direct implementation of the Fourier transform for $n = 1024$ requires 10^6 operations, while the FFT algorithm needs only 10^4 operations (100 times faster). The only condition for applying the algorithm is that the number of points should be equal to 2^p, where p is a positive integer. The principles of the FFT algorithm can be found in many textbooks (see the references at the end of this chapter).

1.2 The inverse Fourier transform

As pointed out in previous sections, a Fourier transform is usually followed by other operations in the frequency domain, after which one usually returns to the original domain by an inverse FT operation. Consider a full sequence of a time-to-frequency-to-time domain operation on a data array of n equispaced data points. By the application of eqn. (2), the n-point discrete Fourier transform $F(u)$ of $f(x)$ is obtained, where $\Delta u = 1/(n\ \Delta x)$ and $u = 0, 1, 2, \ldots, n-1$. The reverse discrete Fourier transform, $f(x) = F^{-1}([F(u)]$, is defined as

$$f(x) = \sum_{u=0}^{n-1} F(u) \exp(j2\pi ux/n) \tag{4}$$

$F(u)$ is a complex number and therefore as every complex number $z = R + jI$, it has a corresponding complex conjugate $z^* = R - jI$. Taking the complex conjugate of both sides of eqn. (4) and dividing them by n gives

$$\frac{1}{n} f^*(x) = \frac{1}{n} \sum_{u=0}^{n-1} F^*(u) \exp(-j2\pi ux/n) \tag{5}$$

By comparing eqns. (5) and (2), one notices that eqn. (5) is analogous to that for the forward transform [eqn. (2)]. Thus, if we input $F^*(u)$ into an algorithm designed to calculate the forward transform, the result is $(1/n)f^*(x)$. Multiplication of that result by n and taking the complex conjugate yields the desired inverse $f(x)$. When $f(x)$ is expected to be real, it is sufficient to take the real part of the result without any further operation. The FFT algorithm can also be used for the calculation of the inverse Fourier transform provided the complex conjugates of the data are determined first.

Example. As an example, the results obtained in Sect. 1 are back-transformed to the time domain. The data are $F(0) = 3.25$, $F(1) = (-2 + j)/4$, $F(2) = -1/4$ and $F(3) = (-2 - j)/4$.
The complex conjugates are $F^*(0) = 3.25$, $F^*(1) = (-2 - j)/4$, $F^*(2) = -1/4$, and $F^*(3) = (-2 + j)/4$.
The forward Fourier transforms of $F^*(u)$ are

$$x = 0;\ \tfrac{1}{4} f^*(0) = \tfrac{1}{4} \sum_{u=0}^{3} F^*(u) \exp(-2\pi ju \cdot 0/4)$$

$$= \tfrac{1}{4}[F^*(0) + F^*(1) + F^*(2) + F^*(3)]$$

$$= \tfrac{1}{4}\left[3.25 + \tfrac{1}{4}(-2 - j) - \tfrac{1}{4} - \tfrac{1}{4}(2 - j)\right]$$

$$= \tfrac{1}{2}$$

References p. 252

$$x=1;\ \tfrac{1}{4}\mathrm{f}^*(1)=\tfrac{1}{4}\sum_{u=0}^{3}\mathrm{F}^*(u)\exp[-2\pi ju(1/4)]$$

$$=\tfrac{1}{4}\{\mathrm{F}^*(0)+\mathrm{F}^*(1)\exp[(-2\pi j(1/4)]$$
$$+\mathrm{F}^*(2)\exp[-2\pi j(2/4)]+\mathrm{F}^*(3)\exp[-2\pi j(3/4)]\}$$

$$=\tfrac{1}{4}\{3.25+\tfrac{1}{4}(-2-j)\exp(-\pi j/2)$$
$$-\tfrac{1}{4}\exp(-\pi j)-\tfrac{1}{4}(2-j)\exp(-3\pi j/2)\}$$

$$=\tfrac{1}{4}\{3.25+\tfrac{1}{4}(-2-j)(-j)-\tfrac{1}{4}(-1)-\tfrac{1}{4}(2-j)(j)\}$$

$$=\tfrac{3}{4}$$

$$x=2;\ \tfrac{1}{4}\mathrm{f}^*(2)=\tfrac{1}{4}\sum_{u=0}^{3}\mathrm{F}^*(u)\exp[-2\pi ju(2/4)]$$

$$=\tfrac{1}{4}\{\mathrm{F}^*(0)+\mathrm{F}^*(1)\exp[-2\pi j1(2/4)]$$
$$+\mathrm{F}^*(2)\exp[-2\pi j\cdot 2(2/4)]-\mathrm{F}^*(3)\exp[-2\pi j\cdot 3(2/4)]\}$$

$$=\tfrac{1}{4}\{3.25+\tfrac{1}{4}(-2-j)\exp(-\pi j)$$
$$-\tfrac{1}{4}\exp(-2\pi j)-\tfrac{1}{4}(2-j)\exp(-3\pi j)\}$$

$$=\tfrac{1}{4}\{3.25+\tfrac{1}{4}(-2-j)(-1)-\tfrac{1}{4}(1)-\tfrac{1}{4}(2-j)(-1)\}$$

$$=1$$

$$x=3;\ \tfrac{1}{4}\mathrm{f}^*(3)=\tfrac{1}{4}\sum_{u=0}^{3}\mathrm{F}^*(u)\exp[-2\pi ju(3/4)]$$

$$=\tfrac{1}{4}\{\mathrm{F}^*(0)+\mathrm{F}^*(1)\exp[(-2\pi j(3/4)]$$
$$+\mathrm{F}^*(2)\exp[-2\pi j\cdot 2(3/4)]-\mathrm{F}^*(3)\exp[-2\pi j\cdot 3(3/4)]\}$$

$$=\tfrac{1}{4}\{3.25+\tfrac{1}{4}(-2-j)\exp(-3\pi j/2)$$
$$-\tfrac{1}{4}\exp(-3\pi j)-\tfrac{1}{4}(2-j)\exp(-9\pi j/2)\}$$

$$=\tfrac{1}{4}\{3.25+\tfrac{1}{4}(-2-j)(0+j)-\tfrac{1}{4}(-1)-\tfrac{1}{4}(2-j)(0-j)\}$$

$$=1$$

Multiplication by $n=4$ and taking the complex conjugate gives f(0) = 2, f(1) = 3, f(2) = 4, f(3) = 4, which is exactly the original data set. Translation to $g(x)$ is carried out by calculating the discretizing interval $\Delta x=1/(n\,\Delta u)=1/(4)=0.25$. It should be noted that the value of x_0 cannot be retrieved from the Fourier transform.

1.3 Properties of the Fourier transform

1.3.1 Periodicity and symmetry

At the beginning of the discussion, we pointed out that the FT of an n-point $(0, n-1)$ discretized signal is defined for $u=0$ to $n-1$. Figure 4 displays the real and imaginary parts of the FT of an exponential function with $n=32$. As discussed in Sect. 1, the values of the transforms at $u=17$ to 31 represent the same frequencies as the points $u=15$ to 1, $u=0$ still represents the DC term and $u=16$ represents the maximal frequency, f_{max}, in the data. Thus, all information of the FT is present in the points $u=0$, $u=16$ and $u=1$ to 15 or, in general, in the interval

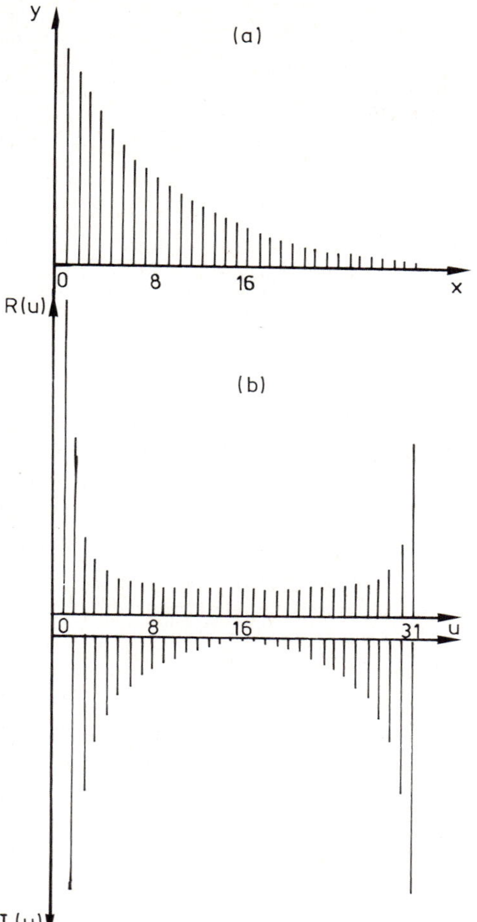

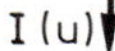

Fig. 4. (a) Exponential signal with $n = 32$ and (b) its FT. R(u) and I(u) are the real and imaginary parts, respectively.

(0, $n/2$), which contains the first $(n/2) + 1$ data points. Because this is true for any signal in the time domain, some computer programs return only the DC term plus $n/2$ complex points for an input array consisting of n real points. Such a partial FT must be symmetrically expanded before back transforming. Some computer algorithms do this automatically and others leave it to the user.

1.3.2 Distributivity

The distributivity property means that the Fourier transform of the sum of two functions equals the sum of the Fourier transform of the separate functions. Thus $F[f1(x) + f2(x)] = F[f1(x)] + F[f2(x)]$

References p. 252

Signal-to-noise enhancement (or filtering) in the Fourier domain is based on that property. If one assumes that the noise (n) is additive to the signal (s), the measured signal $m(x)$ is equal to $s(x) + n(x)$. Therefore

$$F[m(x)] = F[s(x)] + F[n(x)]$$

or

$$M(u) = S(u) + N(u)$$

Assuming that both Fourier spectra $S(u)$ and $N(u)$ contribute at specific frequencies, the true signal, $s(x)$, can be recovered from $M(u)$ after elimination of $N(u)$ (see Sect. 2.2).

1.3.3 Shift

When one wishes to display all n values of the FT with the DC term as a point of symmetry in the centre of the figure, one has to move the origin to the centre and change the frequency span which is originally from 0 to f_{max} and back to 0 into ($-f_{max}$ to $+f_{max}$), where $f_{max} = 1/(2\Delta x)$. This can be accomplished by multiplying $f(x)$, $x = 0, 1, \ldots, n-1$, by $(-1)^x$ before taking the transform. When back transforming Fourier data with the origin at the centre point, the results should thereafter be multiplied by the same factor $(-1)^x$ in order to obtain the right sign in the time domain.

The fact that the discretizing interval, Δu, is the inverse of the measurement time $(=1/n\,\Delta x)$ has consequences for the application of routines for a fast Fourier transform. Because they require the number of data points to be a power of 2, it follows that the spectrum in the time domain has to be extended (e.g. with zeros) or shortened to meet that requirement. This has consequences for the discretizing interval obtained, as this virtually expands or shortens the measurement time.

Consider the discrete signal shown in Fig. 2. If we simply add four zeros to the data, the following situation is obtained: Δx remains unchanged and so does the span of the frequency range, f_{max}. However, n now equals 8 instead of 4. As a consequence, $\Delta u = 1/n\Delta x$ has been halved! Thus, the frequency spectrum is discretized in twice as many points over the same frequency span. The effect is demonstrated on the Fourier transform of a step function (16 points) expanded with zeros to a total of, respectively, 64 and 256 data points and an unchanged data interval (Fig. 5). This demonstrates that one should be very careful in defining the frequency scale of the FT, after appending zeros at the end of the original data sequence (also called zero filling), in order to obtain a smaller discretizing interval in the frequency domain. Zero filling is useful to smooth the appearance of the FT and to reduce the error in the accuracy of estimating the positions of the spectral peaks.

1.3.4 Convolution

As a rule, a measurement is an imperfect reflection of reality. Noise and other blur sources degrade the signal. In the particular case of spectrometry, a major source of degradation is the peak broadening caused by the fact that the bandwidth

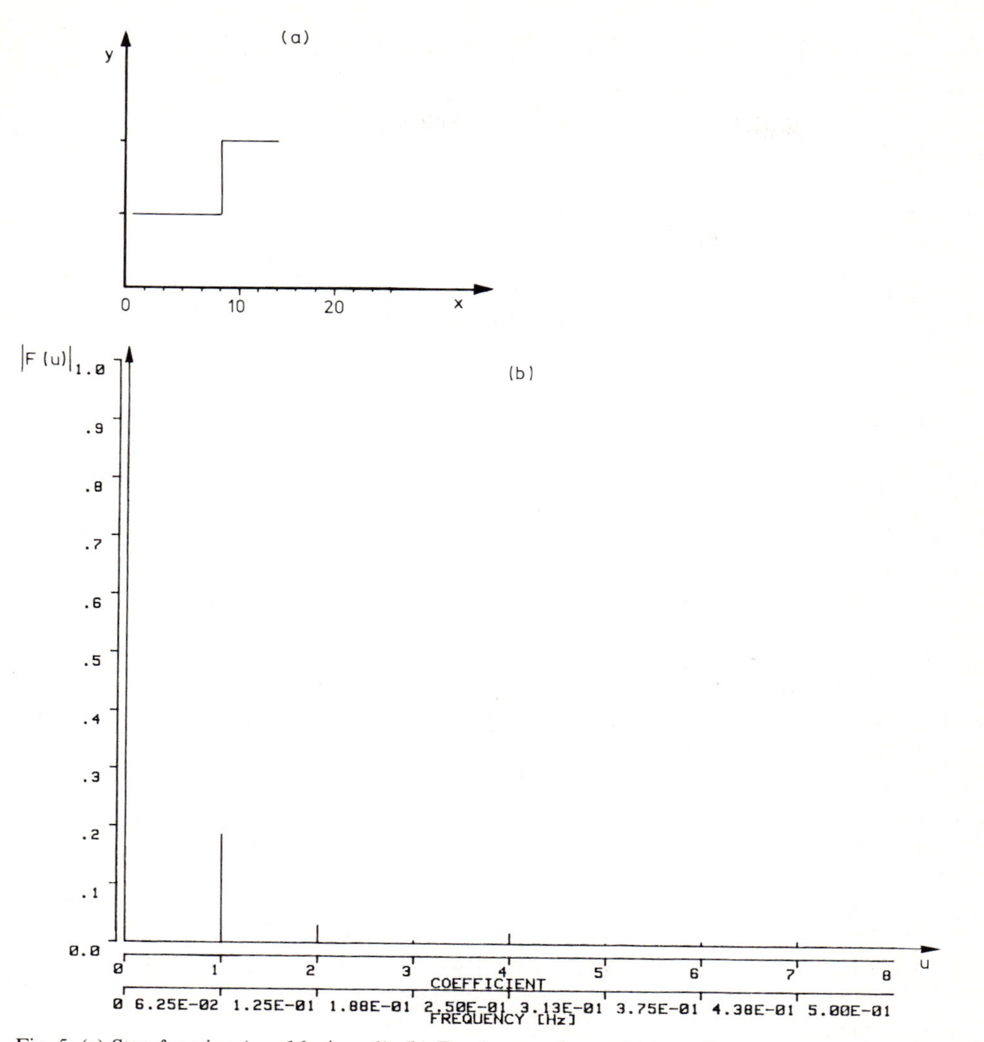

Fig. 5. (a) Step function ($n = 16$, $\Delta x = 1$). (b) Fourier transform of (a). (c) Fourier transform of (a) with 48 zeros added, $n = 64$. (d) Fourier transform of (a) with 240 zeros added, $n = 256$.

of a monochromator is limited. When a spectrophotometer is tuned on a wavelength, radiation with a given intensity profile, the so-called slit function, will pass the exit slit and reach the detector (Fig. 6). Under certain conditions, the shape of the slit function is a triangle, characterized by its half height width, $w_{1/2}$, called the spectral band width. When measuring a "true" absorbance peak with a half-height width not very much larger than the spectral band width, a disturbed peak shape will be observed, resulting in a poorer resolution for overlapping peaks. This is

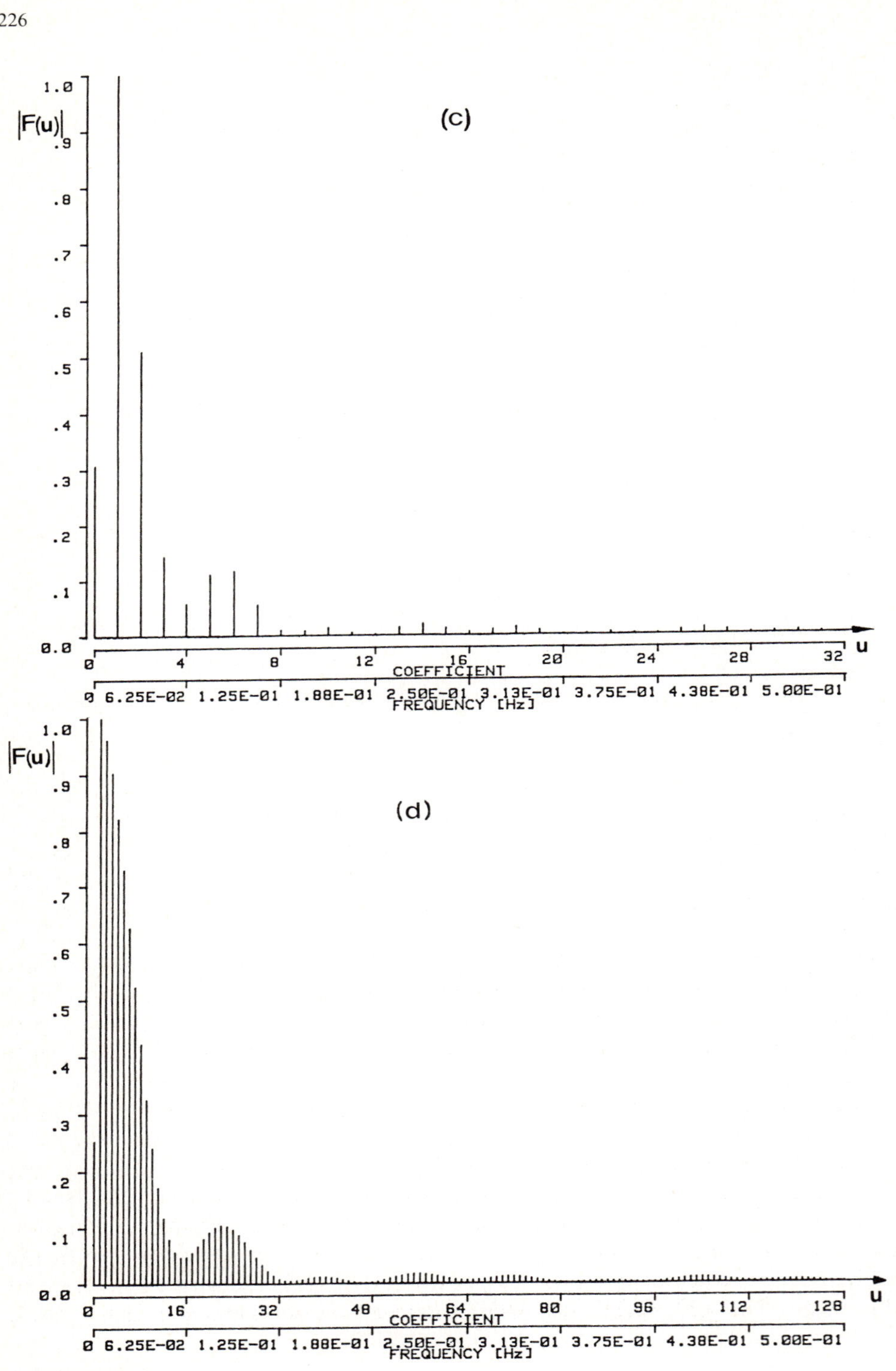

Fig. 5 (continued).

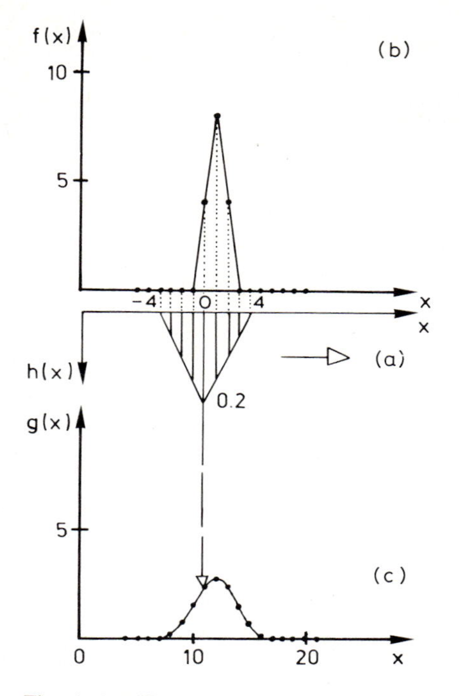

Fig. 6. (a) Slit function h(x), $w_{1/2} = 4\Delta x$. (b) Absorption peak f(x), $w_{1/2} = 2\Delta x$. (c) g(x) is the convolution of (b) with (a).

called convolution. Although, in principle, convolutions can be calculated in the wavelength or time domain, it is preferable to transform the data to the Fourier domain before calculation. The inverse operation, i.e. enhancing the resolution by eliminating the disturbances caused by the slit function (generally denoted as the point-spread function) is called deconvolution. It is only feasible in the Fourier domain.

Because the concept of convolution is rather difficult to understand mathematically, the operation will be clarified taking the example from spectrophotometry. For simplicity, we will consider the absorbance peak to be triangular. It is given by f(x) in Fig. 6(b) (x is the wavelength). The slit function of the spectrometer is also triangular [Fig. 6(a)] and the x axis is discretized at regular intervals. Let us calculate the observed absorbances in points along this axis by moving the slit function h(x) over the x axis.

At position 8, no radiation should normally be measured since f(8) = 0. However, some radiation may reach the detector since the slit function is not infinitely small: radiation from a distance less than 4 units away contributes to some extent. This means that radiation corresponding to $x = 11$ reaches the $x = 8$ position. How large the fraction is, is determined by the value of the slit function. It is equal to $f(11) \cdot h(11 - 8) = f(11) \cdot h(3) = 4 \times 0.05 = 0.2$ and the signal at location 8 is thus, g(8) = 0.2. At position 9, still no radiation should be measured but some radiation intended for positions 11 and 12 gets through and, of course, a larger fraction from 11 than from 12. From 11 it is f(11) · h(11 − 9) and from 12 f(12) · h(12 − 9) so that

$g(9) = f(11) \cdot h(2) + f(12) \cdot h(3) = 0.8$

Further, one computes that

$g(10) = f(10)h(0) + f(11)h(1) + f(12)h(2) = 1.6$

$g(11) = f(10)h(-1) + f(11)h(0) + f(12)h(1) + f(13)h(2) + f(14)h(3) = 2.4$

$g(12) = f(10)h(-2) + f(11)h(-1) + f(12)h(0) + f(13)h(1) + f(14)h(2) = 2.8$

$g(13) = f(10)h(-3) + f(11)h(-2) + f(12)h(-1) + f(13)h(0) + f(14)h(1) = 2.4$

$g(14) = f(10)h(-4) + f(11)h(-3) + f(12)h(-2) + f(13)h(-1) + f(14)h(0) = 1.6$

$g(15) = f(10)h(-5) + f(11)h(-4) + f(12)h(-3) + f(13)h(-2) + f(14)h(-1) = 0.8$

$g(16) = f(10)h(-6) + f(11)h(-5) + f(12)h(-4) + f(13)h(-3) + f(14)h(-2) = 0.2$

The plot of the convolution shown in Fig. 6(c) clearly demonstrates the signal broadening and resulting intensity drop. In general, one can state that, instead of measuring the true value $f[x(i)]$, one measures an average of the absorbance at values around $f[x(i)]$ weighted over (convoluted with) the slit function. The measured signal is given by

$$g[x] = \sum_{i=0}^{m} f(x)h[x(i) - x] \tag{6}$$

where m is at least equal to the number of points defining $h(x)$. Equation (6) is the discrete representation of the continuous convolution integral

$$\begin{aligned} g[x] &= f[x] * h[x] \\ &= \int_{-\infty}^{+\infty} f(x)h[x(i) - x]\, dx \end{aligned}$$

Examples of broadening processes in analytical chemistry are, for example, Doppler and collisional broadening in atomic spectral lines, finite resolving power of the monochromator, the asymmetry introduced by electronic filtering, structural effects in the specimen in X-ray diffraction, etc.

The convolution theorem states that the convolution in the time (or, here, wavelength) domain translates into a simple scalar product in the frequency domain and vice versa. This gives

$$g(x) = f(x) * h(x) \Leftrightarrow F(u) \cdot H(u) = G(u) \tag{7}$$

where $F(u)$ is the Fourier transform of $f(x)$ and $G(u)$ that of $g(x)$.

From this result, it follows that the convolution of the two triangles in our example could also have been calculated in the Fourier domain

$$\begin{matrix} f(x) \overset{\text{FT}}{\rightarrow} F(u) & \\ & \rightarrow F(u) \cdot H(u) \rightarrow G(u) \overset{\text{RFT}}{\rightarrow} g(x) \\ h(x) \overset{\text{FT}}{\rightarrow} H(u) & \end{matrix}$$

Example of convolution in the Fourier domain.

One can calculate the convolution of the triangle and the point-spread function in the Fourier domain. The FT of a triangle is displayed in Fig. 7(a) and that of the point-spread function in Fig. 7(b). Multiplication of the transforms gives the convolution of both functions [Fig. 7(c)]. Of course, one should pay careful attention to multiply values of the Fourier transform at corresponding frequencies. Therefore the digitizing interval, Δx, and the number of data, n, in both signals $f(x)$ and $h(x)$ should be equal. After carrying out a reverse transform, the convolution is obtained in the original wavelength domain [Fig. 7(d)].

One observes that, in comparison with the convolution in the wavelength domain, the data are now shifted by 4 data points to the right. This is one less than half the

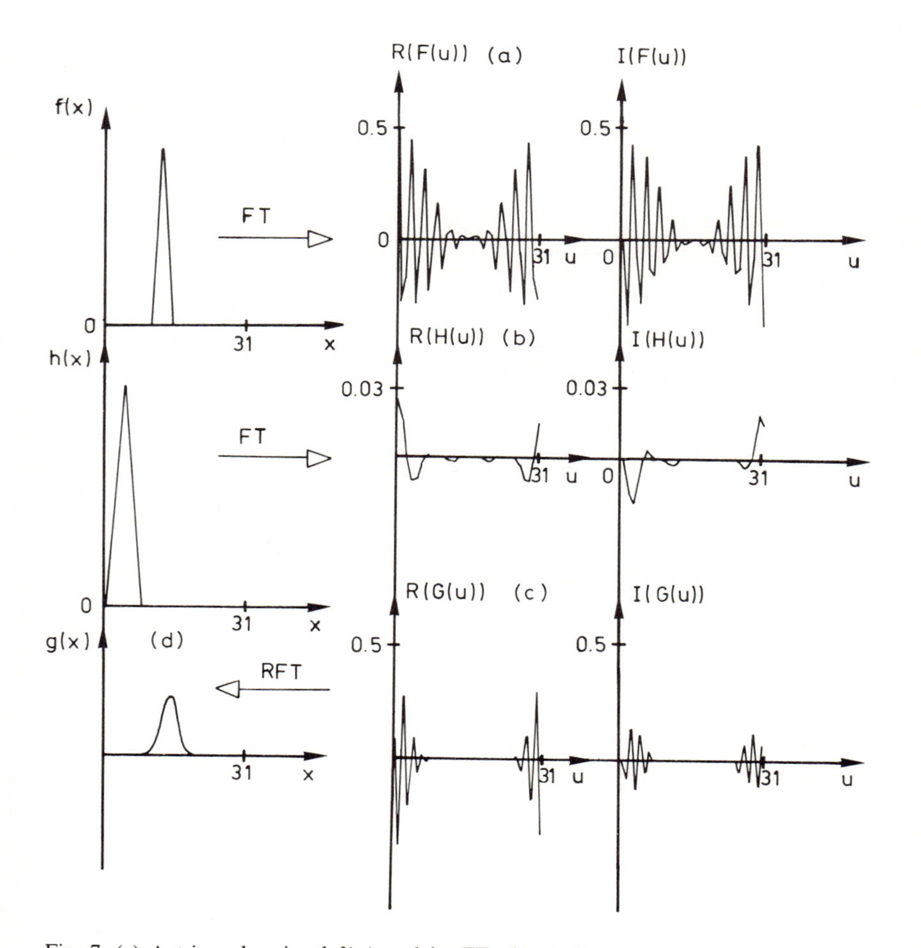

Fig. 7. (a) A triangular signal $f(x)$ and its FT. (b) A slit function $h(x)$ (triangular) and its FT. (c) Multiplication of the FT of (a) with that of (b). (d) The inverse FT of (c).

number of non-zero points in h(x). This is important to know in order to retrieve the convoluted data in the obtained 32 data points. Special precautions are necessary to avoid certain numerical errors in the calculations (wrap-around error). The discussion of the origin of these errors is beyond the scope of this text (see suggested reading). However, one should know that wrap-around errors (also called leakage errors) may be overcome to some extent by appending zeros in f(x) and h(x) before transformation. When f(x) and h(x) are discretized into sampled arrays of size A and B, respectively, both f(x) and h(x) should be extended with zeros to a size of at least $A + B$. Of course, if $(A + B) \neq 2^P$, more zeros should be appended in order to be able to use a fast Fourier transform.

2. Analytical applications

Fourier transformations are mostly applied in analytical chemistry for the digital processing of spectra and for the characterization of analytical signals. In spectrum processing, they are applied in order to reduce complex operations in the original (wavelength) domain to relatively easy operations in the other domain. This is shown schematically in Fig. 8. The application of Fourier transforms for the characterization of analytical signals is an example where characteristics are more easily detected in one domain than in the other.

Two frequently applied operations on spectra are filtering, i.e. maximizing of the signal-to-noise ratio, and deconvolution (resolution enhancement or signal restoration).

2.1 Deconvolution (signal restoration)

Restoration is the estimation of the undistorted signal (spectrum), f(x), from the observed signal, g(x), using the knowledge of the point-spread function, h(x). This inverse operation of convolution is called deconvolution. While in the time domain it is possible to calculate a convolution in a rather straightforward way, the calculation of the inverse operation is much more complicated.

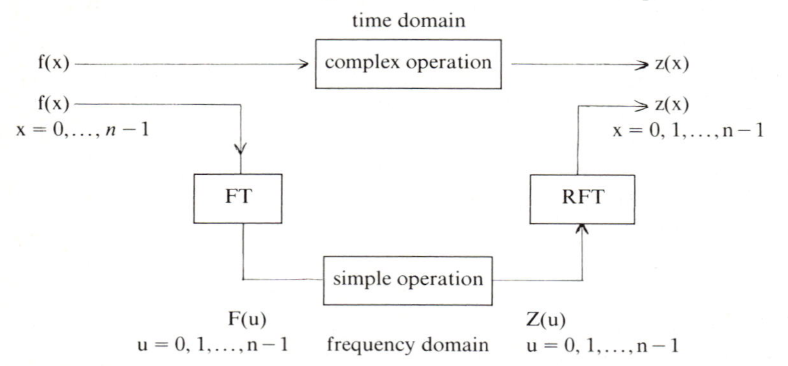

Fig. 8. Comparison of operations in the time and frequency domains.

From eqn. (7) it follows that

$$F(u) = \frac{G(u)}{H(u)}$$

whereafter the reverse transform of $F(u)$ gives $f(x)$. A deconvolution operation can be carried out according to the scheme

$$\begin{matrix} g(x) \overset{FT}{\rightarrow} G(u) \\ \\ h(x) \overset{FT}{\rightarrow} H(u) \end{matrix} \rightarrow G(u)/H(u) \rightarrow F(u) \overset{RFT}{\rightarrow} f(x)$$

This operation is also called inverse filtering. The procedure is more complex than one would think at first sight. When noise is present in the data, as is generally the case in analytical practice, the measured signal is a convolution of the "true" signal with a given function, $h(x)$, to which noise is added. Thus

$$g(x) = f(x) * h(x) + n(x)$$

Applying the property of additivity of Fourier transforms, the transform $G(u)$ of $g(x)$, is given by

$$G(u) = F(u)H(u) + N(u)$$

The application of a deconvolution procedure results in

$$\frac{G(u)}{H(u)} = F(u) + \frac{N(u)}{H(u)}$$

or

$$\hat{F}(u) = F(u) + \frac{N(u)}{H(u)}$$

where $F(u)$ is the FT of the true but unknown signal $f(x)$ and $\hat{F}(u)$ is the estimate of the FT of $f(x)$.

As an example of this operation, the convoluted triangle [Fig. 7(d)] was deconvoluted for the point-spread function in a trial to recover the undistorted signal. Obviously, deconvolution retrieves the original sharp signal [Fig. 9(b)], but not without disturbances. The reason is that, in the presence of noise, the term $N(u)/H(u)$ may introduce trouble because, at high frequencies, $N(u)$ may still be large (because noise contains high frequencies) whereas $H(u)$ is small. Consequently, the division $N(u)/H(u)$ may produce large values at high frequencies, which would mean that the original spectrum contains high frequencies (which is noise), leading to invalid results for the deconvolution. A remedy is to carry out the division to the point where noise begins to dominate the result. The effect of noise on the data is demonstrated by deconvoluting the convoluted triangle but now after superposition of Gaussian distributed noise (4% relative standard deviation) [Fig. 9(c), (d)]. Strong side lobes appear and deteriorate the result considerably. There-

References p. 252

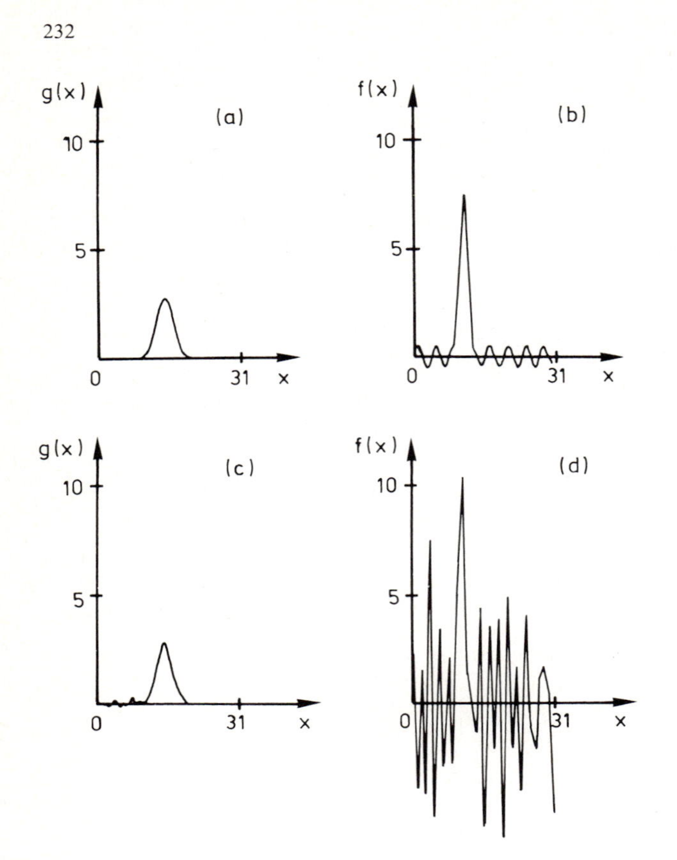

Fig. 9. (a) Gaussian peak convoluted with a triangular point-spread function (psf) ($w_{1/2}\text{psf}/w_{1/2}$ peak = 2). (b) Restoration of (a) for the effect of the psf. (c) Peak (a) + noise N (0.4%). (d) Restoration of (b) for the effect of the psf.

fore, many other possibilities have been proposed ranging from pseudo-deconvolution techniques (peak sharpening) to complex filters.

A necessary condition for deconvolution is the knowledge of the shape of the point-spread function, $h(x)$. In some instances, $h(x)$ can be determined experimentally by measuring a narrow signal or impulse with a bandwidth which is at least 10 times smaller than the point-spread function. An interesting feature of deconvolution is the ability to recover two (or more) overlapping peaks. This depends on the ratio between the bandwidths of the point-spread function and the signal and upon the separation between the peaks and their intensity ratio. A discussion of this point can be found in ref. 3. Figure 10 displays the results of the deconvolution of a composite profile of two Gaussian peaks, demonstrating the ability of deconvolution to restore a signal.

In the case where the point-spread function is not known, pseudo-deconvolution techniques have to be used to enhance resolution. In spectrometry, for example, the

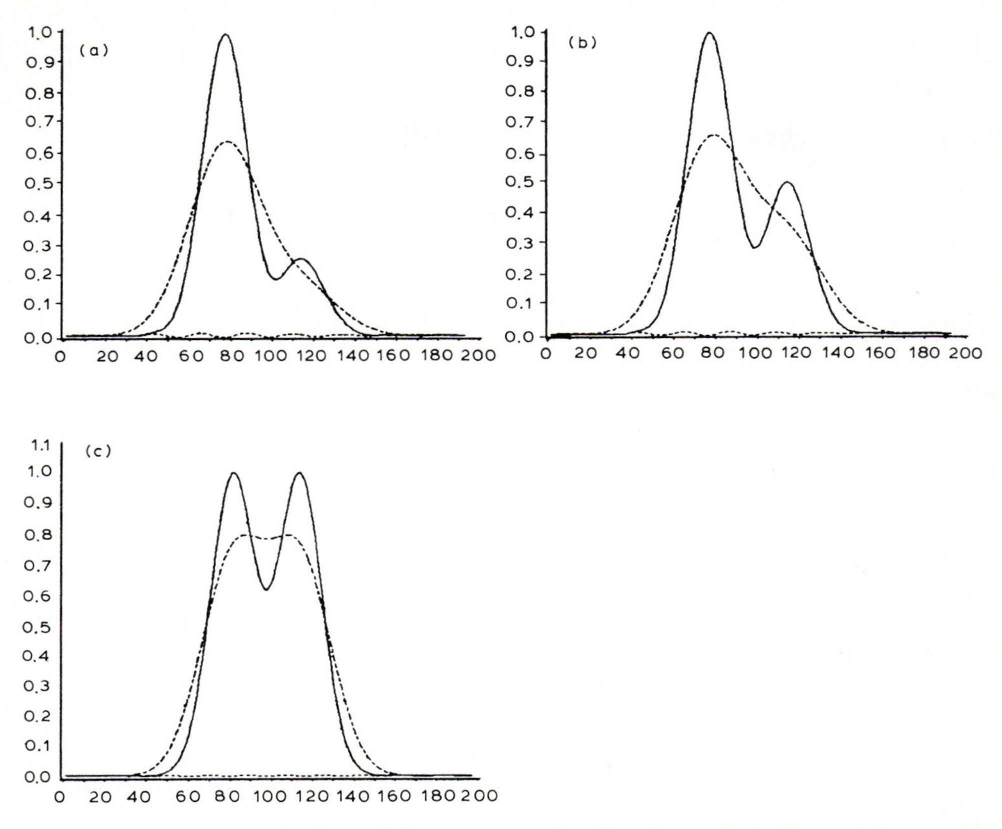

Fig. 10. Ability to restore the signal of the two overlapping Gaussian peaks by deconvolution. -----, Measured peak; ———, restored signal; - - - - - -, difference between the true and restored signals. The ratio between the half-height widths of the psf and signal are (a), (b) 1.25 and (c) 1.0.

point-spread function is symmetrical. Without knowing the exact shape of the point-spread function one can enhance the resolution by subtraction of the second derivative from the measured spectrum $g(x)$

$$f(x) = ag(x) - (1-a)g''(x) \qquad (0 < a < 1)$$

Because the second derivative of any clock-shaped peak is negative between the two inflection points (second derivative is zero) and positive elsewhere (see Fig. 23), the subtraction results in a higher top and narrower wings, which results in a better resolution. Pseudo-deconvolution methods for asymmetric point-spread functions, such as the broadening effect of a slow detector response, are not available.

2.2 Signal and noise characterization

Spectral analysis can be a useful exploratory diagnostic tool for the characterization of analytical signals and their noise. Figure 15 of Chap. 14 and Fig. 13 below,

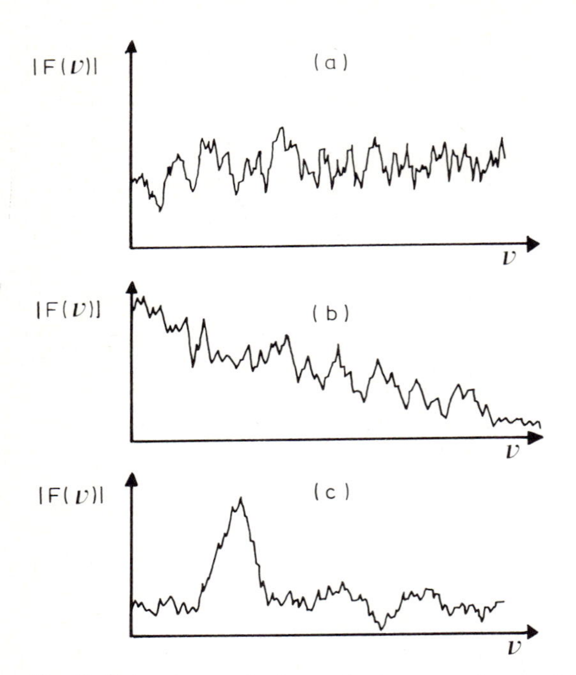

Fig. 11. Noise characterization in the frequency domain. (a) "White" noise; (b) flicker or $1/f$ noise; (c) interference noise.

show, respectively, the autocorrelation function and the power spectrum of a flame ionization detector. These plots give useful information for taking the proper measures to improve the signal-to-noise ratio. To get some feeling for the sort of spectra that may be found when characterizing noise, we display in Fig. 11 three spectra with obviously different characteristics (all spectra are scaled in the same way). Spectrum (a) is the power spectrum of "white noise" which contains all frequencies to the same extent. Examples of white noise are shot noise in photomultiplier tubes and thermal noise occurring in resistors. In spectrum (b), the power of the signal is inversely proportional to the frequency. This type of noise is often called $1/f$ noise (f is the frequency) and originates mostly from fluctuations of the environmental humidity, power supply, temperature, vibrations, quality of chemicals, etc. This type of noise is also very common in analytical systems, where it is called drift. As an example, we display the power spectrum of the noise of an HPLC detector (Fig. 12).

Spectrum (c) in Fig. 11 has the characteristic that the power in some specific areas of the spectrum contains a maximum. A very common source of this type of noise is the 60 Hz power line. This indicates an underlying periodic waveform in the signal. Most noise encountered in real situations will be a mixture of the noise types described. Spectrum analysis is a useful tool in assessing the frequency characteris-

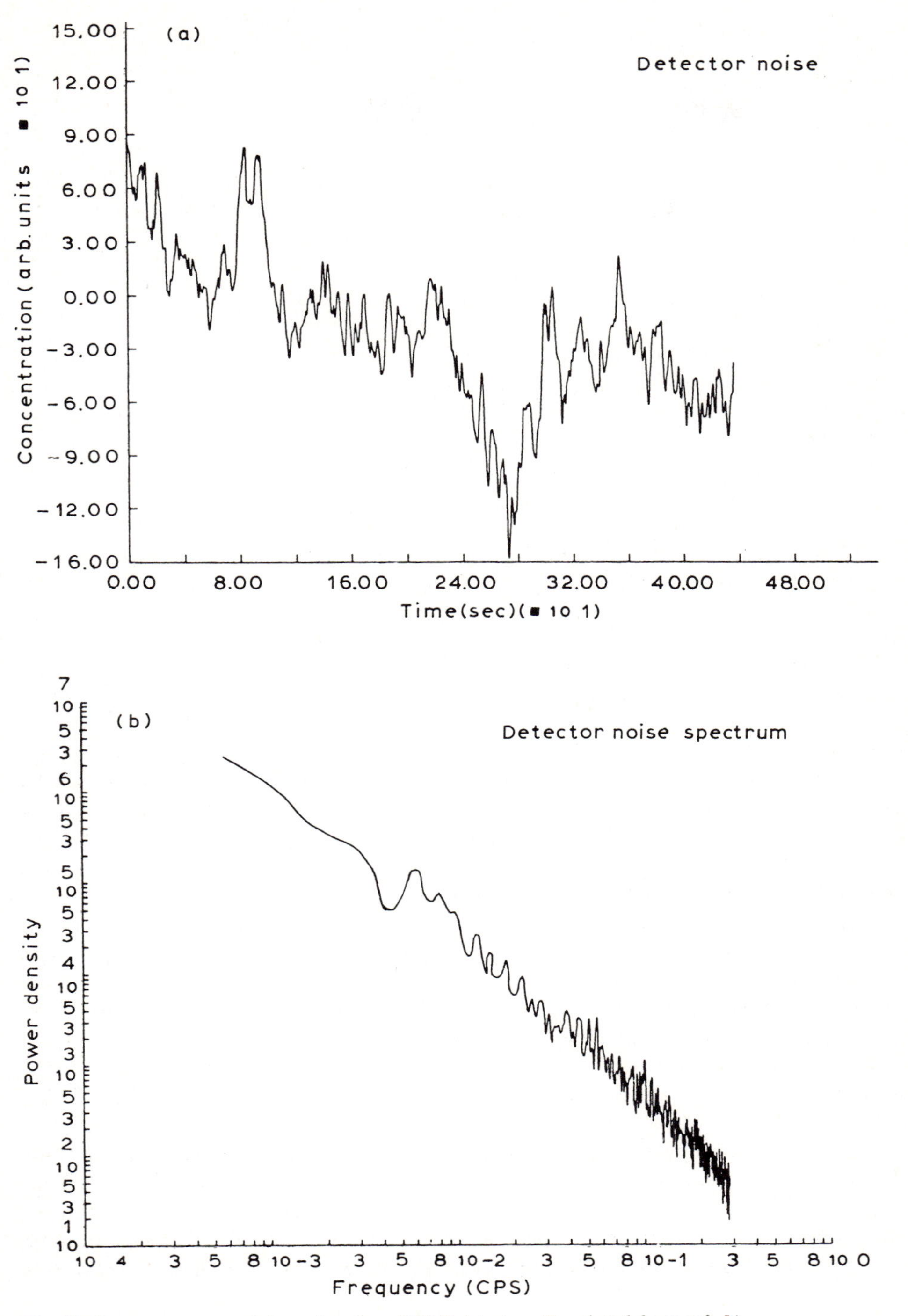

Fig. 12. Power spectrum of the noise of an HPLC detector. (Reprinted from ref. 5.)

tics of these types of signal. The information present in the power spectrum can also be obtained by calculating the autocovariance function or autocorrelation function of the signal. As explained in the previous section, the autocorrelation function of a signal gives a quantitative measure for the correlation between the signal values at different time intervals. The smaller the time constant, the higher the relative content of high frequencies in the signal. Intuitively, one can understand that there is an exact mathematical relationship between the autocorrelation function and the power spectrum of a signal.

2.3 Filtering

Analytical instruments usually contain the necessary electronics to keep the signal-to-noise ratio (S/N) as high as possible, within the constraints of price-to-performance ratio. For example, in AAS, the radiation of a hollow cathode lamp is modulated by a light chopper so as to minimize the $1/f$ noise present in the system, or "notch" filters are used to reject specific frequencies (e.g. 60 Hz interference). However, when measuring at low concentration levels, some additional enhancement of the S/N ratio is often required because some noise is always passed by an electronic device. In other instances, the instrument is provided with a variable read-out time constant or damping (see Chap. 11), the application of which may cause undesirable distortions (see Sect. 3). Further enhancement of the S/N ratio can be obtained by programming software filters, which is more versatile than designing the corresponding hardware filters. When the signal information is located at a frequency or group of frequencies which are well separated from the frequencies of the noise, software filters can be designed which pass a selected range of frequencies and cut the frequencies containing noise. Because information on the frequency content of a signal is obtained in the Fourier domain, it is obvious that filters should be defined in that domain. An additional advantage is that software filters can be tuned by applying them repeatedly on the same signal. Figure 13 shows the power spectrum of the output of a flame ionization detector. The noise contains two principal frequencies, 2 and 10 cps. This information enables an appropriate software filter for the enhancement of the signal-to-noise ratio to be designed. A filter which cuts frequencies higher than approximately 6 cps will reduce the power of the noise by about one half.

The general expression for a filter in the Fourier domain is

$$\mathrm{G}(u) = \mathrm{F}(u) \cdot \mathrm{H}(u) \tag{9}$$

which means that the FT of the signal is multiplied by a filter function, $\mathrm{H}(u)$, whereafter the FT is back-transformed.

The shape of the filter function, $\mathrm{H}(u)$, is chosen as a function of the desired properties of the filtering operation. It is recommended to shift the origin of the data in the Fourier domain to the centre of the data row (Sect. 1.3.3). The filter can then be applied symmetrically around this new origin ($u = 0$) (see Fig. 14). Current methods in analytical chemistry are mainly divided into two groups: methods

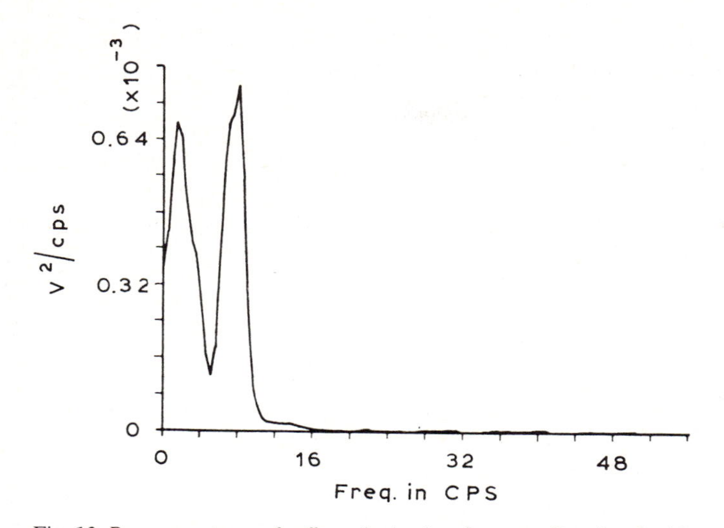

Fig. 13. Power spectrum of a flame ionization detector. (Reprinted with permission from ref. 5.)

applied in the time domain and methods applied in the frequency domain. It is difficult to decide which filter to use. A rule of thumb is that the frequency domain is preferable when the frequency characteristics of signal and noise are very different. A very obvious example is the removal of 50 or 60 Hz modulated noise caused by the power line. No filters in the time domain (Sect. 3.2) can filter this noise adequately, while this is very easily accomplished in the frequency domain. A filter which passes the low frequencies and cuts the high frequencies, is called a low-pass filter.

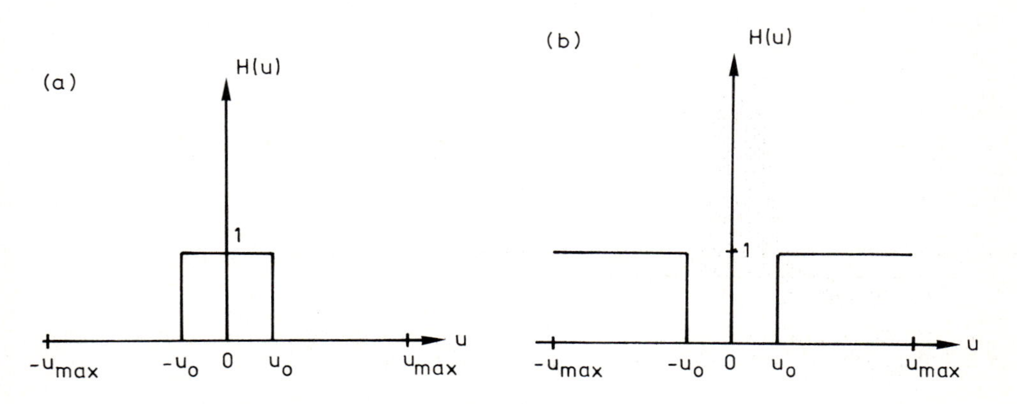

Fig. 14. The shape of (a) a low-pass filter and (b) a high-pass filter in the frequency domain. $u = 0$ is the centre of the frequency axis.

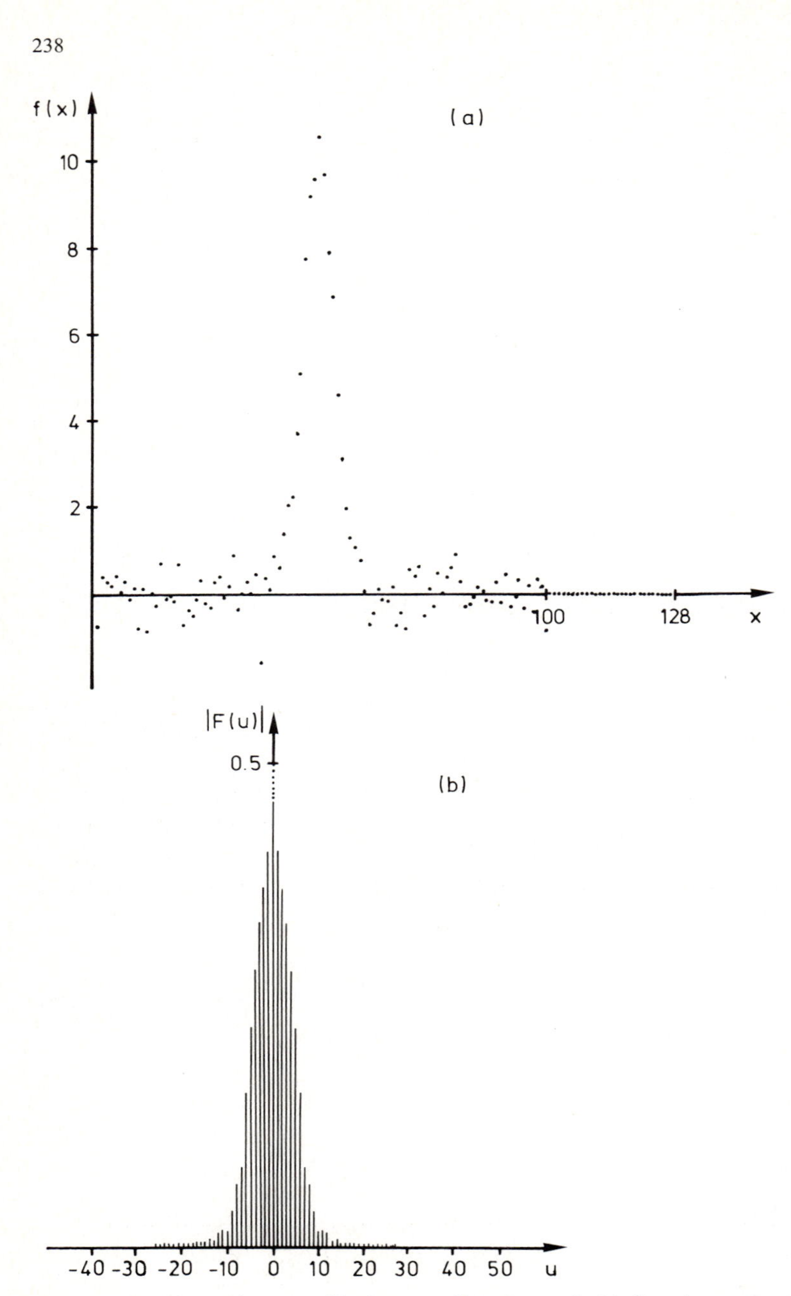

Fig. 15. The effect of low-pass filtering on a Gaussian peak. (a) Gaussian peak; $w_{1/2} = 8\,\Delta x$; (b) FT of (a); (c) signal (a) filtered with $u_0 = 16$; (d) signal (a) filtered with $u_0 = 32$.

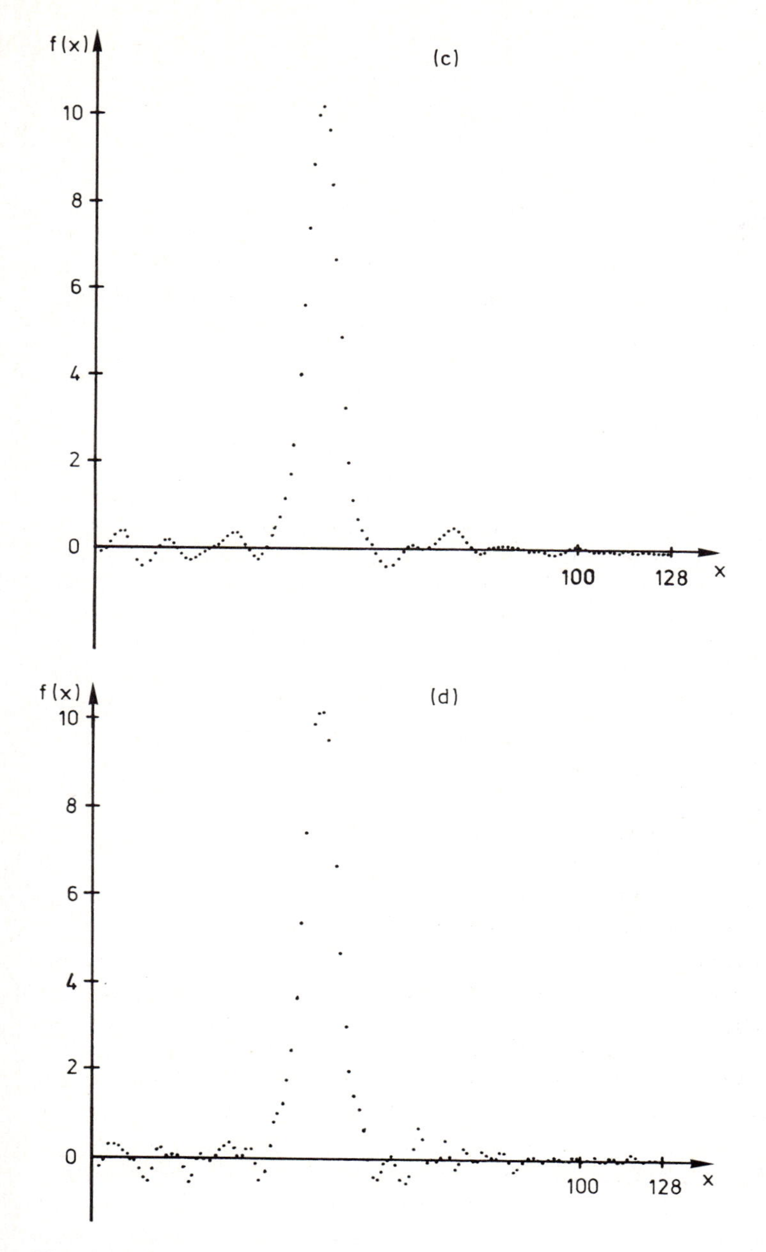

Fig. 15 (continued).

References p. 252

The simplest low-pass filter is the ideal filter with the form

$$H(u) = \begin{cases} 1 & |u| \leqslant u_0 \\ 0 & |u| > u_0 \end{cases}$$

As an example, Fig. 15 shows the effect of the ideal filter on a noisy Gaussian signal, discretized into 100 points and extended with 28 zeros to 128 data points. The effect is clearly visible. For $u = 16$, all high-frequency noise has been removed, introducing some side lobes instead. Furthermore, we notice that, contrary to some methods in the time domain, no asymmetry is introduced on a symmetrical signal by smoothing in the frequency domain and that the top of the peak is not shifted.

A second class of filters, of less importance in analytical chemistry, are the high-pass filters of the form

$$H(u) = \begin{cases} 0 & |u| \leqslant u_0 \\ 1 & |u| > u_0 \end{cases}$$

Many other shapes can be chosen for the low- and high-pass filters: exponential, trapezoidal, etc. The choice is highly dependent on the specific experimental conditions, the desired amount of filtering and the degree of signal distortion that can be tolerated. It is not possible to present a hard rule for determining the filter shape and the simplest approach is an empirical one. Generally, exponential and trapezoidal filters perform better than cut-off filters because an abrupt truncation of the Fourier spectrum may introduce spurious side lobes after back transformation. The problem of choosing filter shapes is discussed in more detail by Horlick [4] with references to more mathematical treatments of the subject.

2.4 Digitization of analytical signals

With experimental data, continuous signals are sampled at finite intervals. As discussed in Sect. 1, the sampling interval between the data points determines the maximum observable frequency, f_{max}. The relationship between sample interval and f_{max} is given by the Nyquist sampling theorem, which states that $f_{max} = 1/(2\,\Delta x)$. If Δx is not small enough, the high frequency information present in the data is lost and may disturb the results at other frequencies by folding. This means that the sampling frequency has to be adjusted to the properties (time constant) of the signal. Applications of Nyquist's (also referred as Shannon's) sampling theorem to the Gaussian elution profile of a chromatogram, leads to the conclusion that approximately 8 samples per 6 σ-width are needed to preserve all the frequency information present in a Gaussian peak. If the peak elutes in 1 s, a sampling frequency of 8 Hz is required to retain all the information in the signal. Reconstruction of the originally measured signal from the digitized signal could be made via the frequency spectrum, but it is evident that this would require lengthy calculations. It is common practice to choose a smaller digitizing interval than the Nyquist interval, permitting the recovery of the signal through regression for the interpolation between the sampled values. Such a higher digitizing rate also enables correction for noise in the data.

3. Noise reduction by direct smoothing of the data

3.1 Noise removal

Some types of noise are very easily distinguished from the true signal. For example, it may occur that the signal is disturbed by sharp spikes of one or more deviating data points. Figure 16(a) shows an absorption peak measured with a furnace atomic absorption spectrometer. Spikes are observed on the signal. Removal of the spikes by filtering in the frequency domain, as described in the previous section, is not appropriate in this case. No low- or high-pass filter in the frequency domain can be designed to remove the spikes because their contribution in the Fourier spectrum is too weak to be isolated. On the other hand, the spikes are easily detectable in the original data and are also easily removed.

Spikes are detected by comparing the value of each data point with its neighbours. If the difference with its neighbours is larger than a given threshold value, it is decided that the data point is a spike. They are removed, without affecting the rest of the signal, by replacing the spike by the average value of its neighbours. The results of spike removal by this type of interpolation are shown in Fig. 16(b).

The method just described has several tuning parameters that can be adjusted to

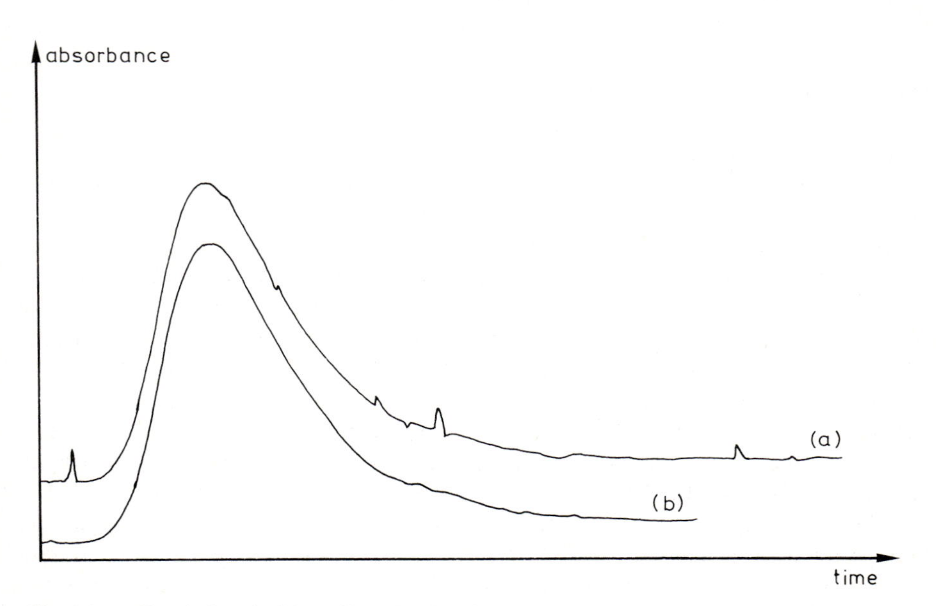

Fig. 16. (a) Signal disturbed by spikes. (b) Signal after removal of spikes.

References p. 252

affect the result: these parameters include the number of neighbours taken into account and the threshold amount by which a data point is allowed to differ from its neighbours. The threshold might be defined as some multiple of the estimated standard deviation of the noise.

3.2 Smoothing

Smoothing methods are not based on the identification and removal of the noise, as described in the preceding section for noise removal. Instead, smoothing reduces the noise by applying some averaging to the signal.

3.2.1 Moving average

One of the simplest ways to smooth data is by a moving average. In this procedure, one chooses a window of a fixed odd number of points and calculates the average of all data points inside the window. The central point in the window is thereafter replaced by that calculated average. Next, the window is moved one data point by dropping the last point and adding the next to the window. The mean value of all points inside the window is recalculated and replaces the central point. This operation is mathematically represented as

$$y_{s,i} = \sum_{j=-m}^{j=+m} y_{i+j}/(2m+1) \tag{10}$$

where the index i indicates the index of the data points and $2m + 1$ is the size of the window. The effect of a 9-point moving average is shown in Fig. 20. In addition to a moving average, various other types of smoothing, discussed below, can be applied, each having its very typical characteristics.

3.2.2 Exponential averaging

Exponential averaging is obtained by calculating the weighted average of the points in a moving window of m data points. The last point in the window (i.e. the point i to be smoothed) is given the greatest weight and each preceding point is attributed a lower weight determined by the shape of the exponential function [$\exp(-j/T)$] with a smoothing constant T. Points following the last point are given zero weight. This filter smooths point i by using data points that precede this point. A worked out example is given in Table 1. This procedure is identical to a real time smoothing with electronic RC filters where, at a time i, no data points $i + 1$, etc. are available. Because of the asymmetry of this smoothing function, a unidirectional distortion is introduced in the smoothed data, as with the RC filter in an actual instrument [Fig. 17(b)]. Besides the desired lowering of the noise, the effect of exponential averaging is a lowering and shift of the peak maximum. The effect is similar to the measurement of a peak with an instrument with a too large response time, as has been discussed in Chap. 11. Thus, an exponential smoothing of a peak with a smoothing constant T has the same effect as measuring that peak with an

TABLE 1

Example of an exponential smoothing filter. Window = 5 data points; $T = 3$ (data points)

The convolutes are given by $\exp(-|j|/T)$

j	-4	-3	-2	-1	0
$\exp(-\|j\|/T)$	0.26	0.37	0.51	0.72	1

Consider the 15-point data series [a]

1	2	3	4	5	6	7	8	9	10	11	12	13	14	15	16	17	18
0.3	0.1	0.4	0.2	0.7	0.9	1.4	1.3	1.0	0.8	0.3	0.3	0.4	0.1	.	.	.	.
4	3	2	1	0													
				0.41													
	$j=4$	3	2	1	0												
					0.59												
		$j=4$	3	2	1	0											
						0.903											
			$j=4$	3	2	1	0										
							1.08										
				$j=4$	3	2	1	0									
								1.11									
					$j=4$	3	2	1	0								
									1.03								
						$j=4$	3	2	1	0							
										0.78							
							$j=4$	3	2	1	0						
											0.57						
								$j=4$	3	2	1	0					
												0.46					
									$j=4$	3	2	1	0				
													0.30				

[a] Bold faced numbers are the smoothed data.

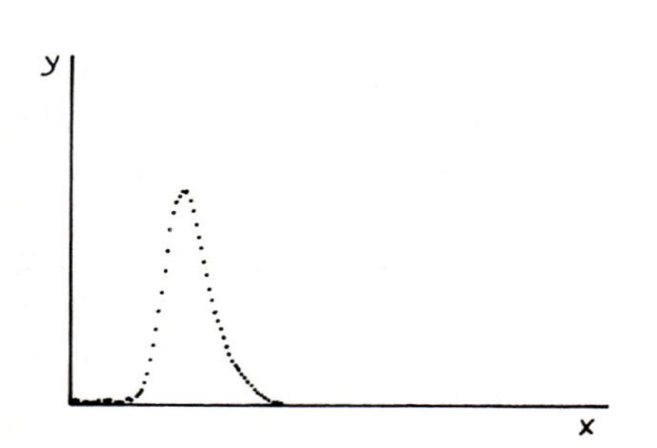

Fig. 17. The effect of exponential averaging on a Gaussian peak.

References p. 252

instrument with a time constant, T_x, equal to T. The intensity drop as a function of the ratio of T and the peak width, σ, can therefore be read from Fig. 6 in Chap. 11.

3.2.3 Polynomial smoothing with Savitzky–Golay filters

In many instances, it is not necessary to carry out the smoothing in real time, i.e. the point i should not be smoothed in the time before data point $(i+1)$ is measured, but can be postponed until a certain number of data points has been collected after i. This allows us to use a symmetric convolution function around the data point to be smoothed. This data point is the central point in a window of points used to calculate the smoothed value. An advantage of symmetric convolutes is the avoidance of a shift of the peak position and the preservation of the peak symmetry.

Consider a window of 7 data points. The central point, 4, is the data point we want to smooth. The window therefore consists of 3 points on the left-hand side and 3 points on the right-hand side of the data point to be smoothed. Polynomial smoothing consists of fitting a polynomial model through the datapoints $(x_{i-3}, y_{i-3}), (x_{i-2}, y_{i-2}), (x_{i-1}, y_{i-1}), \ldots, (x_{i+1}, y_{i+1}), (x_{i+2}, y_{i+2}), (x_{i+3}, y_{i+3})$ by the least-squares method described in Chap. 13. The value of the middle point in the window, y_i, is then replaced by the corresponding value of the model, $\hat{y}_i$. When the model is adequate to describe the deterministic part of the signal, then $\hat{y}_i$ is a better estimate of the true value than y_i. A 6th-order polynomial exactly fits the 7 data points. Thus, the value of the polynomial in the central point, $\hat{y}_i$, is exactly the measured value y_i. Therefore, substitution of y_i by $\hat{y}_i$ has no effect. However, when a lower order polynomial is fitted through the 7 data points (Fig. 18), it is assumed that the polynomial describes the underlying deterministic part of the data. Substitution of the central point, y_i, by the value of the polynomial, $\hat{y}_i$, therefore returns a point which contains less noise. The fit with a zero-order polynomial (a constant) assumes that all variation in the data in the window has to be ascribed to noise and the true relationship is a horizontal line (= a constant). By substituting

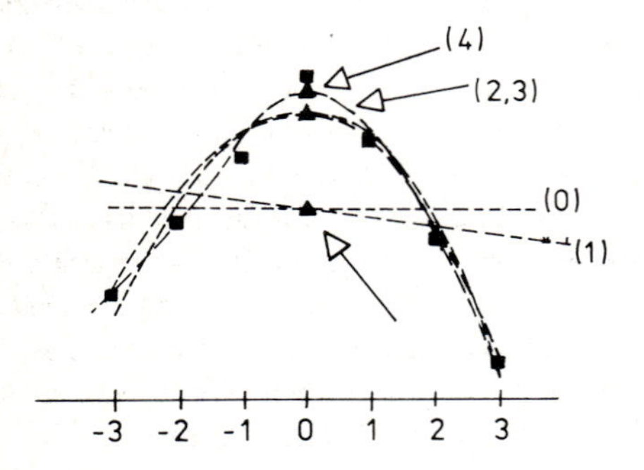

Fig. 18. Polynomial smoothing: a window of seven data points fitted with polynomials of various degrees, 0, 1,...,4. ▲ is the value of the polynomial in the central point.

TABLE 2

Convolutes for quadratic and cubic smoothing

(Reprinted from ref. 7.)

Points	25	23	21	19	17	15	13	11	9	7	5
−12	−253										
−11	−138	−42									
−10	−33	−21	−171								
−09	62	−2	−76	−136							
−08	147	15	9	−51	−21						
−07	222	30	84	24	−6	−78					
−06	287	43	149	89	7	−13	−11				
−05	322	54	204	144	18	42	0	−36			
−04	387	63	249	189	27	87	9	9	−21		
−03	422	70	284	224	34	122	16	44	14	−2	
−02	447	75	309	249	39	147	21	69	39	3	−3
−01	462	78	324	264	42	162	24	84	54	6	12
00	467	79	329	269	43	167	25	89	59	7	17
01	462	78	324	264	42	162	24	84	54	6	12
02	447	75	309	249	39	147	21	69	39	3	−3
03	422	70	284	224	34	122	16	44	14	−2	
04	387	63	249	189	27	87	9	9	−21		
05	322	54	204	144	18	42	0	−36			
06	287	43	149	89	7	−13	−11				
07	222	30	84	24	−6	−78					
08	147	15	9	−51	−21						
09	62	−2	−76	−136							
10	−33	−21	−171								
11	−138	−42									
12	−253										
NORM	5175	8059	3059	2261	323	1105	143	429	231	21	35

the central point by this average value and by moving the window over the data by dropping one at the left and picking up one at the right each time, the moving average filter is obtained. By adapting the order of the polynomial with respect to the number of points in the window, various degrees of smoothing are obtained.

When a spectrum has been recorded over 1000 data points and a 7-point smoothing is applied, a 7-point window is moved 993 times in order to treat all the data. Consequently, an nth-order (with $0 < n < 6$) polynomial has to be fitted 993 times through 7 points, which becomes quite laborious if one wants to repeat the calculations with changing window sizes and degrees of the polynomial. However, values of c_j ($j = -m, +m$) have been tabulated [7] with which the data points in the window have to be multiplied and summed in order to obtain the value of the central point, smoothed with a polynomial of a given degree [eqn. (11)]. Mathematically, polynomial smoothing is obtained by

$$y_{s,i} = \sum_{j=-m}^{+m} c_j y_{j+i}/\text{NORM} \quad (11)$$

References p. 252

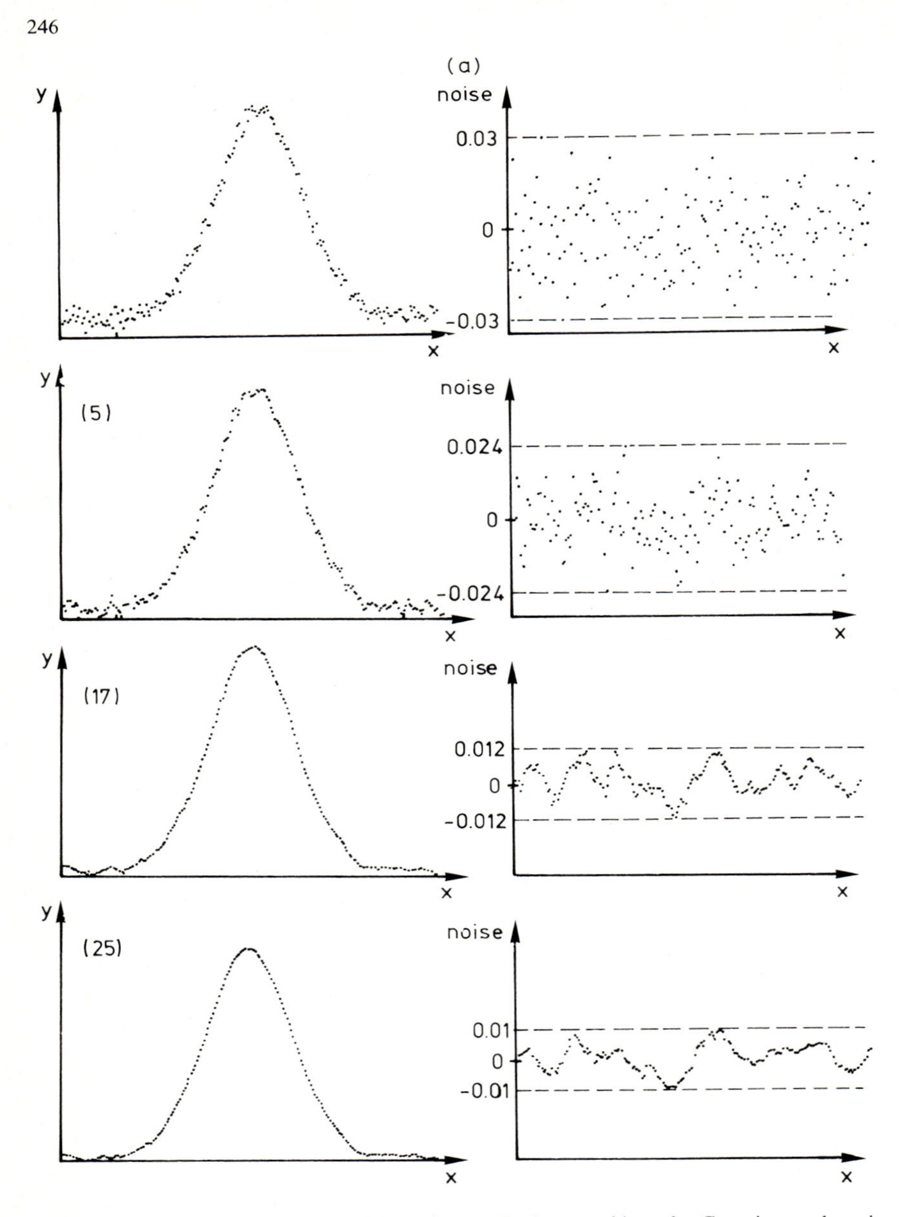

Fig. 19. (a) Five-point, 17-point, and 25-point quadratic smoothing of a Gaussian peak, noise: N (0,3%). (b) S/N after smoothing as a function of the window size.

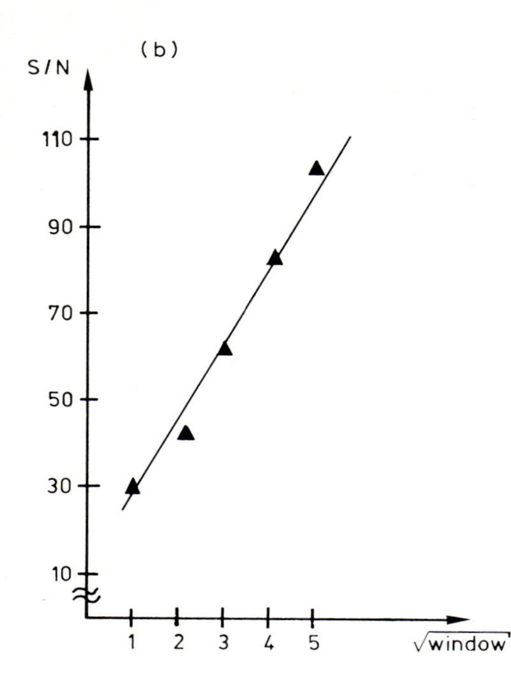

Fig. 19 (continued).

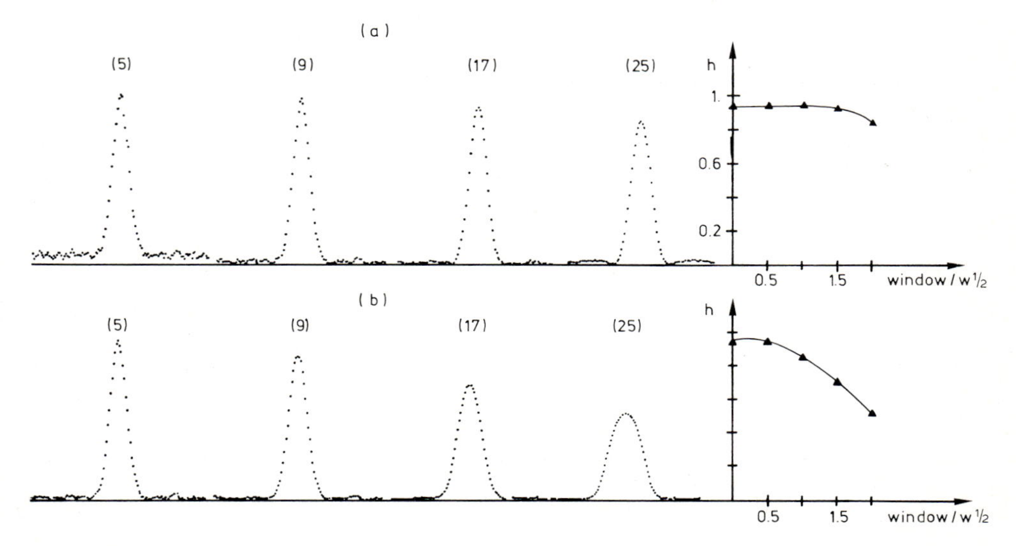

Fig. 20 Distortion, h/h_0, of a Gaussian peak as a function of the ratio between window size and half-height width for (a) polynomial smoothing and (b) moving average. The window size is indicated in parentheses.

TABLE 3

Polynomial smoothing: a worked example [a]

Data from Table 2.

7-point quadratic convolutes: −2 3 6 7 6 3 −2 NORM = 21

1	2	3	4	5	6	7	8	9	10	11	12	13	14	15	16	17
0.3	0.1	0.4	0.2	0.7	0.9	1.4	1.3	1.0	0.8	0.3	0.3	0.4	0.1			
−2	3	6	7	6	3	−2										
(−0.6+0.3+2.4+1.4+4.2+2.7−2.8)/21																
			↓													
			0.36 (smoothed datapoint 4)													
1	2	3	4	5	6	7	8	9	10	11	12	13	14	15	16	17
0.3	0.1	0.4	0.2	0.7	0.9	1.4	1.3	1.0	0.8	0.3	0.3	0.4	0.1	0.3		
	−2	3	6	7	6	3	−2									
	(−0.2 1.2 1.2 4.9 5.4 5.2 −2.6)/21															
				↓												
				0.67 (smoothed datapoint 5)												
					0.981											
						1.24										
							1.26									
								1.06								
									0.69							
										0.30						

[a] Bold faced numbers are the smoothed data.

where NORM is a normalizing factor, $y_{s,i}$ is the smoothed data point i, and $c(j = -m, +m)$ are the tabulated values for a given window size $(2m + 1)$ and the order of the polynomial.

Equation (11) is equivalent to eqn. (6) for the calculation of a convolution by moving a window with weighting factors or convolutes over the signal. Consequently, the signal broadening and weakening introduced by convolution is also observed when smoothing data. On the other hand, a convolution as indicated by eqn. (6) also smooths the data. The equivalence between smoothing, convolution, and filtering is further discussed at the end of this chapter. Because the values of the centre points of kth- and $(k + 1)$th-order polynomials (k is even) when fitted through an odd number of points are equal (Fig. 18), the set of convolutes are the same and the set up to 25 points is shown in Table 2 with the proper normalizing factors. It should be stressed that polynomial smoothing requires a constant digitizing interval. A worked example of a 7-point quadratic smoothing is given in Table 3.

Figure 19 shows a Gaussian peak with random and Gaussian distributed noise smoothed by fitting a quadratic polynomial in a moving window of, respectively, 5, 9, 17 and 25 points. In order to judge the effect of the smoothing on the S/N ratio, plots of the differences between the smoothed and true noise-free spectra are provided. The width of the Gaussian peak has been chosen such that any peak distortion is avoided.

It can be concluded that the signal-to-noise ratio improves with the square root of the window size. A 25-point smoothing gives an approximate 5-fold improvement. Figure 20 shows that the ratio of the window size and the peak width determines the introduced distortion of the data. This distortion is a broadening and lowering of the signal introduced by a second-degree polynomial smoothing of a Gaussian peak. It should be noted at this point that these conclusions apply for the cases in which the noise is white or contains mainly frequencies that are much higher than those in the signal.

The larger the window size, the better the signal-to-noise ratio becomes, but the larger the distortion of the peak. From Fig. 20, it follows that the window size should not be chosen larger than 1.5 times the half-height width of the peaks. The figure also demonstrates that smoothing by a moving average can only be used to smooth relatively broad signals. Other signals such as titration curves are also distorted by smoothing. For each particular signal shape, the effect of the window size and the degree of the polynomial should be determined.

An obvious advantage of polynomial smoothing over filtering in the frequency domain is its simplicity because it is directly applied on the measurements and does not require Fourier transforms and inverse Fourier transforms.

The effect of polynomial smoothing can be compared with filtering in the frequency domain by calculating the filter which corresponds with a given set of smoothing convolutes. The multiplication of the $(2m+1)$ values in a moving

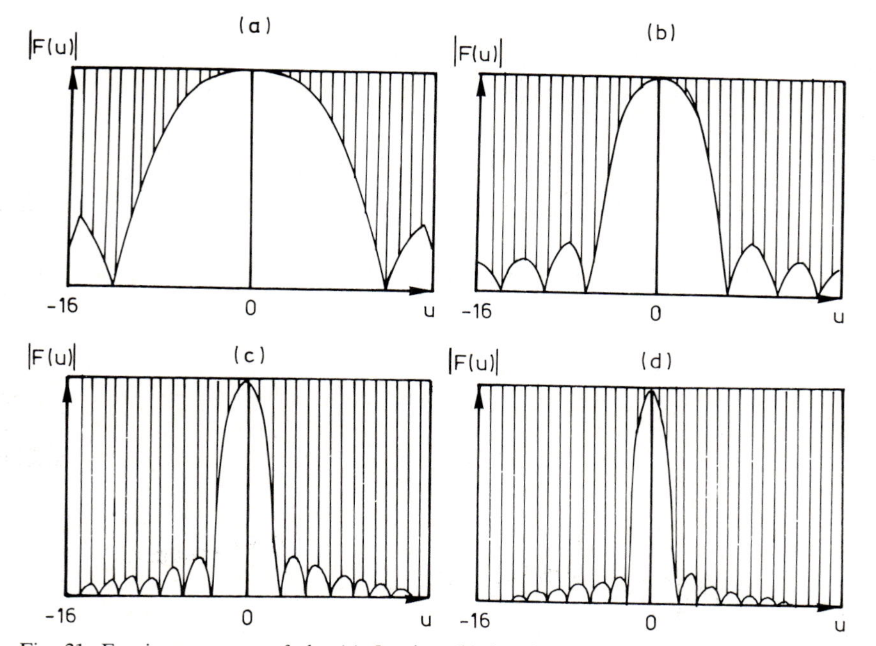

Fig. 21. Fourier spectrum of the (a) 5-point, (b) 9-point, (c) 17-point and (d) 25-point second-order convolutes (zero filled to $n = 128$). $f_{max} = 0.5$ Hz, $\Delta u = 7.8 \times 10^{-3}$, $\Delta x = 1$.

References p. 252

TABLE 4

Convolutes for the one-step calculation of the smoothed second derivative

(Reprinted from ref. 7.)

Points	25	23	21	19	17	15	13	11	9	7	5
−12	92										
−11	69	77									
−10	48	56	190								
−09	29	37	133	51							
−08	12	20	82	34	40						
−07	−3	5	37	19	25	91					
−06	−16	−8	−2	6	12	52	22				
−05	−27	−19	−35	−5	1	19	11	15			
−04	−36	−28	−62	−14	−8	−8	2	6	28		
−03	−43	−35	−83	−21	−15	−29	−5	−1	7	5	
−02	−48	−40	−98	−26	−20	−48	−10	−6	−8	0	2
−01	−51	−43	−107	−29	−23	−53	−13	−9	−17	−3	−1
00	−52	−44	−110	−30	−24	−56	−14	−10	−20	−4	−2
01	−51	−43	−107	−29	−23	−53	−13	−9	−17	−3	−1
02	−48	−40	−98	−26	−20	−48	−10	−6	−8	0	+2
03	−43	−35	−83	−21	−15	−29	−5	−1	7	5	
04	−36	−28	−62	−14	−8	−8	2	6	28		
05	−27	−19	−35	−5	1	19	11	15			
06	−16	−8	−2	6	12	52	22				
07	−3	5	37	19	25	91					
08	12	20	82	34	40						
09	29	37	133	51							
10	48	56	190								
11	69	77									
12	92										
Norm	26910	17710	33649	6783	3876	6188	1001	429	462	42	7

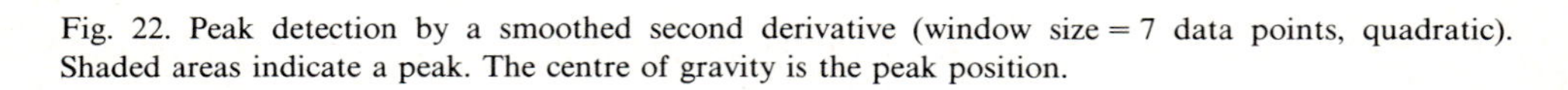

Fig. 22. Peak detection by a smoothed second derivative (window size = 7 data points, quadratic). Shaded areas indicate a peak. The centre of gravity is the peak position.

window of data points with the convolutes given in Table 2 is a convolution of the measured data with these convolutes.

In Sect. 1.3.4, it was explained that a convolution in the time domain is a multiplication in the frequency domain. Consequently, the Fourier transform of the convolutes given in Table 2 (5, 7, 9... or 25 points with as many zeros appended as is necessary to obtain the same number of data points as in the spectrum) represents

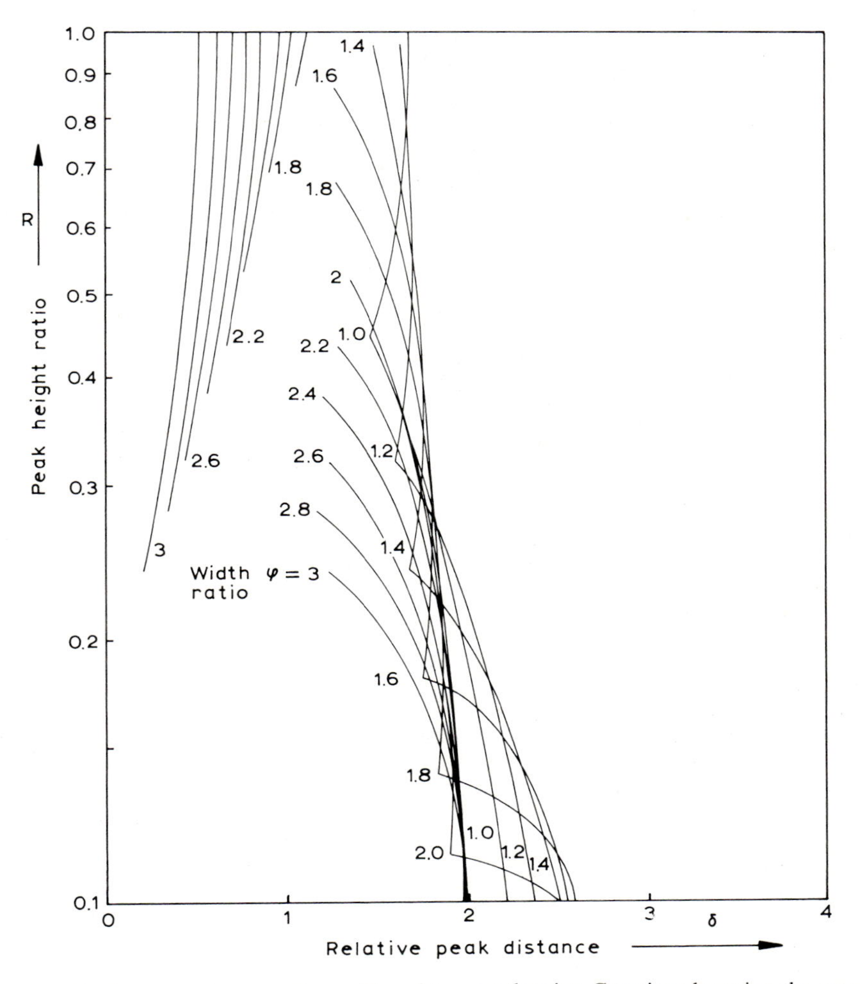

Fig. 23. Recovery of composite profiles of two overlapping Gaussians by using the second derivative. $\delta = d/[(1/w_{1/2})_1 + (1/w_{1/2})_2]$; $R = h_1/h_2$; $\rho = (w_{1/2})_1/(w_{1/2})_2$. (Reprinted with permission from ref. 8.)

References p. 252

the shape of the polynomial filter in the frequency domain. As expected, this corresponds to a low-pass filter as discussed in Sect. 2.2. The larger the window size, the lower are the contributions of the high frequencies, which is equivalent to a smaller cut-off frequency (Fig. 21).

No strict rules can be given as to whether to apply smoothing (in the time domain) or filtering (in the frequency domain) on the data. Disturbances with a periodic nature (60 Hz power line) cannot be removed by smoothing, but are easily removed by filtering. The lower the signal-to-noise ratio and the larger the difference between the frequency content of the signal and noise, the more one prefers filtering methods. It should be noted that smoothing and differentiation of the data can be carried out in one step by choosing the proper set of convolutes c_i $(i = -m, +m)$ in eqn. (11). In Table 4, 5-, 7-..., 25-point convolutes are tabulated for the calculation of the smoothed second derivative of the data in one step. From this second derivative, peak positions (x_p) can be determined by the calculation of the centre of gravity of each negative region, which is considered to represent one peak (Fig. 22), viz.

$$x_p = \sum_{x \in \text{neg.area}} x_i y_i / \sum y_i$$

Figure 23 shows how well peak positions can be detected in overlapping peak systems. If the noise in the analytical signal is sufficiently low, the application of higher derivatives may also detect peaks with a larger overlap.

References

1 I.G. McWilliam and H.C. Bolton, Instrumental peak distortion: I. Relaxation time effects, Anal. Chem., 41 (1969) 1755.

2 E. Grushka, Characterization of exponentially modified Gaussian peaks in chromatography, Anal. Chem., 44 (1972) 1733.

3 A. den Harder and L. de Galan, Evaluation of a method for real time deconvolution, Anal. Chem., 46 (1974) 1464.

4 G. Horlick, Digital data handling of spectra utilizing Fourier transforms, Anal. Chem., 44 (1972) 943.

5 H.C. Smit and H.L. Walg, Base-line noise and detection limits in signal-integrating analytical methods, applications to chromatography, Chromatographia, 8 (1975) 311.

6 T. Lub, H.C. Smit and H. Poppe, Correlation high-performance liquid-chromatography technique for improving detection limit applied to analysis of phenols, J. Chromatogr., 149 (1978) 721.

7 A. Savitzky and M.J.E. Golay, Smoothing and differentiation of data by simplified least-squares procedures, Anal. Chem., 36 (1964) 1627.

8 B.G.M. Vandeginste and L. de Galan, Critical evaluation of curve fitting in infrared spectrometry, Anal. Chem., 47 (1975) 2124.

Recommended reading

Transforms and filtering

R. Brereton, Fourier Transforms. Use, theory and applications to spectroscopic and related data, Chem. Lab., 1 (1986) 17.

G. Horlick, Fourier transform approaches to spectroscopy, Anal. Chem., 43(8) (1971) 61A.

B.R. Lam, R.C. Wiebolt and T.L. Isenhour, Practical computation with Fourier transforms for data analysis, Anal. Chem., 53 (1981) 889A.

G.M. Hieftje, Signal to noise enhancement through instrumental techniques. I. Signals, noise and signal to noise enhancement in the frequency domain, Anal. Chem., 44(6) (1972) 81A.
R.B. Lam and T.L. Isenhour, Equivalent width criterion for determining frequency domain cut offs in Fourier transform smoothing, Anal. Chem., 53 (1981) 1179.
H.L. Walg and H.C. Smit, A user-oriented software Fourier spectrum display for analytical purposes, Anal. Chim. Acta, 103 (1978) 43.
P.R. Griffiths (Ed.), Transform Techniques in Chemistry, Heyden, London 1978.
A.G. Marshall (Ed.), Fourier, Hadamard and Hilbert Transforms in Chemistry, Plenum Press, New York, 1982.
J.L. Grant, Resolution enhancement of X-ray photoelectron spectra by iterative deconvolution, Intell. Instrum. Comput., (1985) 4.
D. Binkley and R. Dessy, Data manipulation and handling, J. Chem. Educ., 56 (1979) 148.

Polynomial smoothing

G.M. Hieftje, Signal to noise enhancement through instrumental techniques. II. Signal averaging, boxcar integration, and correlation techniques, Anal. Chem., 44(7) (1972) 69A.
M.U.A. Bromba, Application hints for Savitzky–Golay digital smoothing filters, Anal. Chem., 53 (1981) 1583.
M.U.A. Bromba and H. Ziegler, Digital filter for computationally efficient smoothing of noisy spectra, Anal. Chem., 55 (1983) 1299.
S. Johannathan and R.C. Patel, Digital filters for noise reduction in optical kinetic experiments, Anal. Chem., 58 (1986) 421.
C.G. Enke and T.A. Nieman, Signal-to-noise ratio enhancement by least-squares polynomial smoothing, Anal. Chem., 48 (1976) 705A.

IR Fourier transform spectroscopy

R. Geick, Fourier methods in analytical chemistry: IR Fourier transform spectroscopy, Z. Anal. Chem., 288 (1977) 1.
F.C. Srong, III, How the Fourier transform infrared spectrophotometer works, J. Chem. Educ., 56 (1979) 681.

Time series analysis

C. Chatfield, The Analysis of Time Series: An Introduction, Chapman and Hall, London, 1984.
G.E.P. Box and G.M. Jenkins, Time Series Analysis, Forecasting and Control, Holden-Day, San Francisco, 1970.

Chapter 16

Response Surfaces and Models

1. Introduction

The concepts developed in this chapter and the next three chapters are general and can be applied not only to analytical chemical systems (e.g. electrochemistry, spectroscopy, chromatography) but also to the systems which are being measured by the analytical chemist (e.g. pharmaceutical manufacturing, polymer production, water quality).

While these concepts offer valuable insights for improving and understanding the responses from systems, it is well to keep in mind that we might never be able to discover the *true* behavior of a system. Uncertainties associated with the experimental measurements and with the mathematical description of the system will almost always exist. Thus, we will usually have a more or less "fuzzy" picture of how the system truly behaves. It follows that if we are to have a sufficiently clear view of how the system behaves, we must maximize our fundamental understanding of the system and minimize the uncertainties associated with our experimental measurements. These dual goals are achieved by the simultaneous application of precise measurement techniques, good chemical models, and proper experimental design.

2. Response surfaces

A *response surface* is the graph of a system response plotted as a function of one or more of the system factors (i.e. "independent variables", see Chap. 13). Response surfaces offer the chemometrician a convenient means of visualizing how various factors affect his or her measurement system.

Consider the familiar analytical example shown in Fig. 1, a photometric method for the classical determination of manganese as permanganate. The amount of light transmitted through the system depends upon the concentration of MnO_4^- in the sample.

2.1 True response surfaces

According to Beer's law (a mathematical model), if the intensity of the light source is absolutely stable, if the light is monochromatic, if there is no stray light in the system, if the MnO_4^- does not undergo chemical or physico-chemical changes as a function of concentration, and if there are no other absorbing species present in solution, then the intensity of transmitted light is given by

$$I = I_0 \times 10^{-\epsilon b C} \tag{1}$$

Reference p. 269

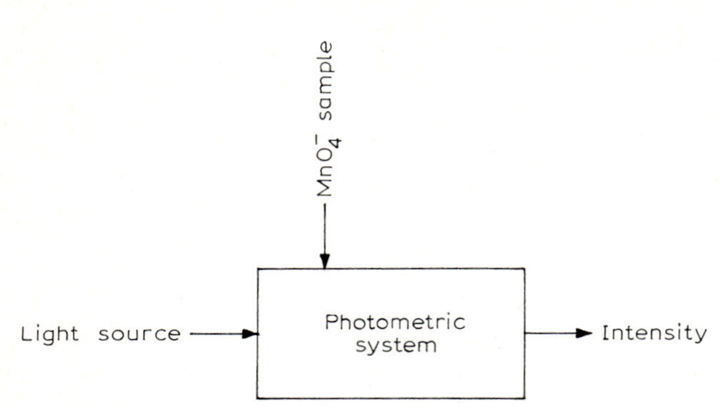

Fig. 1. Systems theory diagram of a photometric method for the classical determination of manganese as permanganate.

and the transmittance, T, is given by

$$T = \frac{I}{I_0} = 10^{-\epsilon b C} \tag{2}$$

where I is the intensity or power of the transmitted light, I_0 is the intensity or

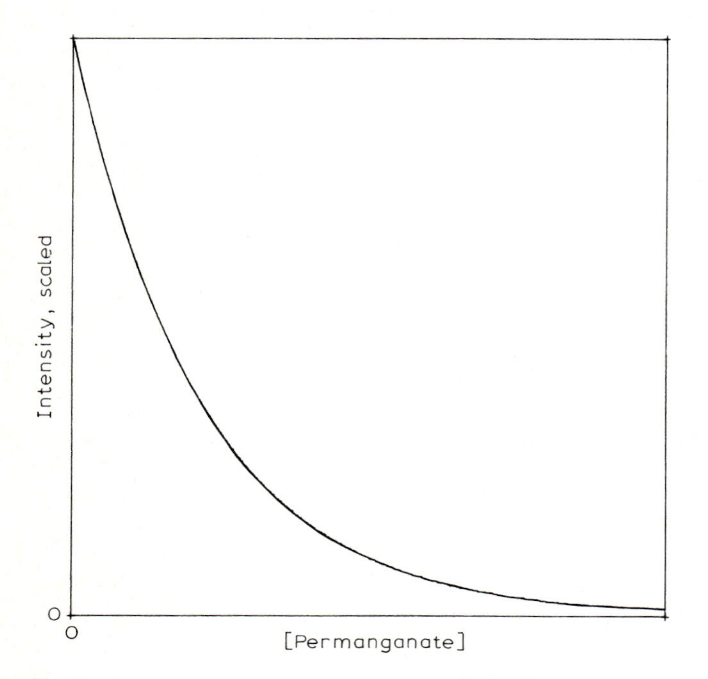

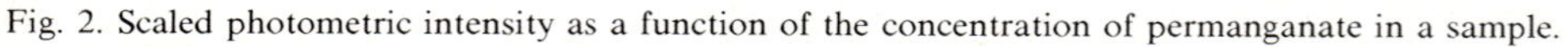
Fig. 2. Scaled photometric intensity as a function of the concentration of permanganate in a sample.

power of the light source, ϵ is the molar absorptivity in L mol^{-1} cm^{-1}, b is the path length through the sample in cm, and C is the concentration of MnO_4^- in the sample in mol l^{-1}.

In this system, the intensity of transmitted light, I, is the system response and the concentration of MnO_4^- in the sample, C, is the system factor. The true response surface (i.e. the graph of the system response plotted as a function of the single system factor) is shown in Fig. 2. Note that, for this true response surface, the physical understanding is exact and there is no experimental uncertainty.

2.2 *Measured response surfaces*

Although many system diagrams are drawn in a manner similar to Fig. 1, such diagrams are incomplete because they do not explicitly take into account a separate but necessary *measurement system*. This is a subtle but important point. In the photometric determination of MnO_4^-, for example, transmitted intensity is the important physical property of interest, but we seldom measure it directly. Instead, with modern photometric instruments, transmitted intensity is converted to an electrical current, then to a voltage, and finally to an analog or digital meter readout from which we visually "observe" the value of the transmitted intensity. A correct view of the double-beam photometric measurement system is given in Fig. 3.

It is clear that the measurement system itself can introduce a set of uncertainties and biases quite separate from those introduced by the photometric system. For example, if electrical line voltage fluctuations affect the measurement system in such a way that the readout fluctuates, the originally well-defined response surface of Fig. 2 becomes uncertain again.

In general, *measured response surfaces are "fuzzy" because of uncertainties in the primary system itself and because of uncertainties introduced by the associated measurement system.* It is usually desirable that this latter source of uncertainty should not

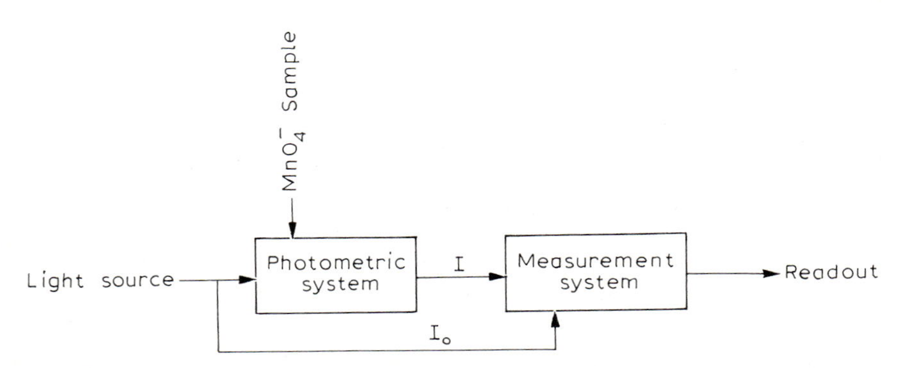

Fig. 3. Systems theory diagram showing the relationship between the photometric system and the measurement system for a double-beam photometer.

be greater than the former; hence the desire for precise (i.e. low uncertainty) analytical chemical methods.

2.3 Estimated response surfaces and models

Consider the set of experimental points shown in Fig. 6. It is probably unrealistic to try to write a mechanistic model (see Chap. 12) to explain these data. Our knowledge of metallurgy would be quickly exceeded. It is possible, however, to propose an empirical model that might provide a good description of the data. One possible empirical model might be the equation of a parabola

$$y_{1i} = b_0 + b_1 x_{1i} + b_{11} x_{1i}^2 + e_{1i} \tag{3}$$

where y_{1i} represents the system response (tensile strength) for experiment i, x_{1i} represents the system factor (percent manganese in this case), b_0 is an offset parameter, b_1 is a first-order parameter, b_{11} is a second-order parameter, and e_{1i} is a residual or deviation between what is observed and what is predicted by the model. This is a single-factor model of a single-factor response surface. Matrix least-squares techniques (see Chap. 13) can be used to fit this model to the data shown in Fig. 6.

As an example, suppose the data in Fig. 6 have the following numerical values (**D** is an "experimental design matrix" that has a number of rows equal to the number of experiments and a number of columns equal to the number of factors; each element of the **D** matrix thus represents the value of a particular factor in a particular experiment).

$$\mathbf{D} = \begin{bmatrix} 3 \\ 4 \\ 4 \\ 4 \\ 5 \\ 5 \\ 6 \\ 6 \\ 6 \\ 7 \end{bmatrix} \quad \mathbf{y} = \begin{bmatrix} 58 \\ 87 \\ 93 \\ 91 \\ 105 \\ 98 \\ 88 \\ 90 \\ 89 \\ 61 \end{bmatrix}$$

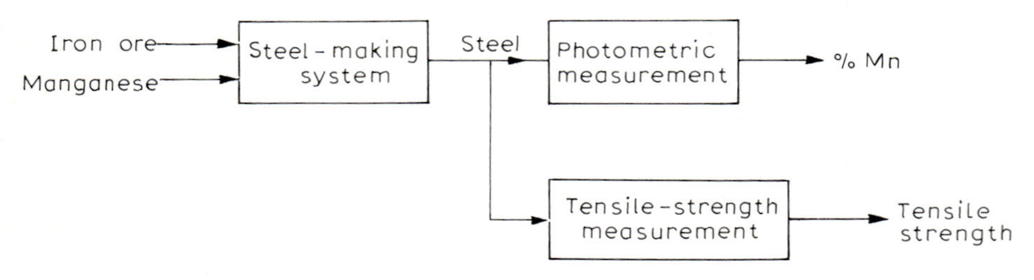

Fig. 4. Systems theory diagram showing the relationships among a steel-making system and two measurement systems.

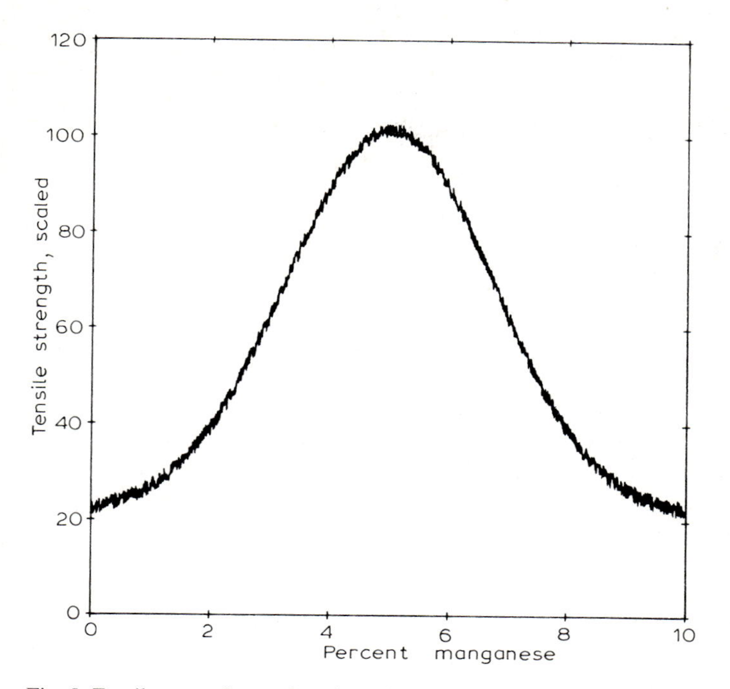

Fig. 5. Tensile strength as a function of the percent manganese in steel (hypothetical data).

We can fit eqn. (3) by defining an **X** matrix, the columns of which contain the coefficients of the model parameters for each experiment.

$$\mathbf{X} = \begin{bmatrix} 1 & 3 & 9 \\ 1 & 4 & 16 \\ 1 & 4 & 16 \\ 1 & 4 & 16 \\ 1 & 5 & 25 \\ 1 & 5 & 25 \\ 1 & 6 & 36 \\ 1 & 6 & 36 \\ 1 & 6 & 36 \\ 1 & 7 & 49 \end{bmatrix} \quad (\mathbf{X}'\mathbf{X}) = \begin{bmatrix} 10 & 50 & 264 \\ 50 & 264 & 1460 \\ 264 & 1460 & 8388 \end{bmatrix}$$

$$(\mathbf{X}'\mathbf{X})^{-1} = \begin{bmatrix} 32.155 & -13.183 & 1.2826 \\ -13.183 & 5.5062 & -0.53438 \\ 1.2826 & -0.54348 & 0.054348 \end{bmatrix}$$

$$(\mathbf{X}'\mathbf{y}) = \begin{bmatrix} 860 \\ 4302 \\ 22534 \end{bmatrix}$$

$$\mathbf{b} = (\mathbf{X}'\mathbf{X})^{-1}(\mathbf{X}'\mathbf{y}) = \begin{bmatrix} -158.41 \\ 103.40 \\ -10.326 \end{bmatrix}$$

The resulting equation is

$$y_1 = -158.41 + 103.40x_1 - 10.326x_1^2 \tag{4}$$

This model is plotted as the solid line in Fig. 6. The agreement between the model

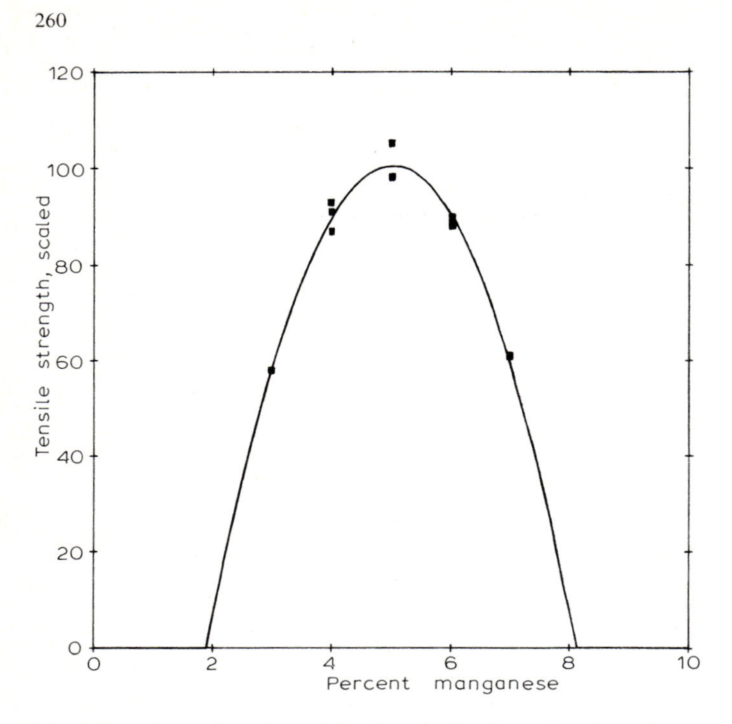

Fig. 6. Experimental results and fitted model for data taken from the system described in Fig. 4.

and the data is good; any deviations seem to be accounted for by the uncertainties that exist in the measurements themselves and not by any lack of fit between the model and the data.

Presumably, the results shown in Fig. 6 came from the same system represented by Figs. 4 and 5. Although the model represented by eqn. (4) is adequate over the domain of acquired data shown in Fig. 6, serious discrepancies between what this model predicts and the actual behavior of the system (Fig. 5) would exist both at lower and at higher percent manganese. This discrepancy, or lack of fit, is a source of uncertainty when trying to use an inadequate model to describe the true behavior of systems.

In general, mechanistic models should be used if systems are adequately understood and there is some guarantee that the systems will not deviate greatly from their expected behavior. Otherwise, mechanistic models might be seriously misleading for predictive purposes; unbiased empirical models might prove to be better choices. Empirical models must be used when systems are not well understood and mechanistic models are not possible.

Full second-order polynomial models [e.g. eqn. (4)] are very useful empirical models. We will use these models in much of the next few chapters because of their versatility in describing a wide variety of naturally occurring response surfaces over limited domains of the factors.

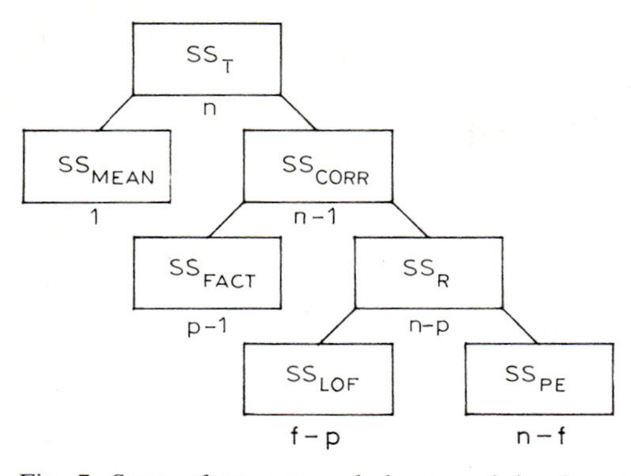

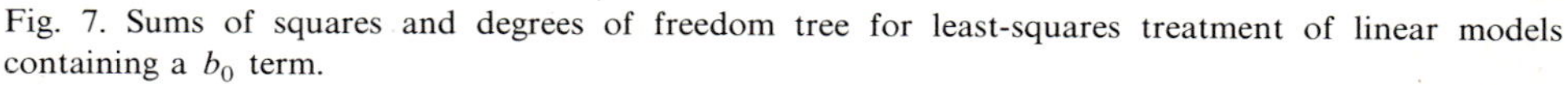
Fig. 7. Sums of squares and degrees of freedom tree for least-squares treatment of linear models containing a b_0 term.

The statistical basis of judging the adequacy of models has been given in Chap. 5 and is summarized here as the sum of squares and degrees of freedom tree shown in Fig. 7. The coefficient of multiple determination (R^2) and the F-test for the significance of regression, F_{REG}, offer useful means of evaluating how good the model is. The F-test for lack of fit, F_{LOF}, is useful as a means of deciding if a better model can reasonably be expected to be found.

We can calculate the values of the coefficient of multiple determination (i.e. the sum of squares due to regression divided by the sum of squares corrected for the mean), the F-ratio for the significance of regression, and the F-ratio for the lack of fit for the results of the previous example and can calculate at what levels the F-values are significant.

$$R^2 = \frac{SS_{FACT}}{SS_{CORR}} = \frac{1962.24}{2018.00} = 0.97237$$

$$F_{REG} = \frac{SS_{FACT}/(p-1)}{SS_R/(n-p)} = \frac{1962.24/2}{55.7578/7} = 123.17$$

$$F_{LOF} = \frac{SS_{LOF}/(f-p)}{SS_{PE}/(n-f)} = \frac{10.5911/2}{45.1667/5} = 0.5862$$

where n is the number of experiments in the set, f is the number of distinctly different factor combinations, and p is the number of parameters in the model.

The F-ratio for regression is significant at the 99.99965% level of confidence. The F-ratio for lack of fit is significant at the 41% level of confidence, i.e. there is no reason to be concerned about the lack of fit, the model is probably adequate.

3. Two-factor response surfaces and models

Many systems exhibit responses that are functions of not one but two factors. Examples of such systems are absorbance as a function of both the determinand

and an interferent, tensile strength of steel as a function of both manganese and cobalt concentrations, and liquid chromatographic retention time as a function of both pH and ion-pairing reagent concentration. In this section, we show the application of empirical models to the two-factor case. The concepts developed here can be easily expanded to cover the multifactor case.

3.1 First-order models

Let us consider again the photometric measurement technique described in Sect. 2.1, but rewrite Beer's Law in the form

$$A_j = \epsilon_j b C_j \tag{5}$$

where A_j is the absorbance ($= -\log[I/I_0]$) of compound j at a given wavelength λ, ϵ_j is the molar absorptivity of j at λ, and C_j is the molar concentration of j. If an interfering substance, k, also absorbs radiation at this wavelength, then the absorbance caused by this interfering substance, A_k, may be written

$$A_k = \epsilon_k b C_k \tag{6}$$

where ϵ_k is the molar absorptivity of the interferent and C_k is the concentration of the interferent.

It is a characteristic of photometric absorption that the total absorbance, A_t, is the sum of the individual absorbances of all absorbing species

$$A_t = A_j + A_k = \epsilon_j b C_j + \epsilon_k b C_k \tag{7}$$

This model may be written in the form

$$y_{1i} = b_1 x_{1i} + b_2 x_{2i} + e_{1i} \tag{8}$$

where $y_{1i} = A_t$, $b_1 = \epsilon_j b$, $b_2 = \epsilon_k b$, $x_{1i} = C_j$ and $x_{2i} = C_k$. A response surface for this model is shown in Fig. 8. Suppose we have obtained experimental data at $x_{11} = 1$, $x_{21} = 1$ and $x_{12} = 2$, $x_{22} = 2$.

$$\mathbf{D} = \begin{bmatrix} 1 & 1 \\ 2 & 2 \end{bmatrix} \quad \mathbf{y} = \begin{bmatrix} 4 \\ 7 \end{bmatrix}$$

If we try to fit the model represented by eqn. (8) to this data, then

$$\mathbf{X} = \begin{bmatrix} 1 & 1 \\ 2 & 2 \end{bmatrix} \quad (\mathbf{X}'\mathbf{X}) = \begin{bmatrix} 5 & 5 \\ 5 & 5 \end{bmatrix}$$

The determinant of $(\mathbf{X}'\mathbf{X})$ is equal to zero and the $(\mathbf{X}'\mathbf{X})^{-1}$ matrix is undefined. Therefore, the model cannot be fit to this data. An explanation for this is that the data points (and the origin) all lie in a straight line (i.e. they are co-linear). But the model is that of a plane. Thus, an infinite number of planes could pass equally well through the line of data and no unique solution exists.

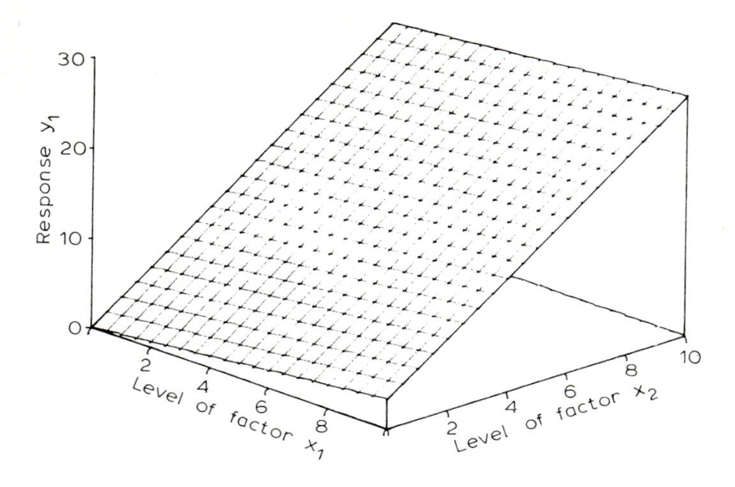

Fig. 8. Two-factor response surface for the model $y_1 = \frac{1}{3}x_1 + \frac{7}{3}x_2$. See eqn. (8).

Let us try a different set of data obtained at (1, 2) and (2, 1).

$$\mathbf{D} = \begin{bmatrix} 1 & 2 \\ 2 & 1 \end{bmatrix} \quad \mathbf{y} = \begin{bmatrix} 5 \\ 3 \end{bmatrix}$$

$$\mathbf{X} = \begin{bmatrix} 1 & 2 \\ 2 & 1 \end{bmatrix} \quad (\mathbf{X}'\mathbf{X}) = \begin{bmatrix} 5 & 4 \\ 4 & 5 \end{bmatrix}$$

$$(\mathbf{X}'\mathbf{X})^{-1} = \begin{bmatrix} 5/9 & -4/9 \\ -4/9 & 5/9 \end{bmatrix} \quad (\mathbf{X}'\mathbf{y}) = \begin{bmatrix} 11 \\ 13 \end{bmatrix}$$

$$\mathbf{b} = (\mathbf{X}'\mathbf{X})^{-1}(\mathbf{X}'\mathbf{y}) = \begin{bmatrix} 1/3 \\ 7/3 \end{bmatrix}$$

The two data points plus the origin are not co-linear and the fitted plane is uniquely defined. The equation of the plane is

$$y_1 = \tfrac{1}{3}x_1 + \tfrac{7}{3}x_2$$

Note that, in Fig. 8, the slope of the response surface with respect to the factor x_1 is always equal to b_1, independent of the values of x_1 and x_2. Similarly, the slope of the response surface with respect to the factor x_2 is always equal to b_2. This is a proper interpretation of eqn. (8): b_1 and b_2 are the partial derivatives of the response y_1 with respect to x_1 and x_2, respectively.

$$\frac{\partial y_1}{\partial x_1} = b_1 \tag{9}$$

$$\frac{\partial y_1}{\partial x_2} = b_2 \tag{10}$$

A more generally useful form of the two-factor first-order polynomial model includes a b_0 term.

$$y_1 = b_0 + b_1 x_1 + b_2 x_2 \tag{11}$$

With this model, the response is not constrained to go through the origin. Addition of the b_0 term may be thought of as giving the response surface the degree of freedom required to move up or down in a "vertical" direction.

3.2 First-order models with interaction

In many systems, the effect of one factor will depend upon the value (or *level*) of another factor. This phenomenon is called *factor interaction* and may be written mathematically as

$$\frac{\partial y_1}{\partial x_1} = b_1 + b_{12} x_2 \tag{12}$$

That is, when the value of x_2 is equal to zero, the slope of the response surface with respect to x_1 is equal to b_1, as before. However, when x_2 is not equal to zero, the slope of the response surface with respect to x_1 depends upon the value of x_2. Substituting this value $(\partial y_1/\partial x_1)$ for the effects of x_1 in eqn. (11) gives

$$y_{1i} = b_0 + (b_1 + b_{12} x_{2i}) x_{1i} + b_2 x_{2i} + e_{1i} \tag{13}$$

or

$$y_{1i} = b_0 + b_1 x_{1i} + b_2 x_{2i} + b_{12} x_{1i} x_{2i} + e_{1i} \tag{14}$$

This is the equation of a full two-factor first-order model with interaction. It is interesting to note that eqn. (14) can be rearranged to give

$$y_{1i} = b_0 + b_1 x_{1i} + (b_2 + b_{12} x_{1i}) x_{2i} + e_{1i} \tag{15}$$

Thus, it is also true that

$$\frac{\partial y_1}{\partial x_2} = b_2 + b_{12} x_1 \tag{16}$$

as expected; that is, the effect of the factor x_2 also depends upon the value of the factor x_1.

As an example, suppose we want to investigate the effect of two variables, chart speed x_1 and recorder sensitivity x_2, on the precision of measuring chromatographic peaks. Assume we have gathered the following set of factor levels and corresponding responses.

$$\mathbf{D} = \begin{bmatrix} -1 & -1 \\ -1 & +1 \\ +1 & -1 \\ +1 & +1 \end{bmatrix} \quad \mathbf{y} = \begin{bmatrix} 3 \\ 3 \\ 3 \\ 7 \end{bmatrix}$$

The factor levels have been "coded" so that a real chart speed of, say, 1 cm min^{-1} is given a coded value of -1 and a chart speed of 5 cm min^{-1} is given a coded value of $+1$. Similarly, coded recorder sensitivities of -1 and $+1$ might correspond to 1 mV full scale and 5 mV full scale, respectively. The responses (uncoded) might represent the standard deviation of peak heights from 10 repetitive injections. Let us fit the model given by eqn. (14) to this data.

$$\mathbf{X} = \begin{bmatrix} +1 & -1 & -1 & +1 \\ +1 & -1 & +1 & -1 \\ +1 & +1 & -1 & -1 \\ +1 & +1 & +1 & +1 \end{bmatrix} \qquad (\mathbf{X}'\mathbf{X}) = \begin{bmatrix} 4 & 0 & 0 & 0 \\ 0 & 4 & 0 & 0 \\ 0 & 0 & 4 & 0 \\ 0 & 0 & 0 & 4 \end{bmatrix}$$

$$(\mathbf{X}'\mathbf{y}) = \begin{bmatrix} 16 \\ 4 \\ 4 \\ 4 \end{bmatrix} \qquad (\mathbf{X}'\mathbf{X})^{-1} = \begin{bmatrix} 1/4 & 0 & 0 & 0 \\ 0 & 1/4 & 0 & 0 \\ 0 & 0 & 1/4 & 0 \\ 0 & 0 & 0 & 1/4 \end{bmatrix}$$

$$\mathbf{b} = (\mathbf{X}'\mathbf{X})^{-1}(\mathbf{X}'\mathbf{y}) = \begin{bmatrix} 4 \\ 1 \\ 1 \\ 1 \end{bmatrix}$$

A graph of the response surface for the fitted model

$$y_1 = 4 + 1x_1 + 1x_2 + 1x_1x_2 \qquad (17)$$

is given in Fig. 9 where the concept of interaction is clearly evident. An interpreta-

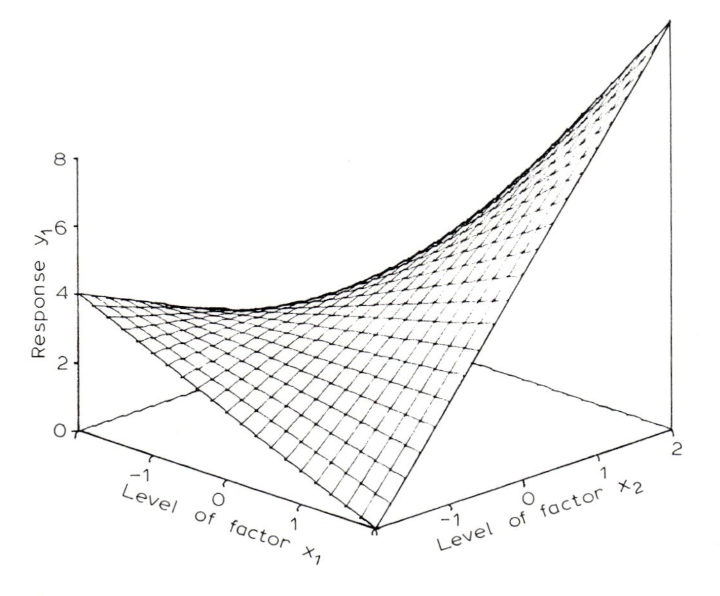

Fig. 9. Two-factor response surface for the model $y_1 = 4 + 1x_1 + 1x_2 + 1x_1x_2$. See eqn. (14).

Reference p. 269

tion of the figure is that, at low recorder sensitivity (-1 level of x_1), the chart speed does not have much of an effect on precision.

$$\begin{aligned} y_1 &= 4 + 1x_1 + 1x_2 + 1x_1x_2 \\ &= 4 + 1(-1) + 1x_2 + 1(-1)x_2 \\ &= 3 \end{aligned} \tag{18}$$

However, at high recorder sensitivity ($+1$ level of x_1), high chart speed ($+1$ level of x_2) gives worse precision than does low chart speed (-1 level of x_2).

$$\begin{aligned} y_1 &= 4 + 1x_1 + 1x_2 + 1x_1x_2 \\ &= 4 + 1(+1) + 1x_2 + 1(+1)x_2 \\ &= 5 + 2x_2 \end{aligned} \tag{19}$$

3.3 Full second-order polynomial models

The concept of interaction can also be applied to individual factors. For example, in Fig. 6, the effect of percent manganese (factor x_1) depends upon the value of that same factor. That is

$$\frac{\partial y_1}{\partial x_1} = b_1 + b_{11}x_1 \tag{20}$$

If a second factor were involved, it might also be true that the effect of this second

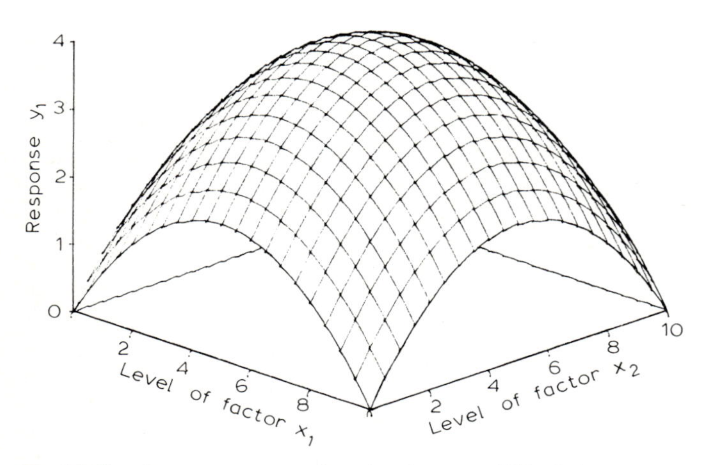

Fig. 10. Two-factor response surface showing a possible relationship between square wave polarographic sensitivity, voltage increment, and drop time. Scaled equation: $y_1 = 0 + 0.8x_1 + 0.8x_2 - 0.08x_1^2 - 0.08x_2^2$.

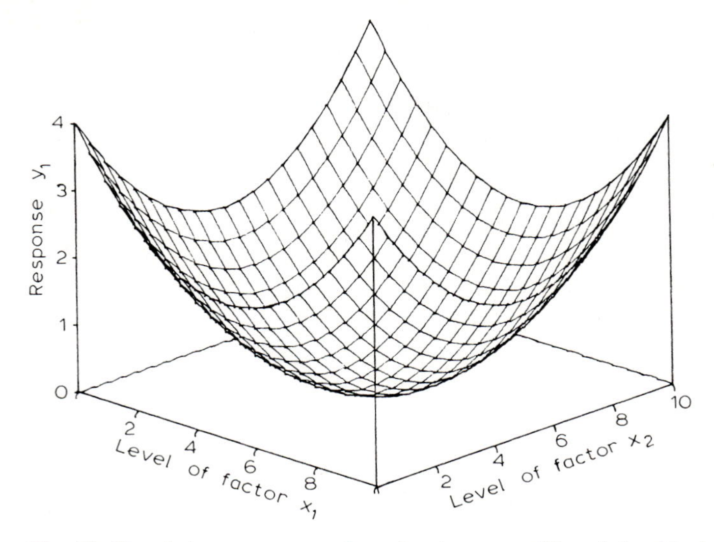

Fig. 11. Two-factor response surface showing a possible relationship between the relative effect of bromine interference in the determination of chlorine, pH, and temperature. Scaled equation: $y_1 = 4 - 0.8x_1 - 0.8x_2 + 0.08x_1^2 + 0.08x_2^2$.

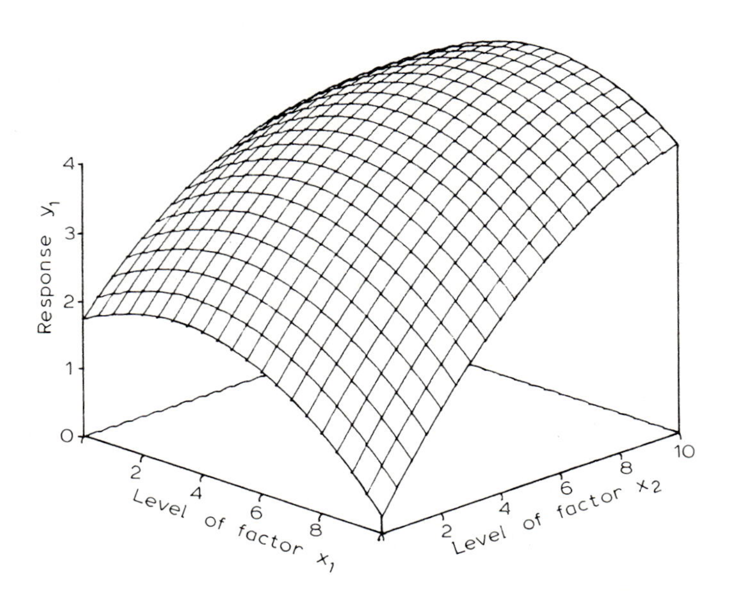

Fig. 12. Two-factor response surface showing a possible relationship between sample throughput efficiency of a laboratory, pay rate, and number of personnel. Scaled equation; $y_1 = 1.75 + 0.25x_1 + 0.50x_2 - 0.04x_1^2 - 0.03x_2^2 + 0.02x_1x_2$.

Reference p. 269

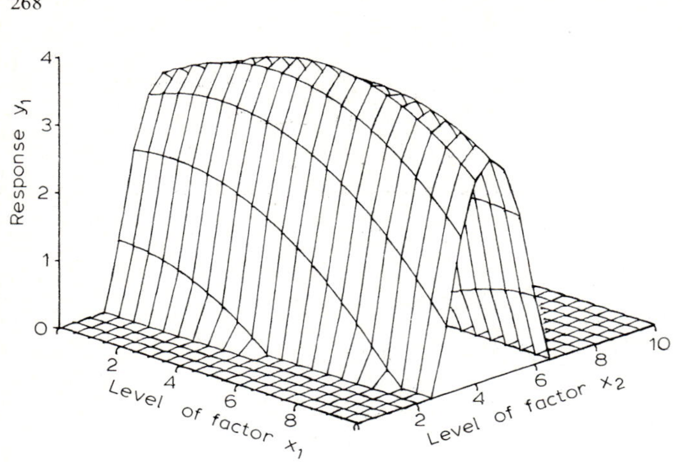

Fig. 13. Two-factor response surface showing a possible relationship between turbidity, temperature, and pH. Scaled equation: $y_1 = -9 - 0.4x_1 + 7x_2 - 0.04x_1^2 - 1x_2^2 + 0.2x_1x_2$.

factor depends upon its value

$$\frac{\partial y_1}{\partial x_2} = b_2 + b_{22}x_2 \tag{21}$$

Substitution of these *self-interaction* effects into the model expressed by eqn. (14) gives

$$y_{1i} = b_0 + (b_1 + b_{11}x_{1i})x_{1i} + (b_2 + b_{22}x_{2i})x_{2i} + b_{12}x_{1i}x_{2i} + e_{1i} \tag{22}$$

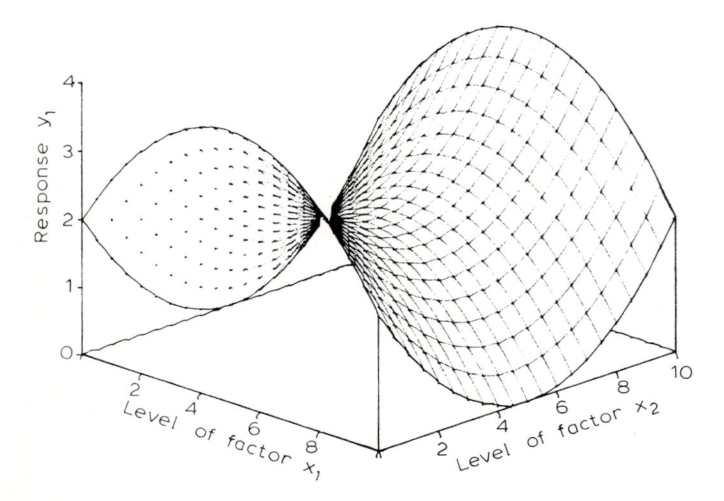

Fig. 14. Two-factor response surface showing a possible relationship between analytical recovery and the concentrations of two different reagents. Scaled equation: $y_1 = 2 + 0.8x_1 - 0.8x_2 - 0.08x_1^2 + 0.08x_2^2$.

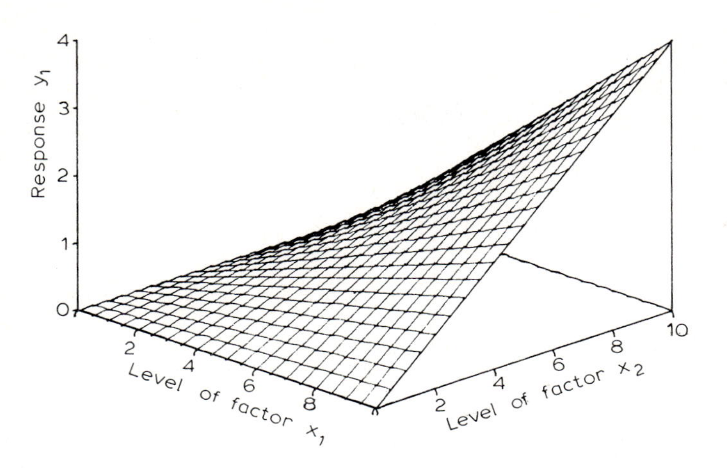

Fig. 15. Two-factor response surface showing a possible relationship between reaction rate and two different reactants. Scaled equation: $y_1 = 0.04x_1x_1$.

or

$$y_{1i} = b_0 + b_1x_{1i} + b_2x_{2i} + b_{11}x_{1i}^2 + b_{22}x_{2i}^2 + b_{12}x_{1i}x_{2i} + e_{1i} \tag{23}$$

Equation (23) represents a full two-factor second-order polynomial model. Such equations are exceptionally versatile for use as empirical models in many systems over a limited domain of the factors [1].

Figures 10–15 show examples of response surfaces that can be fit by the model of eqn. (23). These figures might represent, for example, the following analytical chemical response surfaces: Fig. 10, square wave polarographic sensitivity as a function of voltage increment and drop time; Fig. 11, relative effect of bromine interference in the determination of chlorine as a function of pH and temperature; Fig. 12, sample throughput efficiency of a laboratory as a function of pay rate and number of personnel; Fig. 13, turbidity as a function of temperature and pH; Fig. 14, analytical recovery as a function of the concentrations of two different analytical reagents; and Fig. 15, reaction rate as a function of two different reactants.

Reference

1 G.E.P. Box and K.B. Wilson, On the experimental attainment of optimum conditions, J. R. Stat. Soc. Ser. B, 13 (1951) 1.

Recommended reading

T.B. Barker, Quality by Experimental Design, Dekker, New York, 1985.

L. von Bertalanffy, General System Theory. Foundations, Development, Applications, Braziller, New York, 1968.

G.E.P. Box, W.G. Hunter and J.S. Hunter, Statistics for Experimenters. An Introduction to Design, Data Analysis, and Model Building, Wiley, New York, 1978.

W.G. Cochran and G.M. Cox, Experimental Designs, Wiley, New York, 1957.
O.L. Davies (Ed.), Design and Analysis of Industrial Experiments, Hafner, New York, 2nd edn., 1956.
S.N. Deming and S.L. Morgan, Experimental Design: A Chemometric Approach, Elsevier, Amsterdam, 1987.
A.J. Duncan, Quality Control and Industrial Statistics, Irwin, Homewood, IL, 1959.
R.A. Fisher, The Design of Experiments, Hafner, New York, 1971.
P.M.W. John, Statistical Design and the Analysis of Experiments, MacMillan, New York, 1971.
J. Mandel, The Statistical Analysis of Experimental Data, Wiley, New York, 1964.
W. Mendenhall, Introduction to Linear Models and the Design and Analysis of Experiments, Duxbury, Belmont, CA, 1968.
M.G. Natrella, Experimental Statistics, National Bureau of Standards Handbook 91, U.S. Government Printing Office, Washington, DC, 1963.
G.W. Snedecor and W.G. Cochran, Statistical Methods, Iowa State University Press, Ames, IA, 6th edn., 1967.
G.M. Weinberg, An Introduction to General Systems Thinking, Wiley, New York, 1975.
E.B. Wilson, Jr., An Introduction to Scientific Research, McGraw-Hill, New York, 1952.
W.J. Youden, Statistical Methods for Chemists, Wiley, New York, 1951.

Chapter 17

Exploration of Response Surfaces

1. Introduction

The goal of much research and development in analytical chemistry is to improve and understand the systems with which we are working.

As Box and Wilson [1] have pointed out, this is largely an experimental endeavor involving the exploration and exploitation of response surfaces. Unless good mechanistic models are available to us before we begin the project, we must learn how the system behaves by carrying out a series of designed experiments and then build our models from the experimental results.

In this chapter, we point out the strengths and weaknesses of some experimental designs that can be used to explore response surfaces in analytical chemistry.

2. Two-level factorial designs

Factorial designs are a class of experimental *designs* that are generally very economical, that is they offer a large amount of useful information from a small number of experiments. When the number of experiments that can be carried out is limited, then factorial designs offer an efficient way to obtain maximum information from these experiments.

Factorial designs involve a certain number of *levels* (or values) of each of the *factors* (or variables) of interest. Thus, if we were interested in the effects of both pH and temperature on chromatographic retention, we might want to consider a two-factor factorial design; if we were interested in the effects of pH, temperature, and mobile phase modifier concentration, we might want to consider a three-factor factorial design.

The simplest factorial designs are those that involve just two levels of each of the factors. Figure 1 shows the locations of experiments associated with a two-level two-factor factorial design and Fig. 2 shows the locations of experiments associated with a two-level three-factor factorial design.

Factorial designs are conveniently designated as a base raised to a power, e.g. 2^2 and 2^3. The base is the number of levels associated with each factor (two in this section) and the power is the number of factors in the study (two or three for Figs. 1 and 2, respectively). It is convenient that the designation for factorial designs is equal to the number of different factor combinations (experimental conditions) in the set of experiments, e.g. $2^2 = 4$ experimental conditions for Fig. 1 and $2^3 = 8$ factor combinations for Fig. 2.

Two-level factorial designs can be used to fit models that contain first-order effects (e.g. b_1x_1) and interaction effects (e.g. $b_{12}x_1x_2$). Two-level factorial designs

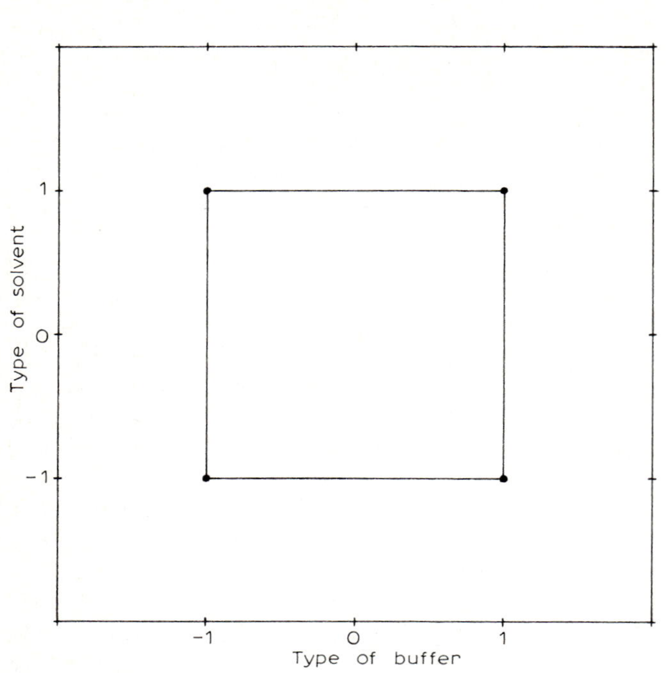

Fig. 1. A two-level two-factor factorial design involving qualitative factors.

cannot be used to fit models that contain second- or higher-order terms in a single factor (e.g. $b_{11}x_1^2$); this limitation is caused by the fact that curvature in a single factor can only be detected by experiments at three or more levels of that factor and

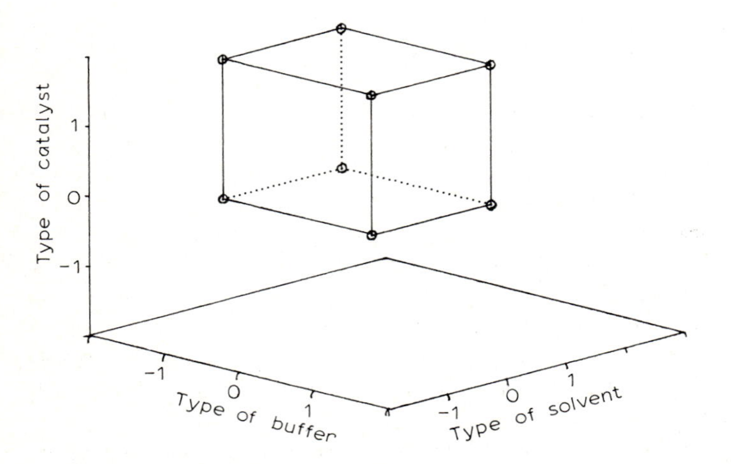

Fig. 2. A two-level three-factor factorial design involving qualitative factors.

two-level factorial designs (by definition) do not meet this requirement. Two-level multifactor factorial designs can be used to estimate higher-order interaction effects (e.g. $b_{123}x_1x_2x_3$). One important area in which such interactions are found frequently is in kinetics where, for example, three-component rate expressions exist, e.g. Rate $= k[A][B][C] = b_{123}x_1x_2x_3$. However, third- and higher-order interactions tend to be rare in most areas of analytical chemistry.

3. Qualitative factors

Factors can be divided into two categories, quantitative and qualitative. Temperature, pressure, concentration, and time are all examples of quantitative factors. Each has a meaningful numerical value that represents the extent or amount of that factor. Type of buffer, type of solvent, and source of supply are all qualitative factors. Any numerical value that might be assigned to any of these factors would not be meaningful in the sense of indicating the extent or amount of that factor.

Consider the following experimental problem. Suppose we want to find out if the use of phosphate buffer instead of acetate buffer (qualitative factors) makes a difference to the sensitivity of an analytical method. Suppose, further, that we want to find out if the use of methyl ethyl ketone instead of acetone as a solvent makes a difference in a later part of the same method. And, finally, suppose we want to find out if there is an interaction between these two factors. Let x_1 be the factor "type of buffer", let x_2 be the factor "type of solvent", and let the response, y_1, be the scaled absorbance for a fixed amount of analyte. The model

$$y_{1i} = b_0 + b_1x_{1i} + b_2x_{2i} + b_{12}x_{1i}x_{2i} + e_{1i} \tag{1}$$

will allow us to estimate the effects we are interested in: b_1 will be the effect of the type of buffer, b_2 will be the effect of the type of solvent, b_{12} will be the interaction effect between the two factors, and b_0 is simply an offset term that avoids forcing the model to go through the origin.

A two-level two-factor factorial design would be appropriate for this problem. The experimental design has four factor combinations and the model has four parameters (this meets the general requirement of experimental designs that the number of factor combinations must be equal to or greater than the number of parameters in the model). If we do not replicate the experiments, there will be no degrees of freedom for residuals ($n - p = 0$; see Fig. 7 in Chap. 16); therefore, let us carry out replicate experiments at each of the four factor combinations so there will be four degrees of freedom for the residuals. Note, however, that these residuals are estimates of purely experimental uncertainty only; there are still no degrees of freedom for estimating lack of fit (see Chap. 16). The only way to obtain degrees of freedom for lack of fit would be to carry out experiments at additional factor combinations. Thus, fitting the model represented by eqn. (1) to data from a two-level two-factor factorial design will appear to give us a "perfect" fit.

One problem remains to be solved before we can actually carry out the experiments and treat the data: how can we use x_1 to express the type of buffer and how can we use x_2 to express the type of solvent? One way might be to give the acetate buffer the value 0 and the phosphate buffer the value 1; similarly, methyl ethyl ketone could be given the value 0 and acetone the value 1. However, if we use the values -1 and $+1$ instead of 0 and $+1$ in each case, we will obtain an $(\mathbf{X}'\mathbf{X})$ matrix that is trivial to invert. This is the scheme used in the labeling of Fig. 1. Given the data

$$\mathbf{D} = \begin{bmatrix} -1 & -1 \\ -1 & -1 \\ -1 & +1 \\ -1 & +1 \\ +1 & -1 \\ +1 & -1 \\ +1 & +1 \\ +1 & +1 \end{bmatrix} \quad \mathbf{y} = \begin{bmatrix} 2.8 \\ 3.2 \\ 6.1 \\ 5.9 \\ 4.4 \\ 4.6 \\ 5.7 \\ 5.3 \end{bmatrix}$$

let us fit the model represented by eqn. (1).

$$\mathbf{X} = \begin{bmatrix} +1 & -1 & -1 & +1 \\ +1 & -1 & -1 & +1 \\ +1 & -1 & +1 & -1 \\ +1 & -1 & +1 & -1 \\ +1 & +1 & -1 & -1 \\ +1 & +1 & -1 & -1 \\ +1 & +1 & +1 & +1 \\ +1 & +1 & +1 & +1 \end{bmatrix} \quad (\mathbf{X}'\mathbf{X}) = \begin{bmatrix} 8 & 0 & 0 & 0 \\ 0 & 8 & 0 & 0 \\ 0 & 0 & 8 & 0 \\ 0 & 0 & 0 & 8 \end{bmatrix}$$

$$(\mathbf{X}'\mathbf{X})^{-1} = \begin{bmatrix} 1/8 & 0 & 0 & 0 \\ 0 & 1/8 & 0 & 0 \\ 0 & 0 & 1/8 & 0 \\ 0 & 0 & 0 & 1/8 \end{bmatrix}$$

$$(\mathbf{X}'\mathbf{y}) = \begin{bmatrix} 38 \\ 2 \\ 8 \\ -4 \end{bmatrix} \quad \mathbf{b} = (\mathbf{X}'\mathbf{X})^{-1}(\mathbf{X}'\mathbf{y}) = \begin{bmatrix} 4.75 \\ 0.25 \\ 1.00 \\ -0.50 \end{bmatrix}$$

The resulting equation is

$$y_1 = 4.75 + 0.25x_1 + 1.00x_2 - 0.50x_1x_2 \tag{2}$$

The results are plotted as a response surface in Fig. 3.

The interpretation of the results of eqn. (2) might be as follows. The effect of changing the buffer from acetate (-1) to phosphate $(+1)$ gives, on the average in this example, an increase of 0.50 units in response (the value of $b_1 = 0.25$ in eqn. (2) is the change per unit in x_1; $\Delta x_1 = 2$ units in going from -1 to $+1$, so the net change in response is $0.25 \times 2 = 0.50$. The effect of changing the solvent from methyl ethyl ketone (-1) to acetone $(+1)$ gives, on the average in this example, an increase of 2.0 units in response.

The interaction effect in this example is relatively large and may be interpreted by rearranging the fitted model as

$$y_1 = 4.75 + (0.25 - 0.50x_2)x_1 + 1.00x_2 \tag{3}$$

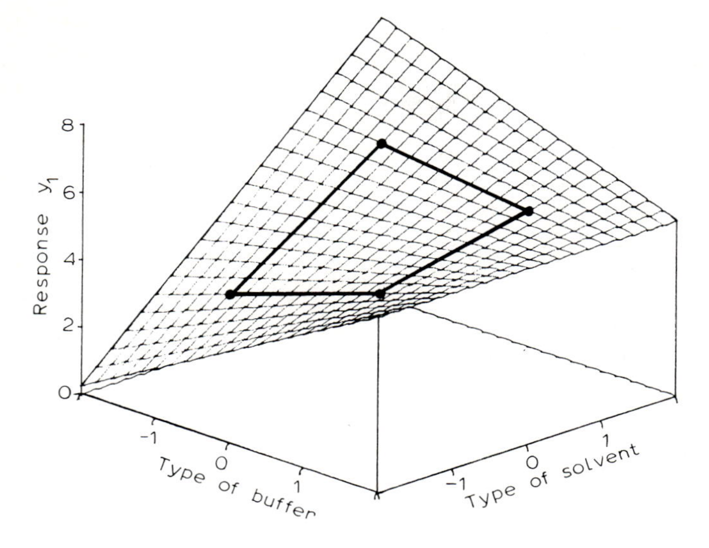

Fig. 3. Response surface corresponding to the fitted model represented by eqn. (2).

The effect of x_1 is thus seen to depend upon the level of x_2. When $x_2 = -1$ (methyl ethyl ketone), the coefficient of x_1 in eqn. (3) is 0.75 and the effect of changing from acetate to phosphate is to *increase* the response by 1.50 units. When $x_2 = +1$ (acetone), the coefficient of x_1 in eqn. (3) is -0.25 and the effect of changing from acetate to phosphate is now to *decrease* the response by 0.50 units. A similar interpretation can be given for the effect of x_2 and its dependence on the value of x_1; this is left as an exercise for the reader.

4. Cautions about the chosen system of coding

The interpretation of b_0 deserves comment. This parameter is the expected response *at the origin of the coordinate system*. If we had used a different set of coded numerical values for buffer and solvent levels (say 0 and +2 instead of -1 and +1), the estimated value of b_0 would have been quite different. It is also true that the estimated values of b_1, b_2, and b_{12} would also be different. For example, given the re-coded data

$$\mathbf{D} = \begin{bmatrix} 0 & 0 \\ 0 & 0 \\ 0 & 2 \\ 0 & 2 \\ 2 & 0 \\ 2 & 0 \\ 2 & 2 \\ 2 & 2 \end{bmatrix} \qquad \mathbf{y} = \begin{bmatrix} 2.8 \\ 3.2 \\ 6.1 \\ 5.9 \\ 4.4 \\ 4.6 \\ 5.7 \\ 5.3 \end{bmatrix}$$

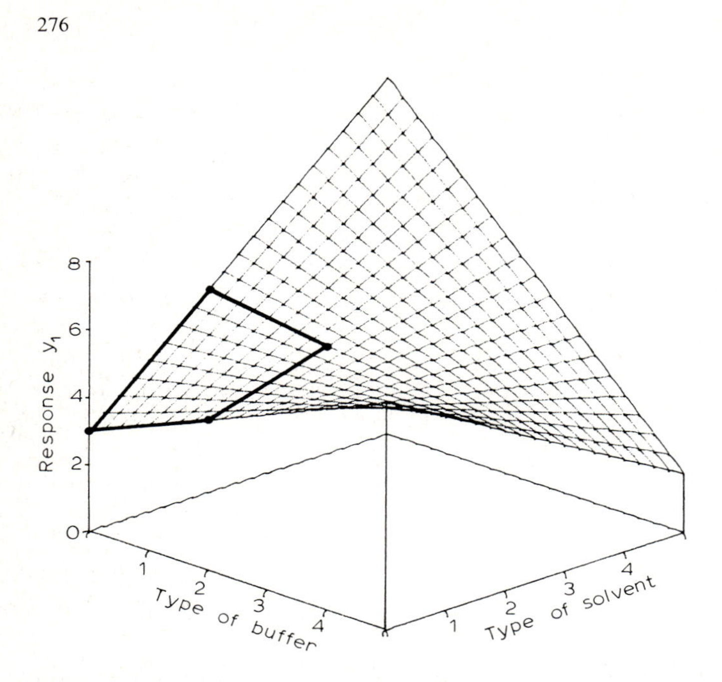

Fig. 4. Response surface corresponding to the fitted model represented by eqn. (4).

when we fit the model represented by eqn. (1) we now obtain

$$\mathbf{X} = \begin{bmatrix} +1 & 0 & 0 & 0 \\ +1 & 0 & 0 & 0 \\ +1 & 0 & 2 & 0 \\ +1 & 0 & 2 & 0 \\ +1 & 2 & 0 & 0 \\ +1 & 2 & 0 & 0 \\ +1 & 2 & 2 & 4 \\ +1 & 2 & 2 & 4 \end{bmatrix}$$

The fitted model is now

$$y_1 = 3.00 + 0.75x_1 + 1.50x_2 - 0.50x_1x_2 \quad (4)$$

Compare this equation with eqn. (2). The results are plotted in Fig. 4. Compare this figure with Fig. 3. It is evident that two different systems of coding result in two different descriptive equations and two different response surfaces for the same set of data.

Although these equations and figures appear to be quite different, their interpretation results in identical conclusions about the factor effects and their interactions. Equation (4) can be rearranged to the same form as eqn. (3)

$$y_1 = 3.00 + (0.75 - 0.50x_2)x_1 + 1.50x_2 \quad (5)$$

When $x_2 = 0$ (methyl ethyl ketone for this coding system), the coefficient of x_1 in eqn. (5) is again 0.75 and the effect of changing from acetate to phosphate is identical to what we found with the other system of coding [eqn. (3)]. When $x_2 = +2$ (acetone for this system of coding), the coefficient of x_1 in eqn. (5) is again -0.25 and the effect of changing from acetate to phosphate is identical to what we found with the other system of coding [eqn. (3)]. Thus, even though eqns. (3) and (5) are different (because of the different systems of coding), they contain the same information and describe the same response surface *with respect to the experimental points* (compare Figs. 3 and 4).

It should be obvious by now that the interpretation of the parameter estimates (b_0, b_1, b_2, and b_{12}) by themselves, one at a time, can be misleading if factor interaction is present in the model. One of the most useful means of evaluating the effects of individual factors is to group the coefficients of these factors in a manner similar to that shown in eqns. (3) and (5).

Whatever system of coding we use (-1 and $+1$, or 0 and $+2$), the arbitrary values assigned to the xs have no chemical or physical meaning: the x values correspond to abstract factors, not to real factors. It makes no sense to ask what response we might get if x_1 were equal to 3: our fitted model *will* blindly make a prediction about the response we would get if x_1 were equal to 3, but there is no meaningful way of choosing a buffer that has an "x_1 factor level" of 3. Thus, in the case of qualitative factors, factorial designs can be used to fit models that have *descriptive* character, but we should not attempt to derive *predictive* information from these descriptive models.

5. Quantitative factors

When working with *quantitative* factors, factorial designs can be used to fit models that have predictive character as well as descriptive character. Suppose that, instead of investigating the type of buffer (a qualitative factor) we were to investigate the pH of the solution (a quantitative factor). And suppose that instead of investigating the type of solvent (a qualitative factor) we were to investigate the dielectric constant of the solution (a quantitative factor). We could again use data from a two-level two-factor factorial design (see Fig. 5) to fit the model of eqn. (1) but, in this case, the x values correspond to real chemical and physical factors.

Given the data

$$\mathbf{D} = \begin{bmatrix} 8 & 72 \\ 8 & 72 \\ 8 & 78 \\ 8 & 78 \\ 10 & 72 \\ 10 & 72 \\ 10 & 78 \\ 10 & 78 \end{bmatrix} \quad \mathbf{y} = \begin{bmatrix} 5.1 \\ 4.9 \\ 6.3 \\ 6.3 \\ 5.4 \\ 5.6 \\ 7.7 \\ 7.9 \end{bmatrix}$$

References p. 290

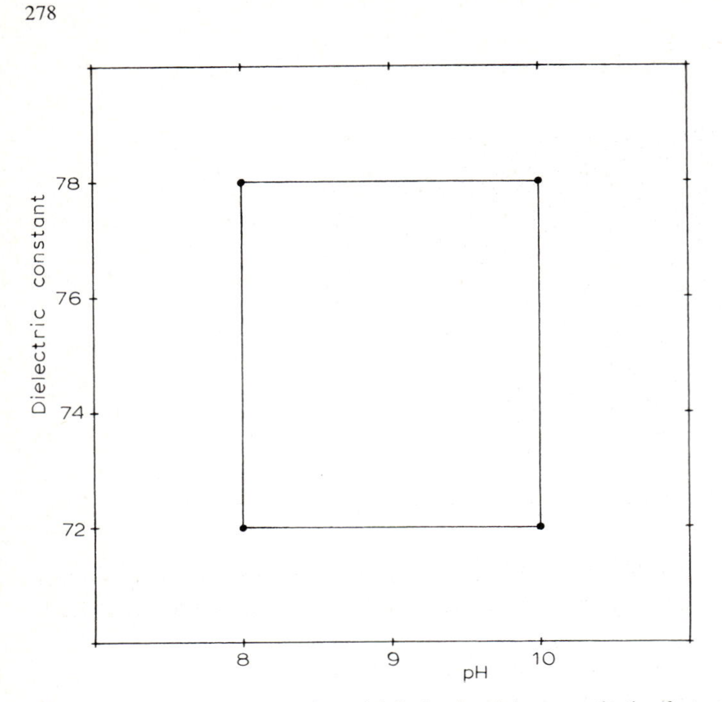

Fig. 5. A two-level two-factor factorial design involving quantitative factors.

let us fit the model represented by eqn. (1).

$$\mathbf{X} = \begin{bmatrix} 1 & 8 & 72 & 576 \\ 1 & 8 & 72 & 576 \\ 1 & 8 & 78 & 624 \\ 1 & 8 & 78 & 624 \\ 1 & 10 & 72 & 720 \\ 1 & 10 & 72 & 720 \\ 1 & 10 & 78 & 780 \\ 1 & 10 & 78 & 780 \end{bmatrix}$$

The fitted model is

$$y_1 = 35.4 - 5.75x_1 - 0.450x_2 + 0.0833x_1x_2 \tag{6}$$

A plot of this model is shown in Fig. 6.

From the model itself [eqn. (6)] it appears that increasing the value of x_1 (pH) causes the response to decrease $(-5.75x_1)$ and increasing the value of x_2 (dielectric constant) also causes the response to decrease $(-0.450x_2)$. Yet the plot of the model *over the domain of the experiments* shows just the opposite behavior (see Fig. 7). This apparent discrepancy is caused by our failure to take the interaction terms into account when examining the x_1 and x_2 effects. For the case of the x_1 effect, rearranging eqn. (6) gives

$$y_1 = 35.4 + (-5.75 + 0.0833x_2)x_1 - 0.450x_2 \tag{7}$$

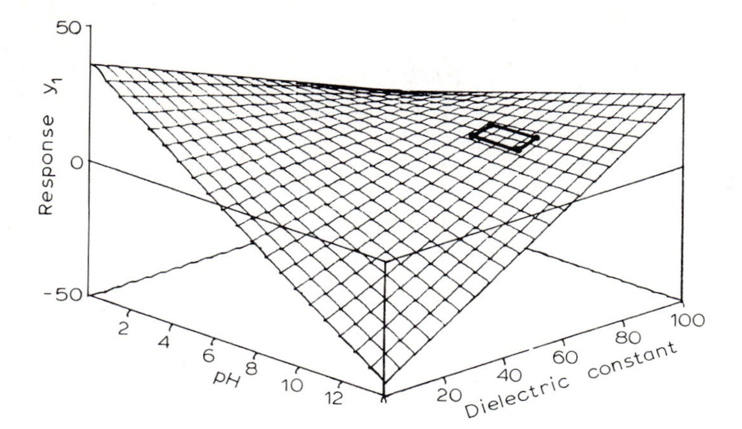

Fig. 6. Response surface corresponding to the fitted model represented by eqn. (4).

When x_2 is equal to 72 and 78, the coefficient of x_1 is 0.25 and 0.75, respectively; thus, over the domain of our experiments, increasing the pH causes the response to increase.

For the case of the x_2 effect, rearranging eqn. (6) gives

$$y_1 = 35.4 + (-0.450 + 0.0833x_1)x_2 - 5.75x_1 \tag{8}$$

When x_1 is equal to 8 and 10, the coefficient of x_2 is 0.217 and 0.383, respectively; thus, over the domain of our experiments, increasing the dielectric constant causes the response to increase.

With quantitative, real factors like pH and dielectric constant, it is now meaningful to ask what response we might get if pH were equal to, say, 9 and the dielectric

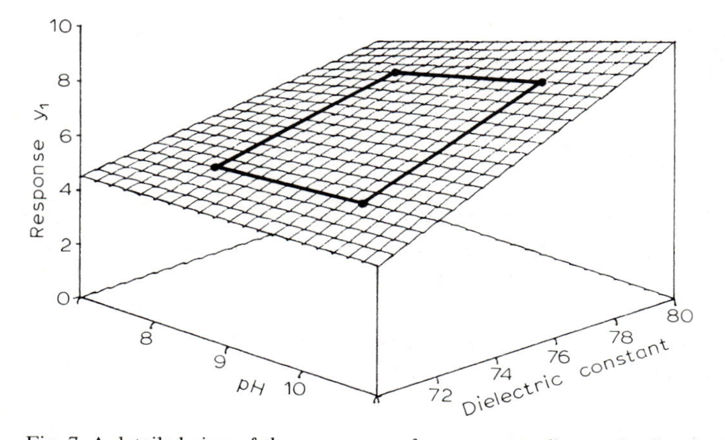

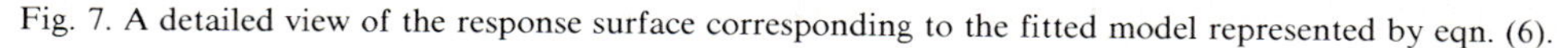

Fig. 7. A detailed view of the response surface corresponding to the fitted model represented by eqn. (6).

References p. 290

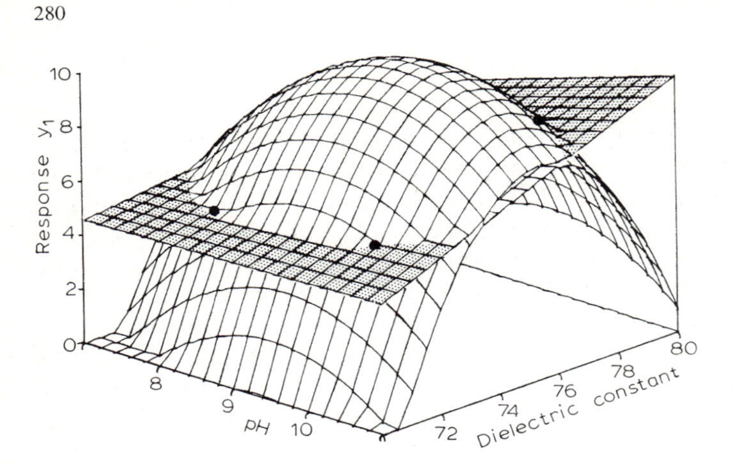

Fig. 8. The fitted model (shaded surface) and the actual response surface (unshaded surface).

constant were equal to, say, 75. For these conditions, the model of eqn. (6) [or eqns. (7) or (8)] predicts a response of 6.15. But will we actually get this response (or a response close to a value of 6.15) if we carry out an experiment at pH = 9 and dielectric constant = 75?

If our fitted model is of the correct form and if the parameters have been estimated reasonably precisely, then there will be very little difference between values predicted by the model and values actually obtained experimentally from the system at factor combinations not too far removed from our original area of experimentation. The statistical way of deciding if the model is of the correct form is to look at the sum of squares due to lack of fit (SS_{LOF}) and compare it with the sum of squares due to purely experimental uncertainty (SS_{PE}), taking into account the degrees of freedom of each sum of squares. Unfortunately, for the model we have been considering [eqn. (6)] and the experimental design we have used (Fig. 5), there are no degrees of freedom for estimating lack of fit. Thus, although our model does a good job of describing the data we have obtained, we still have no idea how good it will be as a means of estimating the response at factor combinations not already investigated.

In view of this, we should not be surprised that we might actually obtain at pH = 9 and dielectric constant = 75 a response equal to 10.0, not 6.15 as eqn. (6) predicts. Figure 8 illustrates a possible reason for the poor predictions of the model we have been using. In Fig. 8, the actual response is represented by the unshaded surface and the response corresponding to the model of eqn. (6) is represented by the shaded surface. It can be seen that both surfaces intersect at the factor combinations corresponding to the two-level two-factor factorial design we have used, but the two surfaces do not agree very well at other factor combinations. In particular, they do not agree at $x_1 = 9$ and $x_2 = 75$. If we had augmented the original factorial design with additional factor combinations, we would have been able to detect this lack of fit of eqn. (6) to the resulting experimental data.

6. Fractional replication

One of the disadvantages of factorial designs is the very large number of experiments required when working with more than two or three factors. As the number of factors increases, the number of experiments required for a two-level factorial design increases in the geometric progression 2, 4, 8, 16, 32,.... Thus, *full* factorial designs in k factors (i.e. experimental designs including all 2^k factor combinations) are seldom used when k is greater than two or three.

For relatively simple models, not all of the 2^k factor combinations are required to be able to estimate the parameters. Recall that the only requirement is that f (the number of factor combinations) is greater than or equal to p (the numer of parameters in the model). If, for example, it is required to fit simple first-order models without interaction, then only $k+1$ parameters are required; using 2^k factorials to estimate these $k+1$ parameters quickly becomes highly inefficient as k increases.

One technique for decreasing the number of experiments while still utilizing the basic layout of the factorial designs is to use *fractional replication.* Using this technique, a geometrically balanced subset of the 2^k factor combinations is employed. The concept is illustrated in Figs. 9 and 10 for the three-factor case. In Fig. 9, the $2^3 = 8$ experiments of a full two-level three-factor factorial design are represented. From these eight experiments, we can choose two subsets, each of which forms a tetrahedron. One of these tetrahedra is shown by dotted, diagonal lines drawn across each face of the cube of experiments. The other tetrahedron can be imagined by drawing the opposite diagonals on each face. Each of these subsets contains 1/2 (or 2^{-1}) of the original number of points and each is called a $(1/2)2^3$ (or $2^{(3-1)}$) fractional factorial design.

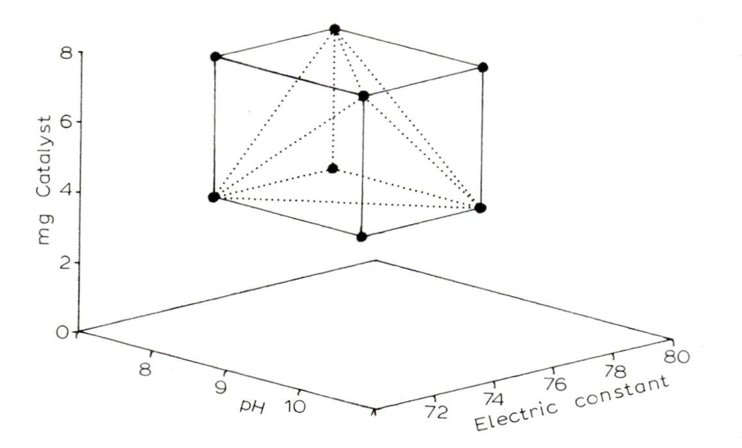

Fig. 9. A two-level three-factor factorial design showing one tetrahedral subset that can be used for fractional replication.

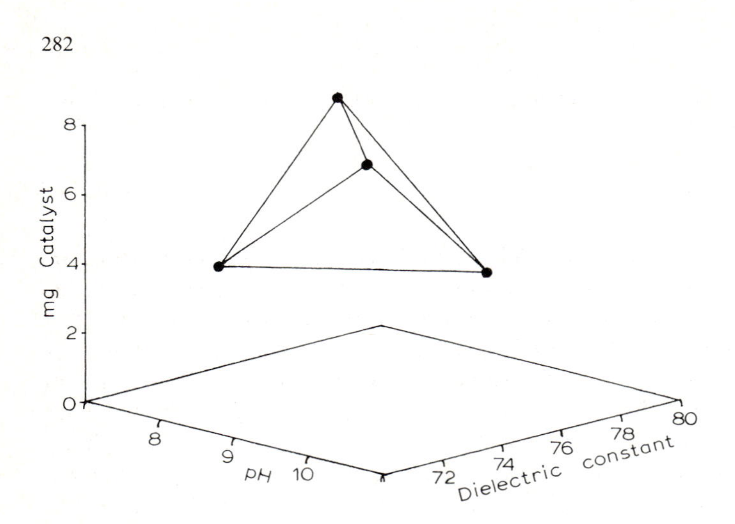

Fig. 10. A $(1/2)2^3$ fractional factorial design.

Figure 10 contains the tetrahedron suggested by Fig. 9. These four experimental conditions can be used to estimate the parameters of the model

$$y_{1i} = b_0 + b_1 x_{1i} + b_2 x_{2i} + b_3 x_{3i} + e_{1i} \tag{9}$$

For example, if we measure the scaled absorbance for a fixed amount of analyte as a function of pH, dielectric constant, and milligrams catalyst, we might obtain the data

$$\mathbf{D} = \begin{bmatrix} 8 & 72 & 4 \\ 8 & 78 & 8 \\ 10 & 72 & 8 \\ 10 & 78 & 4 \end{bmatrix} \quad \mathbf{y} = \begin{bmatrix} 0.5 \\ 7.5 \\ 3.5 \\ 8.5 \end{bmatrix}$$

The model represented by eqn. (9) may be fit using

$$\mathbf{X} = \begin{bmatrix} 1 & 8 & 72 & 4 \\ 1 & 8 & 78 & 8 \\ 1 & 10 & 72 & 8 \\ 1 & 10 & 78 & 4 \end{bmatrix}$$

The fitted model is

$$y_1 = -80.5 + 1.00x_1 + 1.00x_2 + 0.250x_3 \tag{10}$$

If, in fact, the response surface is first-order (i.e. a three-dimensional plane), then we would predict that increasing the pH, increasing the dielectric constant of the solvent, and increasing the amount of catalyst would all produce increases in response. Unfortunately, this type of fractional factorial design is unable to detect curvature (higher-order factor effects) or twist (factor interaction) in the response surface. If these higher-order and interaction effects *are* present, they will *confound* (confuse) the first-order effects in models such as eqn. (9). For example, what we think is caused only by the b_1x_1 effect might, in fact, be caused partially or wholly by a $b_{23}x_2x_3$ interaction.

7. Central composite designs

In Chap. 16 it was suggested that full second-order polynomial models are exceptionally versatile for use as empirical models in many systems over a limited domain of the factors. *Central composite designs* are very useful for obtaining data that can be used to fit these full second-order polynomial models [2]. They are called *composite* because they are a juxtaposition of a *star design* with $2k+1$ factor combinations (see Fig. 11) and a two-level k-factor factorial design with 2^k factor combinations (see Fig. 5) to give a total of 2^k+2k+1 factor combinations. They are called *central* if the centers of the star and factorial designs coincide; otherwise they are called *non-central*. Figure 12 shows a central composite design in two factors; note that it has $2^2+(2\times 2)+1=9$ experimental conditions. Figure 13 shows a central composite design in three factors; note that it has $2^3+(2\times 3)+1=$ 15 factor combinations. The center point of central composite designs is usually replicated a total of three or four times to estimate purely experimental uncertainty.

The number of parameters in a full second-order polynomial model in k factors is $(k+1)\cdot(k+2)/2$. Table 1 compares the number of factor combinations in

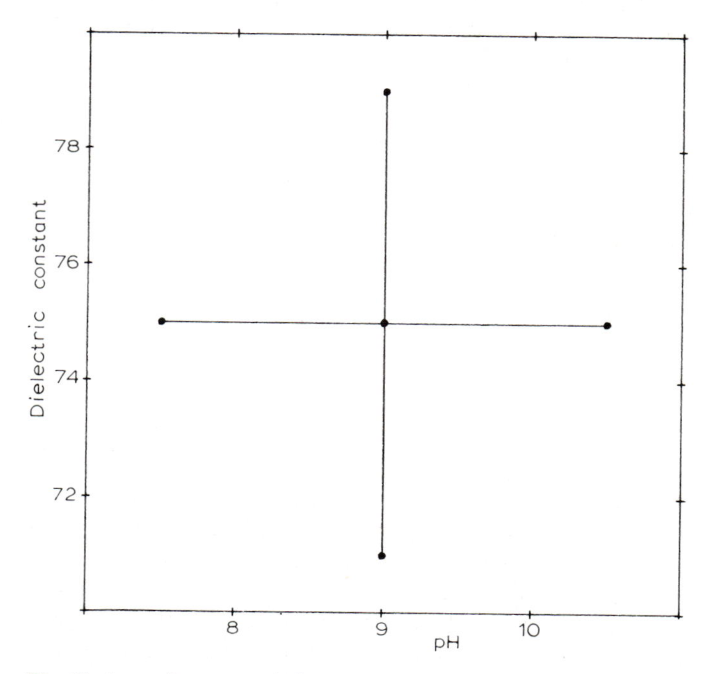

Fig. 11. A two-factor star design.

References p. 290

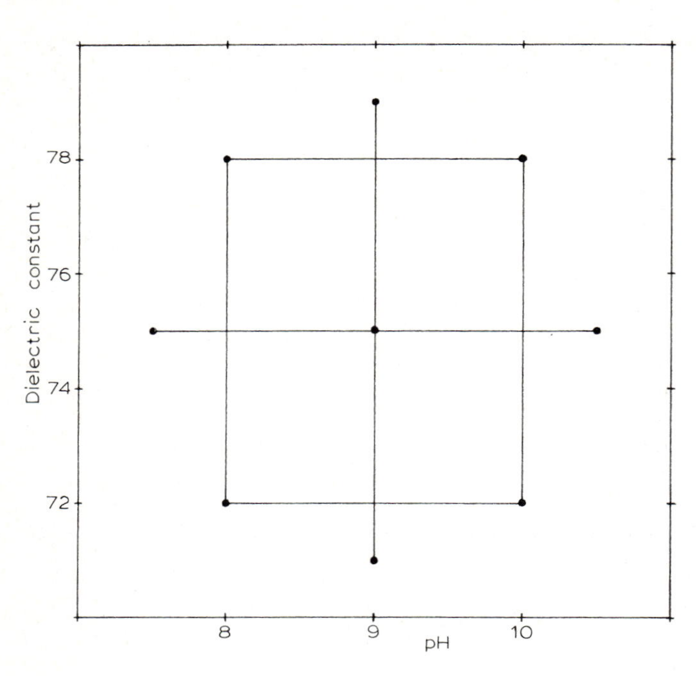

Fig. 12. A two-factor central composite design.

central composite designs with the number of parameters in full second-order polynomial models. The efficiency (the number of parameters to be estimated divided by the number of factor combinations) of central composite designs is quite high for up to about five or six factors and then begins to decrease rapidly.

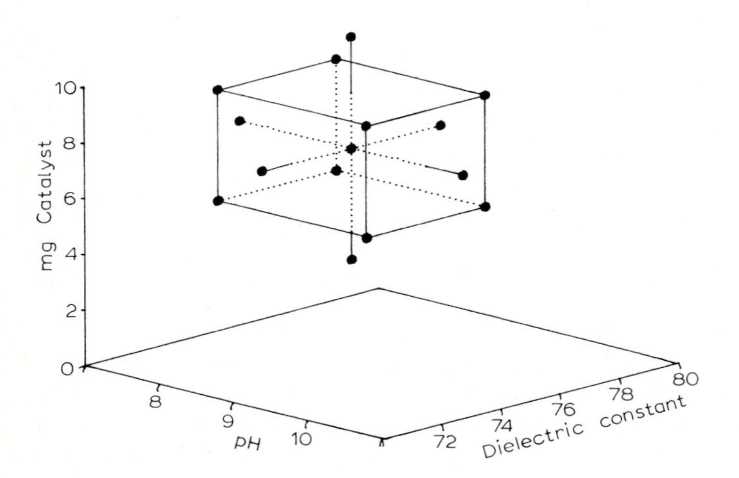

Fig. 13. A three-factor central composite design.

As an example of the use of a central composite experimental design to estimate the parameters of a full second-order model, consider the following data set that might be associated with Fig. 12.

$$\mathbf{D} = \begin{bmatrix} 8 & 72 \\ 8 & 78 \\ 10 & 72 \\ 10 & 78 \\ 7.5 & 75 \\ 10.5 & 75 \\ 9 & 71 \\ 9 & 79 \\ 9 & 75 \\ 9 & 75 \\ 9 & 75 \\ 9 & 75 \end{bmatrix} \qquad \mathbf{y} = \begin{bmatrix} 5.3 \\ 7.2 \\ 6.5 \\ 8.3 \\ 8.0 \\ 9.8 \\ 4.0 \\ 6.5 \\ 10.7 \\ 10.8 \\ 10.7 \\ 10.6 \end{bmatrix}$$

Again, the response might represent the scaled absorbance for a fixed amount of analyte. The model represented by

$$y_{1i} = b_0 + b_1 x_{1i} + b_2 x_{2i} + b_{11} x_{1i}^2 + b_{22} x_{2i}^2 + b_{12} x_{1i} x_{2i} + e_{1i} \tag{11}$$

may be fit using the $\mathbf{X}$ matrix

$$\mathbf{X} = \begin{bmatrix} 1 & 8 & 72 & 64 & 5184 & 576 \\ 1 & 8 & 78 & 64 & 6084 & 624 \\ 1 & 10 & 72 & 100 & 5184 & 720 \\ 1 & 10 & 78 & 100 & 6084 & 780 \\ 1 & 7.5 & 75 & 56.25 & 5625 & 562.5 \\ 1 & 10.5 & 75 & 110.25 & 5625 & 787.5 \\ 1 & 9 & 71 & 81 & 5041 & 639 \\ 1 & 9 & 79 & 81 & 6241 & 711 \\ 1 & 9 & 75 & 81 & 5625 & 675 \\ 1 & 9 & 75 & 81 & 5625 & 675 \\ 1 & 9 & 75 & 81 & 5625 & 675 \\ 1 & 9 & 75 & 81 & 5625 & 675 \end{bmatrix}$$

The fitted model is

$$y_1 = -2006 + 15.7x_1 + 51.5x_2 - 0.802x_1^2 - 0.341x_2^2 - 0.00833x_1x_2 \tag{12}$$

and is plotted in Fig. 14. A sum of squares and degrees of freedom tree is given in Fig. 15. The F-ratio for the significance of regression is 3328.3 and is significant at the 100.000% level of confidence; it is highly probable that one or more of the parameters in the model is not equal to zero. The F-ratio for lack of fit is 0.0896 and is significant at the 4% level of confidence; it is highly improbable that we could find a better model.

References p. 290

TABLE 1

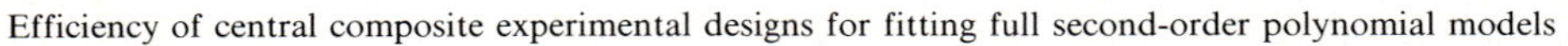

Efficiency of central composite experimental designs for fitting full second-order polynomial models

Experimental factors, k	Parameters, $p=(k+1)(k+2)/2$	Factor combinations, $f=2^k+2k+1$	Efficiency, p/f
1	3	5	0.60
2	6	9	0.67
3	10	15	0.67
4	15	25	0.60
5	21	43	0.49
6	28	77	0.36
7	36	143	0.25
8	45	273	0.16
9	55	533	0.10
10	66	1045	0.06

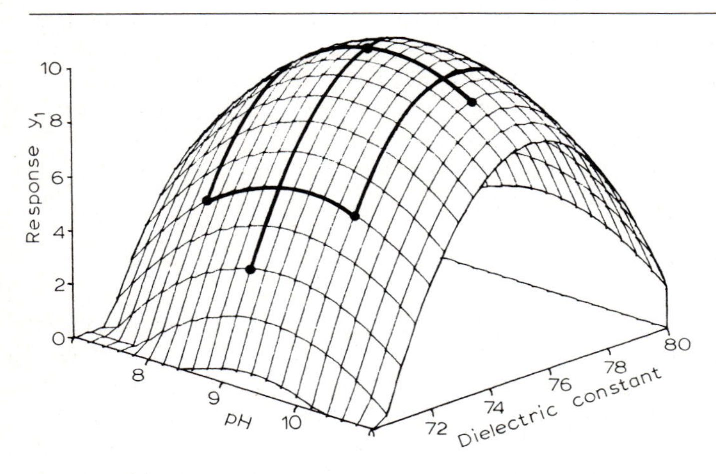

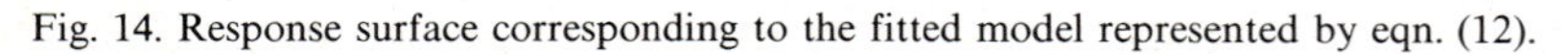

Fig. 14. Response surface corresponding to the fitted model represented by eqn. (12).

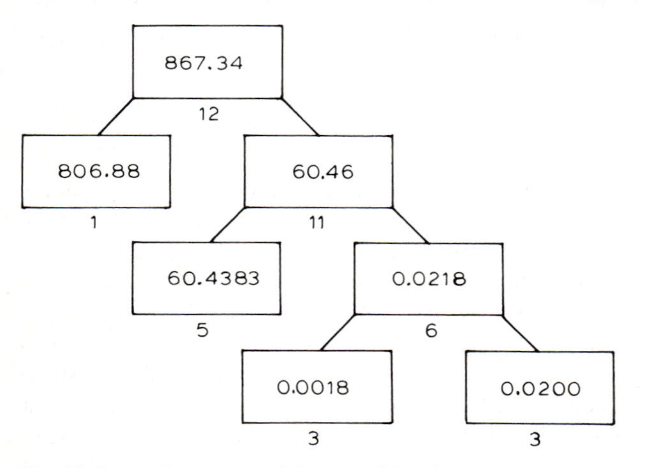

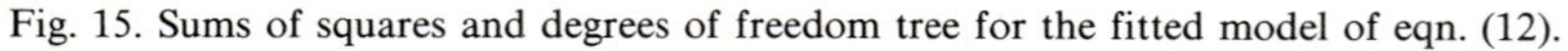

Fig. 15. Sums of squares and degrees of freedom tree for the fitted model of eqn. (12).

8. General comments on experimental designs

Central composite experimental designs can often be used to rapidly optimize an analytical method, especially if the number of factors is not large and if all of the required experiments can be carried out in parallel (i.e. all at the same time) instead of sequentially (i.e. one after the other). The center point is usually chosen to correspond to the chemist's "best guess" of where the optimum might be. The spacing of the other factor combinations also requires judgement on the part of the experimenter. If the factor levels are chosen too close together, then the differences in response will be small and the parameter estimates will not be very significant (i.e. we simply will not know very well what the response surface looks like). However, if the factor levels are chosen too far apart, then the *true* response surface might be quite different from the *estimated* response surface if the model is inadequate (i.e. there might be serious lack of fit between the model and the true response surface, similar to that suggested by Fig. 8). Just what "too close together" and "too far apart" mean in practice is best decided by the experimenter. For example, changing the pH by only ± 0.1 unit would probably result in factor levels that are "too close together" in most svstems; changing the pH by ± 5 units would probably result in factor levels that are "too far apart"; changing the pH by ± 1 unit might be about right for many situations [3].

9. Exploring response surfaces

The experimental designs discussed in this chapter can be used in a sequential manner to explore response surfaces in analytical chemistry. Initially, two-level highly fractional factorial designs are used to determine which of many possible factors are the most important. After the most important factors have been chosen, full two-level factorial designs are used to determine the effects and interactions of these factors so that a move in factor space can be made to improve the response of interest. Finally, when it is felt that adequate conditions have been achieved, central composite designs are used to map out the operating region to determine factor tolerances and factor compensation.

9.1 Screening experiments

As we have seen, classical factorial-type experimental designs require enormous numbers of experiments for large numbers of factors. Thus there is a strong driving force to use only a few factors in a given study. As a result, it is usually necessary to screen a large number of possible factors to select those that appear to be the most significant, i.e. those factors that have the greatest effect upon the response.

Highly fractional two-level factorial designs, known as "saturated fractional factorial designs", are often used for this purpose [4]. The saturated fractional factorial designs result in the sequence $2^{(3-1)}$ (see Fig. 9), $2^{(7-4)}$, $2^{(11-7)}$, ... for determining three main effects in four experiments, seven main effects in eight experiments, fifteen main effects in sixteen experiments, etc. This sequence is highly efficient for screening a number of factors that is one less than a power of two (i.e. 3, 7, 15, 31, 63, 127, 255, ...) but it is inconvenient for other reasonable numbers of factors (e.g. 11 or 19).

A different type of fractional factorial design, known as Plackett–Burman designs [5], can be used when the number of factors is one less than a multiple of four (i.e. 3, 7, 11, 15, 19, 23, 27, 31, ...). The smallest number of factors for which the Plackett–Burman designs exist but the saturated fractional factorial designs do not is 11. Other numbers of factors for which the Plackett–Burman designs are useful are 19, 23, 27, 35, 39, ...

If the number of factors in a real system does not correspond to one less than a multiple of four, dummy factors can be added to achieve the required number. Factor levels are usually coded as -1 and $+1$. The model used for these screening experiments must be first-order in each factor

$$y_1 = b_0 + b_1 x_1 + b_2 x_2 + b_3 x_3 + \cdots \tag{13}$$

Screening experiments are not foolproof. The saturated fractional factorial designs and the Plackett–Burman designs give minimum variance estimates of the first-order factor effects, but they also assume that there are no factor interactions or other higher-order effects present in the system. Thus, it is true that large values of b in the fitted model might mean that the associated factor is important, i.e. varying that factor will have a large effect on the response. It is *not* true, however, that small b values in the fitted model necessarily mean that the associated factor is insignificant.

If the chosen factor levels are too close together, even a highly significant factor will not be able to produce a change in response large enough to be noticed. If the chosen factor levels are too far apart, they might lie on either side of an optimum in such a way as to give approximately the same response and make the factor appear to be insignificant. If factor interaction exists, the main effects could be missed entirely. (As we shall see in the next chapter, alternative optimization strategies exist that do not require preliminary screening of factors.)

9.2 Evolutionary operation

After the three or four most important factors have been identified by the screening experiments, full two-level factorial designs can be constructed. The information obtained from these designs can be used to determine a direction to move in the three- or four-dimensional factor space so that an increase in response is highly probable. If the system is potentially dangerous, risky, hazardous, etc., then the factorial designs should be small and the moves should be conservative. If none of these concerns are present, then the designs can be larger and the moves can be bolder.

This type of approach is generally known as an "evolutionary operation", or EVOP [6]. It has been used successfully in the chemical process industry for many years. However, it should be noted that there are significant differences between the experimental environments of an industrial chemical process and an analytical research laboratory.

In the chemical process industry, the process is probably making money or it would not be in operation. Box's insight was to view this operation as a continuous set of experiments, each of which is normally carried out under identical conditions. Box recognized that a better way of utilizing this experimental situation would be to make small changes in the operating conditions, not enough to degrade the product, but enough to understand the factor effects and their interactions. Based on this knowledge, the set points of the process could be changed to new, slightly different values and the experimental process repeated. As this EVOP is being carried out, the process continues to make good product but makes it more efficiently and at better profit.

Unfortunately, research analytical chemists cannot afford the time and the number of experiments required to carry out this industrial type of slow evolutionary operation. Instead, because most (but not all) analytical processes are not hazardous, bolder moves can usually be made with the result that the EVOP strategy often arrives in the region of optimum results more rapidly.

9.3 Mapping the region of the optimum

Once in the region of the optimum, central composite experimental designs are useful for understanding higher-order factor effects and factor interactions. The resulting information is useful for two important reasons: specifying factor tolerances and taking advantage of factor compensation.

Factor tolerances indicate the domain within which a factor must be controlled to maintain a given reproducibility in response. As an example, wavelength of maximum absorption is an important factor in spectrophotometric methods of analysis. If the absorption band is broad with respect to wavelength, then relatively large changes in wavelength can be tolerated about the optimum; the effect upon absorption will be small and the method will show little variability. If, however, the absorption band is sharp with respect to wavelength, then the same relatively large changes in wavelength will cause drastic changes in absorption and the analytical method will show great variability.

Although broad absorption maxima are preferred to narrow absorption maxima, either can be used successfully as long as the corresponding factor tolerances are specified. In the case of a broad absorption maximum, a factor tolerance of ± 20 nm might give a precision of $\pm 2\%$ uncertainty in absorption measurements. In the case of a narrow absorption maximum, a factor tolerance of ± 3 nm might be necessary to give the same precision of $\pm 2\%$ uncertainty in the absorption measurements. If these tolerances are not specified for a given method, the persons implementing the method might not apply sufficient control to the factor and too much uncertainty might result. (Note that a knowledge of factor tolerances is also important to avoid spending too much time and money on factors that do not need

References p. 290

tight control. It would be wasteful of the analytical chemist's time to require setting the wavelength to ± 3 nm when ± 20 nm would be sufficient.)

Factor interaction exists when response surfaces are tilted with respect to the factor axes (see, for example, Fig. 13 in Chap. 16). Such response surfaces generally have an oblique ridge of optimal response. A knowledge of this ridge can be useful for making economic trade-offs among factors.

Suppose, for example, that the factors in Fig. 13 of Chap. 16 do not correspond to temperature and pH but rather represent mg NaCl (factor x_1) and mg $AuCl_3$ (factor x_2). There is a clear optimum at a certain number of milligrams of sodium chloride and a corresponding number of milligrams of gold(III) chloride. Although this might represent an optimum in analytical sensitivity (or organic yield, or some other measure of response), it does not necessarily represent an *economic* optimum.

To obtain a less expensive but still *adequate* method, we might decrease the number of milligrams of sodium chloride (factor x_1) *and* decrease the number of milligrams of gold(III) chloride (factor x_2). Although the method will no longer give strictly optimal analytical sensitivity, we have been able to move along the ridge in the response surface and keep the analytical sensitivity high *while decreasing the overall cost of the method.*

When factors interact in this way, they are said to be "compensating" factors, that is we can change one factor and compensate for the loss in response by an appropriate change in another factor.

References

1 G.E.P. Box and K.B. Wilson, On the experimental attainment of optimal conditions, J. R. Stat. Soc., Ser. B, 13 (1951) 1.

2 G.E.P. Box and D.W. Behnken, Some new three level designs for the study of quantitative variables, Technometrics, 2 (1960) 445.

3 D.M. Steinberg and W.G. Hunter, Experimental design. Review and comment, Technometrics, 26 (1984) 71.

4 S. Addleman and O. Kempthorn, Some main effects plans and orthogonal arrays of strength two, Ann. Math. Stat., 32 (1961) 1167.

5 R.L. Plackett and J.P. Burman, The design of optimum multifactorial experiments, Biometrika, 33 (1946) 305.

6 G.E.P. Box, W.G. Hunter and J.S. Hunter, Statistics for Experimenters. An Introduction to Design, Data Analysis, and Model Building, Wiley, New York, 1978.

Recommended reading

T.B. Barker, Quality by Experimental Design, Dekker, New York, 1985.

G.E.P. Box and J.S. Hunter, Multifactor experimental designs for exploring response surfaces, Ann. Math. Stat., 28 (1957) 195.

G.E.P. Box and J.S. Hunter, The 2^{k-p} fractional factorial designs. Part I, Technometrics, 3 (1961) 311.

J.A. Cornell, Experiments with Mixtures. Designs, Models, and the Analysis of Mixture Data, Wiley, New York, 1981.

C. Daniel, Applications of Statistics to Industrial Experimentation, Wiley, New York, 1976.

O.L. Davies (Ed.), Design and Analysis of Industrial Experiments, Hafner, New York, 2nd edn., 1956.
C.R. Hicks, Fundamental Concepts in the Design of Experiments, Holt, Rinehart and Winston, New York, 1973.
W. Mendenhall, Introduction to Linear Models and the Design and Analysis of Experiments, Duxbury, Belmont, CA, 1968.
E.B. Wilson, Jr., An Introduction to Scientific Research, McGraw-Hill, New York, 1952.

Chapter 18

Optimization of Analytical Chemical Methods

1. Introduction

In the last chapter, we saw how data from structured experimental designs can be used to fit models that describe the nature of an analytical response as a function of several analytical variables over a limited region of the factor space. Because a major goal of much research and development in analytical chemistry is to improve the response, optimization is often required. In favorable cases, the information from structured experimental designs can point directly to the optimal combination of factors. In other cases, especially those situations involving relatively large numbers of factors, this approach to optimization is not feasible (see, for example, Chap. 17, Table 1).

In this chapter, we point out how the sequential application of simple experimental designs and simple models can be used to achieve optimum performance from systems involving a large number of factors. Once this optimum has been achieved, it is useful to apply multifactor screening designs (Chap. 17, Sect. 9.1) and then to apply more complicated experimental designs (e.g. central composite designs) for the fitting of more complicated models (e.g. full second-order polynomial models) for the most important factors in the region of the optimum. The information thus obtained can be used to set tolerances on the factor levels in the region of the optimum so that good reproducibility can be obtained and to discover any factor interactions that might exist so that further economic gains might be realized (Chap. 17, Sect. 9.3).

The use of simple designs at the beginning of an optimization study is justified by the fact that, if we are far from the optimum, the most important information we need is that which indicates the direction in which to move in order to improve the response. In the beginning stages of analytical methods development, it usually is not worth the time and expense required to describe fully and precisely a portion of the response surface we are not ultimately going to be interested in. Only when we get into the region of the optimum is it important to have a more complete knowledge of the true response surface.

2. Sequential single-factor-at-a-time strategy

Figure 1 shows an analytical response (e.g. sensitivity) as a function of the two factors pH and dielectric constant. In this representation, the elliptical lines represent contours of constant response; e.g. all combinations of pH and dielectric constant that lie on the outermost ellipse will give an identical analytical response. As the elliptical contours get closer to the center, the analytical response increases.

References p. 304

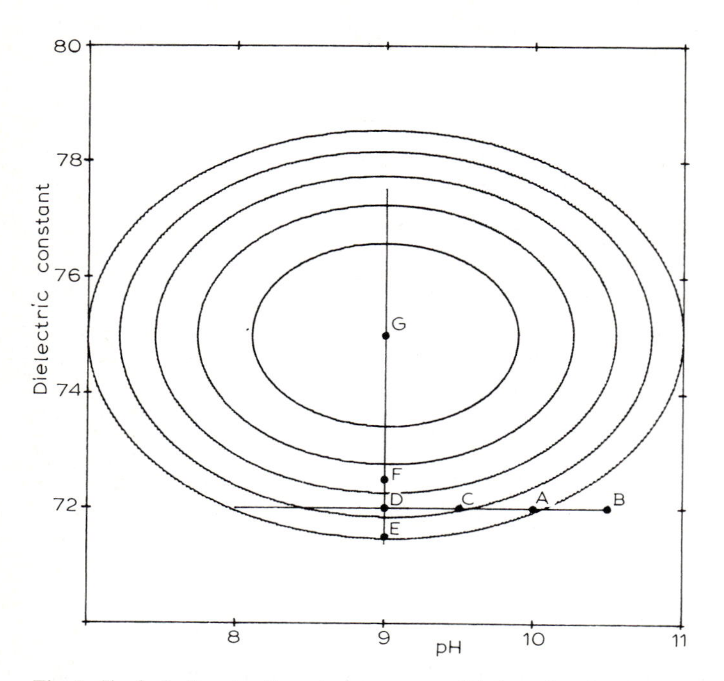

Fig. 1. Single-factor-at-a-time strategy on a well-behaved surface.

In this example, there is no interaction between the two factors pH and dielectric constant.

The sequential single-factor-at-a-time approach requires all factors but one to be held constant while a univariate search is carried out on the factor of interest. In the example shown here (Fig. 1), dielectric constant, x_2 is held fixed at 72 and the pH, x_1, is varied first. If we start our search at pH = 10.0 (point A) and carry out another experiment at pH = 10.5 (point B), we will immediately sense that increasing the pH causes the analytical response to decrease and that this is therefore not a good direction in which to move. Carrying out a third experiment at pH = 9.5 (point C) will provide a more desirable analytical response. We will find that further experiments decreasing the pH continue to improve the response until we get to point D where pH = 9.0. Further decreases in pH would cause a decrease in the analytical response.

We can proceed with the single-factor-at-a-time algorithm by now holding pH constant at its optimum value (pH = 9) and varying the other factor, x_2, the dielectric constant. If we decrease x_2 to 71.5 (point E) we will find that the analytical response will decrease and that this is therefore an undesirable direction in which to move. Increasing x_2 to 72.5 (point F) gives a more desirable analytical response. Continued increases in x_2 will cause further increases in response until $x_2 = 75$ (point G), beyond which the analytical response will decrease.

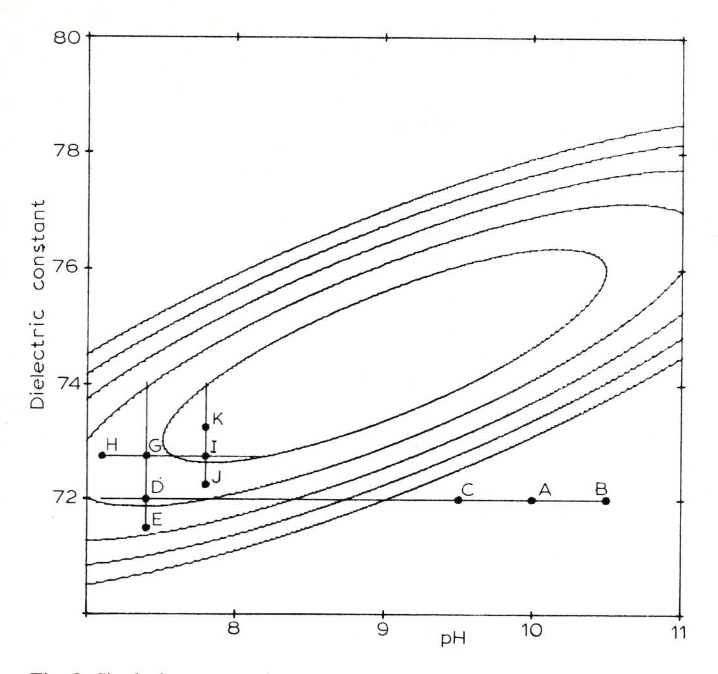

Fig. 2. Single-factor-at-a-time strategy on a response surface exhibiting a diagonal ridge.

If we carry out an experiment in any direction from G, we will find that the analytical response is less than the analytical response at G. Thus, in this example, the single-factor-at-a-time strategy has successfully found the optimum. The single-factor-at-a-time strategy appears to be simple, efficient, and effective.

Consider, however, the analytical response surface shown in Fig. 2. In this example, the two factors interact, i.e. the effect of one factor depends upon the value of the other factor. Starting a single-factor-at-a-time search at $x_1 = 10.0$, $x_2 = 72$ (point A) will give behavior that is similar to what we found before: i.e. increasing the pH to 10.5 (point B) will cause the analytical response to decrease; decreasing the pH (point C) will cause the analytical response to increase until we arrive at pH = 7.4 (point D); further decreases in pH will cause the analytical response to decrease. Now, holding the pH constant at its optimum value (pH = 7.4 in this case), we can vary the other factor, x_2, the dielectric constant. If we decrease x_2 to 71.5 (point E), we will find a decrease in the analytical response; increasing x_2 will give more desirable results. Continued increases in x_2 cause further increases in analytical response until we arrive at point G. Note that point G, however, is obviously not at the true optimum of the response surface.

The complication in this case is that the ridge in our response surface does not lie parallel to the factor axes; instead, it is oblique with respect to the factor axes. Such response surfaces are the rule rather than the exception in analytical chemistry.

References p. 304

When we hold all factors but one constant, we move parallel to a factor axis and locate an optimum that lies near this diagonal ridge. When we vary another factor, we again move parallel to a factor axis but, in general, this direction will not point to the true optimum of the response surface. In Fig. 2, we see that the optimum value for x_1 no longer depends only on x_1 but, in this case, depends on x_2 as well. The optimum in x_2 also depends on the level of x_1. Thus, one cycle through each of the factors using the single-factor-at-a-time strategy does not ensure that we will locate the optimum of the response surface; we might have been able to *improve* the response but we have not been able to *optimize* it.

Further improvements might be possible by reiterating the procedure (see points H, I, J, and K in Fig. 2) but moving along the ridge in this way is a time-consuming and expensive process. In those cases where the steepness of the ridge, the extent of obliqueness, and the resolution of the experiments are unfavorable, it is possible to become stranded on the ridge and not be able to move on toward the optimum with this single-factor-at-a-time strategy.

3. Box-type EVOP

In 1957, Box [1] suggested an alternative optimization technique that would not be stranded by ridges (Chap. 17, Sect. 9.2). Because the diagonal ridge of the response surface arises as a result of the interaction terms (e.g. $b_{12}x_1x_2$), it was proposed that a simple multifactor model containing interaction terms might be used to describe accurately a local region of the response surface. For the two-factor case, the model is

$$y_{1i} = b_0 + b_1x_{1i} + b_2x_{2i} + b_{12}x_{1i}x_{2i} + e_{1i} \tag{1}$$

Two-level two-factor factorial designs (Chap. 17, Sect. 2) are appropriate for obtaining information to be used to fit this model. The model can then be used to estimate the direction of steepest ascent and a new experimental design can be carried out in that direction. The process can be repeated as many times as necessary until the optimum is finally reached. The progress of such a scheme is shown in Fig. 3. The procedure is not stranded by ridges and will eventually find the optimum of the response surface.

In the original paper, Box discussed the similarities of such a procedure to Darwinian evolution: concepts such as *natural variation, survival of the fittest*, and *adaptation toward an optimum* are all present in each case. As a result, this optimization procedure was called an *evolutionary operation*, or EVOP for short. It has been successfully applied to improve yield in many large-scale industrial chemical processes where the continual generation of product can be viewed as a framework for doing an almost infinite number of experiments.

When applied in the analytical research and development laboratory, however, Box-type (factorial) EVOP has two major disadvantages. The first disadvantage is the very large number of experiments required in each factorial design: 2^k where k is the number of factors being optimized. This is not usually a problem in the

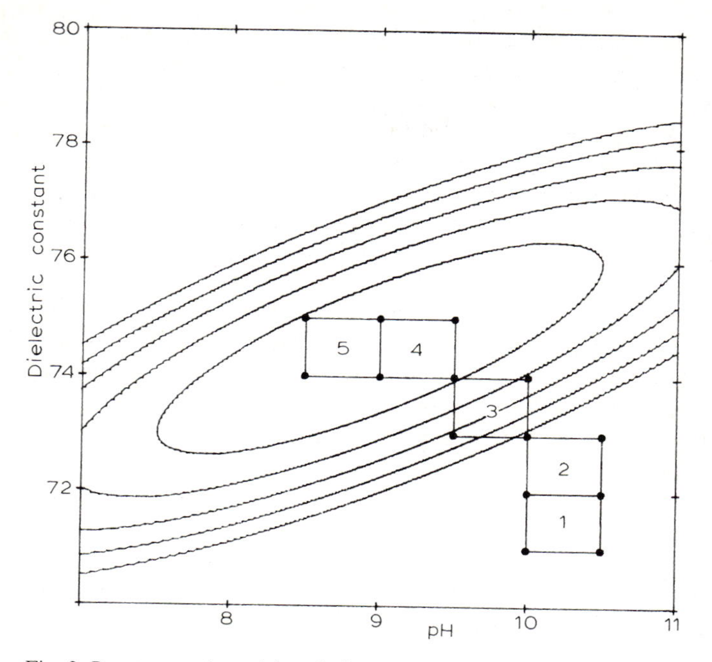

Fig. 3. Box-type or factorial evolutionary operation (EVOP).

industrial environment. By the time industrial processes are put into full production, their behavior is quite well known from studies at the pilot plant level and from studies at the laboratory bench level. Thus, most industrial EVOP procedures use only a few factors and 2^k is small. In contrast, analytical chemical projects at the early stages of research and development are usually not well understood. Thus, the required experimentation must involve many factors and 2^k will often be very large.

The second disadvantage is that each "move" toward the optimum requires at least $2^k/2$ and frequently as many as $2^k - 1$ new experiments. In industrial environments, this is of little consequence: if the plant is operating, it is probably already profitable and the length of time required to achieve further improvement is not too important. In analytical research and development environments, however, where rapid optimization is desired, the necessity of carrying out a large number of experiments becomes an overwhelming burden.

Because of these disadvantages, factorial EVOP has not been used very frequently in analytical research and development environments.

4. Simplex EVOP

In 1962, Spendley et al. [2] introduced the sequential simplex method as an alternative EVOP technique that is much more efficient than factorial EVOP. (The

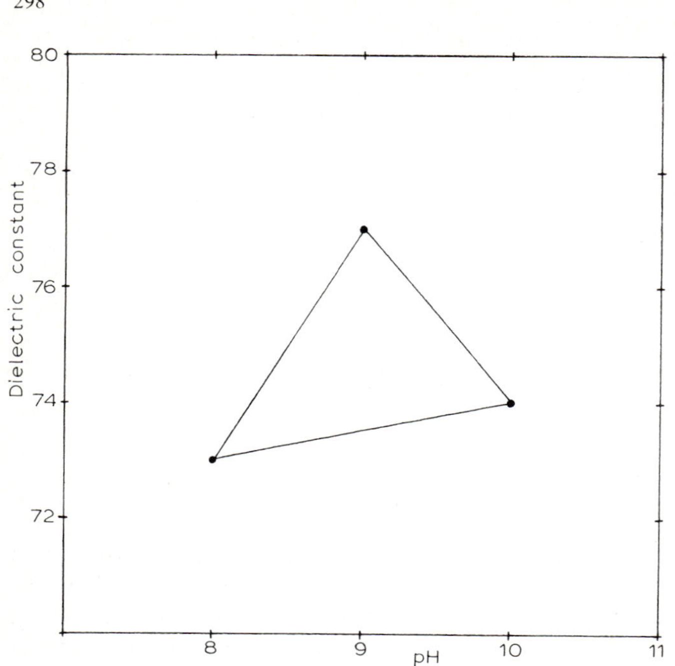

Fig. 4. A two-factor simplex experimental design.

"sequential simplex" method should not be confused with the "simplex tableau" technique in linear programming [3] or with "simplex mixture" designs in formulations problems [4–6].) The efficiency of the sequential simplex method is gained by the smaller number of experiments required by the initial experimental design (only $k + 1$, vs. 2^k for the factorial method) and the smaller number of experiments per move (only 1 or 2, vs. as many as $2^k - 1$ for the factorial method).

Although the simplex method is model-independent (it requires no fitting of models to data), the experimental design is nevertheless based upon a simple k-factor first-order model

$$y_{1i} = b_0 + b_1x_{1i} + b_2x_{2i} + \ldots + b_kx_{ki} + e_{1i} \tag{2}$$

The minimum number of factor combinations (experiments) required to fit this model is $k + 1$. Such a pattern of experiments forms what is known in mathematics as a simplex: a simplex is a geometric figure having a number of vertexes (corners) equal to one more than the number of factors. A simplex in two dimensions has three vertexes (sets of experimental conditions, or factor combinations) and is a triangle. A simplex in three dimensions has four vertexes and is a tetrahedron. Examples of simplex designs for two- and three-factor systems are shown in Figs. 4 and 5, respectively.

When a model of the form of eqn. (2) is fitted to data from a simplex design, the parameters b_1–b_k represent the partial derivatives of the response with respect to

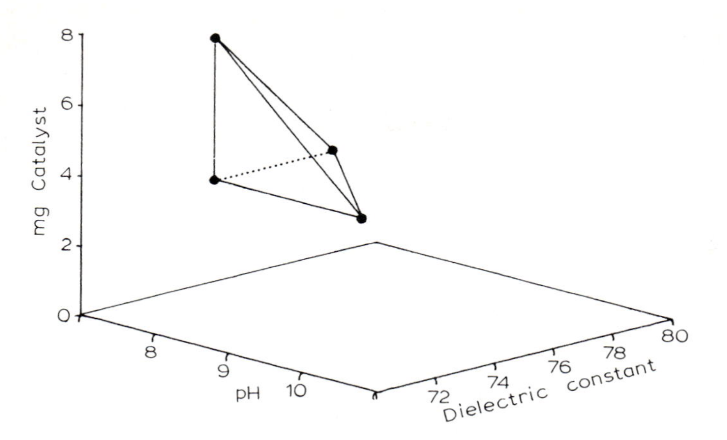

Fig. 5. A three-factor simplex experimental design.

each of the factors; i.e. they represent the slope of the response y_1 with respect to each x. Thus, the model of eqn. (2) can be thought of as a tangential plane (or hyperplane) touching the response surface at a point corresponding to the center of the experimental design. If desired, the model could be fitted to the simplex vertexes and the parameter estimates b_1–b_k could be used to indicate the direction of steepest ascent.

In 1965, Nelder and Mead [7] modified the original simplex procedure of Spendley et al. [2] to give the simplex the ability to accelerate in favorable directions and decelerate in unfavorable directions. Since that time, other modifications of the basic simplex procedure have been proposed, including the "super modified simplex" of Routh et al. [8]. The simplex algorithm discussed in this chapter is essentially that of Nelder and Mead [7] with a modification originally introduced by King [9].

5. The fixed-size simplex algorithm

The fixed-size sequential simplex method of Spendley et al. [2] is a logical algorithm consisting simply of reflection rules. These rules can be understood by referring to Fig. 6 and the initial simplex labeled BNW. (Figure 7 shows a set of possible moves for a three-factor simplex.)

In the initial simplex BNW, vertex B was measured and found to have the best response, vertex W has the worst response, and vertex N represents the next-to-the-worst response. P is the centroid of the face remaining when the worst vertex W is eliminated from the full simplex.

Reflection is accomplished by extending the line segment WP beyond P to generate the new vertex R.

$$R = P + (P - W) \tag{3}$$

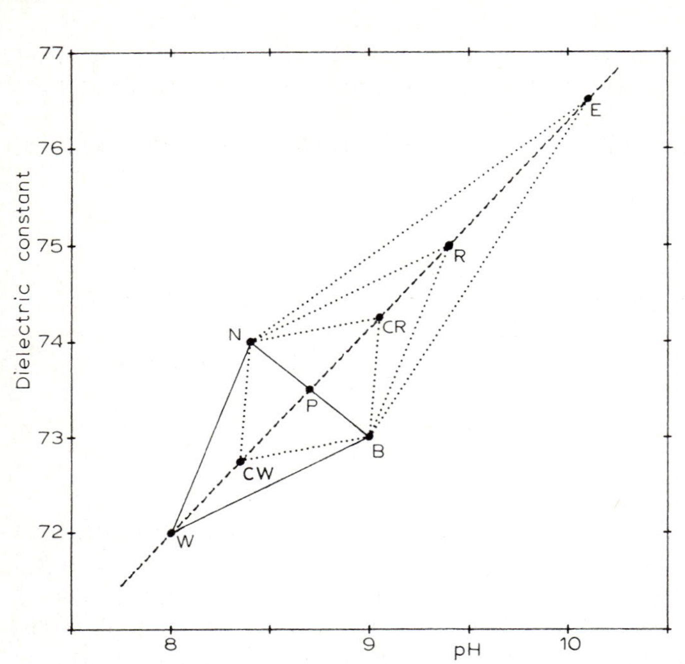

Fig. 6. Possible simplex moves in two dimensions.

An experiment is then carried out under the experimental conditions defined by R, and the response is noted. The vertexes in the new simplex BNR are now ranked to yield the designations BNW and the algorithm is repeated.

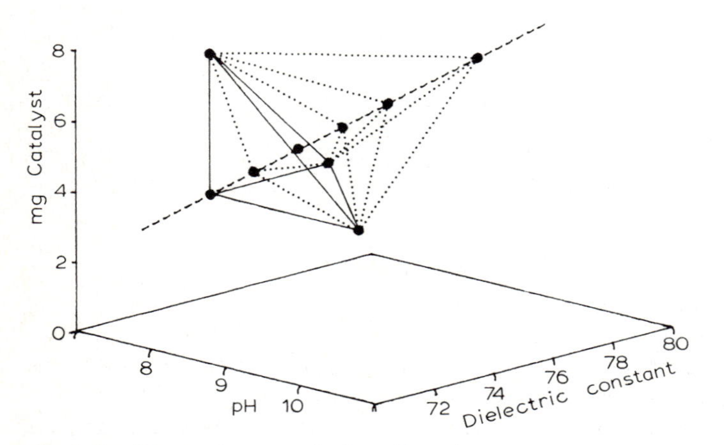

Fig. 7. Possible simplex moves in three dimensions.

One additional rule must be added to prevent the simplex from oscillating about a ridge: if the new vertex is the worst vertex, do not reject it but, instead, reject the next-to-the-worst vertex.

Examples of the progress of the fixed-size simplex may be found in the literature.

6. The variable-size simplex algorithm

The modified sequential simplex method of Nelder and Mead [7] is a logical algorithm consisting of reflection, expansion, and contraction rules. These rules can be understood by referring to Fig. 6 and the reflected simplex labeled BNR. (Figure 7 shows a set of possible moves for a three-factor simplex.) Three possibilities exist for the measured response at R.

(a) The response at R is more desirable than the response at B. An expansion is indicated and the new vertex E is generated.

$$E = R + (P - W) \tag{4}$$

An experiment is then carried out under the experimental conditions defined by E and the response is noted. If the response at E is better than the response at B, E is retained to give the new simplex BNE. If the response at E is *not* better than at B, the expansion is said to have failed and BNR is taken as the new simplex. The algorithm then iterates using the new simplex, either BNE or BNR. Its vertexes are ranked from best (B) to worst (W).

(b) If the response at R is between that of B and N, neither expansion nor contraction is recommended and the algorithm continues with the new simplex BNR.

(c) If the response at R is less desirable than the response at N, a step in the wrong direction has been made and the simplex should be contracted. One of two possible vertexes must be generated.

(i) If the response at R is worse than the response at N but not worse than the response at W, the new vertex should lie closer to R than to W.

$$C_r = P + \frac{P - W}{2} \tag{5}$$

The algorithm then continues using the new simplex BNC_r.

(ii) If the response at R is worse than the previous worst vertex W, then the new vertex should lie closer to W than to R.

$$C_w = P - \frac{P - W}{2} \tag{6}$$

The algorithm then continues using the new simplex BNC_w.

The modification introduced by King states that, when the algorithm continues for the next iteration, the vertex to be rejected in the then current simplex is the vertex that was the next-to-the-worst vertex (N) in the old simplex. This greatly simplifies a set of rules and procedures that otherwise would make the simplex rather inefficient in certain instances.

References p. 304

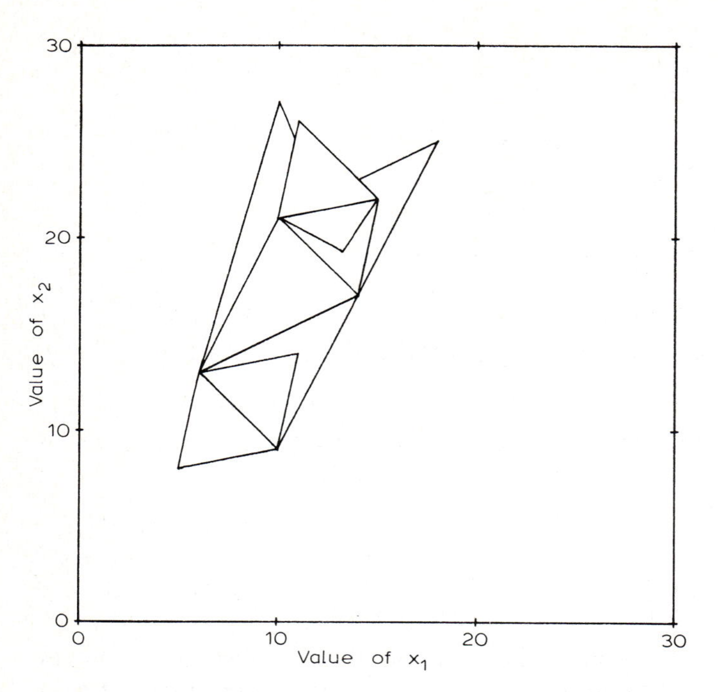

Fig. 8. Examples of simplex moves.

The calculated vertex coordinates shown in Fig. 8 are listed in Table 1. This type of calculation can be done easily with row and column additions as shown here for the first move in Fig. 8.

TABLE 1

Example of simplex calculations

Vertex	Value of x_1	Value of x_2	Response	Formed by [a]
1	5	8	0.075	I
2	10	9	0.417	I
3	6	13	0.437	I
4	11	14	0.500	R
5	14	17	0.535	E
6	10	21	0.586	R
7	10	27	0.408	E
8	18	25	0.454	R
9	15	22	0.514	C_r
10	11	26	0.462	R
11	13.25	19.25	0.576	C_w

[a] I = initial simplex vertex; R = reflection; E = expansion; C_r = contraction on the R side; C_w = contraction on the W side.

$$B = [6 \quad 13]$$
$$N = [10 \quad 9]$$
$$[16 \quad 22] = SUM$$
$$[8 \quad 11] = P = SUM/k$$
$$W = [5 \quad 8]$$
$$[3 \quad 3] = P - W$$
$$R = [11 \quad 14] = P + (P - W)$$

The next move is an expansion and can be calculated as

$$E = [14 \quad 17] = R + (P - W)$$

If a contraction had been required instead, it could have been calculated as one of the two examples

$$C_r = [9.5 \quad 12.5] = P + \frac{P - W}{2}$$

$$C_w = [6.5 \quad 9.5] = P - \frac{P - W}{2}$$

False high results caused by experimental uncertainty might mislead the simplex. This can be detected and corrected if the response of a vertex appearing as best in $k + 1$ successive simplexes is re-evaluated. The average response can be taken as the response of that vertex from then on.

If a vertex lies outside the boundaries of one or more of the factors, a very undesirable response is assigned to that vertex. The simplex will then be forced back inside the boundaries.

The simplex is halted when the step size becomes less than some predetermined value (e.g. 1% of the domain of each variable), or when the differences in response approach the value of the experimental uncertainty, or when adequate response has been achieved.

7. Joint use of simplex and classical experimental designs

The sequential simplex method of optimization and classical experimental designs are complementary techniques that are both useful in the overall strategy of analytical chemical methods development.

Historically, classical experimental designs have often been used to help answer, in order, three questions commonly encountered in chemical research.

(1) Does an experimentally measured response depend on certain factors?

(2) What equation does the dependence best fit?

(3) What are the optimum levels of the important factors?

As Driver [10] has pointed out, "The questions are so related that it is possible for the experimenter not to know which one he wishes to answer and, in particular,

References p. 304

many people try to answer question 2 when in fact they need the answer to the narrower question 3."

When optimization is the desired goal, questions of significant factors and functional relationships are usually of interest only in the area of the optimum. The simplex method offers a means of rapidly and efficiently finding the optimum, *after which* the functional relationship and then the significant factors can be determined using classical experimental designs.

In some cases, however, the time required for a given experiment is long (e.g. days or weeks to get a result). In such cases, the sequential simplex method is clearly undesirable. Instead, classical experimental designs are preferable because, even though they might require a large total number of experiments, the experiments can be carried out simultaneously; the complete set of results is available after only one experimental time period.

References

1 G.E.P. Box, Evolutionary operation. A method for increasing industrial productivity, Appl. Stat., 6 (1957) 81.
2 W. Spendley, G.R. Hext and F.R. Himsworth, Sequential applications of simplex designs in optimization and evolutionary operation, Technometrics, 4 (1962) 441.
3 G.B. Dantzig, Linear Programming and Extensions. Princeton University Press, Princeton, NJ, 1963.
4 J.A. Cornell, Experiments with Mixtures. Designs, Models, and the Analysis of Mixture Data, Wiley, New York, 1981.
5 H. Scheffe, Experiments with mixtures, J. R. Stat. Soc. Ser. B, 20 (1958) 344.
6 J.W. Gorman and J.E. Hinman, Simplex-lattice designs for multicomponent systems, Technometrics, 4 (1962) 463.
7 J.A. Nelder and R. Mead, Simplex method for function minimization, Comput. J., 7 (1965) 308.
8 M.W. Routh, P.A. Swartz and M.B. Denton, Performance of the super modified simplex, Anal. Chem., 49 (1977) 1422.
9 P.G. King, Automated Development of Analytical Methods, Ph.D. Dissertation, Emory University, Atlanta, GA, 1974.
10 R.M. Driver, Chem. Br., 6 (1970) 154.

Recommended reading

W.K. Dean, K.J. Heald and S.N. Deming, Simplex optimization of reaction yields, Science, 189 (1975).
S.N. Deming and S.L. Morgan, Teaching the fundamentals of experimental design, Anal. Chim. Acta, 150 (1983) 183.
S.N. Deming and L.R. Parker, Review of simplex optimization in analytical chemistry, Crit. Rev. Anal. Chem., 7 (1978) 187.

Chapter 19

Optimization of Chromatographic Methods

1. Introduction

Chromatographic systems differ significantly from the types of system discussed in the previous three chapters because of the widespread existence of multiple optima. The intentional variation of system conditions (e.g. temperature or pH) can often cause peaks to "cross" one another, i.e. to change their elution order as a function of system conditions. Conditions for which two or more peaks of interest are eclipsed clearly represent minimum performance from the system (no separation). Conditions for which all peaks are separated from each other represent maximum performance. In chromatographic systems, there are often many sets of conditions that give rise to local maxima. The problem, then, is to predict the location in factor space of these local maxima and to choose the best (or global) maximum, often within certain constraints such as greatest permissible analysis time. Once this desirable domain of factor space has been approximately located, local optimization procedures can be used to find the exact location of the maximum.

Although research is still being directed toward further improvements in stationary phases, most practising analysts are able to improve their separations by adjusting only the temperature and flow rate in gas chromatography and, in addition, the composition of the eluant in liquid chromatography. As a result, multifactor studies are required for the systematic optimization of chromatographic methods.

2. Multifactor designs in chromatography

Figure 1 is an example of a "four-level two-factor" full factorial design in liquid chromatography. Suppose the investigator is interested in finding out how the factors pH and concentration of "ion-interaction" reagent affect the retention time of a hydrocinnamic acid ($C_6H_5CH_2CH_2COOH$) sample in a reversed-phase "ion-pair" liquid chromatographic system. The investigator might prepare 16 eluants, each containing 780 ml distilled water; 10 ml 1 M acetic acid to be used as a buffer; 0.0, 1.5, 3.0, or 5.0 mM octylamine hydrochloride (the so-called "ion-pairing agent" or ion-interaction reagent, IIR); sufficient NaOH to adjust the pH to 3.6, 4.4, 5.2, or 6.0; and sufficient methanol to bring the total volume to 1.00 l. The investigator could then use each of these eluants with a reversed-phase column to measure the retention time of a hydrocinnamic acid sample.

The usefulness of factorial-type designs is shown in Fig. 2 in which retention time is plotted as a function of pH for different values of [IIR], and in Fig. 3 in which

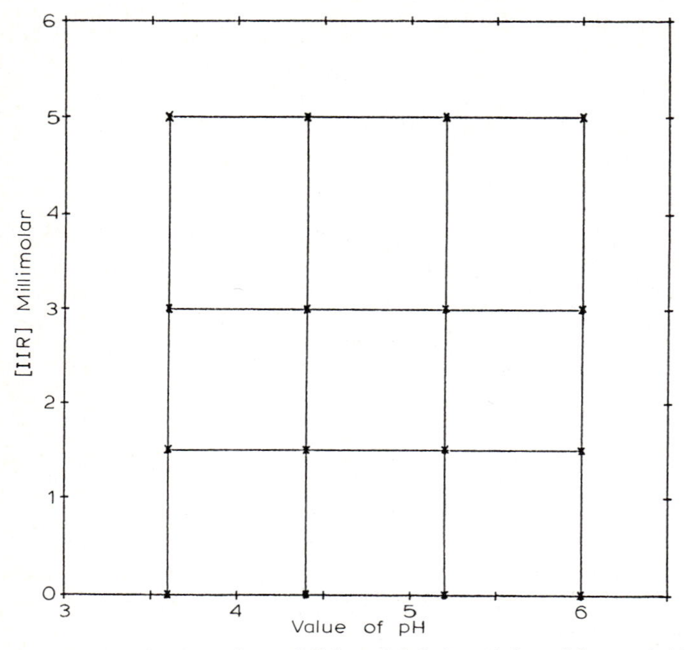

Fig. 1. A four-level two-factor full factorial design. (Adapted from ref. 1.)

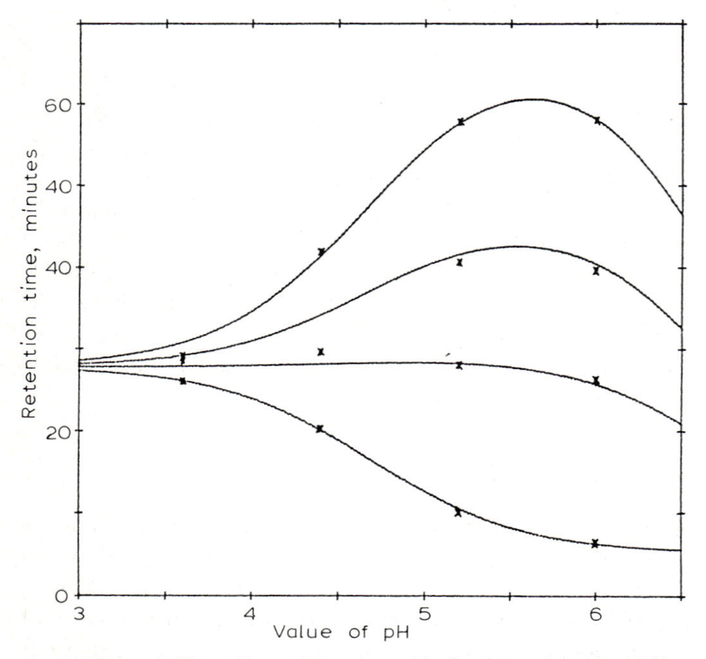

Fig. 2. Effect of pH on the retention time of hydrocinnamic acid at different values of [IIR]. Lower curve to upper curve: 0.0, 1.5, 3.0, and 5.0 mM [IIR]. (Adapted from ref. 1.).

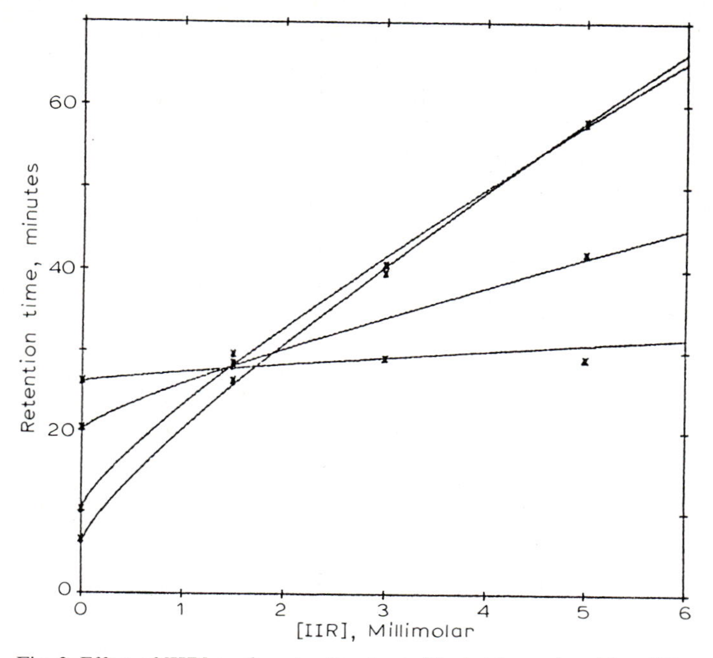

Fig. 3. Effect of [IIR] on the retention time of hydrocinnamic acid at different values of pH. At [IIR] = 0, upper curve to lower curve; pH = 3.6, 4.4, 5.2, and 6.0. (Adapted from ref. 1.)

retention time is plotted as a function of [IIR] for different values of pH. (Figures 2 and 3 contain the same information viewed from two different orthogonal directions.) It is seen that the effect of pH does depend upon the concentration of IIR and that the effect of the concentration of IIR does depend upon the pH.

Note that, if the pH had been varied at [IIR] = 1.5 mM only, it might be concluded that [IIR] has no effect on the retention time (see Fig. 2). Similarly, if [IIR] had been varied at pH = 3.6 only, it might be concluded that pH has no effect on the retention time (see Fig. 3). While these conclusions would be valid at the single value of [IIR] studied and at the single value of pH studied, it is clear from Figs. 2 and 3 that the conclusions are not general and should not be assumed to apply to other values of [IIR] and pH as well.

3. Semi-empirical models

Information gained from multifactor experimental designs such as that shown in Fig. 1 is often very useful for obtaining a much better mechanistic understanding of the separation process. It is evident from the multifactor results shown in Figs. 2 and 3 that both pH and [IIR] have important effects on the retention time of hydrocinnamic acid. Thus, any model that attempts to explain the chromatographic behavior of hydrocinnamic acid in a reversed-phase "ion-pair" system must include the effects of both pH and [IIR] in a mathematical way that describes the real interaction between them.

A semi-empirical model that explains the results shown in Figs. 2 and 3 is

$$t_R = f_{HA} t_{HA} + f_A t_A + f_A f_{HS} b[\mathrm{IIR}]^{1/c} \quad (1)$$

where t_R is the observed retention time of hydrocinnamic acid; f_{HA} is the fraction of hydrocinnamic acid existing in the conjugate acid form, $f_{HA} = [H^+]/([H^+] + K_{HA})$, where K_{HA} is the acid dissociation constant of hydrocinnamic acid; t_{HA} is the retention time of hydrocinnamic acid in its protonated conjugate acid form (i.e. at low pH in the absence of any IIR); f_A is the fraction of hydrocinnamic acid existing in the conjugate base form, $f_A = K_{HA}/([H^+] + K_{HA})$; f_{HS} is the fraction of octylamine existing in the protonated conjugate acid form, $f_{HS} = [H^+]/([H^+] + K_{HS})$, where K_{HS} is the acid dissociation constant of octylammonium ion; t_A is the retention time of hydrocinnamic acid in its unprotonated conjugate base form (i.e. at high pH in the absence of any IIR); and b and c are empirical constants arising from the use of a Freundlich adsorption isotherm. The third mathematical term of the model, $f_A f_{HS} b[\mathrm{IIR}]^{1/c}$, expresses the interaction between pH and [IIR]. Further details of this model and its mechanistic basis may be found in the literature [2]. This non-linear model was fitted to the experimental data points shown in Figs. 2 and 3 using the variable-size simplex method (see Chap. 18); the lines in Figs. 2 and 3 were drawn from the fitted model.

The response surface shown in Fig. 4 represents an alternative view of the information contained in Figs. 2 and 3. The surface was drawn from the fitted equation, eqn. (1), for hydrocinnamic acid. The interaction between pH and [IIR] is

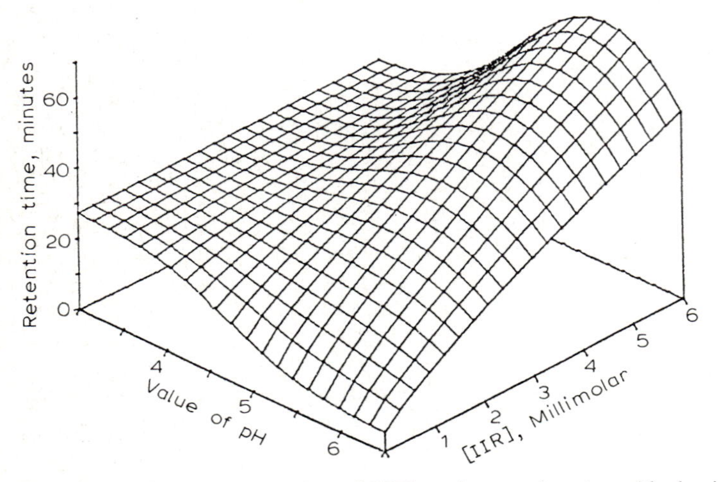

Fig. 4. Multifactor effects of pH and [IIR] on the retention time of hydrocinnamic acid. (Adapted from ref. 2.)

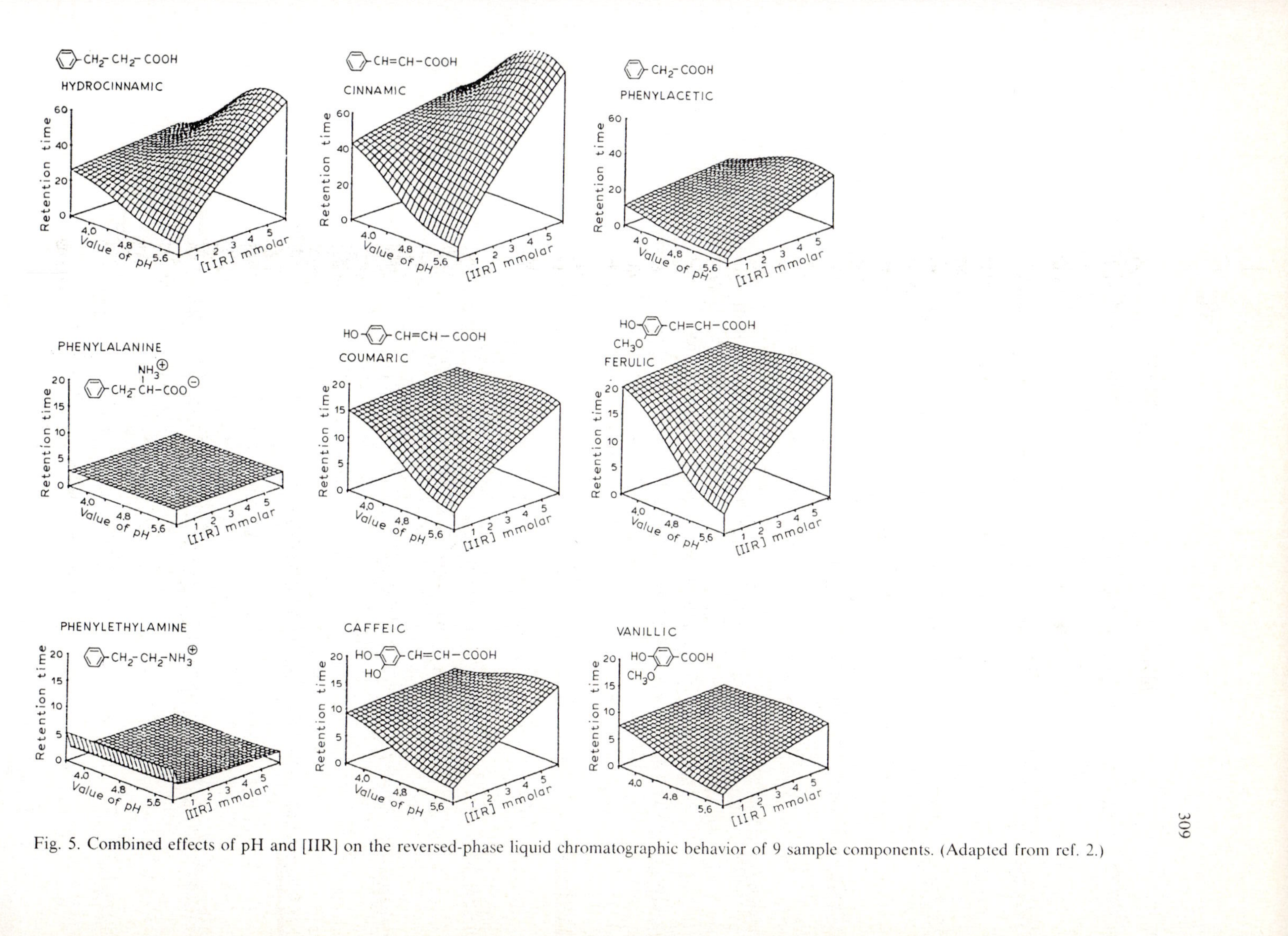

Fig. 5. Combined effects of pH and [IIR] on the reversed-phase liquid chromatographic behavior of 9 sample components. (Adapted from ref. 2.)

References p. 316

clearly evident. Figure 5 shows the behavior of hydrocinnamic acid and 8 other sample compounds: *trans*-cinnamic, phenylacetic, *trans*-*p*-coumaric, *trans*-ferulic, *trans*-caffeic, and vanillic acids, phenylethylamine (a weak base), and phenylalanine (a zwitterionic compound). We will use the predicted behavior of these 9 compounds in the following discussion.

4. Multifactor chromatographic optimization

One of the major goals of chromatographic methods development is to obtain adequate separation of all components of interest in a reasonable analysis time. This goal is achieved (or approached as closely as necessary) by adjusting accessible chromatographic factors (pH, [IIR], temperature, organic modifier concentration, etc.) to give the desired response.

As mentioned earlier, optimization of chromatographic systems is unusually difficult because of the ever-present possibility of "elution order reversal". This is illustrated in Fig. 6 in which the retention time response surfaces for *trans*-ferulic acid and phenylacetic acid have been superimposed. The intersection of the two surfaces occurs at combinations of pH and [IIR] that give identical retention times for these two components, i.e. under these conditions of pH and [IIR], *trans*-ferulic and phenylacetic acid will not be separated. For those combinations of pH and [IIR] that occur to the "left" of this intersection, phenylacetic acid will elute first followed by *trans*-ferulic acid. For those combinations of pH and [IIR] that occur to

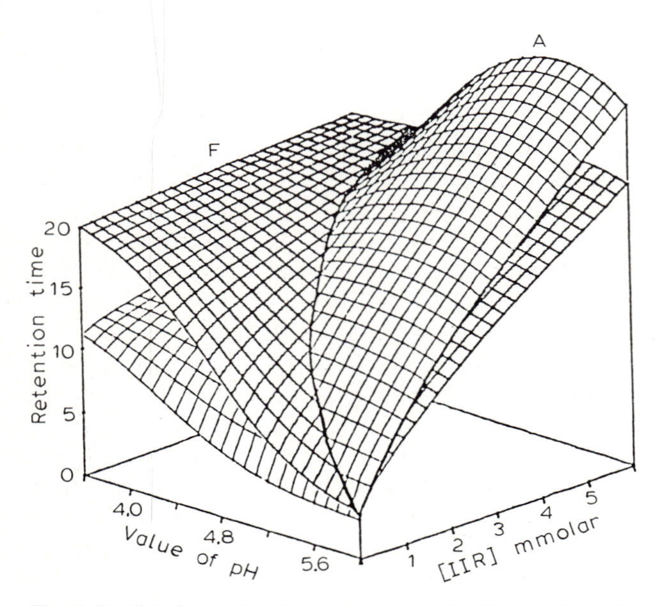

Fig. 6. Predicted retention behavior of *trans*-ferulic acid (F) and phenylacetic acid (A) as a function of pH and [IIR]. (Adapted from ref. 2.)

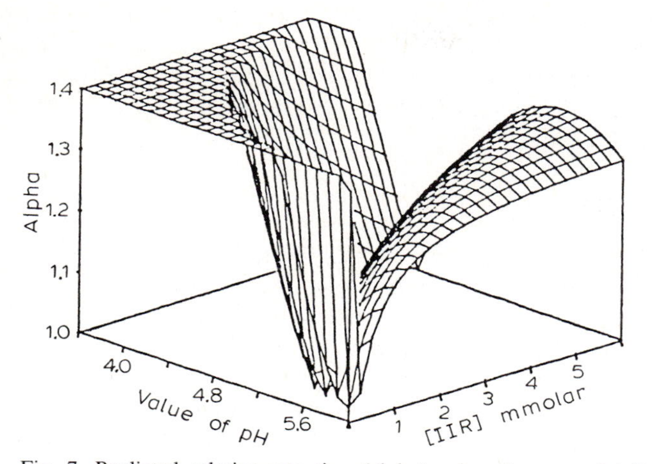

Fig. 7. Predicted relative retention (alpha) values for *trans*-ferulic acid and phenylacetic acid as a function of pH and [IIR]. Values of α greater than 1.4 set equal to 1.4. (Adapted from ref. 2.)

the "right" of this intersection, the elution order will be reversed: *trans*-ferulic acid would elute first followed by phenylacetic acid. Thus, there are two broad domains of experimental conditions that will give acceptable separation of the two compounds; there is a narrow domain between them that will give unacceptable separation.

The "window diagram" technique pioneered by Laub and Purnell [3] for single-factor optimization can be applied to the present multifactor case to make these ideas quantitative. Because relative retention, α, is a better measure of separation than is the difference in retention times, the two-dimensional "alpha diagram" shown in Fig. 7 can be produced by "dividing" the higher surface by the lower surface shown in Fig. 6 at all combinations of pH and [IIR]. (In fact, the capacity factor surfaces must first be formed by subtracting the time equivalent of the void volume from each surface in Fig. 6 and dividing each of the resulting surfaces by the time equivalent of the void volume; the ratios of these capacity factor surfaces then give the relative retention surface shown in Fig. 7.) Other measures of chromatographic performance can also be used for the vertical axis: separation, resolution, percent overlap, etc. The two domains giving acceptable separation are evident in Fig. 7 as the higher parts of the surface; the unacceptable domain occurs in the "valley" of the figure.

All other pairs of compounds in the nine-component chromatographic sample can be treated in the same way, resulting in a number of diagrams similar to Fig. 7. Each of these surfaces corresponds to a single line in Laub and Purnell's original window diagrams. If all of these alpha surfaces are superimposed and if only those parts of the surfaces that are first encountered above the pH–[IIR] plane are plotted, the resulting surface shown in Fig. 8 is the two-dimensional equivalent of the Laub and Purnell windows, a response surface that looks like a mountain range.

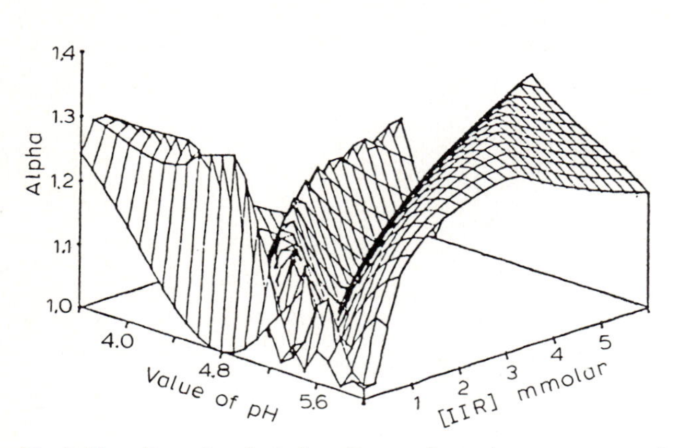

Fig. 8. Two-dimensional window diagram for a nine-component mixture. (Adapted from ref. 2.)

Figure 8 can be interpreted as follows: any point on the surface represents the best separation of the two worst separated pairs of compounds at that particular combination of pH and [IIR]. If the investigator is willing to define optimum chromatography as the best relative retention of the worst separated pairs of compounds, then optimum chromatography occurs at the top of the tallest "mountain" in Fig. 8. If the investigator were to prepare an eluant with the corresponding pH and [IIR], then the best separation possible for this system should be obtained.

How closely the actual chromatographic performance matches that predicted by the window diagram techniques depends upon both the accuracy of the models used in constructing the window diagrams and the precision and accuracy of the data that has been used to fit these models. Thus, it is important that high-quality experimental data be obtained and that good chromatographic models be used.

Any optimum obtained in this way is still conditional. If a third factor were included, say the percent of methanol in the eluant, then the shape of Fig. 8 would, in general, change as this third factor was varied. The "optimum chromatography" might get better, or worse, as the percent methanol is changed. Three- and higher-factor systems are challenging in this regard and offer even more power for obtaining improved separations.

5. Ruggedness

There are three domains in Fig. 8 that give approximately the same locally optimum results: the domain near pH = 4.0, [IIR] = 0.5 mM (near the front left side); the domain near pH = 4.4, [IIR] = 3.0 mM (along the ridge in the middle); and the domain near pH = 5.8 [IIR] = 3.0 mM (near the right side of the figure). Although all three optima would yield approximately the same quality of seapration, the domain on the right is preferable because it is more "rugged".

Ruggedness is a measure of a system's lack of sensitivity to small changes in operating conditions. The domain in the middle of Fig. 8 is very sensitive to small

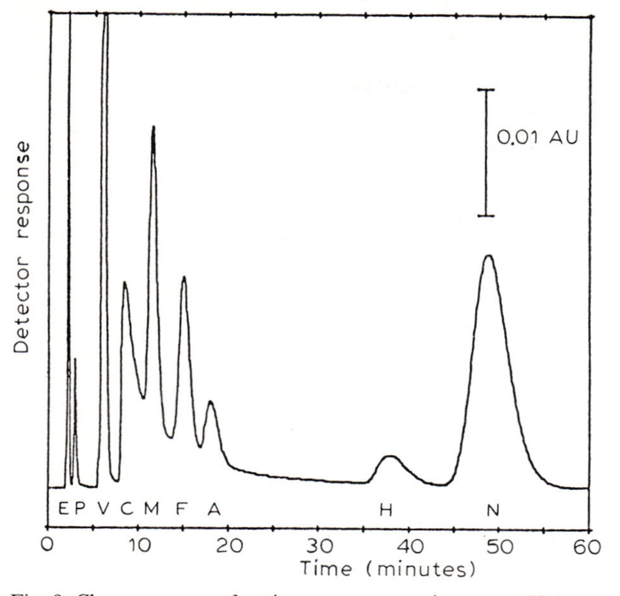

Fig. 9. Chromatogram of a nine-component mixture at pH 5.8 and [IIR] = 3.2 mM. Ultraviolet detector, 254 nm. E = Phenylethylamine, P = phenylalanine, V = vanillic acid, C = *trans*-caffeic acid, M = *trans-p*-coumaric acid, F = *trans*-ferulic acid, A = phenylacetic acid, H = hydrocinnamic acid, N = *trans*-cinnamic acid. (Adapted from ref. 2.)

changes in pH and in the concentration of IIR. If a technician were to prepare an eluant that was in error by a small fraction of a pH unit, the resulting separation would probably be useless: at least two peaks would be nearly merged. The same situation exists for the domain at the left of Fig. 8, small changes in pH or [IIR] could cause major degradation of the quality of the separation. The domain at the right of Fig. 8 is clearly more rugged and would be preferable for that reason. Figure 9 shows the chromatogram that was obtained in this domain at pH = 5.8, [IIR] = 3.2 mM.

6. Other approaches to multifactor chromatographic optimization

A related approach to multifactor chromatographic optimization is the "overlapping resolution map" (ORM) technique introduced by Glajch et al. [4]. In most applications using ORMs, four solvents have been used to optimize the separation of a set of similar compounds. The four solvents are typically methanol (MeOH), acetonitrile (ACN), tetrahydrofuran (THF), and water [5]. The fourth solvent, water, is used to adjust the solvent strength of each of the other solvents so the resulting mixtures (MeOH–water, ACN–water, and THF–water) are approximately isoeluotropic (i.e. giving about the same retention time for the last eluting component).

References p. 316

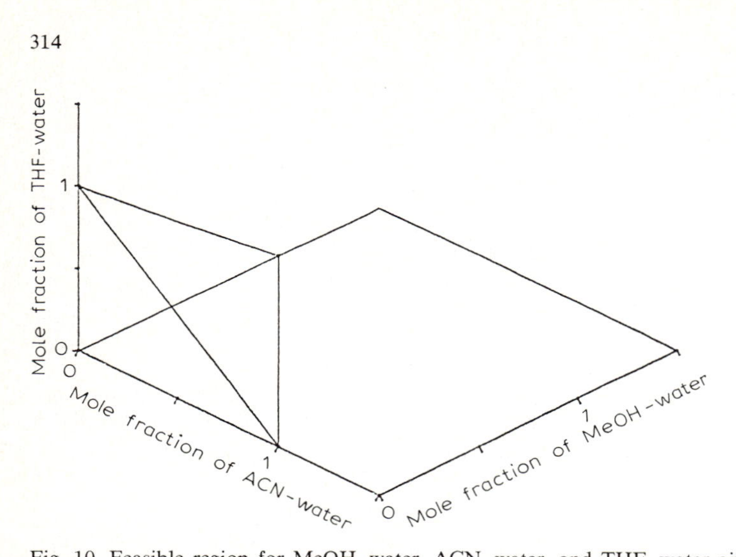

Fig. 10. Feasible region for MeOH–water, ACN–water, and THF–water given the constraint that the sum of the mole fractions must equal unity. (Adapted from ref. 1.)

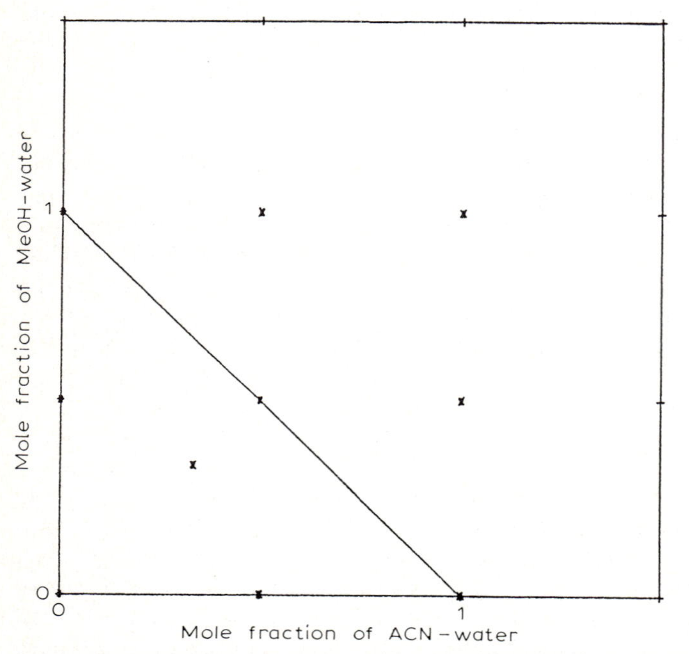

Fig. 11. An alternative view of the feasible region for MeOH–water, ACN–water, and THF–water given the constraint that the sum of the mole fractions must equal unity. A two-factor three-level full factorial experimental design is superimposed. (Adapted from ref. 1.)

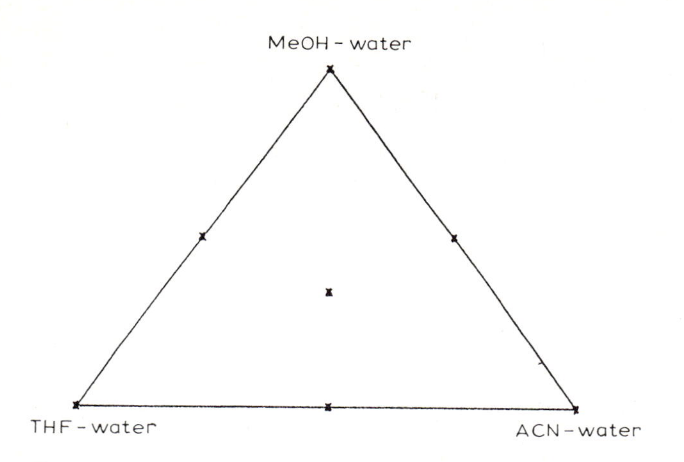

Fig. 12. An alternative representation of the information in Fig. 11. (Adapted from ref. 1.)

Once these solvent mixtures have been defined, water is no longer a variable. Thus, there are only three variables: MeOH–water, ACN–water, and THF–water.

In any experimental design involving mixtures, there is always the constraint that the sum of the mole fractions (or volume fractions) of all components must be unity. In the present case

$$X_{\text{MeOH-water}} + X_{\text{ACN-water}} + X_{\text{THF-water}} = 1 \tag{2}$$

Thus, there are only two *independent* variables (the third is fixed by the values of the first two). This situation is shown graphically in Fig. 10. Only those experiments that fall on the triangular oblique plane are permitted.

An alternative view of this situation is shown in Fig. 11. This view is generated by looking down on the MeOH–water/ACN–water plane from the THF–water axis. Although the mole fraction of THF–water is not shown, it may be obtained from a knowledge of $X_{\text{MeOH-water}}$ and $X_{\text{ACN-water}}$ by difference from eqn. (2). A two-factor three-level factorial design has been superimposed over the feasible region. Only six of the nine experiments (those lying within the feasible region and its boundaries) can actually be carried out. For purposes involving degrees of freedom in model fitting, one additional experiment is carried out in the "middle" of the feasible region; this gives a total of seven experiments.

Figure 12 gives a physical chemistry ternary phase representation of the information in Fig. 11. The top of the triangle corresponds to MeOH–water only ($X_{\text{MeOH-water}} = 1$); the bottom left side corresponds to THF–water only; and the bottom right side corresponds to ACN–water only. Experiments are carried out at the indicated locations to determine the retention times of each component of interest. The results for each component are fitted to a full second-order polynomial model (including a mathematical interaction effect).

$$t_R = b_0 + b_1 X_1 + b_2 X_2 + b_{11} X_1^2 + b_{22} X_2^2 + b_{12} X_1 X_2 \tag{3}$$

References p. 316

where the bs are parameters of the model and X_1 and X_2 represent the mole fractions of MeOH–water and ACN–water, respectively.

The resulting smooth parabolic response surfaces are treated in a way similar to that of the window diagrams discussed in the previous section. The resolution maps can be overlapped to find the set of chromatographic conditions that give the best overall resolution.

7. Sequential simplex optimization

Sequential simplex optimization has been used to optimize multifactor chromatographic systems. Although the sequential simplex technique is a powerful multifactor optimization technique, its use in chromatographic systems is limited because of the existence of multiple optima caused by elution order reversal (see Fig. 8). If started from any domain of factor space (e.g. any domain of pH and [IIR]), the simplex will converge to one of the several local optima. Although there is no guarantee that it will converge to the global optimum (the best of the local optima), its convergence to the global optimum is probable because the global optimum usually spans the largest domain of factor space; thus, it is likely that one of the simplex vertexes will be positioned rather far up the side of the global optimum and the remaining vertexes will converge into this region.

In chromatographic systems where elution order reversal is possible, window diagram techniques should be used first if a sufficient number of experiments can be carried out and if a sufficiently appropriate model can be written to describe the behavior of the system. If the resulting chromatographic performance needs further improvement, a small simplex can then be used to "fine tune" the system and obtain the maximum chromatographic performance.

References

1 S.N. Deming, J.G. Bower and K.D. Bower, Multifactor optimization of HPLC conditions, Adv. Chromatogr., 24 (1984) 35.

2 R.C. Kong, B. Sachok and S.N. Deming, Multifactor optimization of reversed-phase liquid chromatographic separations, J. Chromatogr., 199 (1980) 307

3 R.J. Laub and J.H. Purnell, Criteria for the use of mixed solvents in gas–liquid chromatography, J. Chromatogr., 112 (1975) 71.

4 J.L. Glajch, J.J. Kirkland, K.M. Squire and J.M. Minor, Optimization of solvent strength and selectivity for reversed-phase liquid chromatography using an interactive mixture–design statistical technique, J. Chromatogr., 199 (1980) 57.

5 L.R. Snyder and J. Kirkland, Introduction to Modern Liquid Chromatography, Wiley, New York, 2nd edn., 1979.

Recommended reading

J.C. Berridge, Unattended optimization of reversed-phase high-performance liquid chromatographic separations using the modified simplex algorithm, J. Chromatogr., 244 (1982) 1.

J.A. Cornell, Experiments with Mixtures. Designs, Models, and the Analysis of Mixture Data, Wiley, New York, 1981.

J.A. Nelder and R. Mead, Simplex method for function minimization, Comput. J., 7 (1965) 308.

J.H. Nickel and S.N. Deming, Use of the sequential simplex optimization algorithm in automated liquid chromatographic methods development, Liq. Chromatogr., 1 (1983) 414.

P. Schoenmakers, Optimization of Chromatographic Selectivity, Elsevier, Amsterdam, 1986.

S.N. Deming, J.G. Bower and K.D. Bower, Multifactor optimization of HPLC conditions, Adv. Chromatogr., 24 (1984) 35.

J.C. Berridge, Techniques for the Automated Optimization of HPLC Separations, Wiley, New York, 1985.

Chapter 20

The Multivariate Approach

Analytical data are generated to characterize objects (meteorites, olive oil samples, blood samples from patients, etc.). This characterization is relatively easy when there are only a few analytical data (up to three, see Sect. 2) for each object. In modern computer-aided analytical chemistry, one usually obtains many more data. Instrumental multielement methods, such as neutron activation analysis, or multisubstance methods, such as chromatography, give many more than three results per object. In routine clinical chemistry, a blood sample is analyzed simultaneously, for instance, for twelve variables. The central question of this part of the book is how to extract the information from such an amount of data. In this chapter, some terminology and some intuitive concepts will be introduced with examples. In Chaps. 21–23, these concepts will be discussed in more detail.

1. A multivariate look at clinical chemistry. Multivariate distributions

Clinical chemistry is one of the areas where one usually determines many substances simultaneously and where large amounts of data are produced. Because computers provide extensive facilities for storing and handling these data, the impact of computers in clinical chemistry has been enormous. Although the results obtained for the variables are often related to each other, they are often used either individually (e.g. if a parameter is higher or lower than normal the patient would be suspected of having a certain disease) or successively (e.g. the observation that parameter a is high but b is normal leads to certain conclusions). As the data are obtained simultaneously and concern one sample, it would be preferable to use them simultaneously instead of individually and Fig. 1, illustrates one of the

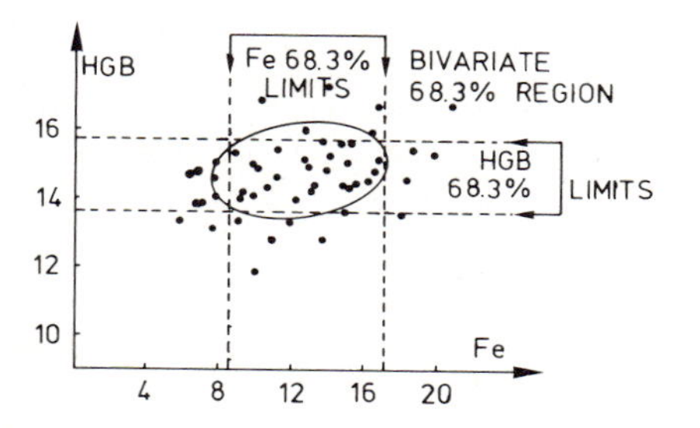

Fig. 1. A bivariate distribution. The parameters HGB (B-haemoglobin) and Fe (S-iron, transferrin bound) were measured for 52 healthy men [1].

References p. 337

advantages of doing so. The ellipsoidal contour line delineates the region expected to contain 68.3% (one standard deviation) of the samples, as determined from a two-dimensional gaussian distribution.

The reason that the main axis of the ellipsoid is not parallel to the abscissa is that the data are correlated. It follows that the correlation coefficient is one of the parameters that characterize a bivariate distribution. If one uses the parameters individually (i.e. two separate univariate gaussian distributions), the normal region would be the rectangle. Several points fall outside the rectangle and would be declared abnormal by the unvariate-thinking person, whereas in fact they are normal, as would be recognized by the bivariate-thinking observer.

This reasoning can be generalized to more dimensions. This is the multivariate normal approach. The starting point of multivariate problems is always a data matrix in which, for n samples (or objects), m data (or variables or features) have been obtained.

	—m variables →			
	—	—	—	—
	—	—	—	—
n samples	—	—	—	—
	—	—	—	—

2. The need for visual information

The human observer is very good at recognizing patterns. He is able to associate a character such as "a" or "b" with a particular sound and to recognize a person with certain characteristics as his neighbor and (hopefully) to distinguish him/her from other people. Just try to describe your neighbor in such a way that a machine could sort her or him out from many others and you will understand how superior the combination of the human brain and eye really is.

Analytical chemists are human and so they take advantage of this faculty in interpreting data they obtain. Let us suppose an analyst determines a single variable (the concentration of one element, x_1) in some samples. He will then draw a graph of the results as in Fig. 2 or, if there are more results, he will make a histogram. From Fig. 2 one concludes that there are two groups of related objects.

In many instances, the analytical chemist will consider that a single variable is not sufficient and that a second or a third is necessary to completely characterize the objects investigated. If he wants to classify meteorites, as in Sect. 4, he will not be happy with the determination of Ni only, but will also want to measure Ge or

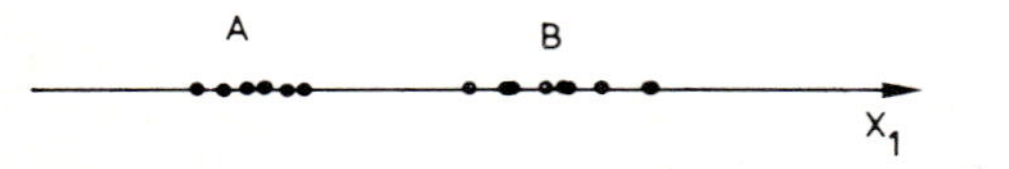

Fig. 2. Plot of the results of one variable (univariate plot).

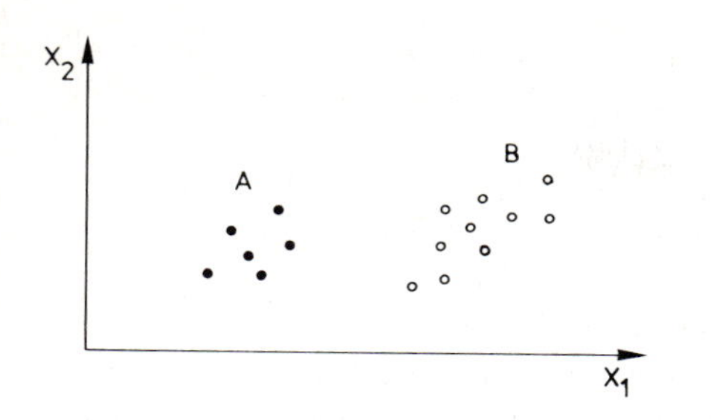

Fig. 3. Plot of the results of two variables (bivariate plot).

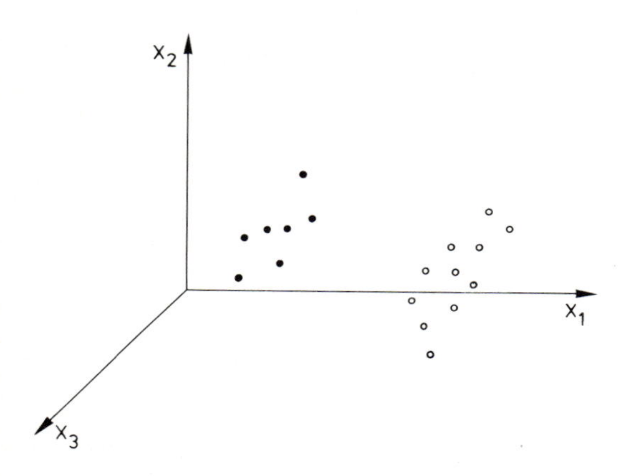

Fig. 4. Plot of the results of three variables.

even Ga. The results for two or three variables might then be those of Figs. 3 and 4. Again, it is concluded without difficulty that two groups of similar objects exist. Note that, while making the figure for two variables (two dimensions) is not at all difficult, it becomes much more difficult to draw it for three variables or dimensions. However, a computer can do this and once it has been done, we have little trouble in discerning groups of similar objects. Clearly, however, three variables constitute the limit to our visual ability. If we want to make use of our visual pattern-recognizing ability, it will be necessary to represent the data in a two- or three-dimensional way. Much of pattern recognition is concerned with the question: how does one condense m-dimensional information in 2 or 3 dimensions. In each of the following examples, we will see that making pictures is a step in the pattern-recognition procedure.

3. Patterns, classes, and distances

In the preceding section, we have used the word "pattern" in an intuitive way. For example, the letter "a" shows characteristics which make it unique compared

References p. 337

with other characters. Apparently, our eye and brain use features to describe such a character and distinguish it from others. This list of features constitutes a pattern. Certain patterns of characteristics will lead us to the conclusion that letter "a" or "b" is present. In the same way, an object characterized by several analytical results may be said to be characterized by a pattern of analytical results. It is hoped that specific patterns can be observed for different kinds of sample.

In Sect. 6, we will discuss how to differentiate between olive oils from different geographical origins. Let us, for the moment, suppose that there are two regions A and B and that for some samples of both regions the concentrations of two fatty acids were measured. The two concentrations define a two-dimensional space and the A and B groups are found to occupy different locations in this space (Fig. 3). One is easily able to distinguish between the two classes. This reasoning can be generalized to more dimensions. If 8 fatty acids are determined, each olive oil sample can be viewed as a point in an 8-dimensional space, each coordinate x_i being the concentration of one of the eight fatty acids. Such a point is conveniently represented by a vector (pattern vector)

$$\mathbf{x} = (x_1, x_2, \ldots, x_i, \ldots, x_8) \tag{1}$$

The vector is composed of m measurement results (8 in this example) constituting a set of m scalar values (coordinates). Each of these patterns is a row of the data matrix (Sect. 1). The m variables define the m-dimensional or m-variate pattern space.

If one were able to observe the pattern space visually, as is possible in the two-dimensional case in Fig. 3, one would note that the points tend to form groups or clusters. In the same way, meteorites with similar characteristics or patterns will tend to form clusters (see Sect. 4). The recognition of similar patterns or the isolation of the clusters is therefore of great analytical interest.

The word similar is important in this context. In Fig. 5, one observes that it has a geometrical meaning. For instance a and b are considered more similar to each other than to c, and d is considered very different from a, b and c. This two-dimensional example shows that objects (or the pattern vectors by which they are characterized) are considered all the more similar when they are near to each other

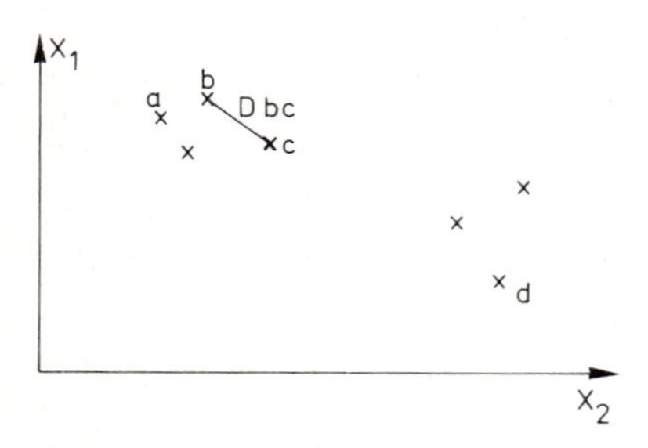

Fig. 5. Distance and similarity. Seven objects are characterized by two variables. a is more similar to b than to c because it is nearer to it; a is very unlike d.

in the pattern space. The distance between objects is therefore a measure of similarity between these objects in the sense that a small distance means a high similarity.

Distance is not the only way of expressing similarity: correlation can also be used for this purpose. More details about similarity measures and distances are given in Sect. 9.

4. The classification of iron meteorites. Clustering

About 600 iron meteorites have been found on earth and have been subjected to exhaustive inorganic analysis. For most of them, one has determined many variables, such as the contents of Ni, Ga, Ge, Ir, Au, etc. About thirteen such parameters are significant for the classification of the meteorites. Such a classification is important because meteorite specialists believe that each group comes from a different celestial body so that this classification permits a better understanding of a part of astronomical history. Classifying is very generally used by human beings to understand the structure of large sets of objects and the interrelationships between these objects. A typical application is the taxonomy of living species.

Figure 6 gives a very small part of the classification of plants. One observes that species are grouped in families, families in classes, etc. This classification was obtained historically by determining characteristics such as the number of cotyledones, the flower formula, etc. In fact, one could have determined for each plant a certain number of characteristics in a data matrix

Characteristics →

— — — —

Plants — — — —

— — — —

— — — —

and applied a mathematical technique to define optimal groupings and, therefore, a mathematically objective classification. This is what is being done by modern specialists in what is called numerical taxonomy. Their principal technique is clustering. Clustering can also be applied to the 600 (meteorites) × 13 (variables) data matrix describing iron meteorites. A small part of this classification is given in Fig. 7.

In both cases, the classification is represented in a tree (called a dendrogram). The dendrogram is a way to visualize the relationships between the objects and it aids the comprehension of the classification obtained. For instance, one observes that meteorites 4, 5, 6, 7, 8, 10, 11 and 39 are closely related and that they also bear some similarity to the groups 9, 12, 13, 14 and 34, 36, 37, 38, 40, 41, 43, 46 but not to meteorites 26 and 3, which are rather different from most other meteorites. Clustering techniques are described in Chap. 22.

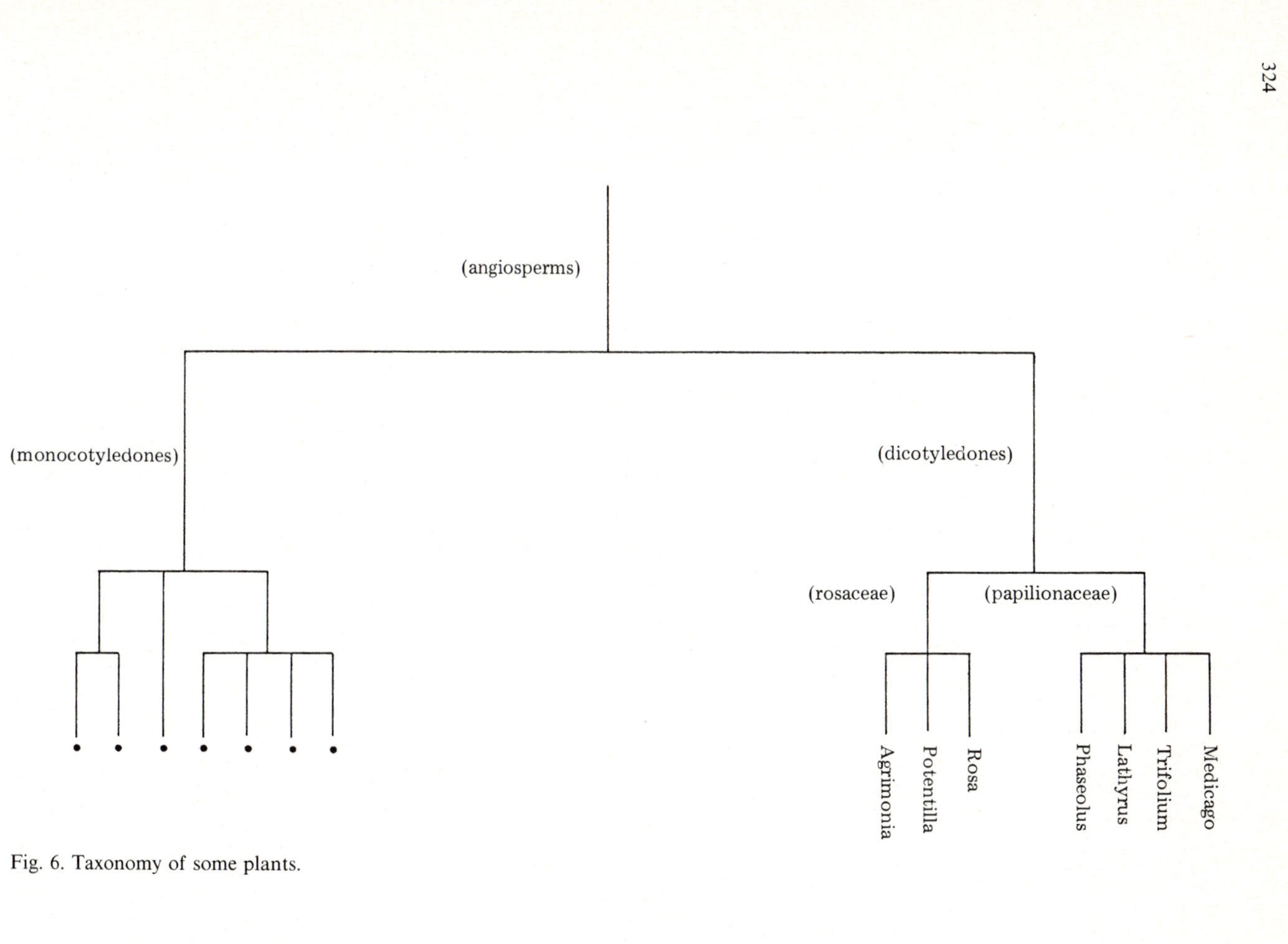

Fig. 6. Taxonomy of some plants.

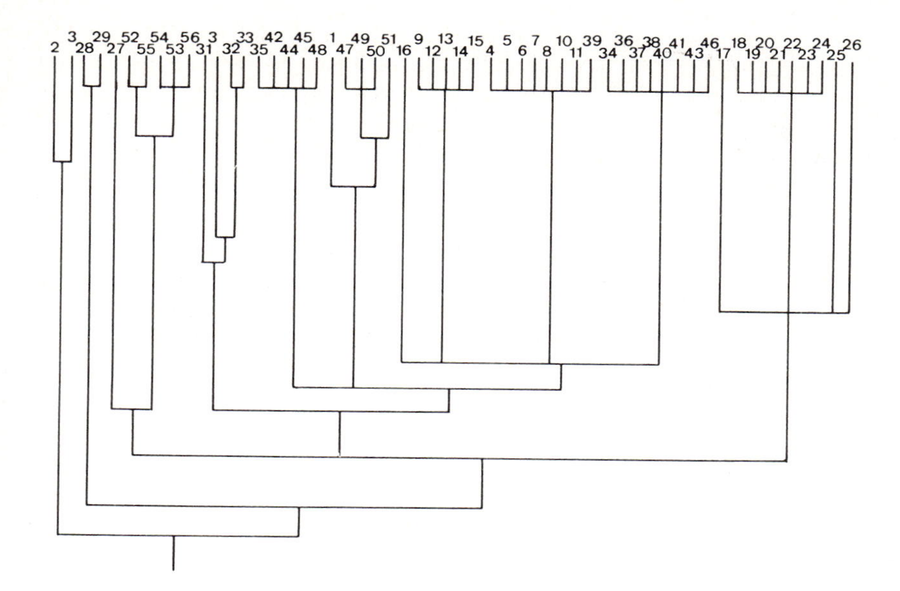

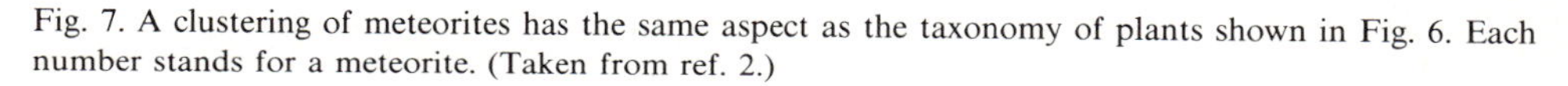
Fig. 7. A clustering of meteorites has the same aspect as the taxonomy of plants shown in Fig. 6. Each number stands for a meteorite. (Taken from ref. 2.)

5. Air pollution. Principal components

In a measuring station located in the Netherlands (see the map in Fig. 8), an air sample was taken every week for 3 years and analyzed by capillary gas chromatography [3]. These analyses yielded about 300 components, 35 of which were considered important as they were found in nearly every sample. At the same time, some 12 meteorological data were collected of which we will only consider the direction of the wind in this example. The data set therefore consists of about 150 samples, each characterized by 35 chemical variables and the wind direction.

The questions asked were

(a) are there subsets of objects with similar chemical characteristics or, in other words, are certain chemical air pollution patterns prevalent?

(b) what is the relationship between chemical air pollution patterns and the wind direction?

These are, as in the preceding section, questions about unknown categories but, unlike the preceding section, these categories are probably rather vague and overlapping. If the classes depend on the direction of the wind, which is a continuous variable, no clear-cut clusters of patterns are expected. Although a dendrogram as determined in the preceding section would yield some information, a much better way is to try and observe as well as possible the 150 samples in the 35-dimensional space of chemical variables.

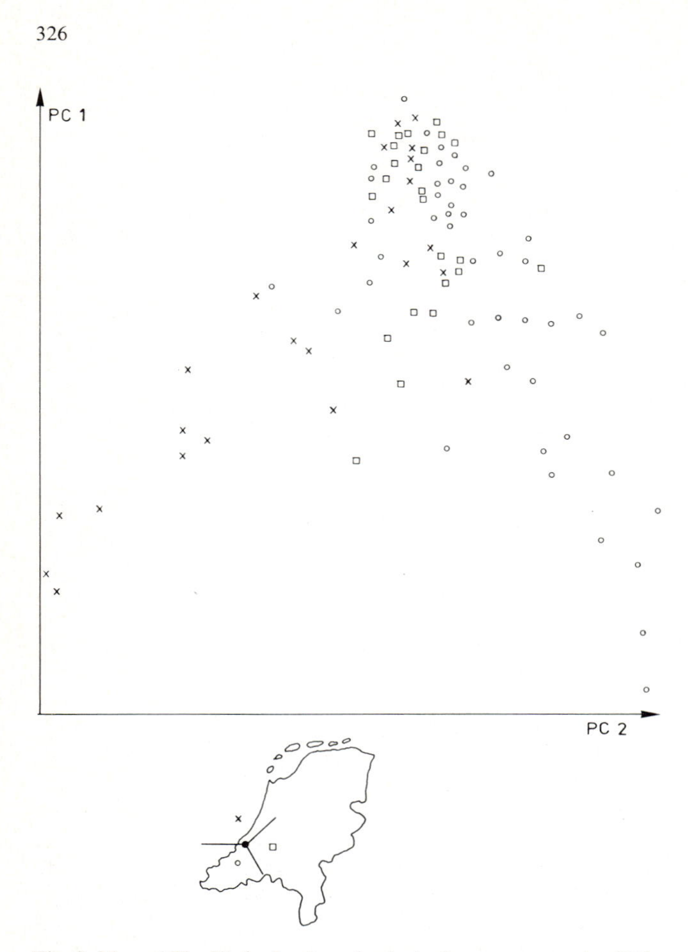

Fig. 8. Map of The Netherlands and principal components plot (PC_1 vs. PC_2) of air samples taken at the point shown on the map. Three possible wind directions are identified on the map by the symbols, ○, ×, and □. The same symbols are used on the plot to identify the prevailing wind direction at the moment the samples were taken.

This cannot be done directly. However, by projecting the data on a well-chosen two-dimensional plane, one may hope to observe at least the salient features of the distribution of the sample. Methods that permit this are called display methods or feature reduction methods (in pattern recognition one often uses the term feature instead of variable). In this section, an introduction will be given to the principal components method. Principal components and related methods are discussed in more detail in Chap. 21.

Before discussing the air pollution situation, let us consider a two-dimensional space (Fig. 9). A feature or dimension reduction in this space can only lead to a one-dimensional representation (i.e. by representing the points on a line). Of course,

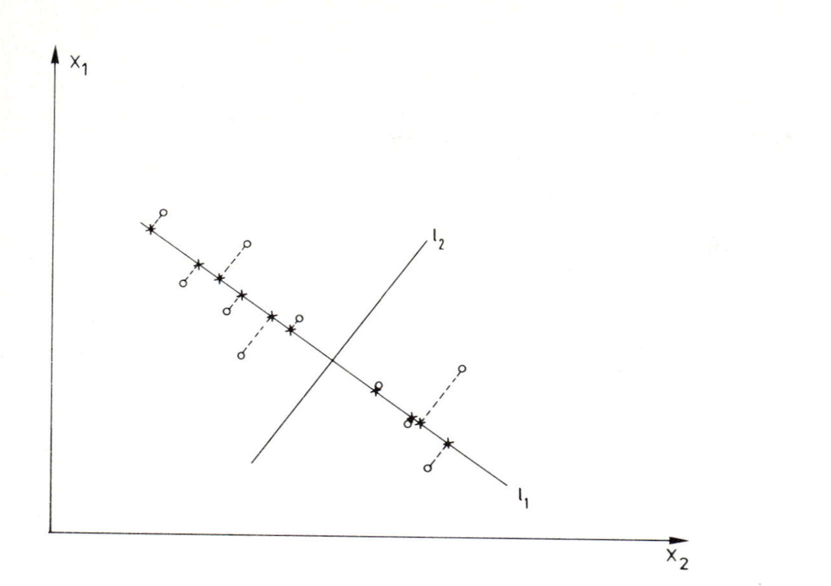

Fig. 9. Objects characterized by two variables, x_1 and x_2, are projected (×) on line l_1. For line l_2, see text.

there is no sense in doing this in practice, but it will be possible to generalize the conclusions obtained to m dimensions.

One can choose several possible directions for the line on which to project the points. Two such directions, l_1 and l_2 are given in Fig. 9. Clearly, direction l_1 is to be preferred to l_2 because the original structure is better preserved. On l_1, one observes that there are two groups of points as in the original two-dimensional case. On l_2, this is no longer possible. Line l_1 is chosen in such a way as to preserve the maximal variation among the points. This is the definition of a principal component.

The principal component is a new variable describing the objects and is a combination of the two old variables. Each object i is now described by the single value u instead of by variables x_{1i} and x_{2i}. The combination is linear

$$u_i = ax_{1i} + bx_{2i} \tag{2}$$

In a three-dimensional situation (Fig. 10), one can reduce the dimensions to one or two. The selection of the first principal component is again carried out in such a way that it retains as much variation as possible. If one wants a second principal component, this will be chosen to be orthogonal to the first and to explain as much variation as possible of that left unexplained by the first principal component. Once one has obtained two principal components, one has, in fact, defined a plane on which the three-dimensional, or by generalization m-dimensional, points are projected. The direction of the plane is chosen to represent the original data structure in

References p. 337

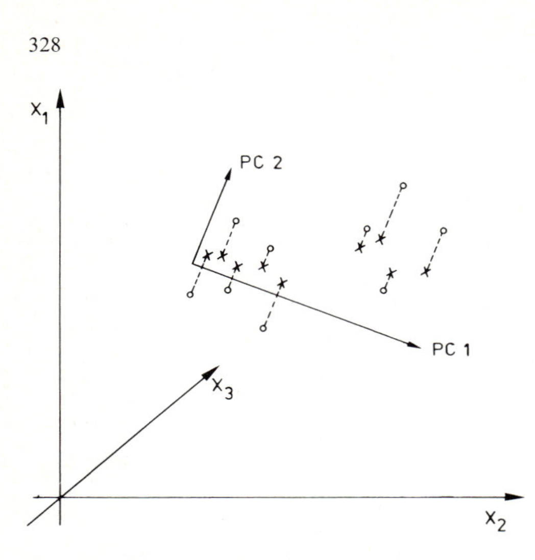

Fig. 10. Objects originally present in a three-dimensional space (○) are projected (×) on the principal components plane PC_1–PC_2.

the best possible way. From a principal components analysis, one therefore obtains two main results, namely

(1) a two-dimensional representation of the data so that relationships among points may be observed and

(2) the magnitude and sign of the coefficients *a* and *b*, called loadings (see Chap. 21, Sect. 3), which give an indication about their significance for determining the data structure.

A third property, which is of no interest in this particular context but which will be used and explained in Chap. 23, is that the principal components result can be considered to be a model of the data.

Let us now consider the results of the principal components analysis of the air pollution data. A plot of PC_1 versus PC_2 is given in Fig. 8. To prepare this figure, a different symbol was given to samples corresponding to three different wind sectors. The first of these blows over an industrial zone, the second over a large town and the third mainly over a huge greenhouse area and highway complex. The figure consists of a kernel, in which all "urban" samples and some "industrial" and "greenhouse" samples are located, and two wings, one of which consists of "industrial" and the other of "greenhouse" samples. The figure by itself makes it clear that the wind direction is a determining factor for the pattern of pollution. A more detailed understanding can be derived by considering the coefficients making up the principal components. Indeed, one may write [see eqn. (2)] for all objects *i*

$$(PC_1)_i = a_1[\text{benzene concn.}]_i + b_1[\text{toluene concn.}]_i + c_1[\text{octane concn.}]_i + d_1[\text{nonane concn.}]_i + \ldots$$

and

$$(PC_2)_i = a_2[\text{benzene concn.}]_i + b_2[\text{toluene concn.}]_i$$
$$+ c_2[\text{octane concn.}]_i + d_2[\text{nonane concn.}]_i + \ldots$$

where $(PC_1)_i$ is the value of sample i along the first principal component and [benzene concn.]$_i$ is the concentration of benzene in sample i.

In this particular example, one could consider PC_1 to be a measure of global pollution and therefore the kernel in the figure can be considered as a normal situation and the two wings, which have higher absolute PC_1 values, as two different situations of higher pollution. The values of the loading coefficients also show which substances contribute most to pollution. Table 1 shows that the highest coefficient value is for *m*-ethylmethylbenzene and the lowest for anisole. Clearly, *m*-ethylmethylbenzene contributes more to PC_1 than anisole.

On PC_2, benzene, toluene and, in general, the aromatics have a negative sign and high values of the coefficient. Decane, undecane and the other aliphatics also have high values but with a positive sign. One concludes that PC_2 differentiates essen-

TABLE 1

Loading coefficients for Vlaardingen (adapted from ref. 3)

The substances are ordered according to the absolute value of the loading for principal component 1. Only coefficients with absolute values higher than 0.2 are shown.

Explained variance	Principal component		
	1 35%	2 16%	3 10%
m-Ethylmethylbenzene	−0.300		
Toluene	−0.288		
1,2,4-Trimethylbenzene	−0.277		
o-Xylene	−0.232	−0.310	
n-Decane	−0.231	0.260	
n-Tridecane	−0.231		−0.243
Ethylbenzene	−0.230	−0.234	
n-Dodecane	−0.227	0.260	
n-Octane	−0.226		
p-Xylene	−0.216	−0.274	
Styrene	−0.202	−0.204	
n-Undecane		0.315	
m-Xylene		−0.226	
Acetophenone		0.247	
n-Nonanal		0.281	
n-Decanal			−0.284
Isopropyl acetate			0.508
Isobutyl acetate		0.200	0.472
2-Ethoxyethyl acetate			0.359
Benzene		−0.289	
Anisole			

tially between aromatic and aliphatic type pollutions.

In conclusion, the analysis showed us that there are three different situations

(a) a normal situation when the wind blows over the urban zone and sometimes when it blows from the other two directions,

(b) a high pollution situation with predominantly aromatic character may arise only when the wind blows from the industrial sector, and

(c) a high pollution situation with predominantly aliphatic character may arise when the wind blows from the greenhouse/highway direction.

Principal components and some related techniques are discussed in Chap. 21.

6. The geographical origin of olive oils. Supervised pattern recognition

This example is concerned with what is called food authentification, i.e. the verification that a foodstuff comes from an alleged origin. In this particular instance, one wanted to derive a rule which would discriminate between olive oils from nine Italian regions. To start with, about 500 olive oils of known origin were analyzed: the concentrations of 8 fatty acids were determined for each. One wanted to know if, in some way, these results could be used to derive a procedure to determine the origin of new samples. In pattern recognition terminology, this question can be rephrased as: use the learning samples (i.e. those with known origin) to derive a classification rule which allows us to classify test samples (i.e. those with unknown origin) in one of nine known classes. In contrast with the examples of Sects. 4 and 5, this is called supervised pattern recognition or supervised learning.

Mathematically, this means that one needs to assign portions of an 8-dimensional space to the nine classes. A new sample is then assigned to the class which occupies the portion of space in which the sample is located. This may be understood more easily by re-reading Sect. 3. The best way to analyze such a problem is to start by trying to visualize it. This can be done by a principal components analysis. The result (for clarity of representation not all the 500 samples are shown) is given in Fig. 11, which immediately shows the classes that can be separated easily and where the difficulties lie. For instance, one does not expect difficulties in distinguishing oils from Calabria and Umbria but, on the other hand, a discrimination between Calabria and Sicily is much more difficult.

A first way of achieving the classification would seem to be (1) to determine PC values [the u values of eqn. (2)] for the new sample to be classified using the PC coefficients derived from the 500 known samples, (2) to plot the new sample in Fig. 11 and to see in which group it falls. This, with some modifications which will be discussed below, is what is done in a method called linear discriminant analysis which is probably the most used supervised pattern recognition method, although it presents some technical disadvantages (see Chap. 23).

In linear discriminant analysis, as in principal components analysis, one tries to reduce the number of features. Let us therefore consider again the two-dimensional space of Fig. 3. Again, one needs to determine a one-dimensional space (a line) on which the points will be projected. However, while principal components selects a

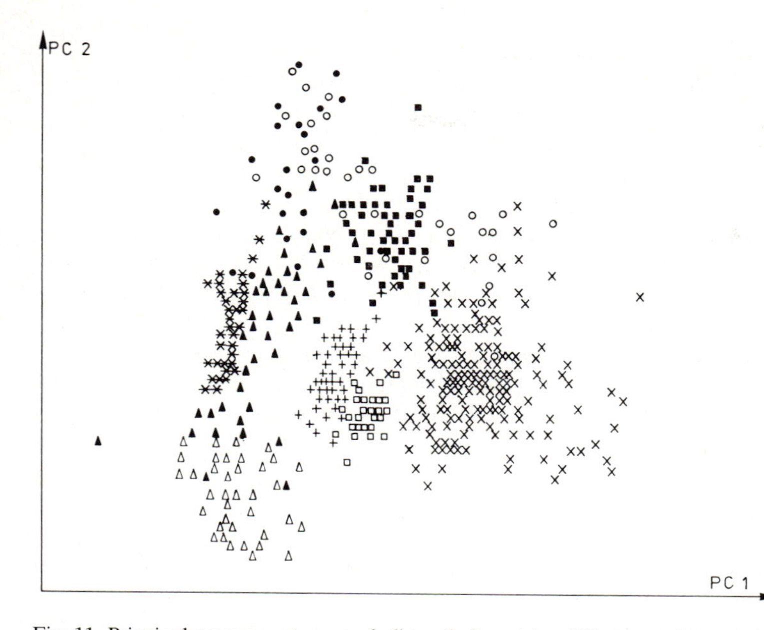

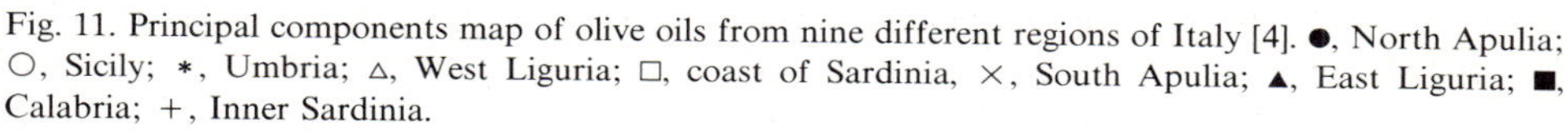

Fig. 11. Principal components map of olive oils from nine different regions of Italy [4]. ●, North Apulia; ○, Sicily; ∗, Umbria; △, West Liguria; □, coast of Sardinia, ×, South Apulia; ▲, East Liguria; ■, Calabria; +, Inner Sardinia.

direction which retains maximal structure in a lower dimension among the data, linear discriminant analysis selects a direction which achieves maximum separation among the given classes. The discriminant function obtained in this way leads to a new variable which, as in principal components, is a linear combination of the original variables. When there are k classes, one can determine $k-1$ discriminant functions.

In Fig. 12, two discriminant functions are plotted against one another for three overlapping classes. A new sample can be allocated by determining its location in the figure.

7. Thyroid status. Supervised pattern recognition

In linear discriminant analysis, allocation can also be achieved in a more formal way by determining the centroid of each group and drawing a boundary half-way between the two centroids.

Linear discriminant analysis and several other supervised methods put the accent on finding optimal boundaries between classes: their first goal is to discriminate. This is not necessarily the best possible approach. An example from clinical chemistry will show this. This example concerns the thyroid gland. People whose thyroid gland functions normally are called euthyroid and patients whose thyroid gland is too active or not active enough are called, respectively, hyperthyroid or hypothyroid. Clinicians want to make a distinction between the three classes. We will call them EU, HYPER, and HYPO. The distinction can be made using five chemical determinations such as serum thyroxine or thyroid-stimulating hormone. This clearly is a multivariate (five-dimensional) situation and linear discriminant analysis can be applied. The resulting plot of discriminant functions is given in Fig. 12.

This plot is typical for many situations of clinical chemistry: it shows a tight normal group (the EU group) and spreading out from it much more disperse abnormal groups (the HYPO and HYPER groups). In fact, this kind of picture is also found in many non-clinical situations too (see, for instance, the air pollution situation of Fig. 8). If one now determines boundaries as described in the preceding section, namely by situating them half-way between the centroids of adjacent classes, some patients of the more disperse abnormal classes are catalogued as members of the more condensed classes.

This classification problem can be solved by developing better boundaries. For instance, using so-called potential methods (see also Chap. 23) leads to the boundaries of Fig. 13.

There are several other problems with discrimination-oriented methods. One of them is that one needs to classify every object in one of the given classes. It is, however, quite possible that an object should not be classified in either class. Returning to the oil example, where a classification between oils from Calabria and Sicily was wanted, one must realize that the sample submitted for classification may, in reality, be for instance a Spanish oil. However, using the discrimination-oriented

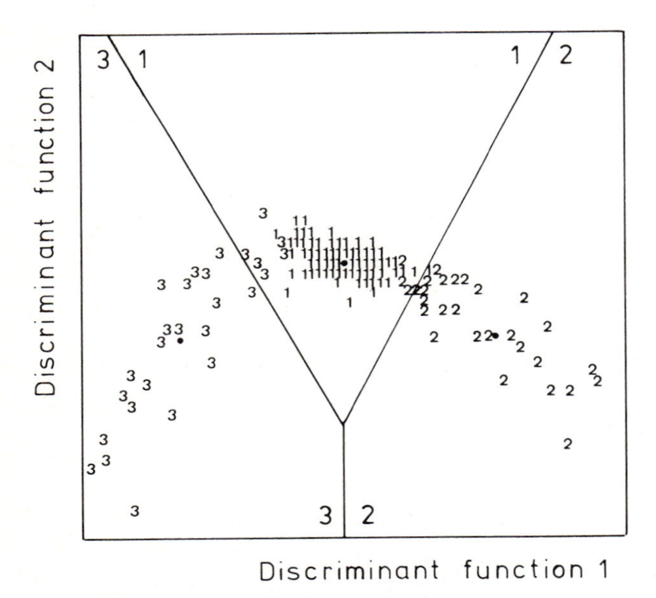

Fig. 12. Linear discriminant analysis plot for three classes with different thyroid status [15].

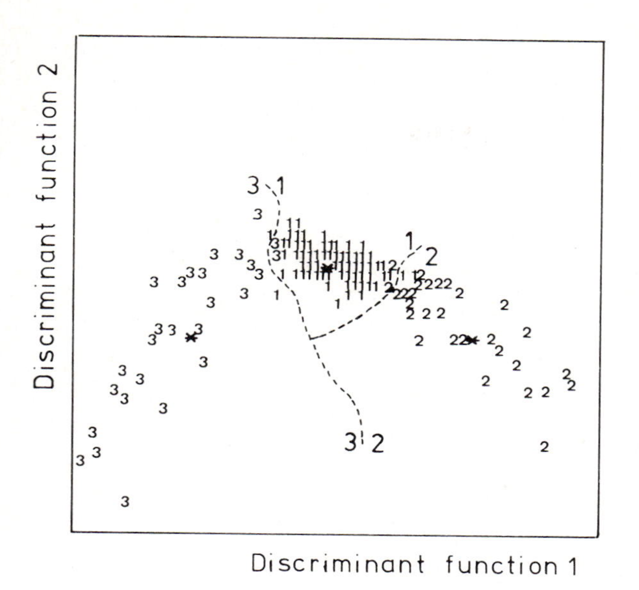

Fig. 13. As Fig. 12 but with a boundary obtained by a potential method [5].

methods, one will classify it necessarily in one of the Italian regions. Often, classification problems are not mere discrimination problems. A medical doctor, for instance is not satisfied with the knowledge that a patient is HYPO, HYPER, or EU. He wants to know, for instance, to what extent he is HYPO, whether the pattern of results is unusual for a HYPO, since this may give an indication of additional factors to be considered, how the HYPO condition reacts to treatment. In fact, he is much more concerned by the relative situation of a patient in a class, than by the discrimination between classes. This is also true in most non-clinical classification problems. Since the pattern recognition method should be a help in understanding the data structure, it must be adapted to the way we think. Apparently, our way of thinking is more to consider the situation of an object in a certain class relative to other objects in the classes and to other classes rather than to merely discriminate between classes. A very different approach to supervised pattern recognition is therefore needed. This consists of making a "model" of each class. Objects which fit the model for a class are considered part of it and objects which do not fit are classified as non-members. In discrimination terms, one could say that the model discriminates between membership and non-membership of a certain class.

The conceptually simplest model, which for reasons explained in Chap. 23 is called UNEQ, is based on the multivariate normal distribution. Returning to Fig. 1, we could say that the ellipsoid contour describing the 95% confidence limit for a bivariate distribution can be considered as a model of a class of healthy patients. Those within its limits are considered healthy and those outside as non-members of

References p. 337

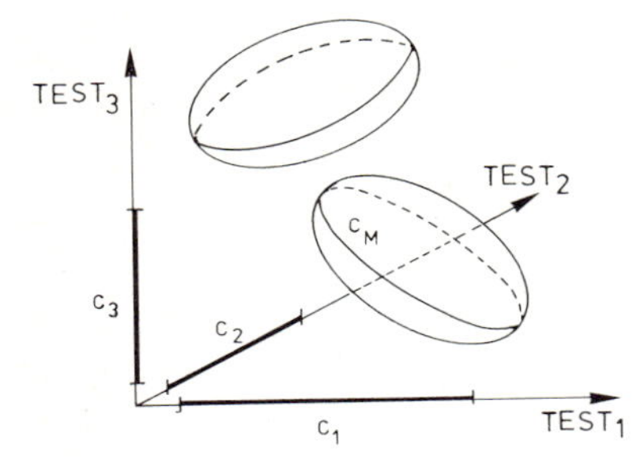

Fig. 14. Class envelopes in three dimensions as derived from the three-variate normal distribution.

the healthy class. The bivariate normal distribution is therefore a model of the healthy class.

In three dimensions (Fig. 14), the model takes the shape of a rugby ball and in *m* dimensions, one must imagine an *m*-dimensional American football or rugby ball. In the figure, two classes are considered and one now observes that four situations can be encountered when classifying an object, namely

(a) the object is only part of class I,

(b) the object is only part of class II,

(c) the object is not a member of class I or II: it is an outlier, and

(d) the object is a member of both classes I and II: it is situated in a zone of doubt.

Moreover, since there is a mathematical model describing the class, it is fairly easy to situate an object inside a class. For instance, one can now easily determine whether an object is close to the center or to the boundary and in which direction.

One concludes that model-oriented methods have several advantages compared with discrimination-oriented methods. Therefore, we recommend it as preferable to discrimination-oriented methods, except, perhaps, in one instance, namely feature reduction.

8. Optimal combination of variables. Feature selection

One can reduce the number of features in two ways. In Sect. 5, the original variables were combined in a smaller number of principal components. This is called feature reduction. Another way is feature selection: from the *m* variables one selects a subset of variables that seem to be the most discriminating. The features obtained therefore correspond to some of the given measurements, while in the display methods the dimensionality reduction is obtained by using all the variables but combining these into a small number of new ones. Feature selection therefore constitutes a means of choosing sets of optimally discriminating variables and, if these variables are the results of analytical tests, this consists, in fact, of the selection of an optimal combination of analytical tests or procedures.

TABLE 2

Classification of patients according to their thyroid status (adapted from ref. 6)

Number of tests considered	Classification success (expressed as selectivity [a])	
	EU/HYPO	EU/HYPER
5	80.0	91.4
4	83.0	88.6
3	86.7	85.7
2	96.7	77.1

[a] See Chap. 26.

Most of the supervised pattern recognition procedures permit the carrying out of stepwise selection, i.e. the selection first of the most important feature, then, of the second most important, etc. The results for the linear-discriminant analysis of the EU, HYPER, HYPO classification of Sect. 7 is given in Table 2. For the HYPER/EU discrimination, the elimination of successively one, two, and three of the variables leads to the expected result that, since there is less information, the classification is less successful. On the other hand, for the HYPO/EU discrimination, a less evident result is obtained. A smaller number of tests leads to an improvement in the classification results. One concludes that the variables eliminated either have no relevance to the discrimination considered, and therefore only add noise, or else that the information present in the eliminated variables was redundant (correlated, see Chap. 9) with respect to the retained variables.

It should be stressed here that feature selection is not only a data manipulation operation, but may have economic consequences. For instance, one could decide on the basis of Table 2, to reduce the number of different tests for a EU/HYPO discrimination problem to only two. A less straightforward problem with which the decision maker is confronted is to decide how many tests to carry out for a EU/HYPER discrimination. One loses some 3% in selectivity by eliminating one test. The decision maker must then compare the economic benefit of carrying out one test less with the loss contained in a somewhat smaller diagnostic success. In fact, he carries out a cost–benefit analysis (see Chap. 10). This is only one of the many instances where an analytical (or clinical) chemist may be confronted with such a situation. Making decisions is discussed in more detail in Chaps. 24–27.

9. Similarity and distance

Essential to the whole of pattern recognition is the concept of similarity. Different measures can be considered according to the situation. This is shown with the aid of Figs. 5, 15, and 16. In Fig. 5, a is considered more similar to b than to c because $D_{ab} < D_{ac}$, where D is the so-called Euclidian distance between a and b. Grouping together of objects with small distances leads to roundish clusters. In Fig. 15, however, one might consider a more similar to b than to c. This leads to the selection of linear clusters.

References p. 337

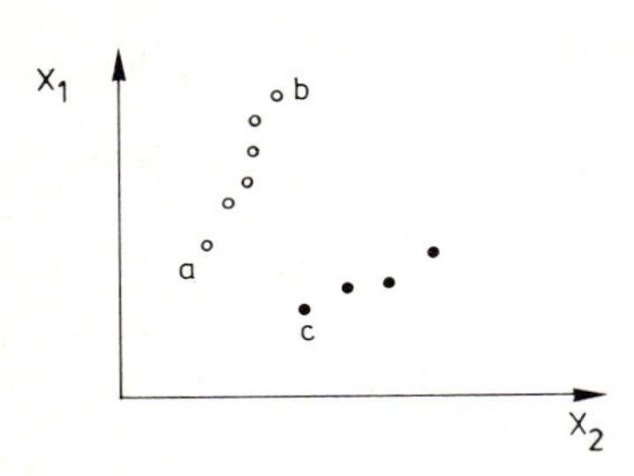

Fig. 15. Similarity of linear clusters. a is more similar to b than it is to c although it is nearer to c.

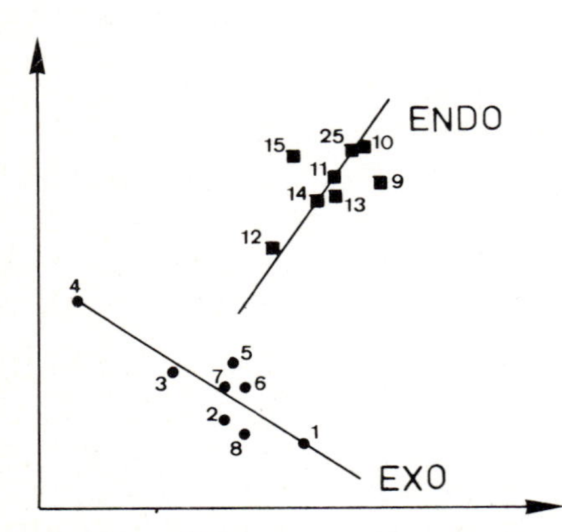

Fig. 16. The ^{13}C NMR spectra of substituted norbornanes lead to linear clusters [7].

Linear clusters often occur in chemistry. Figure 16 shows a two-cluster situation. The clusters are made up of the exo and the endo 2-substituted norbornanes, respectively. The variables are features extracted from the ^{13}C NMR spectrum of these substances. In some instances, and when more than two variables are present, the values of the variables for similar objects show strong correlation (see also Chap.

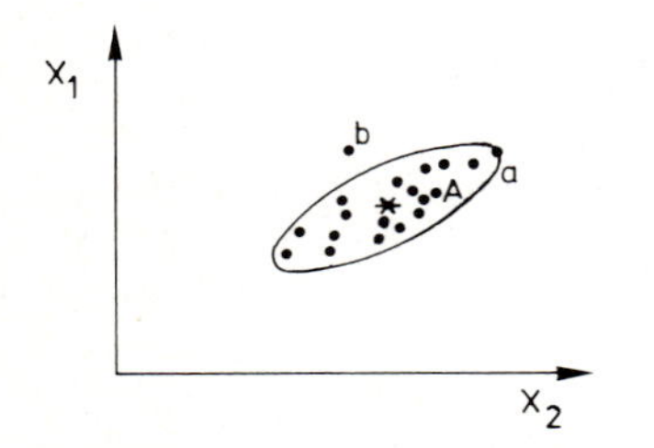

Fig. 17. Distances of objects from a group. a is nearer to group A than is b although b is nearer to the class centroid (*) in geometrical terms.

22). It follows that the correlation coefficient is a similarity measure. It is useful to detect linear clusters.

In Fig. 17, a is considered to be nearer to the center of group A than is b although the distance from the class centroid to a is larger than the distance from the centroid to b. This clearly is due to the fact that the members of A have correlated values of x_1 and x_2. When determining the distance of an object from a group (or, in fact, two groups from each other), it is therefore sometimes necessary to take into account the position of the object considered compared with the direction of the axis going through the group. Such a measure is given by the Mahalanobis distance. Essentially, the Mahalanobis distance is equal to the Euclidian distance with an additional factor taken into account, namely correlation (or covariance). The Euclidian distance is discussed further in Chap. 22 and the Mahalanobis distance in Chap. 23.

References

1 P. Winkel, Patterns and clusters—multivariate approach for interpreting clinical chemistry results, Clin. Chem., 19 (1973) 1329.
2 D.L. Massart, L. Kaufman and K.H. Esbensen, Hierarchical nonhierarchical clustering strategy and application to classification of iron meteorites according to their trace element patterns, Anal. Chem., 54 (1982) 911.
3 J. Smeyers-Verbeke, J.C. Den Hartog, W.H. Dekker, D. Coomans, L. Buydens and D.L. Massart, The use of principal component analysis for the investigation of an organic air pollutants data set, Atmos. Environ., 18 (1984) 2741.
4 M. Forina and C. Armanino, Eigenvector projection and simplified non-linear mapping of fatty acid content of Italian olive oils, Ann. Chim. (Rome), 72 (1982) 127.
5 D. Coomans, D.L. Massart and I. Broeckaert, Potential methods in pattern recognition. Part 4. A combination of ALLOC and statistical linear discriminant analysis, Anal. Chim. Acta, 133 (1981) 215.
6 D. Coomans, I. Broeckaert, M. Jonckheer, P. Blockx and D.L. Massart, The application of linear discriminant analysis in the diagnosis of thyroid diseases, Anal. Chim. Acta, 103 (1978) 409.
7 S. Wold, The analysis of multivariate chemical data using SIMCA and MACUP, Kem. Kemi, 9 (1982) 401.

Recommended reading

Only general books and review or tutorial articles are given here. Books and articles about more specific aspects are given in Chaps. 21–23.

Books

F. Cailliez and J.P. Pagès, Introduction à l'Analyse des Données, Smash, Paris, 1976.
R.O. Duda and P.E. Hart, Pattern Classification and Scene Analysis, Wiley, New York, 1973.
R. Gnanadesikan, Statistical Data Analysis of Multivariate Observations, Wiley, New York, 1977.
A.D. Gordon, Classification Methods for the Exploratory Analysis of Multivariate Data, Chapman and Hall, London, 1981.
D.J. Hand, Discrimination and Classification, Wiley, Chichester, 1981.
M. Kendall, Multivariate Analysis, Griffin, London, 1975.
L. Lebart, A. Morineau and J.P. Fénelon, Traitement des Données Statistiques. Methodes et Programmes, 1982. This book contains many program listings.

F. Morrison, Multivariate Statistical Methods, McGraw-Hill, New York, 1976.

N.N. Nie, C.H. Hull, J.G. Jenkins, K. Steinbrenner and D. Bent, Statistical Package for the Social Sciences (SPSS), McGraw-Hill, New York, 1975.

O. Strouf, Chemical Pattern Recognition, Research Studies Press, Letchworth, 1986.

A.J. Stuper, W.E. Brugger and P.C. Jurs, Computer Assisted Studies of Chemical Structure and Biological Function, Wiley, New York, 1979.

M. Tatsuoka, Multivariate Analysis—Techniques Educational and Psychological Research, Wiley, New York, 1971.

D.D. Wolff and M.L. Parsons, Pattern Recognition Approach to Data Interpretation, Plenum Press, New York, 1983.

K. Varmuza, Pattern Recognition in Chemistry, Springer, Berlin, 1980.

T.Y. Young and T. Calvert, Classification, Estimation and Pattern Recognition, Elsevier, New York, 1974.

Review and tutorial articles

M.P. Derde, D.L. Massart, W. Ooghe and A. Dewaele, Use of pattern recognition display techniques to visualize data contained in complex data-bases, J. Automat. Chem., 5 (1983) 136.

L. Kryger, Interpretation of analytical chemical information by pattern recognition methods—a survey,Talanta, 28 (1981) 871.

K. Varmuza, Pattern recognition in analytical chemistry, Anal. Chim. Acta, 122 (1980) 227.

S. Wold, C. Albano, W.J. Dunn, III, K. Esbensen, S. Hellberg, E. Johansson and M. Sjöström, Pattern recognition: finding and using regularities in multivariate data, in H. Martens and H. Russwurm, Jr., (Eds), Food Research and Data Analysis, Applied Science Publishers, Barking, 1983.

Chapter 21

Principal Components and Factor Analysis

The most important use of principal components (and factor analysis) as shown with the air pollution example in Sect. 20.5, is to represent the n-dimensional data structure in a smaller number of dimensions, usually two or three. This permits one to observe groupings of objects, outliers, etc., which define the structure of a data set.

To understand what happens, first consider a very simple example: 6 samples of polluted air have been analysed for two variables, benzene (y_1) and toluene (y_2) and the results are plotted in Fig. 1. Suppose now that one is not able to visually observe two dimensions, but only a single dimension (which of course is not true). Observation of the complete data structure is then not possible since the data are given in two dimensions; thus one needs to reduce the number of dimensions to one. In pattern recognition one often uses the term feature reduction. A reduction to one dimension means the determination of a line on which the data points must be projected from the two-dimensional space. As there are infinitely many lines on which such a projection is possible a criterion is necessary to define the best line. In principal components analysis (PCA) the direction of the line is chosen in such a way that the projection preserves as well as possible the data structure originally present in the two-dimensional space.

In Fig. 1 there are two groups of points in the original two-dimensional data space. These two groups should also be observed along the one-dimensional line. One can put this in another way: the line must be chosen in such a way that it retains the maximum variation among the data, or more precisely, one wants to maximize variation along the line and to minimize variation around it. Geometrically, this means that the line is directed more or less along the maximal elongation of the cloud of points. Algebraically the spread of the points can be represented by the sum of squares of their distances to their centroid O

$$S^2 = |O1|^2 + |O2|^2 + \ldots + |Oi|^2 + \ldots + |O6|^2$$

where $|Oi|^2$ is the square of the distance between the centroid O and point i. We now introduce a straight line L and consider the projection of the six points on this straight line. Let us call $1', 2', 3', \ldots, 6'$ the projections of the points $1, 2, 3, \ldots, 6$ on the straight line L (see Fig. 2).

Since $|Oi|^2$ can be decomposed in the following way

$$|Oi|^2 = |Oi'|^2 + |ii'|^2$$

the total variation is given by

$$S^2 = |O1'|^2 + |O2'|^2 + \ldots + |O6'|^2 + |11'|^2 + |22'|^2 + \ldots + |66'|^2$$

References p. 369

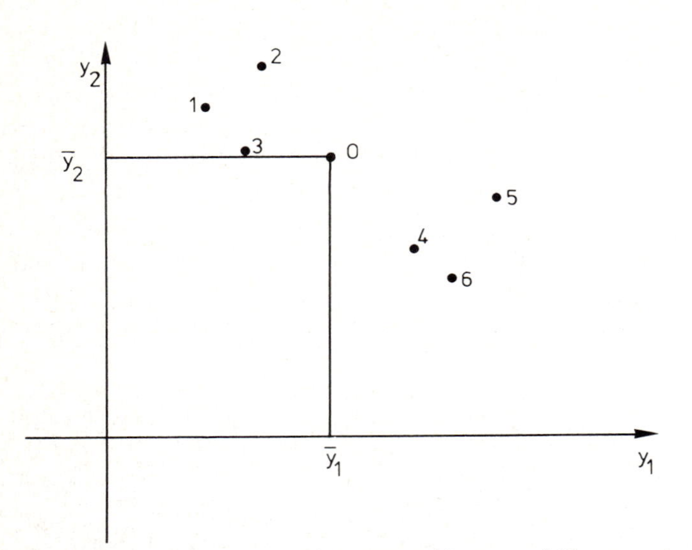

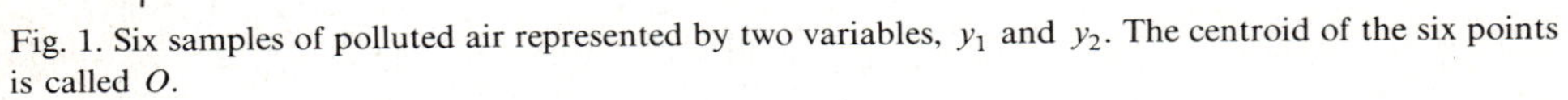
Fig. 1. Six samples of polluted air represented by two variables, y_1 and y_2. The centroid of the six points is called O.

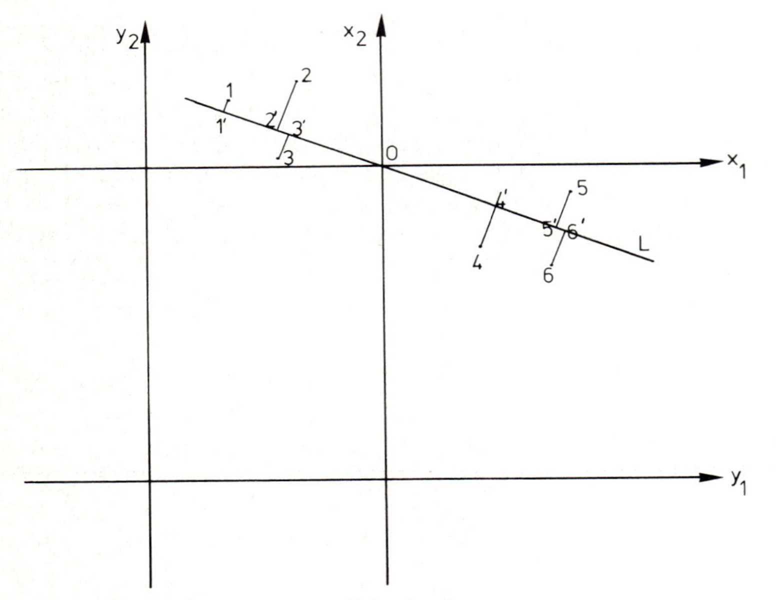

Fig. 2. Reduction from two to one-dimensional space.

The first part of this expression gives the variation along the line and must be maximized while the second is the variation around the line and must be minimized.

From now on, for ease of notation, we will select the centroid O of the points as the origin of the coordinate system. This is achieved by carrying out a translation which can be represented by the equations

$x_1 = y_1 - \bar{y}_1$ and $x_2 = y_2 - \bar{y}_2$

We should note here that this notation is not consistent with previous chapters in which the y variables are used to denote quantities and the x variables concentrations. However, it is adopted here as it would be confusing to introduce another symbol, and in the rest of this chapter only the x variables are used for the actual calculations.

1. The interpretation of principal components plots

Let us consider again the air pollution situation and imagine that the air pollution only originates from two sources, industry (I) and traffic (T). Each produces a distinct pattern of pollution. For instance, I may be characterized by rather high concentrations of substance a and low concentrations of b, while T produces about the same amount of each. The geographic situation is shown in Fig. 3.

At the measuring station, one measures many substances (a, b, $c, \ldots, m$) in samples taken at different times M_1, M_2, etc. One can say that each sample is characterized by a pattern of m polluting substances. Therefore this is an m-dimensional situation and one would like to display samples M_1, M_2, ... using only two new variables: such variables will be called principal components. A figure resembling Fig. 4 would then result. However, in this particular situation one might try to arrive at a two-dimensional representation in a more fundamental way. Indeed, the concentrations of the polluting substances are determined by the contributions of I and T. These contributions vary according to conditions (for instance east wind

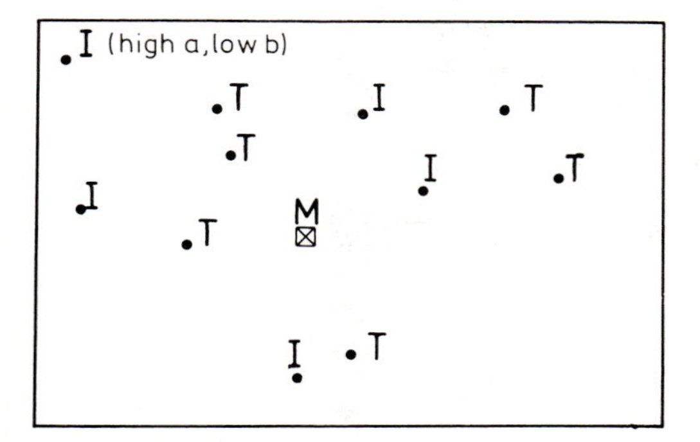

Fig. 3. Air pollution is measured in station M. It is due to industry sources I and traffic sources T.

References p. 369

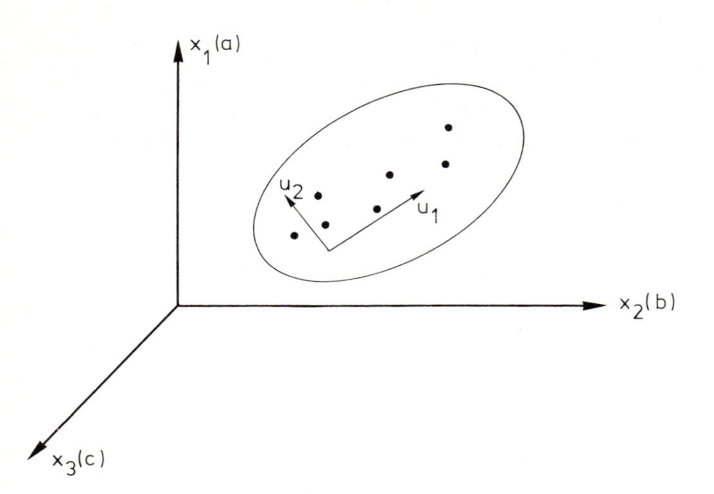

Fig. 4. Data from a three-dimensional space may be reduced to two dimensions.

favors I) and have a more fundamental meaning in the situation investigated. One could hope to obtain a plot of the contribution of I versus the contribution of T. This certainly would make air pollution specialists happy.

For a substance a, one could write

$$\begin{pmatrix}\text{Observed concentration}\\ \text{of } a \text{ in sample } M_1\end{pmatrix} = \begin{pmatrix}\text{concentration of}\\ a \text{ in } I\end{pmatrix} \times \begin{pmatrix}\text{relative importance}\\ \text{of } I \text{ for } M_1\end{pmatrix} + \begin{pmatrix}\text{concentration}\\ \text{of } a \text{ in } T\end{pmatrix} \times \begin{pmatrix}\text{relative importance}\\ \text{of } T \text{ for } M_1\end{pmatrix} + \begin{pmatrix}\text{measurement}\\ \text{error}\end{pmatrix}$$

If the concentrations of a, b, ... m due to I and T were known, the relative importance of I and T could be estimated by solving a linear system of equations. Unfortunately, the concentration of a polluting substance due to a certain source in a measuring station usually cannot be measured directly; for instance, because part of the substance disappears by precipitation, reaction or adsorption on its way to the measuring station. This does not necessarily happen in the same proportion for each substance, so that the pattern due to I and T as perceived at the measuring station is not known. Furthermore, the number of contributing sources is usually not known in advance. Principal components makes it possible to determine this number and a method called factor analysis allows an estimation of the various contributions. This method is discussed in Sect. 5.

It should be noted here that there are two possible representations of the results of a principal components analysis. The first representation considers the old variables as linear combinations of the new ones. In this way, one may consider the old variables to be explained by a (smaller) set of new variables which are called factors. This representation can also be considered as a model for a set of points or samples. Alternatively, one may consider the new variables as linear combinations of the old ones. As will be seen in Sect. 4 this makes it possible to give a graphical representation of the old variables and also of the samples. This second representation was used for the air-pollution example above (see Chap. 20, Sect. 5).

2. Principal components in two dimensions

Consider two variables x_1 and x_2. A general linear transformation transforms the variables x_1 and x_2 into two new variables u_1 and u_2.

Such a general linear transformation can be written as

$$u_1 = a(y_1 - \bar{y}_1) + b(y_2 - \bar{y}_2) = ax_1 + bx_2$$

$$u_2 = c(y_1 - \bar{y}_1) + d(y_2 - \bar{y}_2) = cx_1 + dx_2$$

These equations are represented geometrically in Fig. 5. A data point P, defined by the values of variables x_1 and x_2, is now characterized by the values of the new variables u_1 and u_2. At first sight it may seem useless to replace two variables by two other variables in this way. Suppose, however, that the structure of a set of points shows a clearly preferred direction as in Fig. 6. The new variable u_1 then provides meaningful information concerning the set of data points.

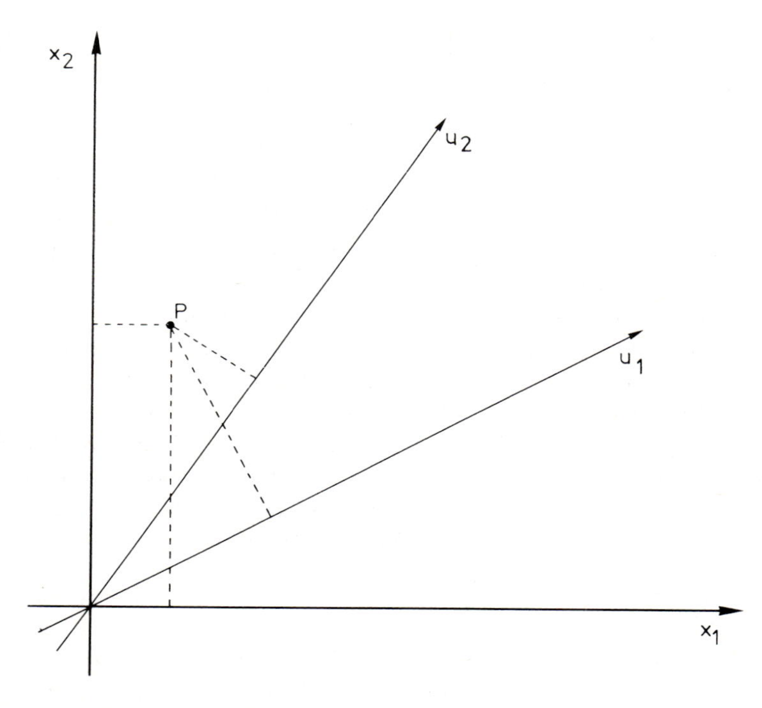

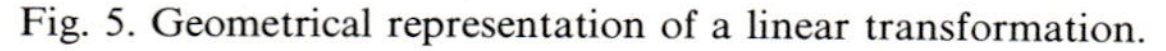

Fig. 5. Geometrical representation of a linear transformation.

References p. 369

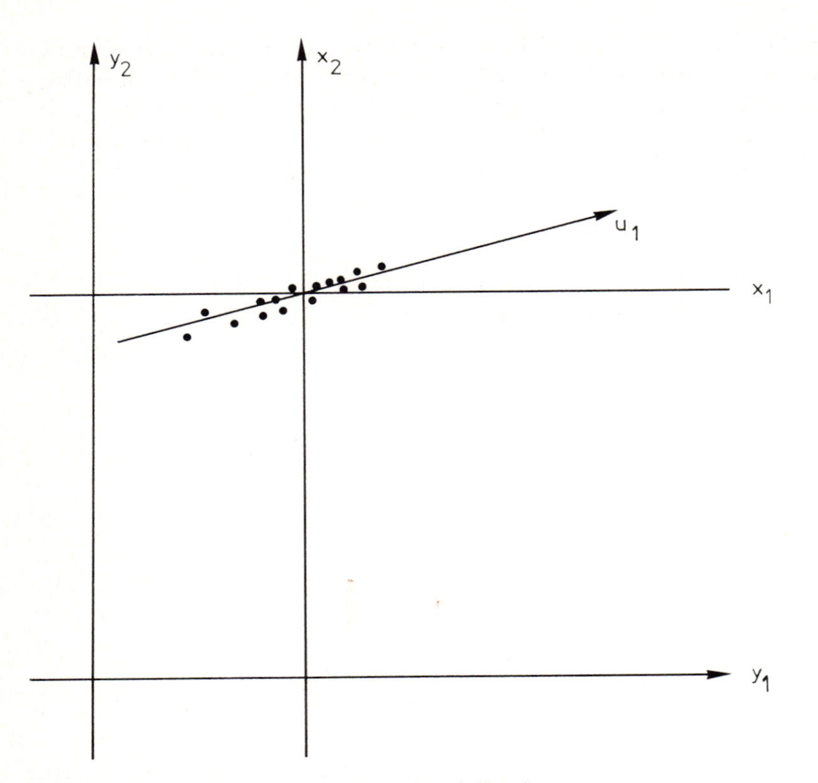

Fig. 6. Set of data points with a clear preferred direction.

In the principal components model new variables are found which give a clear picture of the variability of the data. This is best achieved by giving the first new variable maximum variance: the second new variable is then selected so as to be uncorrelated with the first one. If there are more than two variables the second one should have maximum variance among all variables uncorrelated with the first one, and so on. Geometrically this means that the new variables are orthogonal axes. If there are only two dimensions the condition can be written as

$$a \cdot c + b \cdot d = 0 \tag{1}$$

The variance of u_1 can be made arbitrarily large by multiplying the coefficients a and b by a constant and the same can be done with variable u_2 by multiplying c and d. For example two possible sets of values which satisfy condition (1) are

$$a = \frac{1}{\sqrt{2}} \quad b = \frac{1}{\sqrt{2}} \quad c = \frac{-1}{\sqrt{2}} \quad d = \frac{1}{\sqrt{2}}$$

and

$$a = 2 \quad b = 2 \quad c = \frac{-1}{\sqrt{2}} \quad d = \frac{1}{\sqrt{2}}$$

The second set of values will clearly yield a much larger variance along u_1 than the first. Therefore it is necessary to impose a further restriction (called a normalizing constraint) on the coefficients. Many such restrictions have been considered in the literature. In this introduction we will limit ourselves to the simple restriction of considering variables u_1 and u_2 obtained using the conditions

$$\begin{aligned} a^2 + b^2 &= 1 \\ c^2 + d^2 &= 1 \end{aligned} \tag{2}$$

Condition (1) implies that the new coordinate system (u_1, u_2) is orthogonal and condition (2) that the transformation vectors are of unit length. Therefore the angle θ between u_1 and x_1 is the same as the angle between u_2 and x_2 so that one can also write

$$u_1 = \cos\theta\, x_1 + \sin\theta\, x_2$$

$$u_2 = -\sin\theta\, x_1 + \cos\theta\, x_2$$

so that in the original equation

$$a = d = \cos\theta$$
$$b = -c = \sin\theta$$

Equivalently

$$\begin{pmatrix} u_1 \\ u_2 \end{pmatrix} = \begin{pmatrix} \cos\theta & \sin\theta \\ -\sin\theta & \cos\theta \end{pmatrix} \cdot \begin{pmatrix} x_1 \\ x_2 \end{pmatrix} \tag{3}$$

This relationship which represents a rotation of the coordinate system is shown in Fig. 7.

To maximize the variance of u_1 we first consider the general variance–covariance matrix of the u variables

$$\begin{pmatrix} \sigma_{u_1}^2 & \sigma_{u_1 u_2} \\ \sigma_{u_2 u_1} & \sigma_{u_2}^2 \end{pmatrix}$$

The elements of this matrix can be written in function of the variances and covariances of the x variables. For instance

$$\sigma_{u_1}^2 = \cos^2\theta\, \sigma_{x_1}^2 + \sin^2\theta\, \sigma_{x_2}^2 + 2\sin\theta\cos\theta\, \sigma_{x_1 x_2} = \text{variance along } u_1 \tag{4}$$

Note also that $\sigma_{u_1 u_2}$ and $\sigma_{u_2 u_1}$ are both zero because the new variables are restricted to be uncorrelated. The angle for which $\sigma_{u_1}^2$ is maximum, obtained by setting the derivative of $\sigma_{u_1}^2$ equal to zero, is given by:

$$\theta = \frac{1}{2} \operatorname{Arctan} \frac{2\sigma_{x_1 x_2}}{\sigma_{x_1}^2 - \sigma_{x_2}^2} \qquad \text{if } \sigma_{x_1}^2 > \sigma_{x_2}^2$$

and

$$\theta = 90 + \frac{1}{2} \operatorname{Arctan} \frac{2\sigma_{x_1 x_2}}{\sigma_{x_1}^2 - \sigma_{x_2}^2} \qquad \text{if } \sigma_{x_1}^2 < \sigma_{x_2}^2 \tag{5}$$

References p. 369

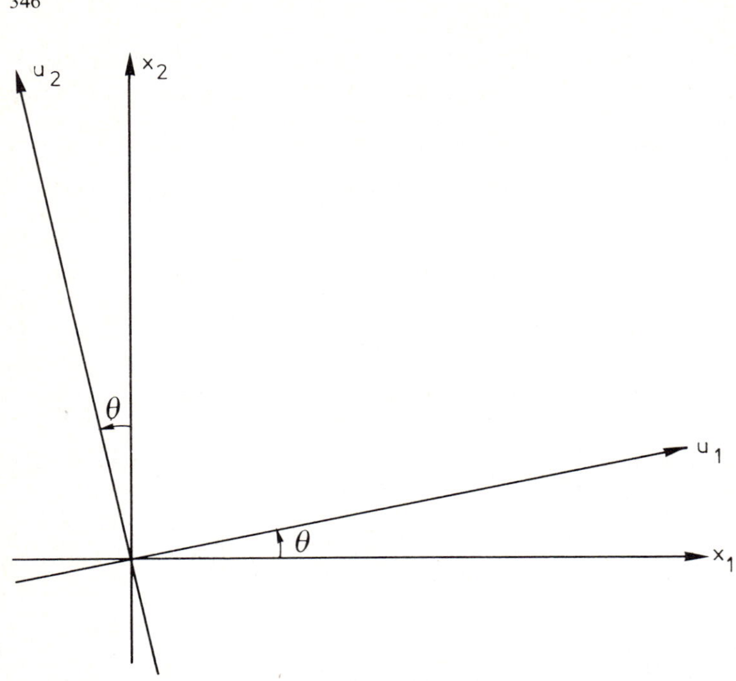

Fig. 7. Geometrical representation of the orthogonal linear transformation given by eqn. (3).

If the two variances are equal, the angle is 45° for positively correlated variables and −45° for negatively correlated variables.

Observe that this derivation can only be carried out for two dimensions. The main reason for giving it is that in this situation it provides an appealing representation of the results.

A numerical example

Consider, once again, the air pollution example of Chap. 20 and suppose eight samples have been measured for two variables, for example the concentrations of benzene and toluene.

Fictive measurement values are given in Table 1. In order to find the principal components by a rotation we first subtract the mean values $\bar{y}_1$ (= 35) and $\bar{y}_2$ (= 14) from each measurement value. The resulting measurements are listed in Table 2 and the points are shown in Fig. 8. We then calculate the two variances and the covariance

$$s_{x_1}^2 = 77.71$$

$$s_{x_2}^2 = 80.86$$

$$s_{x_1 x_2} = 76.29$$

TABLE 1

Measurements for two variables concentration of benzene (y_1) and concentration of toluene (y_2) on eight air samples

y_1	y_2
48	26
44	20
40	24
38	18
32	9
28	6
26	5
24	4

From formula (5) we find that $\theta = 45.59°$ and this allows us to calculate the coefficients a and b (these are the coefficients of the variables x_1 and x_2 in the equation of the first principal component).

$a = 0.6998$
$b = 0.7144$

The linear combination yielding the first principal component is

$u_1 = 0.6998x_1 + 0.7144x_2$

and it is shown in Fig. 9. This equation can also be written in the original variables y_1 and y_2; it then has the form

$$u_1 = 0.6998(y_1 - 35) + 0.7144(y_2 - 14)$$
$$= 0.6998y_1 + 0.7144y_2 - 34.4946$$

It should be noted that each vector which defines a principal component, for example the vector

$$\begin{pmatrix} a \\ b \end{pmatrix} = \begin{pmatrix} 0.6998 \\ 0.7144 \end{pmatrix}$$

is called an eigenvector of the variance–covariance matrix. The variance along this vector is called its eigenvalue. This is explained in more detail in the next section.

TABLE 2

Transformed air sample measurements obtained by subtracting the mean values from the numbers of Table 1 and values of the eight samples for the two principal components

$x_1 = y_1 - \bar{y}_1$	$x_2 = y_2 - \bar{y}_2$	u_1	u_2
13	12	17.67	−0.89
9	6	10.58	−2.23
5	10	10.64	3.43
3	4	4.96	0.66
−3	−5	−5.67	−1.36
−7	−8	−10.61	−0.60
−9	−9	−12.73	0.13
−11	−10	−14.84	0.86

References p. 369

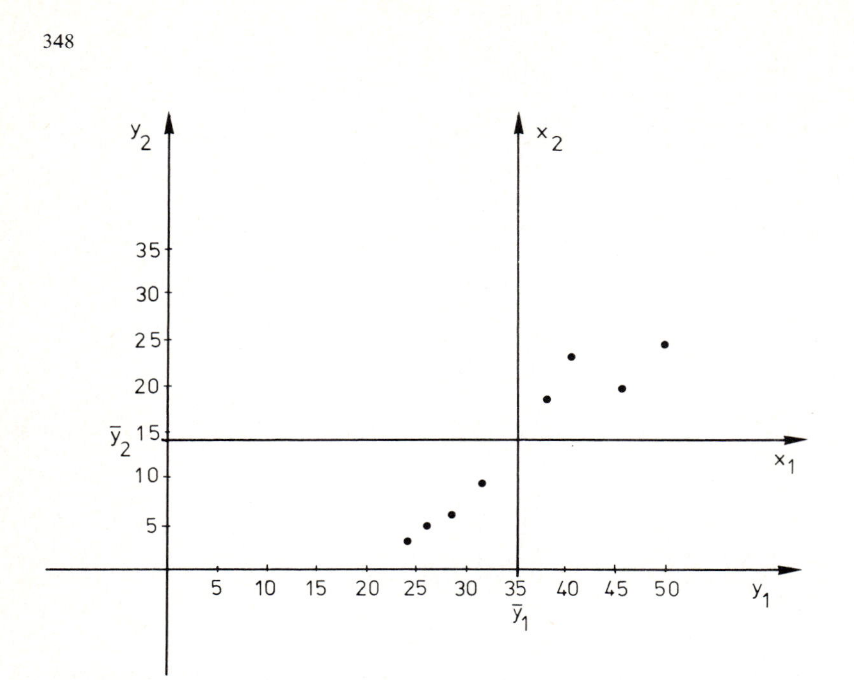

Fig. 8. Graphical representation of the air sample values (see Tables 1 and 2).

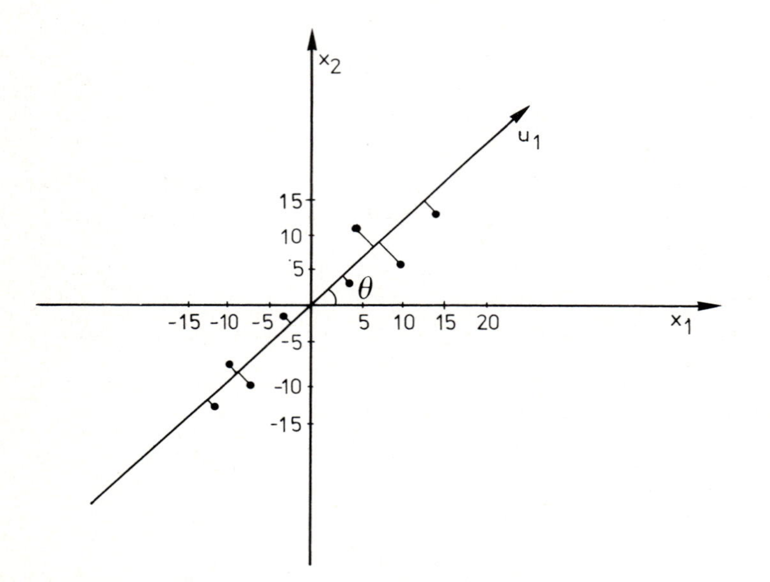

Fig. 9. First principal component for the example of Table 2.

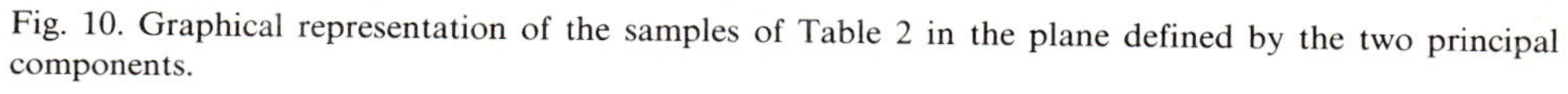
Fig. 10. Graphical representation of the samples of Table 2 in the plane defined by the two principal components.

In two dimensions the second (and last) principal component is uniquely defined. It is the straight line perpendicular to the first one and passes through the origin of the coordinate system (x_1, x_2). It is represented by the equation

$$\begin{aligned} u_2 &= -0.7144x_1 + 0.6998x_2 \\ &= -0.7144(y_1 - 35) + 0.6998(y_2 - 14) \\ &= -0.7144y_1 + 0.6998y_2 + 15.2068 \end{aligned}$$

After having determined the equations for u_1 and u_2 the new coordinates of the 8 samples can be calculated (see Table 2). In Fig. 10 these new coordinates are used to obtain a graphical representation of the data. Note that the first principal component contains almost all of the variance of the data set. Furthermore, the two groups of points which could be distinguished in the original figure (Fig. 8) are still visible here. The first (one-dimensional) principal component is therefore a good representation of the two-dimensional reality.

3. Principal components in *m* dimensions

Let us now consider the same problem in m dimensions. The general transformations used to obtain the new variables u_1, u_2 ... u_m are written as

$$\begin{aligned} u_1 &= v_{11}x_1 + v_{12}x_2 + \ldots + v_{1m}x_m \\ u_2 &= v_{21}x_1 + v_{22}x_2 + \ldots + v_{2m}x_m \\ &\vdots \\ u_m &= v_{m1}x_1 + v_{m2}x_2 + \ldots + v_{mm}x_m \end{aligned} \qquad (6)$$

References p. 369

We call $\mathbf{V}$ the matrix of coefficients v_{rj}

$$\mathbf{V} = \begin{pmatrix} v_{11} & v_{12} & \dots & v_{1m} \\ v_{21} & v_{22} & \dots & v_{2m} \\ \vdots & & & \\ v_{m1} & v_{m2} & \dots & v_{mm} \end{pmatrix}$$

and $\mathbf{u}$ and $\mathbf{x}$ the column vectors of new and old variables

$$\boldsymbol{u} = \begin{pmatrix} u_1 \\ u_2 \\ \vdots \\ u_m \end{pmatrix} \quad \mathbf{x} = \begin{pmatrix} x_1 \\ x_2 \\ \vdots \\ x_m \end{pmatrix}$$

The transformation can then be written as

$$\mathbf{u} = \mathbf{V} \cdot \mathbf{x} \tag{7}$$

The conditions which must be satisfied by the coefficients of the m-dimensional principal components are

(1) for each pair of components u_k and u_r

$$v_{k1}v_{r1} + v_{k2}v_{r2} + \dots + v_{km}v_{rm} = 0 \tag{8}$$

(2) for each component u_r

$$v_{r1}^2 + v_{r2}^2 + \dots + v_{rm}^2 = 1 \tag{9}$$

These conditions can be understood by considering the coefficients of a new variable as a vector in m-dimensional space. For example the first such vector is $(v_{11}, v_{12}, \dots, v_{1m})$, which is the first row of matrix $\mathbf{V}$. Condition (8) means that the product of two of the rows of $\mathbf{V}$ is 0 and therefore that the corresponding vectors are orthogonal. Condition (9) means that this vector must have unit length. Taken together, conditions (8) and (9) are equivalent to

$$\mathbf{V} \cdot \mathbf{V}' = \mathbf{I} \tag{10}$$

where $\mathbf{V}'$ is the transpose of $\mathbf{V}$ and $\mathbf{I}$ is the identity matrix

$$\mathbf{I} = \begin{pmatrix} 1 & 0 & \dots & 0 \\ 0 & 1 & \dots & 0 \\ \vdots & & & \\ 0 & 0 & \dots & 1 \end{pmatrix}$$

In the two-dimensional case condition (10) is automatically satisfied by taking a transformation matrix $\mathbf{V}$ given by

$$\begin{pmatrix} \cos\theta & \sin\theta \\ -\sin\theta & \cos\theta \end{pmatrix}$$

In the m-dimensional case it is not possible to satisfy condition (10) by making use of the same approach as for two dimensions as this would yield much more complex expressions. Similarly to eqn. (4) for the two-dimensional case it can be shown that the variance–covariance matrix of the new variables u_r can be expressed as a function of the variance–covariance matrix of the x variables in the following way.

$$\mathbf{C}_u = \mathbf{V} \cdot \mathbf{C}_x \cdot \mathbf{V}' \tag{11}$$

where $\mathbf{V}'$ is the transpose of $\mathbf{V}$.

If one considers the variance of u_1 the expression to be maximized is

$$\mathrm{Var}(u_1) = \mathbf{v}_1' \cdot \mathbf{C}_x \cdot \mathbf{v}_1 \tag{12}$$

subject to $\mathbf{v}_1' \cdot \mathbf{v}_1 = 1$, where $\mathbf{v}_1'$ is a row vector which represents the transformation by which u_1 is obtained as a linear transformation of the x variables (the first row of matrix $\mathbf{V}$)

$$\mathbf{v}_1' = (v_{11}\ v_{12}\ v_{13}\ \ldots v_{1m}) \tag{13}$$

Because the quantity $\mathbf{v}_1' \cdot \mathbf{C}_x \cdot \mathbf{v}_1$ can be made arbitrarily large by taking larger and larger values in $\mathbf{v}_1'$, constraint (9) should be applied.

This maximization problem is equivalent to

$$\mathbf{C}_x \cdot \mathbf{v} = \lambda\, \mathbf{v} \tag{14}$$

The vector $\mathbf{v}$ found by solving this equation is called an eigenvector of the variance–covariance matrix $\mathbf{C}_x$ and λ is called an eigenvalue. Equation (14) has many solutions, each consisting of an eigenvector and a corresponding eigenvalue. The first solution, represented as a row vector $\mathbf{v}_1'$, will be the first row of matrix $\mathbf{V}$. To calculate the eigenvalue, one multiplies both sides of eqn. (14) by $\mathbf{v}_1'$ which gives

$$\mathbf{v}_1' \cdot \mathbf{C}_x \cdot \mathbf{v}_1 = \lambda_1(\mathbf{v}_1' \cdot \mathbf{v}_1) = \lambda_1 \tag{15}$$

Therefore the maximum of the function is equal to λ_1 and one has to take the largest eigenvalue and the corresponding eigenvector.

The second row of matrix $\mathbf{V}$ is obtained by maximizing the variance of u_2 subject to the restrictions

(1) u_2 and u_1 are uncorrelated

(2) $\sum_{j=1}^{m} v_{2j}^2 = 1$

By the same reasoning as used above for the variable u_1, the second vector $\mathbf{v}_2'$ is obtained as an eigenvector of $\mathbf{C}_x$. Condition (2) is satisfied by taking $\mathbf{v}_2$ as the eigenvector corresponding to λ_2, the second largest eigenvalue of $\mathbf{C}_x$. It follows from the properties of eigenvectors that $\mathbf{u}_2$ and $\mathbf{u}_1$ are uncorrelated.

In the same way eigenvectors corresponding to the further eigenvalues of $\mathbf{C}_x$ yield the subsequent rows of matrix $\mathbf{V}$. This procedure can be continued until m eigenvalues and eigenvectors have been obtained.

The new variables u_r are called *principal components*. They are uncorrelated linear functions of the original variables. The coefficients of the original variables

References p. 369

for a principal component are the coordinates of the corresponding eigenvector. The *loading* of a variable for a principal component is defined as this coordinate multiplied by the square root of the eigenvalue of this principal component. It should be noted here that the coefficients themselves are often called the loadings. The higher the loading of a variable j on a principal component r, the more the variable has in common with this component. The loadings can be interpreted as correlations between the variables and the components. Small (positive or negative) values indicate a weak relationship between the two. It will be seen in the numerical example of this section that the loadings of the variables on the two first principal components can be used to obtain a useful graphical representation of the relation between variables. The value taken by an object for a principal component is called the *score* of the object for this principal component.

It is theoretically possible to determine m principal components. As they are obtained in order of decreasing contribution to the total variance it is usually sufficient to consider the first few principal components and still retain most of the variance. Several methods have been proposed for doing this. A much used method is to select the first p principal components in such a way that they account for at least 80 or 90% of the total variation. By eqns. (12) and (15) λ_1 represents the variance of u_1 and because the u variables are uncorrelated

$$\sum_{r=1}^{m} \lambda_r$$

is the total variation of all the variables. One then selects the p first principal components by using for example the condition

$$\frac{\sum_{r=1}^{p} \lambda_r}{\sum_{r=1}^{m} \lambda_r} \geqslant 0.8 \qquad (16)$$

Another criterion used to select principal components is explained in the numerical example. This problem is discussed further in Sect. 5.3.

In practice the first two or three principal components usually explain an important part of the total variation. In this case a graph of the points (objects) being investigated (in a coordinate system consisting of the first two PCs), gives a good picture of the data set. By this we mean a picture without too much loss of information. This aspect of principal components analysis is called feature reduction.

In the case of a multinormal distribution there is a relationship between the eigenvalues of the variance–covariance matrix and the representation ellipses which can be drawn around the points and which contain predetermined $(1-\alpha)$ percentages of these points (see also Chap. 23, Sect. 6). It can be shown that the ratios of the two axes of these ellipses equal the ratio of the square roots of the eigenvalues.

A feature of the principal components method which is important for its use in practice is that the results obtained are not independent of the units in which the variables are measured. Therefore strictly speaking all variables should be measured on the same scale. When this is not the case, one usually transforms the variables into standardized variables by subtracting the mean value and dividing by the standard deviation. This is equivalent to the use of the correlation matrix $\mathbf{R}_x$ in expression (11) instead of the variance–covariance matrix $\mathbf{C}_x$. However it should be observed that in most applications the objective is to find groups of objects and one is most interested in the relative position of the objects. In such situations, the essential aspect is to remove unimportant variables, which is achieved by both approaches.

A numerical example

We again consider the air pollution example and add the measurements of a third variable y_3. The measurements for 8 samples on all three variables as well as the values corrected for the means are given in Table 3.

A computer program was used to find the eigenvalues and eigenvectors of the variance–covariance matrix. The following results were obtained

$\lambda_1 = 155.61 \qquad u_1 = 0.7000x_1 + 0.7140x_2 + 0.0134x_3$

$\lambda_2 = 24.15 \qquad u_2 = 0.1446x_1 - 0.1600x_2 + 0.9765x_3$

$\lambda_3 = 1.96 \qquad u_3 = 0.6993x_1 - 0.6816x_2 - 0.2152x_3$

For example the first eigenvector is

$(0.7000 \quad 0.7140 \quad 0.0134)$

The sum of the eigenvalues equals 181.72 which is also the sum of the variances of the original variables. By the rule expressed in eqn. (16) the first principal component is sufficient to give an adequate representation of the measurements. Indeed

$$\lambda_1 / \sum_{r=1}^{3} \lambda_r$$

equals 0.86 and therefore 86% of the total variance is accounted for by the first

TABLE 3

Measurements of eight samples on three variables and values corrected for the mean

y_1	y_2	y_3	x_1	x_2	x_3
48	26	17	13	12	4
44	20	15	9	6	2
40	24	8	5	10	−5
38	18	10	3	4	−3
32	9	12	−3	−5	−1
28	6	22	−7	−8	9
26	5	8	−9	−9	−5
24	4	12	−11	−10	−1

principal component.

The correlation matrix is

$$
\begin{array}{c} \\ x_1 \\ x_2 \\ x_3 \end{array}
\begin{array}{c}
\begin{array}{ccc} x_1 & x_2 & x_3 \end{array} \\
\begin{pmatrix} 1 & 0.9624 & 0.1078 \\ 0.9624 & 1 & -0.0462 \\ 0.1078 & -0.0462 & 1 \end{pmatrix}
\end{array}
$$

The three eigenvalues of this matrix are 1.9644, 1.0102, and 0.0254. Observe that the sum equals 3 which is the dimension of the data. It also equals the total variance in the standardized data set. A criterion often used to select principal components is to keep eigenvalues which exceed one. The reason is that the variance of the corresponding principal component is larger than the average for a variable in the data set. Therefore, these principal components can be considered to provide useful information.

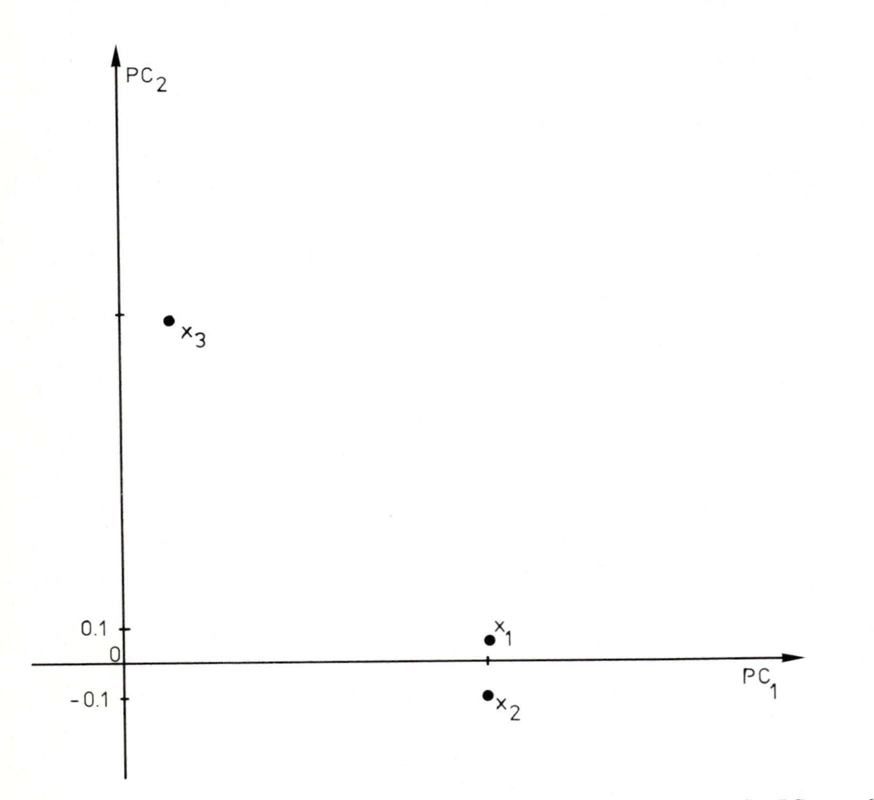

Fig. 11. Example of Table 3: plot of the loadings of three variables in a PC_1–PC_2 coordinate system obtained from the correlation matrix.

In our example PCs 1 and 2 are retained. The proportion of the total variance explained by these variables is

$$\frac{\lambda_1 + \lambda_2}{\lambda_1 + \lambda_2 + \lambda_3} = \frac{1.9644 + 1.0102}{3} = 99.2\%$$

The loadings corresponding to the principal components as determined from the correlation matrix (and not as previously from the variance–covariance matrix) are

	PC_1	PC_2
x_1	0.9926	0.0467
x_2	0.9875	−0.1113
x_3	0.0636	0.9979

Two types of figure are used to represent the results of a principal components analysis. In the first type (see for example Fig. 10) the scores of the objects are used, making it possible to show the relationships between objects. In the second type, for example Fig. 11, the loadings are used in a plot of the two first PCs. From this plot one can for example see that variables x_1 and x_2 are very close and therefore provide similar information (at least for these two principal components) while x_3 is entirely different. Looking at the loadings one sees that variables x_1 and x_2 both score high on the first PC while x_3 scores high on the second PC. In a sense one may conclude that the first PC accounts for variables x_1 and x_2 while the second PC corresponds to x_3. In general, one may conclude that the higher a loading, the more the variable has in common with the principal component. Furthermore as x_1 and x_2 both score highly on PC_1 and are situated near each other they must have a strong positive correlation.

4. Hard modelling versus soft modelling

In Chap. 5 we discussed univariate calibration and described it as a linear regression problem. Multivariate calibration can also be carried out. The usual way of doing this is described in Chaps. 8 and 13. Multivariate calibration is usually carried out in spectrometric applications. Let us therefore first consider its application in UV/VIS spectrophotometry. The usual way of applying multivariate calibration in this case is to select at least as many wavelengths as there are components to be measured. This leads to a set of equations such as eqn. (1) of Chap. 8 which can be solved for the concentrations of components present. This is an instance of what has been called hard modelling [1].

Another approach to the same problem is the use of soft modelling. Principal components play a key role in this way of solving multivariate calibration problems. Suppose that we want to determine the fat content of a foodstuff. A possible way of doing this is to determine the concentration of a particular fatty acid, say stearic acid, and relate this to the fat content. This is univariate calibration since the amount of fat will be estimated from a regression type equation such as

$$\text{Concentration of fat} = b_0 + b_1 \cdot \text{concentration of stearic acid} \quad (17)$$

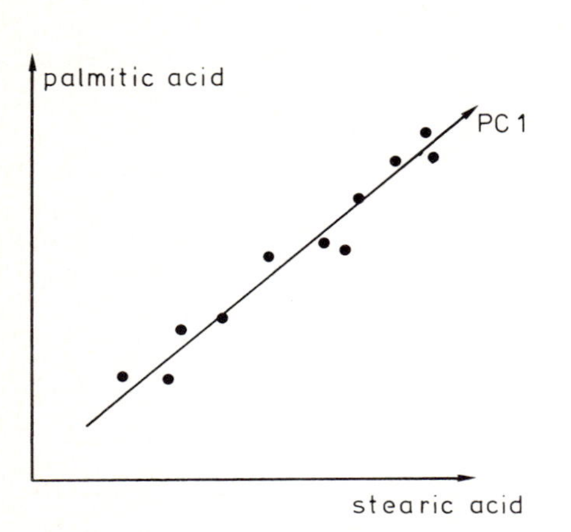

Fig. 12. Concentrations of stearic and palmitic acids in a number of fat samples.

Of course, a still better way is to determine more fatty acids. To keep it simple, we suppose first that only one other such acid is measured, for instance palmitic acid. The calibration problem written as a hard model is then

$$\text{Concentration of fat} = b_0 + b_1 \cdot \text{concentration of stearic acid} + b_2 \cdot \text{concentration of palmitic acid} \quad (18)$$

If a graph is made relating palmitic acid to stearic acid, Fig. 12 would probably be obtained (at least, if the origin of the fat is always the same).

The relationship can be modelled by the regression of palmitic acid on stearic acid or vice versa. Such an approach is called hard modelling. If in this model there are interfering substances in the independent variables it is necessary to use a different approach based on principal components and called soft modelling. The soft model for determining the concentration of fat would then be

$$\text{Concentration of fat} = b_0 + b_1 \text{ (score on PC1)} \quad (19)$$

One notes that a feature reduction method was applied since the two original variables were reduced to a single one. Reducing the number of features means that we accept that there is no other source of variation in the sample affecting the acid concentrations than the amount of fat.

Let us now suppose that there is another source of variation, for instance because there are two kinds of fat present. The ratios of the two acids are different in both kinds of fat. Principal components would then lead to Fig. 13, where PC1, the main source of variation, is concentration related and PC2 fat-source related. By using the scores on PC1 in eqn. (17) instead of the concentration to determine the fat content, one therefore filters out the interfering source of variation. Instead of relating the original variables to the concentrations by regression, one relates a principal component to that concentration. This is called principal components regression.

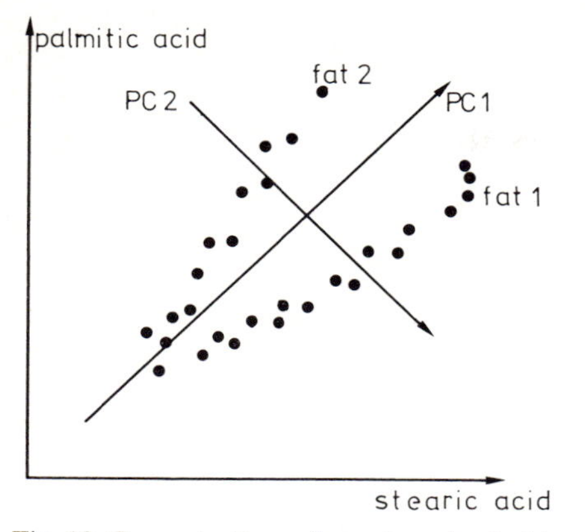

Fig. 13. Concentrations of stearic and palmitic acids in a number of fat samples of two different origins.

Note that PC2 also includes interesting information: it is due to the unknown interfering component.

This principle can also be applied to UV/VIS spectrophotometry. To determine a certain substance X, the normal way is hard modelling using an equation such as (17)

$$C_{MX} = \epsilon'_{MX} A_X + b_0 \qquad (20)$$

where

C_{MX} = molar concentration of X

ϵ'_{MX} = is a constant related to the molar absorptivity coefficient of X. In the univariate situation and with $b_0 = 0$, it would be the reciprocal value of that coefficient

A_X = measured absorption of X

ϵ'_{MX} then plays the role of the regression coefficient b_1 of eqn. (17).

Suppose however, that we measure X at two wavelengths. Replacing stearic acid by $A(\lambda_1)$ and palmitic acid by $A(\lambda_2)$ we would again obtain Fig. 12 and could apply an equation similar to (17) to determine C_{MX}. There are several good reasons for doing this. The first is shown in Fig. 14. The sample designated by x clearly has a different ratio of $A(\lambda_1)/A(\lambda_2)$. x (or the measurement of x) is detected as an outlier.

The most important application is however analogous to Fig. 13. Suppose that unknown to the analyst, there is another substance present in varying amounts. This substance has its own spectrum and affects the result of $A(\lambda_1)$ and $A(\lambda_2)$ in different ways. If substance x were to be measured at only one wavelength this would go unnoticed and systematic errors (Chap. 2) would result. Now a figure such as 13 would result with the two collections of points being due to two different concentrations of x. In fact, the systematic error is filtered out. The importance of this to analytical chemistry is evident. Note also that hard modelling would have allowed this to be achieved if the presence of an interfering substance were known

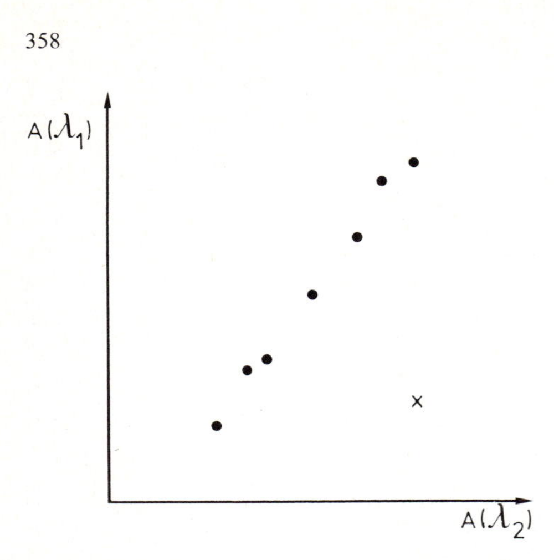

Fig. 14. Detection of outlier by measuring at two wavelengths.

beforehand to the analytical chemist. However, principal components regression allows this to be achieved even when the interfering substance is unknown. An important method which has been applied for this purpose, for instance in NIRA (Near Infrared Absorption) is PLS (Partial Least Squares). It is a generalization of the method described above with some interesting additional features that are however beyond the scope of this chapter. It can be applied to the determination of one or more components in a sample.

5. Factor analysis

5.1 Determination of factors

It was seen in Sect. 2 that the conditions imposed on the coefficients v_{rj} imply that the transformation to obtain the new variables u_r can be written as

$$u_1 = \cos\theta\ x_1 + \sin\theta\ x_2$$

$$u_2 = -\sin\theta\ x_1 + \cos\theta\ x_2$$

In Sect. 3 this was generalized to m dimensions. The transformation is then written as

$$\mathbf{u} = \mathbf{V} \cdot \mathbf{x}$$

and the conditions can be written as

$$\mathbf{V} \cdot \mathbf{V}' = \mathbf{I} \tag{21}$$

A matrix **V** which satisfies condition (21) has the property that its inverse is equal to its transpose

$$\mathbf{V}^{-1} = \mathbf{V}'$$

This implies that

$$\mathbf{x} = \mathbf{V}^{-1} \cdot \mathbf{u} = \mathbf{V}'\mathbf{u} \tag{22}$$

which can also be written in the following extensive way.

$$\begin{aligned} x_1 &= v_{11}u_1 + v_{21}u_2 + \ldots + v_{m1}u_m \\ x_2 &= v_{12}u_1 + v_{22}u_2 + \ldots + v_{m2}u_m \\ &\vdots \\ x_m &= v_{1m}u_1 + v_{2m}u_2 + \ldots + v_{mm}u_m \end{aligned} \tag{23}$$

In this expression the original variables x_j are written in terms of the same number, m, of new variables u_r, which are uncorrelated. The coefficient v_{rj} is the loading of x-variable j on principal component (or factor) r.

Another way of looking at eqn. (23) is to consider each object separately. For example, for an object i one obtains the expression

$$\begin{aligned} x_{i1} &= v_{11}u_{i1} + v_{21}u_{i2} + \ldots + v_{m1}u_{im} \\ x_{i2} &= v_{12}u_{i1} + v_{22}u_{i2} + \ldots + v_{m2}u_{im} \\ &\vdots \\ x_{im} &= v_{1m}u_{i1} + v_{2m}u_{i2} + \ldots + v_{mm}u_{im} \end{aligned}$$

In vector notation this can be written as

$$\mathbf{x}_i' = \mathbf{u}_i'\mathbf{V}$$

The entire data matrix $\mathbf{X}$ of measurements

$$\mathbf{X} = \begin{pmatrix} x_{11} & x_{12} & \ldots & x_{1m} \\ \vdots & \vdots & & \vdots \\ x_{i1} & x_{i2} & \ldots & x_{im} \\ \vdots & \vdots & & \vdots \\ x_{n1} & x_{n2} & \ldots & x_{nm} \end{pmatrix}$$

can therefore be decomposed into a product of two matrices

$$\mathbf{X} = \mathbf{U} \cdot \mathbf{V} \tag{24}$$

5.2 Rotation of factors

The rows of the matrix $\mathbf{U}$ are the "scores" which belong to a given object. The first row of $\mathbf{U}$ contains the scores which belong to the first object, and so on. The score matrix $\mathbf{U}$ can be calculated by multiple linear regression (Chap. 13). By applying eqn. (8) in Chap. 13 one finds that

$$\mathbf{U}' = [\mathbf{V}' \cdot \mathbf{V}]^{-1} \cdot [\mathbf{V}] \cdot \mathbf{X}'$$

Because the eigenvectors are orthogonal: $[\mathbf{V}' \cdot \mathbf{V}] = \mathbf{I}$ (identity matrix), consequently

$$\mathbf{U}' = \mathbf{V}\mathbf{X}' \text{ or } \mathbf{U} = \mathbf{X}\mathbf{V}'$$

In Sect. 3 it has been explained that usually not all m eigenvectors have to be included in $\mathbf{V}$ in order to reconstruct $\mathbf{X}$ within the noise, but that the first p eigenvectors may be sufficient, which gives

$$\mathbf{X} = \mathbf{U}_p \cdot \mathbf{V}_p + \mathbf{E} \quad (25)$$

where $\mathbf{V}_p$ is a matrix which consists of the p first rows of $\mathbf{V}$, $\mathbf{U}_p$ consists of the p first rows of $\mathbf{U}$ and $\mathbf{E}$ is a matrix of errors or residuals.

For $p = 2$, one can plot each object (i) in the $\boldsymbol{v}_1$–$\boldsymbol{v}_2$ space by considering the scores u_{i1} and u_{i2} (the i-th row of $\mathbf{U}_p$) as new coordinates. In some instances one is interested in a physically meaningful interpretation of p, $\mathbf{V}_p$ and $\mathbf{U}_p$. This is the subject of factor analysis.

After selection of p important components or factors the final step of factor analysis is their rotation into terminal factors. In general the exact configuration of the factor structure is not unique: without violating the basic assumptions a factor solution can be transformed into many other factor solutions. This implies that there are many equivalent ways to define the underlying dimensions of a data set. Not all these factor solutions are statistically equivalent: each tells something slightly different about the data. The major option open to the researcher using factor analysis is whether to choose an orthogonal or an oblique rotation of the factors obtained in the previous step. Basically these two types of rotation differ because an orthogonal rotation leads to uncorrelated factors while oblique factors may be correlated.

The basic motivation for using any rotational method is to achieve a simpler and theoretically more meaningful representation of the underlying factors.

The initial solution extracts orthogonal factors in order of decreasing importance. The first factor is usually a general factor with loadings of the same sign on every variable. The second factor tends to be bipolar: about half the variables have high positive loadings and half have high negative loadings. The remaining factors, while often being bipolar are usually quite difficult to interpret.

As an illustration consider Fig. 15. In this figure u_1 and u_2 represent unrotated factors and u_1^r and u_2^r are rotated factors.

On the first unrotated factor all variables load very high, and on the second unrotated factor variables x_1, x_2 and x_3 have positive loadings and variables x_4 and x_5 have negative loadings. Furthermore, for the first unrotated factor the loadings are higher than for the second. After rotation, variables x_1, x_2 and x_3 all load very low for the first factor and high for the second factor while the opposite holds for variables x_4 and x_5. The clustering into two groups of variables is now quite marked for both factors. The rotated solution makes it possible to characterize each variable by a single factor which is conceptually much simpler. Examples show that when there are more than two factors the unrotated factors are even harder to interpret and a rotated solution is necessary.

Another reason often evoked for using a rotated solution is that by removing a single variable the loadings of the unrotated factors may change drastically while this is much less the case for rotated factors.

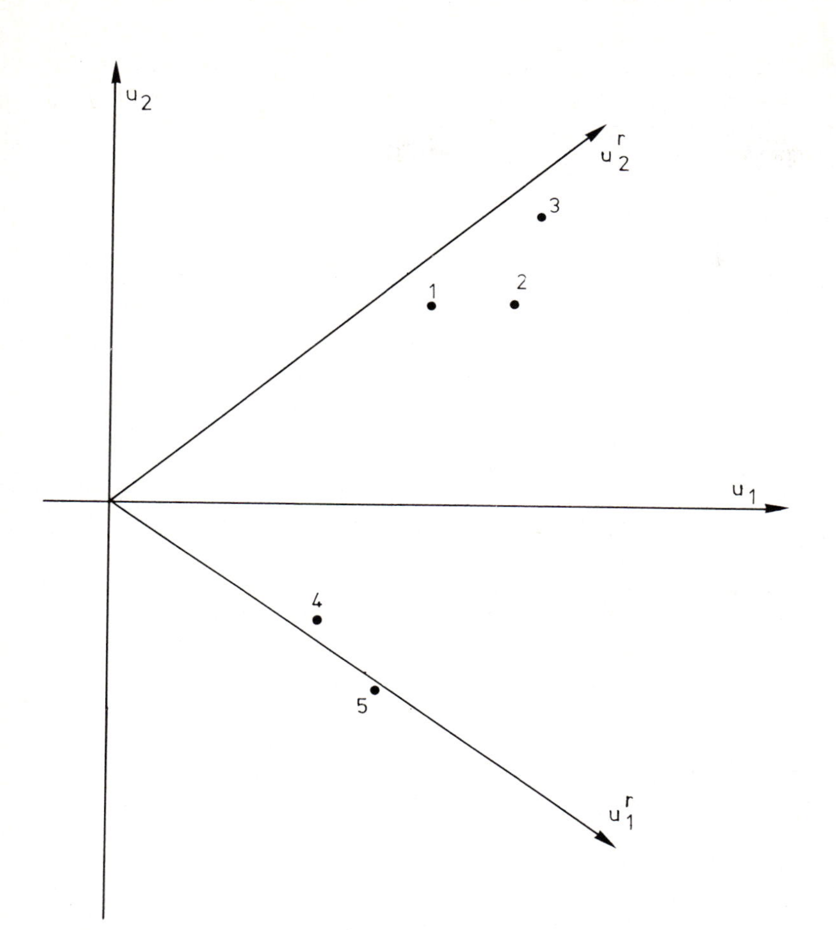

Fig. 15. Graphical representation of the loadings of five variables before and after rotation.

The same general principles apply to oblique rotations as to orthogonal rotations. The main reason for using oblique rotations is that these are in a sense more realistic. Indeed, there is no a priori reason to believe that the underlying dimensions of a data set are unrelated to each other.

The principal objectives of carrying out a particular rotation method are to try to obtain that

(a) many variables have very small loadings on a particular factor: in this way the factor is characteristic for the other variables

(b) few variables score highly on several factors.

Many rotational methods have been proposed to achieve these aims. The most important are Varimax, Quartimax and Equimax among the orthogonal rotations and Oblimin among the oblique rotations. It should be noted that in practice the

References p. 369

transformations and rotations used in Factor Analysis (FA) are related to the chemical constraints of the system being investigated. For example when the estimated loadings represent concentrations the model must of course yield positive values.

5.3 An analytical example

Let us consider again the air pollution situation, in which the air pollution measured at a station M originates from two sources, industry (I) and traffic (T).

Each source produces a distinct pattern of pollution. Such a pattern is built up by the concentrations of m different constituents: e.g. the industrial source generates the pattern

$$\mathbf{I} = \begin{pmatrix} I_1 \\ I_2 \\ \vdots \\ I_m \end{pmatrix} \text{ and the traffic source gives } \mathbf{T} = \begin{pmatrix} T_1 \\ T_2 \\ \vdots \\ T_m \end{pmatrix}$$

In this notation I_1 denotes the concentration of the first substance due to the industry source. At each station (i) a pattern $\mathbf{x}_i$ is measured which is a linear combination of both patterns.

$$\mathbf{x}_i = \begin{pmatrix} x_{i1} \\ x_{i2} \\ \vdots \\ x_{im} \end{pmatrix} = b_{i1}\mathbf{I} + b_{i2}\mathbf{T} \tag{26}$$

The measurements of all constituents at all (say n) stations form a data matrix, $\mathbf{X}$, which is equal to

$$\mathbf{X} = \mathbf{B} \cdot \mathbf{S} \tag{27}$$

The rows of $\mathbf{B}$ indicate the contributions b_{i1} and b_{i2} of the sources $\mathbf{I}$ and $\mathbf{T}$ in each station i. $\mathbf{S}$ is a matrix of which the two rows are respectively $\mathbf{I}$ and $\mathbf{T}$. From eqn. (27) it follows that the data matrix, $\mathbf{X}$, can be decomposed into two other matrices $\mathbf{B}$ and $\mathbf{S}$, which contrary to the decomposition given by eqn. (25), have a physical meaning: namely the contributions of the two pollution sources to each data station.

In many instances these two sources of pollution are unknown and hence neither the pollution patterns $\mathbf{I}$ and $\mathbf{T}$ are available, nor the matrix $\mathbf{S}$. Therefore the decomposition given by eqn. (27) is not directly feasible. By a Principal Components Analysis (PCA), however, an alternative decomposition $\mathbf{X} = \mathbf{U}_p \cdot \mathbf{V}_p + \mathbf{E}$ is always possible. It can be shown that when $\mathbf{X}$ is built up by linear combinations of q physically meaningful patterns, the number of significant eigenvectors in $\mathbf{V}$ is equal to q, and hence $p = q$.

Several methods have been proposed to determine the value of p. The first is given in eqn. (16). Others are based on an evaluation of the statistical properties of $\mathbf{E}$, when increasing p. A more sophisticated method which is not explained here is called cross-validation which is included in software packages for pattern recognition (SIMCA) and multivariate calibration (PLS). In general the matrix $\mathbf{V}_p$ will be different from the matrix $\mathbf{S}$. For instance, the eigenvectors in $\mathbf{V}_p$ are orthogonal. This means that the multiplication of any pair of columns in $\mathbf{V}_p$ will be zero. The vectors which represent the concentration patterns of the m constituents of the industrial source (**I**) and traffice source (**T**) are usually not orthogonal. This can be checked by evaluating the sum of the products of the concentrations of corresponding constituents in both sources. It would be a remarkable coincidence to find this sum equal to zero. Therefore the rows **I** and **T** of matrix **S** are not orthogonal. Because the p rows of **V** are orthogonal, they should be considered to represent "abstract" factors, while the two rows of **S** are "true" factors. By factor analysis one tries to transform the "abstract" factors into "true" factors.

A simple reasoning forms the basis of such a transform. According to eqn. (25) every pollution $\mathbf{x}_i$ is built-up by a linear combination of p (here $p = 2$) eigenvectors

$$\mathbf{x}_i = u_{i1}\mathbf{v}_1 + u_{i2}\mathbf{v}_2$$

According to eqn. (26) it is also true that

$$\mathbf{x}_i = b_{i1}\mathbf{I} + b_{i2}\mathbf{T}$$

If one of the stations would have been situated very near to the traffic source (**T**) and another one very near to the industrial source (**I**), the "true" factors **I** and **T** would have been directly observed and would have been included in **X**. Therefore the "true", but unknown patterns of the sources can also be described as a linear combination of the p (here two) eigenvectors. Thus

$$\mathbf{T} = u_{T1}\mathbf{v}_1 + u_{T2}\mathbf{v}_2 \quad \text{and} \quad \mathbf{I} = u_{I1}\mathbf{v}_1 + u_{I2}\mathbf{v}_2 \tag{28}$$

Equation (28), which is a linear transformation of $\mathbf{v}_1$ and $\mathbf{v}_2$ into **T** and **I**, represents a rotation of the eigenvectors $\mathbf{v}_1$ and $\mathbf{v}_2$ into **T** and **I**. If the results **T** and **I** are still orthogonal, the rotation is called orthogonal, if not, the rotation is oblique. In principle any rotation of $\mathbf{v}_1$ and $\mathbf{v}_2$ is permitted, and infinitely many solutions can be obtained. Therefore the main condition for successful factor analysis is to have the possibility of formulating constraints which bound the solutions into a small region. These constraints have to be formulated using specific knowledge of the particular system being investigated: e.g. when **T** and **I** represent concentrations of substances in the air, all elements of **T** and **I** should be non-negative.

Moreover every pollution pattern $\mathbf{x}_i$ should be built-up by non-negative contributions of **I** and **T**. Thus, u_{T1}, u_{T2}, u_{I1} and u_{I2} in eqn. (28) can only take values for which **T** and **I** have non-negative elements. Moreover, for each pollution pattern $\mathbf{x}_i$ the solution of

$$\mathbf{x}_i = b_{i1}\mathbf{I} + b_{i2}\mathbf{T} = b_{i1}(u_{I1}\mathbf{v}_1 + u_{I2}\mathbf{v}_2) + b_{i2}(u_{T1}\mathbf{v}_1 + u_{T2}\mathbf{v}_2)$$

should yield non-negative contributions b_{i1} and b_{i2}.

A system for which the application of the non-negativity constraints has also proven to produce "true" factors of good quality is HPLC with a UV-VIS diode array detector (HPLC-DAD). In HPLC-DAD a full UV-VIS spectrum is measured

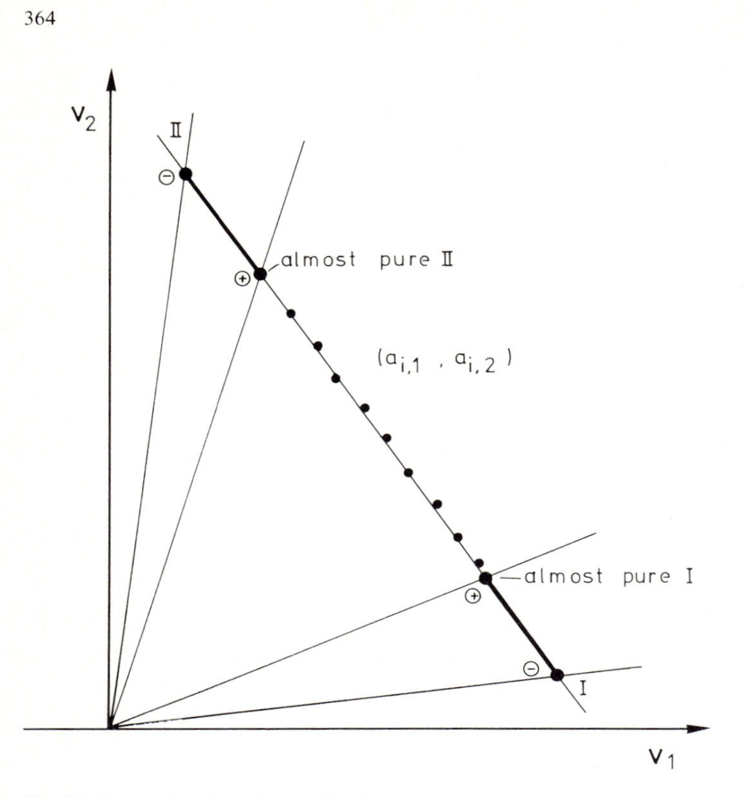

Fig. 16. Representation of normalized spectra (same area) of two component mixtures. (u_{i1}, u_{i2}) are the scores of spectrum (i) in the space defined by the two first eigenvectors of the variance–covariance matrix of the spectra. The solid lines represent the confidence regions of the two fine spectra.

at intervals of 0.1 to 1 s. The rows of the data matrix, **X**, represent spectra, whereas the columns represent chromatograms at different wavelengths.

In some instances when new samples are separated for the first time, peak clusters may appear of unresolved compounds. The questions to solve are then the following: how many compounds are co-eluting in that cluster and what are the spectra of these compounds? The first question is answered by carrying out a PCA of the data matrix **X**, which decomposes into $\mathbf{U}_p \cdot \mathbf{V}_p$.

One can derive that when all spectra (rows of **X**) are normalized to unit area, the sum of the factor scores of each spectrum will be constant: thus $u_{i1} + u_{i2} = 1$. (Note that i corresponds to an object.) By displaying the scores (u_{i1}, u_{i2}) as coordinates in the $\mathbf{v}_1$–$\mathbf{v}_2$ system of orthogonal axes, one can observe that all spectra are represented by points on a straight line (Fig. 16).

Co-eluting compounds are always found in a typical sequence: almost pure I, much compound I + little II, much II + little I, almost pure II. The spectra of almost pure I and II are found at both ends of the line of Fig. 16. From a plot of the

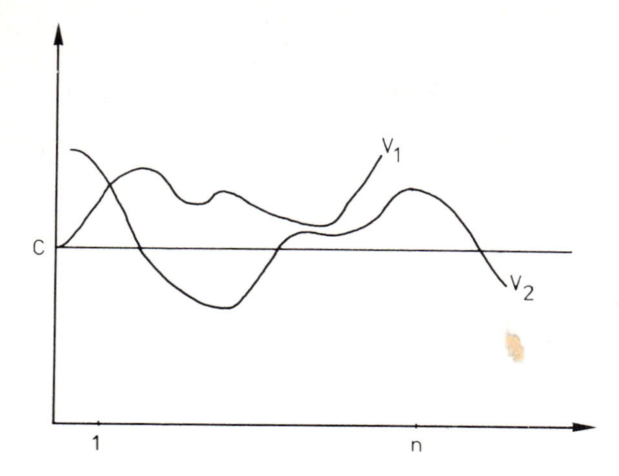

Fig. 17. The loadings of the two first eigenvectors of the variance–covariance matrix of the spectra matrix.

eigenvectors $\mathbf{v}_1$ and $\mathbf{v}_2$ it is obvious that $\mathbf{v}_1$ and $\mathbf{v}_2$ do not represent the "true" spectra, but should be considered to be "abstract" spectra (Fig. 17).

The "true" spectra $\mathbf{S}_1$ and $\mathbf{S}_2$ have to be found by rotating $\boldsymbol{v}_1$ and $\boldsymbol{v}_2$: or

$$\mathbf{S}_1 = u_{\text{I}1}\mathbf{v}_1 + u_{\text{I}2}\mathbf{v}_2$$

$$\mathbf{S}_2 = u_{\text{II}1}\mathbf{v}_1 + u_{\text{II}2}\mathbf{v}_2$$

The scores $(u_{\text{I}1}, u_{\text{I}2})$ and $(u_{\text{II}1}, u_{\text{II}2})$ can be found by applying the following constraints.

(a) As the scores of all mixture spectra are found on a straight line, with the purest spectra at both ends, it is evident that the scores of the pure spectra $\mathbf{S}_1$ and $\mathbf{S}_2$ should be found by extrapolating the line at both ends.

(b) Because $\mathbf{S}_1$ and $\mathbf{S}_2$ should be non-negative, and all elements of $\mathbf{v}_1$ have the same sign while the elements of $\mathbf{v}_2$ have mixed signs, boundaries (I and II) for $(u_{\text{I}1}, u_{\text{I}2})$ and $(u_{\text{II}1}, u_{\text{II}2})$ can be calculated. Namely

$$u_{\text{I}1} = \sqrt{1 - u_{\text{I}2}^2}$$

$$u_{\text{I}2} = -\min_{j=1,\dots,m} \frac{v_{1j}}{\sqrt{v_{1j}^2 + v_{2j}^2}} \quad \text{for } v_{2j} \geqslant 0$$

$$u_{\text{II}1} = \sqrt{1 - u_{\text{II}2}^2}$$

$$u_{\text{II}2} = \min_{j=1,\dots,m} \frac{v_{1j}}{\sqrt{v_{1j}^2 + v_{2j}^2}} \quad \text{for } v_{2j} < 0$$

(c) Because the contributions of each pure spectrum in the mixture spectra should be non-negative, the boundaries (I and II) should be further narrowed.

References p. 369

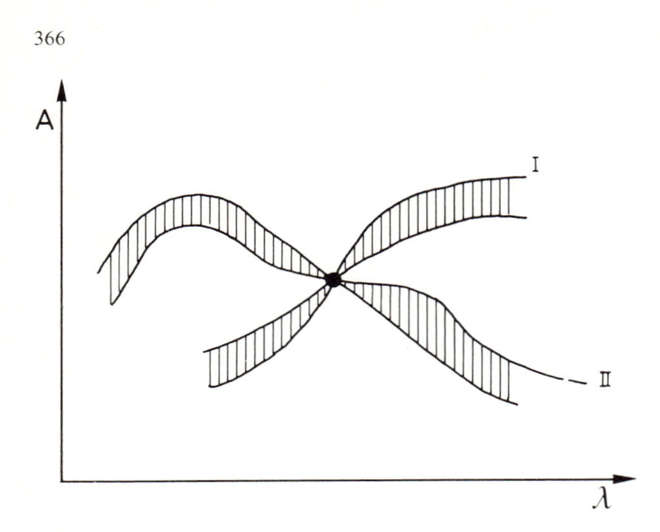

Fig. 18. The almost pure spectra I and II, and the estimated fine spectra I and II. The shaded area represents the confidence region for the true pure spectra.

As a result the range of the scores between which the "true" spectra should be found is given by the solid lines in Fig. 18. Substitution of the four pairs of scores in eqn. (28) gives an upper and lower estimate of both spectra (Fig. 18).

6. Other display methods

6.1 Correspondence factor analysis

Correspondence factor analysis is a technique by which the similarity structure of the objects and of the variables can be represented simultaneously. This makes it possible to plot the objects and the variables in the same graph. The main advantage of the method is that it makes it possible to relate the variables to those objects for which they are particularly meaningful: around a point in the graph representing a variable one finds the objects scoring high on this particular variable. Graphically it makes it possible to combine figures such as 10 and 11 of this chapter. An example of the possible results is shown in Fig. 19.

6.2 Non-linear mapping (multidimensional scaling)

Non-linear mapping (sometimes called multidimensional scaling) is a dimension reducing method which attempts to retain the distances between data points as well as possible.

If the original distances between objects are denoted by d_{ij} (for objects i and j) and the new distances (in a two-dimensional space) by d_{ij}^*, one searches for those d_{ij}^* for which the differences with the d_{ij} are as small as possible. Many algorithms

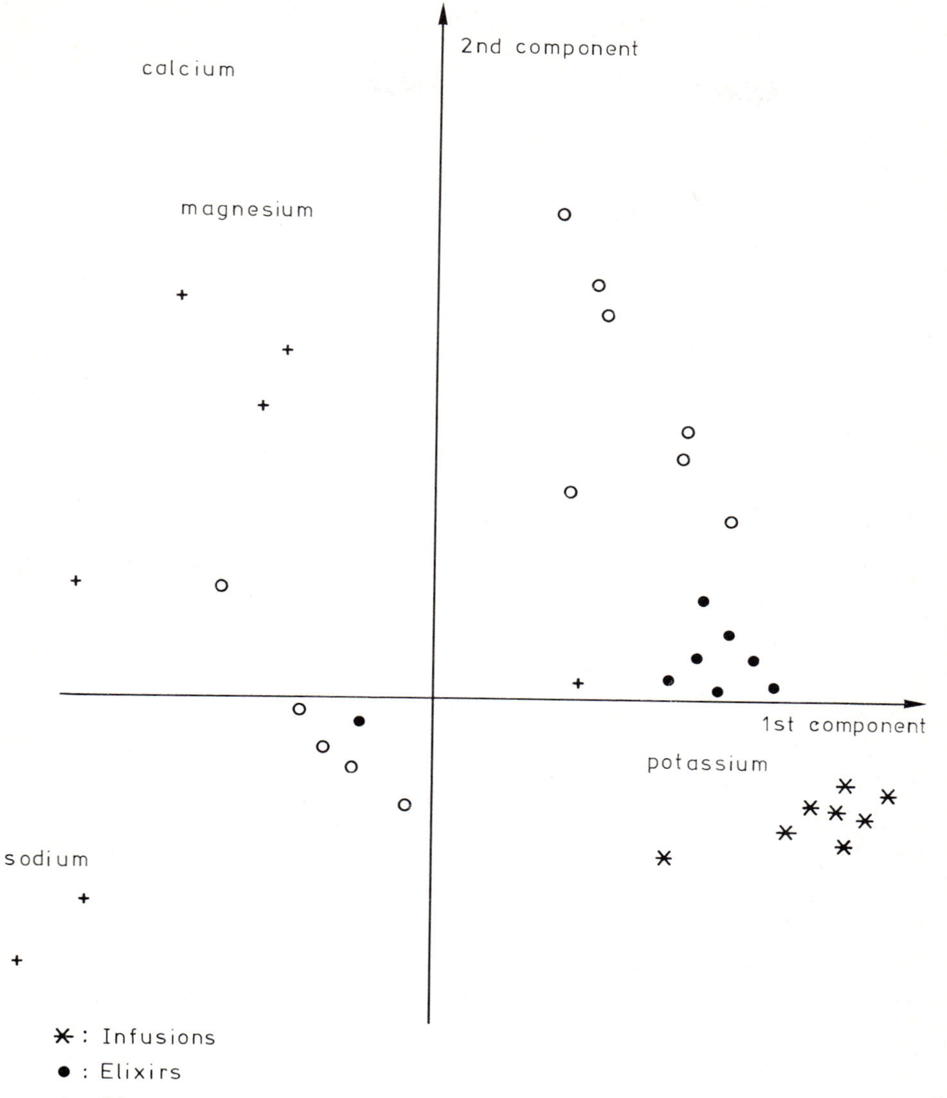

Fig. 19. Example of output obtained from correspondence factor analysis, carried out on a data set consisting of 35 food samples by 4 chemical elements (Ca, Mg, K and Na). One can see that similar food groups appear together. For example, this is the case for the infusions (denoted by $*$). Also these similar objects tend to cluster around significant variables (for example potassium for the infusions). One can also see that the group of algae (denoted by $+$) are all very distant from the variable potassium. Some of them are strongly related to sodium and another group can be found near the variable magnesium [2].

References p. 369

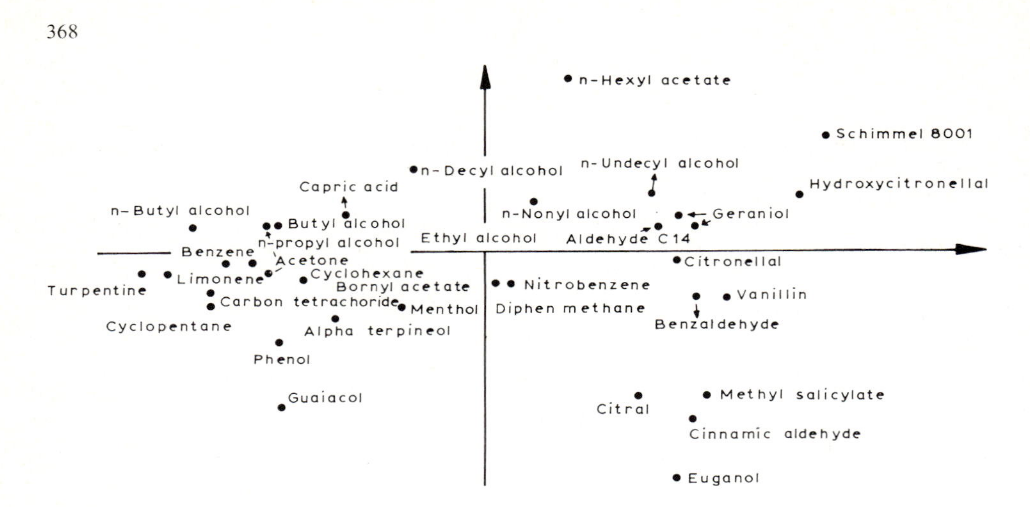

Fig. 20. Representation of the olfactory quality of chemical substances obtained by non-metric multidimensional scaling [4].

have been proposed for multidimensional scaling (see for example the introductory book by Kruskal and Wish). Several of these are based upon the minimization of the so-called mapping error [3]

$$E = \sum_{i<j} \frac{(d_{ij} - d_{ij}^*)^2}{d_{ij}} \tag{29}$$

Non-linear mapping (as well as the other methods discussed in this chapter) assumes data measured on an interval or ratio scale. A similar method called non-metric multidimensional scaling has been proposed for ordinal data. Instead of representing the relationship between objects in a quantitative way one does this by ordering the dissimilarities to obtain a sequence such as

$$d_{i_1j_1} \leqslant d_{i_2j_2} \leqslant \ldots \leqslant d_{i_Nj_N}$$

The method attempts to find new distances which follow the same (or almost the same) ordering between the distances of the objects. This is achieved by considering the ranks of the distances between objects (R_{ij}) and the ranks of the distances between scaled objects (R_{ij}^*) and minimizing a function similar to the mapping error of eqn. (29). This function is called the stress (S) and is given by

$$S = \frac{\sum\limits_{i<j} (R_{ij} - R_{ij}^*)^2}{\sum\limits_{i<j} R_{ij}^2}$$

Non-linear mapping and multidimensional scaling have been widely used in analytical chemistry. An interesting example of the results which can be obtained by this method is given in Fig. 20. Such a map can be used for classification purposes.

References

1 H.A. Martens, Multivariate calibration. Quantitative interpretation of non-selective chemical data, Dr. Technol. Thesis, Technical University of Norway, Trondheim, 1985.
2 A. Thielemans, Vrije Universiteit, Brussels, personal communication.
3 J.B. Kruskal and M. Wish, Multidimensional Scaling, Sage Publications, Beverly Hills, CA, 1978.
4 S.S. Schiffman, Physicochemical correlates of olfactory quality, Science, 185 (1974) 112.

Recommended reading

B.S. Everitt, Graphical Techniques for Multivariate Data, Heinemann, London, 1978.
B.R. Kowalski and C.F. Bender, Pattern recognition. II. Linear and nonlinear methods for displaying chemical data, J. Am. Chem. Soc., 95 (1973) 686.
M.P. Derde, D.L. Massart, W. Ooghe and A. De Waele, Use of pattern-recognition display techniques to visualize the data contained in complex data-bases. A case study, J. Autom. Chem., 5 (1983) 136.

For principal components and factor analysis

W. Lindberg, J.A. Persson and S. Wold, Partial least squares method for spectrofluorimetric analysis of mixtures of humic acid and ligninsulfonate, Anal. Chem., 55 (1983) 643.
J.O. Kim and C.W. Mueller, Introduction to Factor Analysis, Sage Publications, Beverly Hills, CA, 1978.
H. Seal, Multivariate Statistical Analysis for Biologists, Methuen, London, 1964, Chap. 6.
J. Smeyers-Verbeke, J.C. Den Hartog, W.H. Dekker, D. Coomans, L. Buydens and D.L. Massart, The use of principal components analysis for the investigation of an organic air pollutants data set, Atmos. Environ., 18 (1984) 2471.
S. Wold, K. Esbensen and P. Geladi, Principal component analysis, Intell. Lab. Syst., 2 (1987) 37.
S. Wold, C. Albano, W.J. Dunn, III, K. Esbensen, S. Hellberg, E. Johansson, W. Lindberg and M. Sjostrom, Modelling data tables by principal components and PLS: class patterns and quantitative relations, Analusis, 12(10) (1984) 477.
P.K. Hopke, Receptor Modeling in Environmental Chemistry, Wiley, New York, 1985.
E.R. Malinowski and D.G. Howery, Factor Analysis in Chemistry, Wiley, New York, 1980.
C.S. Gutteridge, H.J. MacFie and J.R. Norris, Use of principal components for displaying variation between pyrograms of micro-organisms, J. Anal. Appl. Pyrol., 1 (1979) 67.
G.T. Rasmussen, T.L. Isenhour, S.R. Lowry and G.L. Ritter, Principal component analysis of the infrared spectra of mixtures, Anal. Chim. Acta, 103 (1978) 213.
W. Windig, P.G. Kistemaker and J. Haverkampf, Chemical interpretation of differences in pyrolysis-mass spectra of simulated mixtures of biopolymers by factor analysis with graphical rotation, J. Anal. Appl. Pyrol., 3 (1981) 199.
L. Buydens, D.L. Massart and P. Geerlings, Pharmacological activity of neuroleptic drugs and physicochemical, topological and quantum chemically calculated parameters: a QSAR study, Eur. J. Med. Chem. Chim. Ther., 21 (1986) 35.
S. Wold and K. Anderson, Major components influencing retention indices in GC, J. Chromatogr., 80 (1973) 43.
R. Fellous, L. Lizzani-Cuvelier, R. Luft and D. Lafaye Micheaux, Data analysis in gas liquid chromatography of benzene derivatives, Anal. Chim. Acta, 154 (1983) 191.

For correspondence factor analysis

J.P. Bretandiere, G. Dumont, R. Rej and M. Bailly, Suitability of control materials. General principles and methods of investigation, Clin. Chem., 27 (1981) 798.

J.P. Benzecri et al., L'analyse des Données, Vols. 1 and 2, Dunod, Paris, 1973.
M.O. Hill, Correspondence factor analysis. The method of Hirschfeld revisited, J. R. Stat. Soc. Ser. C, 23 (1974) 340.
P.J. Lewi, Multivariate Data Analysis in Industrial Practice, Research Studies Press, Wiley, Chichester, 1982.
M.J. Greenacre, Theory and Applications of Correspondence Analysis, Academic Press, London, 1984.
M.O. Hill, Correspondence analysis: a neglected multivariate method, Appl. Stat., 23 (1974) 340.

For non-linear mapping and multidimensional scaling

G. Wieten, J. Haverkamp, H.L.C. Meuzelaar, H.W.B. Engel and L.G. Berwald, Pyrolysis mass spectrometry: a new method to differentiate between mycobacteria of the "tuberculosis complex" and other mycobacteria, J. Gen. Microbiol., 122 (1981) 109.
M. Forina and C. Armanino, Eigenvector projection and simplified nonlinear mapping of fatty acid content of Italian olive oils, Ann. Chim. (Rome), 72 (1982) 127.

Other display methods

H. Chernoff, The use of faces to represent points in k-dimensional space graphically, J. Am. Stat. Assoc., 68 (1973) 361.
E.A. Robertson, A.C. Van Steirteghem, J.E. Byrkit and D.S. Young, Biochemical individuality and the recognition of personal profiles with a computer, Clin. Chem., 26 (1980) 30.
P.J. Rousseeuw, Silhouettes: a graphical aid to the interpretation and validation of cluster analysis, Reports of the Department of Mathematics and Information, Technical University of Delft, 1984; J. Comp. Appl. Math., in press.

Chapter 22

Clustering Techniques

Let us consider the meteorite example of Chap. 20. The problem consists of finding groups among some 600 iron meteorites, each of which is characterized by 13 variables (the results of 13 analytical determinations). This can be achieved by grouping (classifying) meteorites with analogous concentration patterns. Let us first attempt such a classification by using only two variables (for instance, Ge and Ni). Fictive concentrations of these two metals for a number of meteorites (called A, B, . . . , J) are shown in Fig. 1. A classification of these meteorites permits one to distinguish first two groups (or classes or clusters), namely ABCDE and FGHIJ. On closer observation, one notes that the first group can be divided into two sub-groups, namely ABC and ED, and that in the second group one can also discern two sub-groups, namely FGHI and J.

There are two ways of representing these data by clustering. The first is depicted by the tree, also called a dendrogram, of Fig. 2 and consists in the elaboration of a hierarchical classification of meteorities. It is hierarchical because large groups are divided into smaller ones (for instance, the group ABCDE splits into ABC and ED). These are then split up again until eventually each group consists of only one meteorite.

The other possibility is to make a table containing different clusterings. A clustering is a partition into clusters. For the example of Fig. 1, this could yield

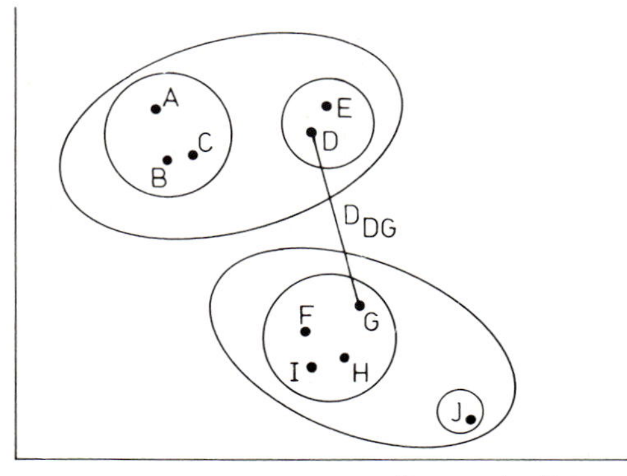

Fig. 1. Concentration of Ni and Ge for ten meteorites (fictive values and samples).

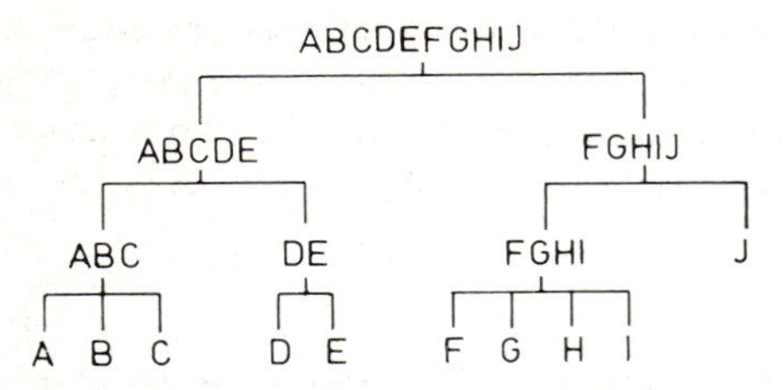

Fig. 2. Dendrogram for the classification of the meteorites represented in Fig. 1.

TABLE 1

A list of clusterings derived from Fig. 1 using MASLOC (see Sect. 3)

Number of clusters	Composition of the clusters
1	A B C D E F G H I J
2	A B C D E – F G H I J
3	A B C – D E – F G H I J
4	A B C – D E – F G H I – J
6	A – B C – D E – F I – H G – J
7	A – B C – D E – F – G – H I – J
10	A – B – C – D – E – F – G – H – I – J

Table 1. Such a table does not necessarily yield a complete hierarchy (e.g. in going from 6 to 7, clusters H and I, separated for cluster 6, are joined again for 7). Therefore the presentation is called non-hierarchical.

1. Measures of (dis)similarity

In Sect. 3 of Chap. 20, the term similarity was introduced. To be able to cluster objects, one must measure their similarity. In Chap. 20, it was already shown that "distance" may be such a measure. In fact, many types of similarity coefficients are applied. We will discuss the two most important, namely the Euclidian distance and

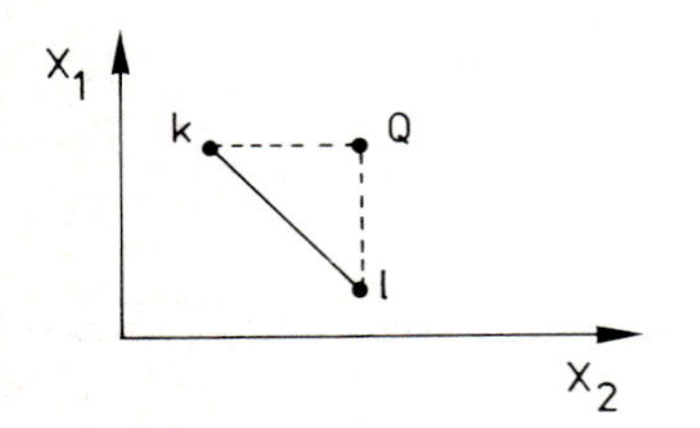

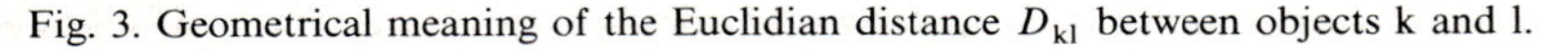

Fig. 3. Geometrical meaning of the Euclidian distance D_{kl} between objects k and l.

the correlation coefficient. The equation for the correlation coefficient has been given in Chap. 14 and the equation for the Euclidian distance between k and l is

$$D_{\mathrm{kl}} = \sqrt{\sum_{j=1}^{m} (x_{\mathrm{k}j} - x_{\mathrm{l}j})^2} \tag{1}$$

where m is the number of variables. The geometrical meaning of the Euclidian distance is understood most easily when picturing it in two dimensions (Fig. 3).

Then, D_{kl} is the hypotenuse of the triangle klQ and

$$\begin{aligned} D_{\mathrm{kl}} &= \sqrt{(|\mathrm{kQ}^2| + |\mathrm{lQ}^2|)} \\ &= \sqrt{(x_{\mathrm{k}1} - x_{\mathrm{l}1})^2 + (x_{\mathrm{k}2} - x_{\mathrm{l}2})^2} \\ &= \sqrt{\sum_{j=1}^{2} (x_{\mathrm{k}j} - x_{\mathrm{l}j})^2} \end{aligned}$$

The difference between the two measures of similarity is explained with the help of Table 2. Table 2 gives the retention indices of five substances on three gas chromatographic stationary phases (SFs) and the question is which of these phases should be considered similar. The similarity measure to be chosen depends on the point of view of the analyst. One point of view might be that he considers similar those SFs that have more or less the same over-all retention, i.e. the same polarity towards a variety of substances. In that case, SF_3 is very dissimilar from both SF_1 and SF_2, while SF_1 and SF_2 are rather similar. The best way to express this is the Euclidian distance. D_{13} and D_{23} are then much higher than D_{12}.

On the other hand, the analyst might not be interested in global retention indices. Indeed, by increasing the temperature for SF_3, he would obtain similar retention indices as for the other two. He will then observe that the relative retention time, i.e. the retention times of the substances compared with each other, are the same for SF_1 and SF_3 and different from SF_2. Chemically, this means that SF_3 has different polarity from SF_1, but the same specific interactions. This is best expressed by using the correlation coefficient as the similarity measure. Indeed, $r_{13} = 1$, indicating complete similarity, while r_{12} and r_{23} are much lower. If, as with Euclidian distance, one would like the numerical value of the similarity measures to increase with

TABLE 2

Retention indices of five substances on three stationary phases in GLC

Stationary phases (SF)	1	2	3	4	5
1	100	130	150	160	170
2	120	110	170	150	145
3	200	260	300	320	340

References p. 383

decreasing similarity, then one should use, for instance, $1 - |r|$. Since the similarity measure to be used depends on the analyst's point of view, it is not possible to recommend one of these two measures. That a good choice is important is, however, evident from the example.

A problem with the Euclidian distance, which does not arise with correlation, is a scale effect. In the meteorite example, the concentration of Ni is of the order of 50 000 p.p.m. and the Ga content of the order of 50 p.p.m. Small relative changes in the Ni content have then, of course, a much higher effect on the Euclidian distance than equally high relative changes of the Ga content. One might also consider two metals M and N, one ranging in concentration from 900 to 1100 p.p.m., the other from 500 to 1500 p.p.m. Concentration changes from one end of the range to the other in N would then be more important in the Euclidian distance than the same kind of change in M. It is probable that the person carrying out the classification will not agree with these numerical consequences and consider them as artefacts. Both problems can be solved by scaling the variables. The most usual way of doing this is using the z-transform, also called autoscaling (see also Sect. 2.6.2 of Chap. 2). One then determines

$$z_{ij} = \frac{x_{ij} - \bar{x}_j}{s_j}$$

where x_{ij} is the value for object i of variable j, $\bar{x}_j$ is the mean for variable j, and s_j is the standard deviation for variable j. One then uses z_{ij} in eqn. (1).

TABLE 3

Example of data matrix

System	Concentrations (arbitrary units)			
	Metal a	Metal b	Metal c	Metal d
A	100	80	70	60
B	80	60	50	40
C	80	70	40	50
D	40	20	20	10
E	50	10	20	10

TABLE 4

Similarity matrix for the distances obtained from Table 3

	A	B	C	D	E
A	0				
B	40.0	0			
C	38.7	17.3	0		
D	110.4	70.7	78.1	0	
E	111.4	72.1	80.6	14.1	0

The similarities between all pairs of objects are measured. This yields the similarity matrix. It is a symmetrical $n \times n$ matrix containing the similarities between each pair of objects. Let us suppose, for example, that the meteorites A, B, C, D, and E in Table 3 have to be classified. Using eqn. (1), one obtains the similarity matrix in Table 4. Because the matrix is symmetrical, only half of this matrix need be used.

2. Hierarchical methods

There is a wide variety of hierarchical algorithms available and it is impossible to discuss all of them here. Therefore, we shall only explain the ones most used, namely the average linkage and the single linkage methods.

In the similarity matrix, one seeks the two most similar objects. When using distance as the similarity measure, this means that one looks for the smallest D_{kl} value. Let us suppose that it is D_{qp}, which means that of all the objects to be classified, q and p are the most similar. They are considered to form a new combined object p*. The similarity matrix is thereby reduced to $(n-1) \times (n-1)$. In average linkage, the similarities between the new object and the others are obtained by averaging the similarities of q and p with these other objects. For example, $D_{kp^*} = (D_{kq} + D_{kp})/2$. In single linkage, D_{kp^*} is expressed as the similarity between the objects and the nearest of the linked objects, i.e. it is set equal to the smallest of the two distances D_{kq} and D_{kp}.

At the same time, one starts constructing the dendrogram by linking together q and p. This process is repeated until all objects are linked in one hierarchical classification system, which is represented by a dendrogram. This procedure can now be illustrated using the data of Tables 3 and 4. The smallest D is 14.1 (between D and E). D and E are combined first and yield the combined object D^*. In the average linkage mode, one may introduce a weighting of the objects when clusters of unequal size are linked. Both weighted and unweighted methods exist.

The successive reduced matrices obtained by average linkage are given in Table 5 and those obtained by single linkage in Table 6.

One observes that, in this particular instance, the only noteworthy difference between the algorithms is the distance at which the last link is made (93.1 for average linkage and 70.7 for single linkage). When larger data sets are studied, the differences may become more pronounced. In general, average linkage is preferred to single linkage. Another method which gives good results (i.e. has been shown to give meaningful clusters) is known as Ward's method.

Single linkage has the advantage of mathematical simplicity, particularly when it is calculated using an operational research technique called the minimal spanning tree. Although at first sight the computations seem to be very different from those in Table 6, exactly the same results are obtained. This can be readily verified by the reader by applying the method to these data (after having read this section). For a good explanation of the method we need a matrix with some more objects. The data matrix is given in Table 7 and the resulting similarity matrix in Table 8.

We may think of these objects as towns, the distances between which are given in

References p. 383

TABLE 5

Successive reduced matrices for the data of Table 4 obtained by average linkage

(a)

	A	B	C	D *
A	0			
B	40.0	0		
C	38.7	17.3	0	
D*	110.9	71.4	79.3	0

D* is the object resulting from the combination of D and E.

(b)

	A	B*	D*
A	0		
B*	39.3	0	
D*	110.9	75.3	0

B* is the object resulting from the combination of B and C.

(c)

	A*	D*
A*	0	
D*	93.1	0

A* is the object resulting from the combination of A and B*.

(d)
The last step consists in the junction of A* and D*. The resulting dendrogram is given in Fig. 4(a).

TABLE 6

Successive reduced matrices for the data of Table 4 obtained by single linkage

(a)

	A	B	C	D*
A	0			
B	40.0	0		
C	38.7	17.3	0	
D*	110.4	70.7	78.1	0

(b)

	A	B*	D*
A	0		
B*	38.7	0	
D*	110.4	70.7	0

(c)

	A*	D*
A*	0	
D*	70.7	0

(d)
The last step consists in the junction of A* and D*. The resulting dendrogram is given in Fig. 4(b).

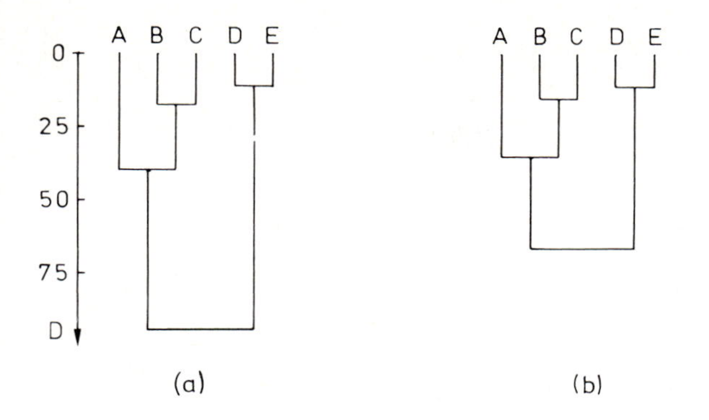

Fig. 4. Dendrograms for the data of Tables 3–6. (a) Average linkage; (b) single linkage.

TABLE 7

Values characterizing objects A, ...G (see Fig. 6)

	x_1	x_2
A	45	24
B	24	43
C	14	23
D	64	52
E	36	121
F	56	140
G	20	148

TABLE 8

Distance between points in Fig. 5
Reprinted from ref. 1 with the permission of the American Chemical Society

	A	B	C	D	E	F	G
A	0						
B	28	0					
C	32	23	0				
D	35	40	60	0			
E	100	80	103	76	0		
F	119	104	128	90	29	0	
G	127	105	126	105	40	35	0

the table, and suppose that the seven towns must be connected to each other by highways (or a production unit to six clients using a pipeline). This must be done in such a way that the total length of the highway is minimal. Two possible configurations are given in Fig. 5. Clearly, (a) is a better solution than (b). Both (a) and (b) are graphs that are part of the complete graph containing all possible links and both

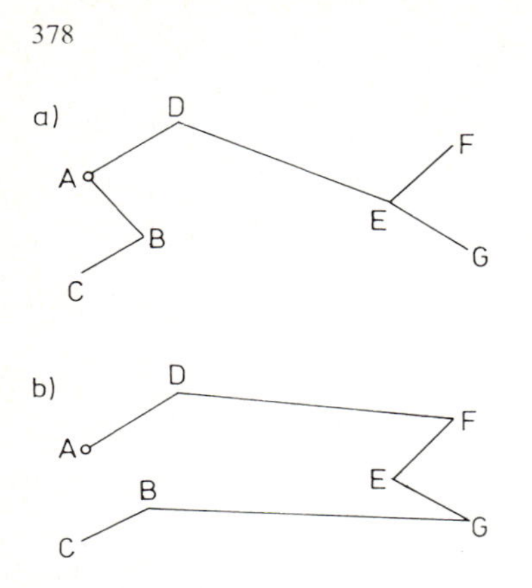

Fig. 5. Examples of trees in a graph. (a) is the minimal spanning tree [1]. Reprinted with the permission of The American Chemical Society.

are connected graphs (all of the nodes are linked directly or indirectly to each other). These graphs are called trees and the tree for which the sum of the values of the links is minimal is called the minimal spanning tree. This minimal spanning tree is also the optimal solution for the highway problem. The terminology used in this chapter comes from graph theory. This is discussed in Chap. 25.

Let us now consider how to find the minimal spanning tree. Several algorithms can be used to achieve this. One of these is Kruskal's algorithm which can be stated as follows: add to the tree the edge with the smallest value which does not form a cycle with the edges already part of the tree. According to this algorithm, one selects first the smallest value in Table 8 (link BC, value 23). The next smallest value is 28 (link AB). The next smallest values are 29 and 30 (links EF and EG). The next smallest value in the table is 32 (link AC). This would, however, close the cycle ABC and is therefore eliminated. Instead, the next link that satisfies the conditions of Kruskal's algorithm is AD and the last one is DE. The minimal spanning tree obtained in this way is that given in Fig. 5(a).

After careful inspection of this figure, one notes that two clusters can be obtained in a formal way by breaking the longest edge (DE). When a more detailed classification is needed, one breaks the second longest edge, etc., until the desired number of classes is obtained. In the same way, clusters are obtained from Fig. 4 by breaking first the lowest link (i.e. the one with highest distance), etc.

The hierarchical methods so far discussed are called agglomerative. Good results can also be obtained with hierarchical divisive methods, i.e. methods that first divide the set of all objects in two so that two clusters result. Then each cluster is again divided in two, etc., until all objects are separated. These methods also lead to a hierarchy. They present certain computational advantages. An example is CLUE [2].

3. Non-hierarchical methods

Let us now cluster the objects of Table 7 with a non-hierarchical algorithm. Instead of clustering by joining objects successively, one wants to determine directly

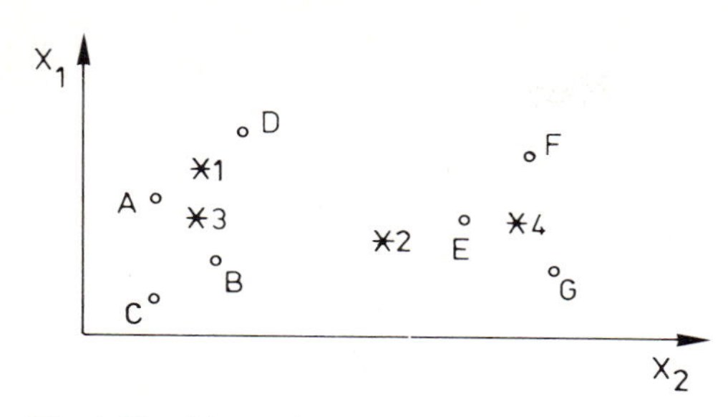

Fig. 6. Non-hierarchical clustering. The meaning of the numbered asterisks is given in the text.

a K-clustering, by which is meant a classification into K clusters. One first does this for 2 clusters. Of course, one is able to see that the correct 2-clustering is (A, B, C, D)(E, F, G). In general, one uses m-dimensional data and it is then not possible to visually observe clusters. In this section, we will also suppose that we are not able to do this. To obtain 2 clusters, one selects 2 "seed points" among the objects and classifies each of the objects with the nearest seed point. In this way, an initial clustering is obtained. For the objects of Table 7, A and B are selected as first seed pionts. In Fig. 6, one observes that this is not a good choice (A and E would have been better) but one should remember that one is supposed to be unable to observe this. D is nearest to A and C, E, F and G are nearest to B (Table 8). The initial clustering is therefore (A, D)(B, C, E, F, G).

For each of these clusters, one determines the centroid (the point with mean values of variables x_1 and x_2 for each cluster). For cluster (A, D), the centroid (*1) has values

$$x_1 = (45 + 64)/2 = 54.5$$

$$x_2 = (24 + 52)/2 = 38$$

and for cluster (B, C, E, F, G) the centroid (*2) is given by

$$x_1 = (24 + 14 + 36 + 56 + 20)/5 = 30$$

$$x_2 = (43 + 23 + 121 + 140 + 148)/5 = 95$$

The two centroids are shown in Fig. 6. In the next step, one reclassifies each object according to whether it is nearest to *1 or *2. This now leads to the clustering (A, B, C, D)(E, F, G). The whole procedure is then repeated: new centroids are computed for the clusters (A, B, C, D) and (E, F, G). These new centroids are situated in *3 and *4 (see Fig. 6). Reclassification of the objects leads again to (A, B, C, D)(E, F, G). Since the new clustering is the same as the preceding one, this clustering is considered definitive.

The method used here is called Forgy's method. It involves the following steps.

(1) Select an initial clustering.

(2) Determine the centroids of the clusters and the distance of each object to these centroids.

(3) Locate each object in the clusters with the nearest centroid.

(4) Compute new cluster centroids and go to step (3). One continues to do this until convergence occurs (i.e. until the same clustering is found in two successive assignment steps).

Instead of using centroids as the points around which the clusters are constructed, one can select some of the objects themselves. These are then called centrotypes. A non-hierarchical method which works in this way is called MASLOC [4]. Returning to the simple example of Fig. 6, suppose that one selects objects A and E as centrotypes. Thus, B, C and D would be classified with A since they are nearer to it than to E, and F and G would be clustered with E. This method is based on an operations research model (Chap. 25), the so-called location model. The points A–G might then be cities in which some central facilities must be located. Determining the clustering (A, B, C, D)(E, F, G) is then equivalent to determining that the sum of the distances from each town to the nearest facility is minimal when the facilities are located in A and E. A, B, C and D will then be served by the facility located in A and E. A, B, C and D will then be served by the facility located in A and E, F and G by the one located in E.

4. Some other methods

The literature abounds with clustering methods and some of these are very sophisticated. For instance, methods based on minimizing the volume of a cluster seem to give very good results but are mathematically very complex. A group of methods which is used quite often is based on the idea of describing high local densities of points. A supervised pattern recognition method based on this idea is described in some detail in Sect. 5 of Chap. 23 and mode analysis, a graph theoretical method, is explained in Sect. 3 of Chap. 25.

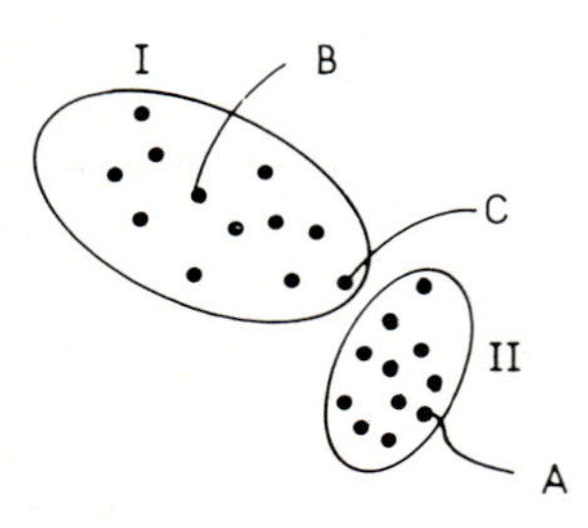

Fig. 7. A non-hierarchical clustering would lead to the two clusters obtained in the figure. Fuzzy clustering would lead to two clusters and a measure of belongingness, p, to those clusters for each object. For example

for object A $p_1(A) = 0$ $p_2(A) = 1$

for object B $p_1(B) = 1$ $p_2(B) = 0$

for object C $p_1(C) = 0.47$ $p_2(C) = 0.53$

All these methods and the methods of the preceding section have one characteristic in common: an object may be part of only one cluster. An important family of clustering techniques, called fuzzy clustering, applies other principles. It permits objects to be part of more than one cluster. This leads to results such as those illustrated by Fig. 7.

Fuzzy clustering has, however, been applied only to a very limited extent in analytical chemistry.

5. Selection of a clustering method

It is a matter of considerable debate which clustering method should be preferred. A simple answer is not possible, but a general selection scheme might be based on Fig. 8. If only an impression of the relationships between the objects is wanted, hierarchical methods are to be preferred because programs are readily

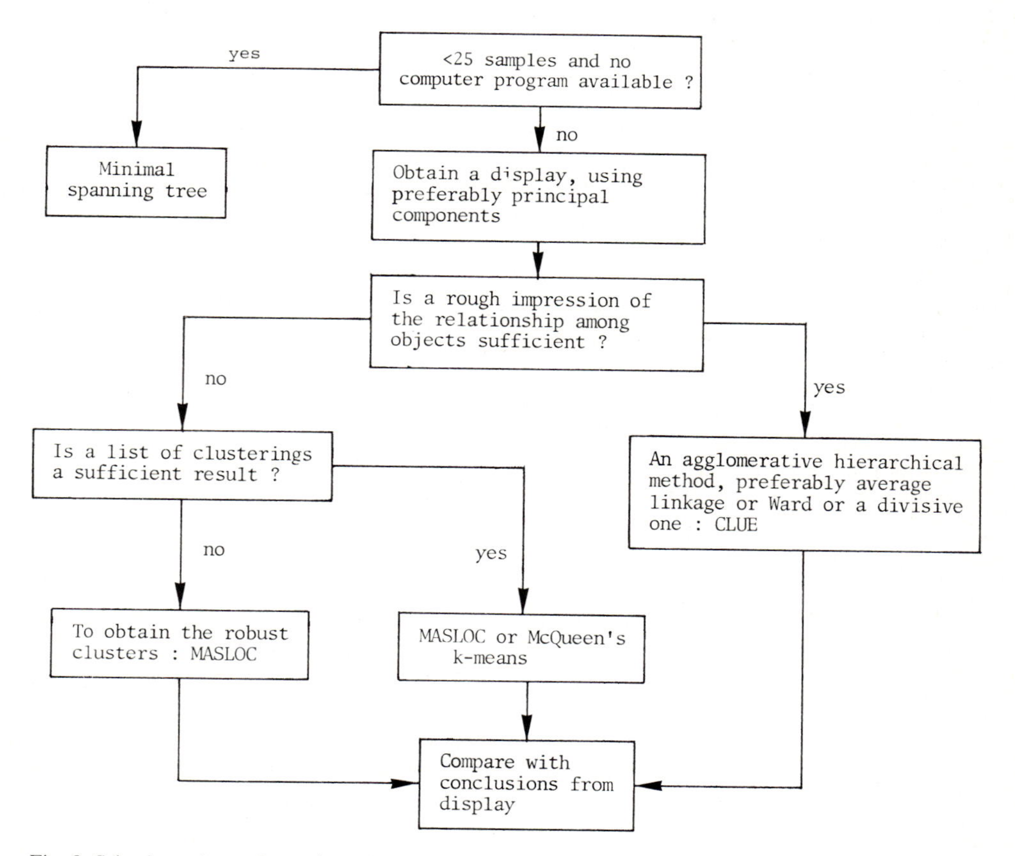

Fig. 8. Selection scheme for a clustering method.

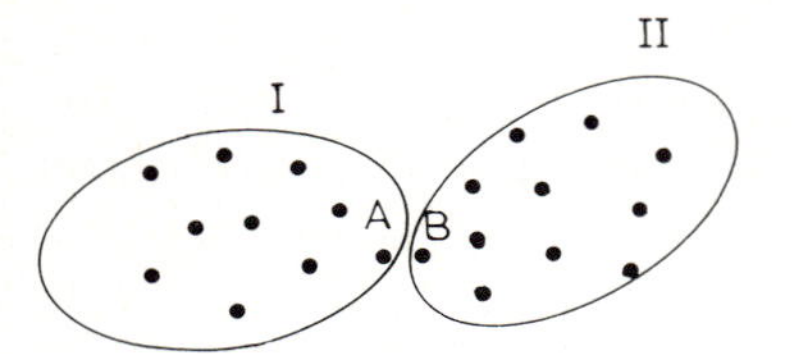

Fig. 9. Non-hierarchical and divisive methods will select clusters I and II. Agglomerative hierarchical methods start by joining A to B because they are nearest to each other and totally meaningless clusters will result.

available in many computer packages and, when the number of objects is not too large, one may even compute a clustering by hand using the minimum spanning tree. An important advantage of hierarchical clustering compared with most non-hierarchical methods is that the output is a dendrogram which allows quick visual inspection. Among the hierarchical methods, average linkage or Ward's method are usually preferred. An alternative is the divisive methods.

If exact results for a certain K-clustering are wanted, a non-hierarchical method is to be preferred because one is not bound by early decisions. A simple example of how disastrous this can be is given in Fig. 9 where an agglomerative hierarchical method would start by linking A and B.

Forgy's method is an attractive and simple one. However, the solutions obtained by Forgy's method may depend on the selection of the initial seed points. MASLOC's results are mathematically exact for less than 50 objects. A higher number of objects requires the use of less rigorous computational procedures, because of the high computer times involved. The results are, however, much better than those obtained with Forgy. A method of the centroid sorting type, which gives better results than Forgy and is less time consuming than MASLOC, is McQueen's K-means method.

An important advantage of MASLOC is that it contains a technique to select what are called robust clusters. This can be explained with Fig. 10. If one applies a non-hierarchical procedure, one asks for a certain number, K, of clusters. In the present case, one could ask for $K = 1, 2, \ldots, 9$ clusters and one would obtain the optimal list of K-clusters for each K. Neither the 2- nor the 4-clustering are really important. The important result is the 3-clustering. However, this is not known beforehand but can be derived if the MASLOC method is used. The following reasoning is applied. In Fig. 10, one observes that E, F and D are separated when one asks for the 2-clustering. When one asks for the 3-clustering, one of the clusters obtained is EDF. Now, in going from the 2-clustering to the 3-clustering, one could expect that one of the two clusters would fall apart to give three clusters. Since E, F and D are joined together in the 3-clustering, this is not the case and one concludes that something unnatural has happened. The 2-clustering is mathematically optimal, but not meaningful. When going from the 3-clustering to higher ones, the same phenomenon is not encountered: the objects separated in this clustering are never joined again. The 3-clusters are therefore meaningful or "robust".

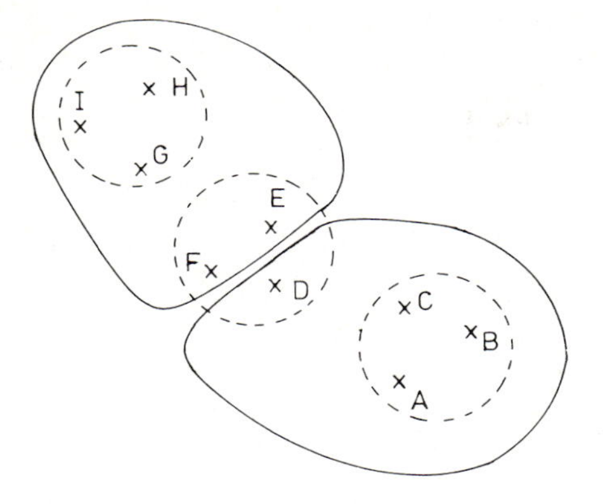

Fig. 10. Nine objects, A–I, are clustered in two and three clusters. The 2-clustering is exact but not meaningful. The 3-clustering is robust.

The result of the clustering procedure depends on which procedure is applied and on the similarity measures used. Each gives a different view of the complex reality in the data set. It is therefore highly recommended that a clustering method is combined with a principal component output and, if possible, that several clustering methods and several types of similarity are used.

Representative examples of hierarchic or non-hierarchic clustering methods are the clustering of stationary GLC phases [4] and meteorites [3]. Other examples can be found in the books cited in ref. 1.

References

1 D.L. Massart and L. Kaufman, The application of operational research in analytical chemistry, Anal. Chem., 47 (1975) 1244A.

2 A. Thielemans, M.P. Derde and D.L. Massart, CLUE, Elsevier Software, Amsterdam, 1985.

3 D.L. Massart, L. Kaufman and K.H. Esbensen, Hierarchical nonhierarchical clustering strategy and application to classification of iron meteorites according to their trace element patterns, Anal. Chem., 54 (1982) 911.

4 D.L. Massart, M. Lauwereys and P. Lenders, The selection of preferred liquid phases after classification by numerical taxonomy techniques, J. Chromatogr. Sci., 12 (1974) 617.

Recommended reading

Books

M.R. Anderberg, Cluster Analysis for Applications, Academic Press, New York, 1973.

B. Everitt, Cluster Analysis, Heinemann, London, 1974.

J. Hartigan, Clustering Algorithms, Wiley, New York, 1975.

M. Jambu, Classification Automatique pour l'Analyse des Données, Vols. 1 and 2, Dunod, Paris, 1978.

D.L. Massart and L. Kaufman, The Interpretation of Analytical Chemical Data by the Use of Cluster Analysis, Wiley, New York, 1983.

P.H.A. Sneath and R.R. Sokal, Numerical Taxonomy, Freeman, San Francisco, 1973.
H. Spath, Cluster Dissection and Analysis, Ellis Horwood, Chichester, 1985.
J. Zupan, Clustering of Large Data Sets, Research Studies Press, Chichester, 1982.

Selected applications

G.W. Adamson and D. Bawden, Comparison of hierarchical cluster analysis techniques for automatic classification of chemical structures, J. Chem. Inform. Comput. Sci., 21 (1981) 204.

D. Bawden, DISCLOSE: an integrated set of multivariate display procedures for chemical and pharmaceutical data, Anal. Chim. Acta, 158 (1984) 363.

P.D. Gaarenstroom, S.P. Perone and J.L. Moyers, Application of pattern recognition and factor analysis for characterization of atmospheric particulated composition in southwest desert atmosphere, Environ. Sci. Technol., 11 (1977) 795.

C. Hansch, S.H. Unger and A.B. Forsythe, Strategy in drug design. Cluster analysis as an aid in the selection of substituents, J. Med. Chem., 16 (1973) 1217.

J.F.K. Huber and G. Reich, Extraction of information on the chemical structure of monofunctional compounds from retention data in gas–liquid chromatography by pattern recognition, J. Chromatogr., 294 (1984) 15.

L. Kaufman, A. Pierreux, P. Rousseeuw, M.P. Derde, M.R. Detaevernier, D.L. Massart and G. Platbrood, Clustering on a microcomputer with an application to the classification of coals, Anal. Chim. Acta, 153 (1983) 253.

A.M. Massart-Leen and D.L. Massart, The use of clustering techniques in the elucidation or confirmation of metabolic pathways, Biochem. J., 196 (1981) 611.

Y. Miyashita, Y. Takahashi, Y. Yotsui, H. Abe and S. Sasaki, Application of pattern recognition to structure–activity problems. Use of minimal spanning tree, Anal. Chim. Acta, 133 (1981) 615.

P. Simon, B.C. Giessen and T.R. Copeland, Categorization of papers by trace metal content using atomic absorption spectrometric and pattern recognition techniques, Anal. Chem., 49 (1977) 2285.

Y. Takahashi, Y. Miyashita, H. Abe, S. Sasaki, Y. Yatsui and M. Sano, A structure–biological activity study based on cluster analysis and the non-linear mapping method of pattern recognition, Anal. Chim. Acta, 122 (1980) 241.

P. Willet, A comparison of some hierarchical agglomerative clustering algorithms for structure–property correlation, Anal. Chim. Acta, 136 (1982) 29.

J. Zupan, Hierarchical clustering of infrared spectra, Anal. Chim. Acta, 139 (1982) 143.

Chapter 23

Supervised Pattern Recognition

1. Decision rules and their validation

Let us consider the thyroid example from Chap. 20, Sect. 7. In its simplest form, it can be represented by Fig. 1. One can imagine that samples from class K (for example, euthyroid EU) should be discriminated from samples from class L (hyperthyroid-HYPER) according to two variables, x_1 (for example, serum tri-iodothyronine, T3) and x_2 (for example, thyroxin stimulating hormone, TSH).

In contrast with Chap. 22, we now know which classes must be considered and our object is to separate those classes and to find rules for deciding in which class a sample of unknown position should be placed. In our example, this means that samples for which one knows to which of the two possible classes (eu- and hyperthyroid) they belong will be used to classify an unknown sample into the EU or HYPER category. Some methods will also make it possible to decide that it belongs to neither.

The accent therefore in this chapter is on discrimination. In Fig. 1, this discrimination can be achieved by drawing a line such as a.

This can be generalized to situations with more variables. An m-dimensional space is then obtained in which the samples, objects or patterns are represented by points. The easiest way of doing this is to characterize them with their pattern vector. In the two-dimensional case in Fig. 1, the sample for which the vector is shown can be denoted by $\mathbf{x}_u = (x_{u1}, x_{u2})$. In general, a sample will be represented in a hyperspace by $\mathbf{x} = (x_1, x_2, \ldots, x_m)$.

In this chapter, which considers supervised learning systems, there are two kinds of sample, namely those which constitute the training or learning set and those which have to be classified (the test set). The training set consists of samples for

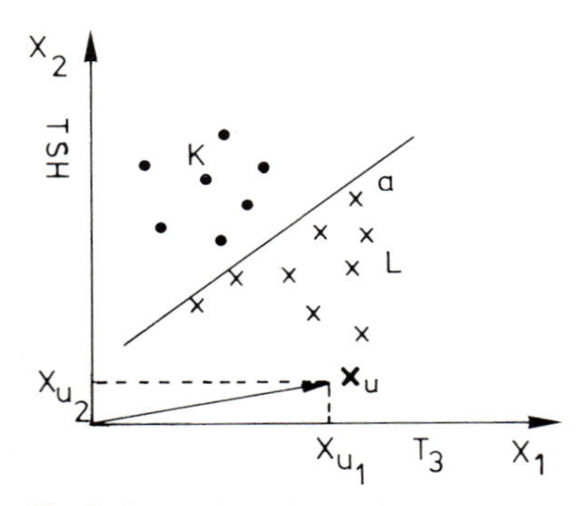

Fig. 1. Separation of two classes, K and L, in two-dimensional space.

which both the pattern vector and the identity are known. In the training or learning step, one develops a decision model (a rule) which allows classification of the unknown samples to be carried out. The decision model of Fig. 1 consists of line a and the rule that objects to the right of it are assigned to class *L* and objects to the left to class *K*. Once a decision rule has been obtained, one still needs to demonstrate that it is a good one. This can be done by observing how successful it is at classifying known samples. One distinguishes between recognition and prediction ability. The recognition (or classification) ability is characterized by the percentage of the members of the training set that are correctly classified. The prediction ability is determined by the percentage of the members of the test set correctly classified by using the decision functions or classification rules developed during the training step. Since the test set consists by definition of samples from unknown origin an artifice is necessary. One determines the prediction ability by developing the decision model on a part of the training set only and uses the other part as a mock test set. This is repeated a few times until all training samples have been used as test samples. If several objects at a time are considered as test samples, one calls this a jackknife method and when only one sample at a time is removed from the training set, one calls this a leave-one-out procedure. If the training set consists of 20 objects, a jackknife method could be carried out as follows. One first deletes objects 1–6 from the training set and develops the classification rules with the remaining objects 7–20. Then, one considers 1–6 as the test set, classifies them with the rules obtained on objects 7–20 and notes how many objects were classified correctly. The whole procedure is then repeated after replacing first 1–6 in the training set but deleting 7–12 and finally on a training set consisting of 1–12. The percentage of successes on the 3 runs together is then called the prediction ability.

When one only determines the recognition ability, there is a risk that one will be deceived into taking an overoptimistic view of the classification result. In particular, classification methods such as linear discriminant analysis and the learning machine, which try to maximize the differences between classes, tend to give an optimistic classification rate if this is calculated from the classification success of the training set only. It is therefore also necessary to verify the prediction ability. If the prediction and the recognition ability are substantially different, this means that the decision rules depend too much on the actual objects in the training set: the solution obtained is not stable and should therefore not be trusted.

As discussed in Chap. 20, we will consider two kinds of pattern recognition method. Most methods explicitly or implicitly try to find a boundary between classes. Of the methods we will discuss in this chapter, statistical linear discriminant analysis (SLDA, Sect. 2), the linear learning machine (LLM, Sect. 3) and density methods such as ALLOC (Sect. 5) are designed to find explicit boundaries between classes while the *K*-nearest neighbour (KNN, Sect. 4) does this implicitly. Other methods such as SIMCA (Sect. 6.2) and UNEQ (Sect. 6.1) put the accent more on similarity within a class than on discrimination between classes.

One also makes a distinction between parametric and non-parametric techniques. In the parametric techniques such as SLDA, UNEQ and SIMCA, statistical parameters of the distribution of the samples are used in the derivation of the decision function (almost always a multivariate normal distribution is assumed). The non-parametric methods such as KNN, LLM and ALLOC are not explicitly based on distribution statistics. The most important disadvantage of parametric methods is that to apply the method correctly statistical requirements must be fulfilled. Why this is not always the case will be discussed in the section on SLDA. The most important advantage for the parametric methods is that probabilities of correct classification can be estimated, although this is possible too with some non-parametric methods such as those of Sect. 5.

2. Statistical linear discriminant analysis

Let us consider again the case of the separation of two classes. This was depicted in Fig. 1, but for convenience it is shown again in Fig. 2(a). The supposedly normal distribution of the two classes to be separated, K and L, is shown in Fig. 2(b) for one of the variables, x_2. The mathematical problem is then to find an optimal decision rule for the classification of these two groups.

Let us consider first the simplest possible case. Two classes, K and L, have to be distinguished using a single variable, x_2 . It is clear that the discrimination will be better when the distance between $\bar{x}_{K2}$ and $\bar{x}_{L2}$ (i.e. the mean values, or centroid, of x_2 for classes K and L) is large and the width of the distributions is small or, in other words, when the ratio of the difference between means to the variance of the distributions is large. Analytical chemists would be tempted to say that the resolution should be as large as possible.

When one considers the situation with two variables, x_1 and x_2, it is again evident that the discriminating power of the combined variables will be good when

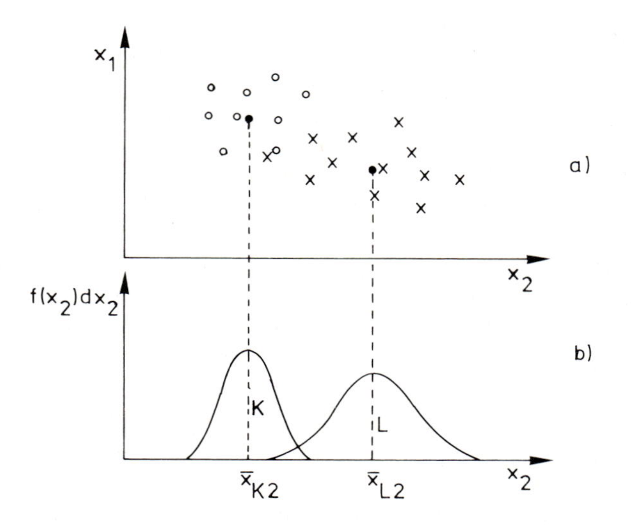

Fig. 2. Two classes, K and L, in (a) a two-dimensional space and (b) their projection on x_2.

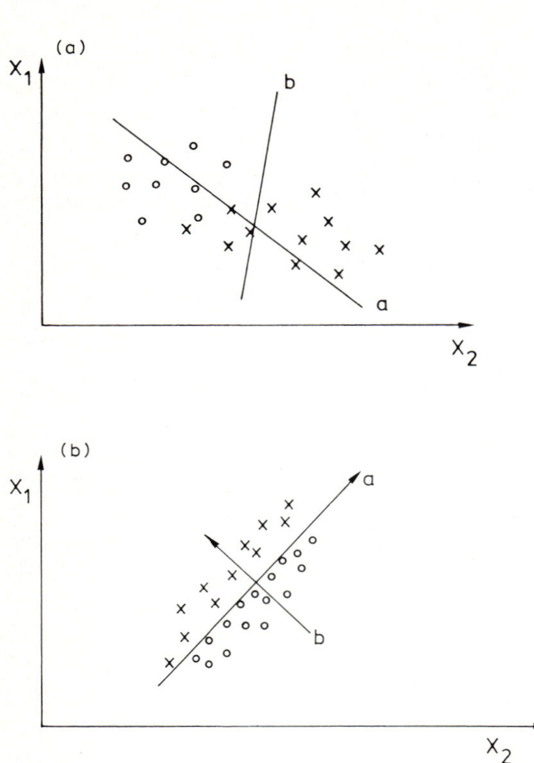

Fig. 3. (a) Line a is a better discrimination function than line b because the projections of the objects belonging to K (○) and L (×) are better separated on a than on b. (b) A situation where the first principal component, a, and the discriminant function, b, differ to a considerable extent.

the centroids of the two sets of samples are sufficiently distant from each other and when the clusters are tight or dense. In mathematical terms this means that the between-class variance is large compared with the within-class variances. On the other hand, it is also clear that if the two variables are highly correlated then the benefit of adding the second variable to the first will be small (see Chap. 9). When the two variables are absolutely correlated ($r = |1|$), the second is of no help and can be eliminated. One therefore finds that the three mathematical parameters that determine the discriminating effect of two chemical variables are the between-class variance, the within-class variance and the correlation between the chemical variables.

In the method of linear discriminant analysis, one seeks a linear function D of the variables x_j

$$D = \sum_{j=1}^{m} w_j x_j \tag{1}$$

which maximizes the ratio between both variances, taking correlation into account; w_j are weights given to the variables and m is the number of variables. To this one should add a constant term $w_{(j+1)}$ if the variables are not standardized prior to the analysis. Geometrically, this means that one looks for a line through the cloud of points, such that the projections of the points of the two groups are separated as much as possible. The approach is comparable to principal components, where one seeks a line that explains best the variation in the data (see Chap. 21). The principal component line and the discrimination often more or less coincide [as would be the case in Fig. 3(a)]. This is not necessarily so as shown in Fig. 3(b).

Let us first consider the situation in which D is determined for two classes K and L in a bivariate situation, i.e. there are only two variables x_1 and x_2. Then

$$D = w_1 x_1 + w_2 x_2$$

The score of an object u along D is then

$$\text{score}(u) = w_1 x_{u1} + w_2 x_{u2}$$

The score of the centroids $\bar{x}_K$ and $\bar{x}_L$ is therefore given by

$$\text{score}(\bar{x}_K) = w_1 \bar{x}_{K1} + w_2 \bar{x}_{K2}$$

$$\text{score}(\bar{x}_L) = w_1 \bar{x}_{L1} + w_2 \bar{x}_{L2}$$

The between-class variance is the square of the distance between the centroids and must be maximal so that

$$[\text{score}(\bar{x}_K) - \text{score}(\bar{x}_L)]^2 = [w_1(\bar{x}_{K1} - \bar{x}_{L1}) + w_2(\bar{x}_{K2} - \bar{x}_{L2})]^2 = \text{maximal}$$

The within-class variance taking correlation between variables into account is equal to (see Chap. 14, Sect. 3.1)

$$\text{var}(w_1 x_1 + w_2 x_2) = w_1^2 s_1^2 + w_2^2 s_2^2 + w_1 w_2 \text{ cov}(1, 2) + w_1 w_2 \text{ cov}(2, 1)$$

and this must be minimal.

Generalizing this to m variables and combining the conditions for between- and within-variance then leads to maximizing.

$$\frac{\left\{ \sum_{j=1}^{m} w_j (\bar{x}_{Kj} - \bar{x}_{Lj})^2 \right\}}{\sum_{j=1}^{m} \sum_{j'=1}^{m} w_j w_{j'} c_{jj'}} \tag{2}$$

where $\bar{x}_{Kj}$ is the mean of variable x_j for class K and $c_{jj'}$ is an element of the pooled or average variance–covariance matrix. The use of a pooled variance–covariance matrix implies that the variance–covariance matrices for both populations are assumed to be not significantly different. The consequences of this are discussed below (Fig. 4).

To clarify the use of the variance–covariance matrix observe that the numerator of eqn. (2) for two variables is (Chap. 14, Sect. 3.1).

$$w_1^2 s_1^2 + w_2^2 s_2^2 + w_1 w_2 \text{ cov}(1, 2) + w_1 w_2 \text{ cov}(2, 1)$$

Maximizing eqn. (2) therefore means minimizing the within-class variances s_1^2 and s_2^2 (i.e. attaching higher weight to the variable with the smaller variance) and minimizing covariance at the same time. These conditions mean, more generally, that variables are given weight or selected (see Sect. 8) according to small variance and small correlation with other variables. The latter condition is in accordance with information theory (Chap. 9).

Differentiation and setting the derivative equal to zero with respect to w_j leads to

$$\bar{x}_{Kj} - \bar{x}_{Lj} = \frac{\sum_{j=1}^{m} w_j(\bar{x}_{Kj} - \bar{x}_{Lj}) \sum_{j'=1}^{m} c_{jj'} w_{j'}}{2 \sum_{j=1}^{m} \sum_{j'=1}^{m} w_j w_{j'} c_{jj'}} \tag{3}$$

$w_{j'}$ can be obtained from

$$w_{j'} = \sum_{j=1}^{m} c_{jj'}^{-1} (\bar{x}_{Kj} - \bar{x}_{Lj}) \tag{4}$$

$c_{jj'}^{-1}$ is an element of the inverse of the pooled variance–covariance matrix.

Once the w values have been obtained, one calculates the values of the discriminant function D (also called discriminant scores) for the centroids of classes K and L as

$$D_K = \sum_{j=1}^{m} w_j \bar{x}_{Kj} \tag{5}$$

$$D_L = \sum_{j=1}^{m} w_j \bar{x}_{Lj} \tag{6}$$

For an individual object u with variable values x_{uj}, the same function

$$D_u = \sum_{j=1}^{m} w_j x_{uj} \tag{7}$$

is obtained and u is classified with K if D_u is situated nearer to D_K than to D_L. Generalization is possible: when there are k classes, up to $k-1$ discriminant functions can be obtained.

To investigate what happens when certain statistical requirements are not fulfilled, we turn to an example we have used earlier in this part of the book, namely the thyroid example. There are now three classes to be considered (EU, HYPO, HYPER) so that we can obtain 2 discriminant functions. In contrast with the principal component axes discussed in the previous chapter, these discriminant functions are not orthogonal. They can, however, be used to display the samples in an optimal discriminating way, by plotting the first discriminant score against the second one. The result was already given in Chap. 20, Fig. 13.

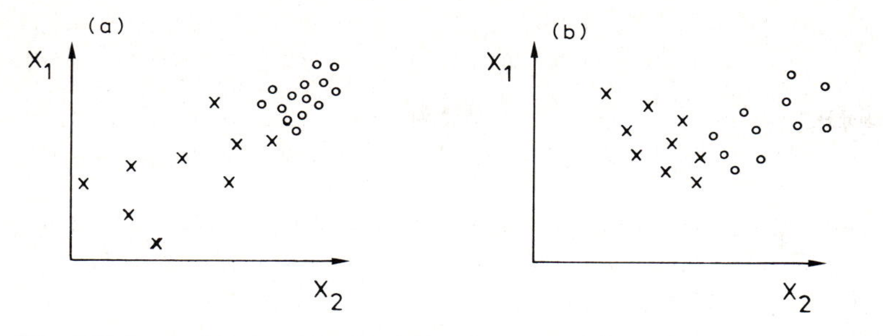

Fig. 4. Unfavourable situations for SLDA. Difference in (a) dispersion and (b) direction.

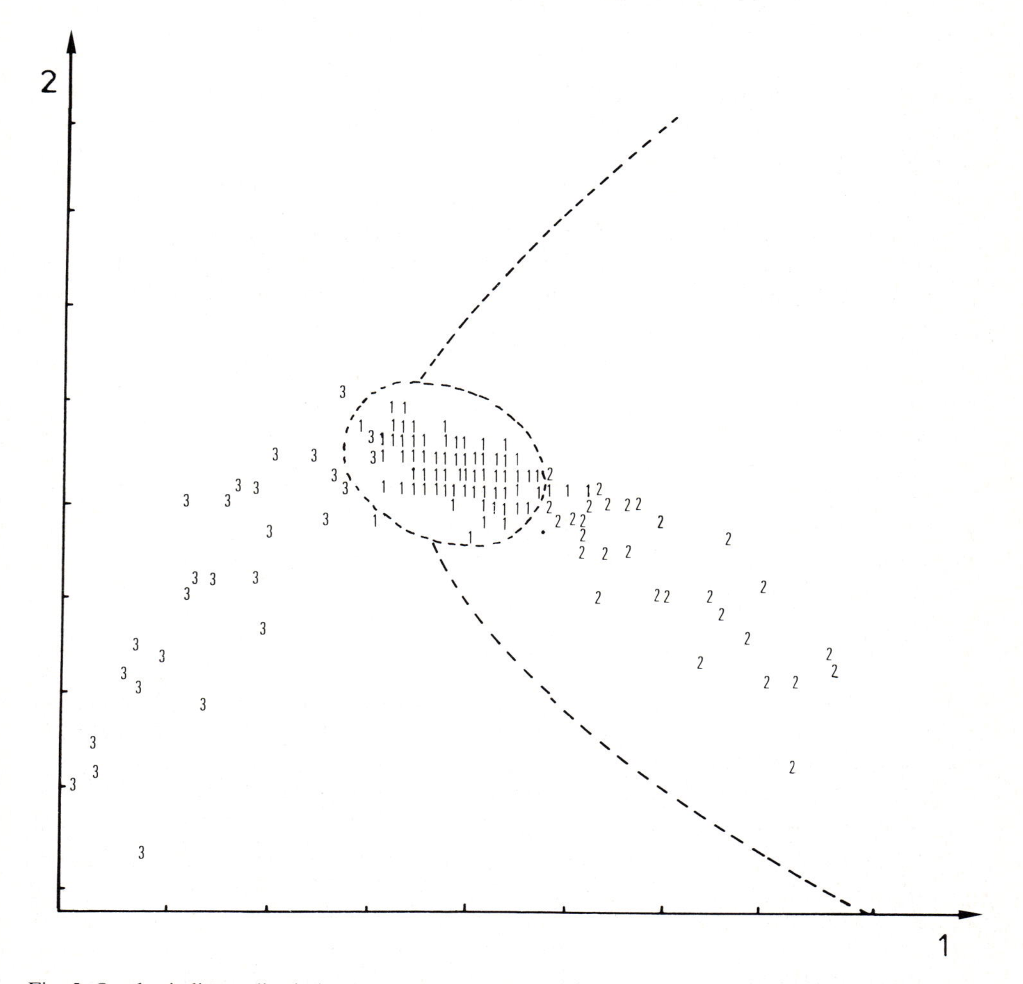

Fig. 5. Quadratic linear discrimination boundaries plotted in the discriminant map of Fig. 13, Chap. 20 [1].

References p. 412

One observes that the HYPER and HYPO classes are much more diffuse than the EU group. The boundaries obtained by discriminant analysis are also given in the figure and clearly they are not optimal since part of the HYPO and HYPER samples are located in the EU group. Because the boundary is situated halfway between the centroids of a diffuse and a dense group, part of the diffuse group is included in the EU group.

SLDA is a parametric method, which means that it is based on certain statistical assumptions. Clearly it is not optimal for a data set with the characteristics displayed here. Indeed, one of the requirements for optimal use of SLDA was equality between the variance–covariance matrices of the groups to be separated. To understand this we must go back to Chap. 14, Sect. 5 and Chap. 13, Sect. 3.

Equal covariance is obtained when the correlations and variances are equal. Equal variance means that the classes must have equal dispersion. In Chap. 14, Sect. 3 it is shown that the regression coefficient of a linear model fitted through the data points (i.e. the direction of a group!) is related to the correlation coefficients. Therefore the requirement of equal covariance really means that classification by SLDA can only be optimally applied when the dispersion of the classes is equal and when they have the same direction in the pattern space (see also Fig. 4).

When applying SLDA, visual inspection of the shape and the direction of the groups in a principal components plot is therefore indicated. When this indicates a clear deviation from the conditions in which SLDA may be used, one can expect to obtain a better result by using quadratic discriminant functions, at the price, however, of increased mathematical sophistication. Figure 5 gives the quadratic boundaries displayed on the same data points as Chap. 20, Fig. 13.

It should be borne in mind that the difficulties due to unfavourable statistical characteristics only influence the boundaries and therefore the classification by SLDA, but not the determination of the discriminant functions. These can therefore be used perfectly well to display the data such as in Fig. 13.

3. The linear learning machine

The linear learning machine (LLM) also tries to locate a linear boundary between classes. In this sense it is also sometimes described as a linear discriminant method. Books on multivariate statistics often reserve the same term for the technique described in the preceding section. To avoid confusion between the general and the restricted use of the term linear discriminant analysis, we have called the technique of the preceding section statistical linear discriminant analysis.

To understand LLM we need to consider again Fig. 1. One wishes to find a decision line that separates the two classes K and L. This can be achieved by a trial and error procedure, in which one uses the errors of the preceding trial to plan a more intelligent trial. One starts with a decision boundary (here a line) chosen at random or, if this is possible, from prior experience. Figure 1 would for instance give rise to Fig. 6 where a_1 is a test boundary chosen at random. One then verifies whether all dot samples are located on the same side of a_1 and all cross samples on

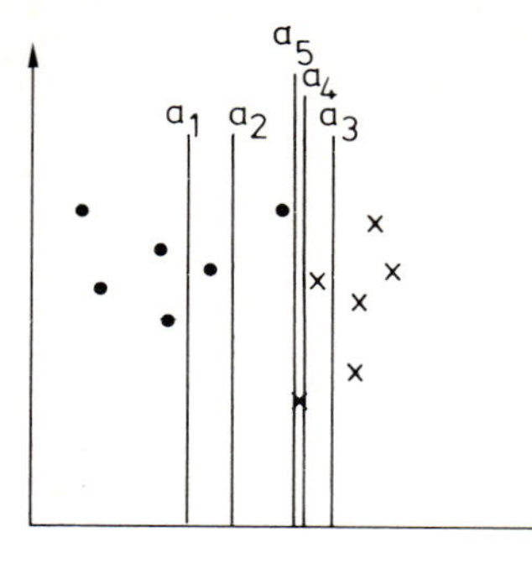

Fig. 6. The linear learning machine strategy: $a_1 - a_5$ are successive decision boundaries.

the other side. When one comes to point 5 one notes that a_1 separates this sample from the other dot samples so that the boundary is not correct. The boundary is then displaced so as to bring 5 to the same side as the other dot samples, for instance by constructing a_2 situated at the same distance from 5 but on the other side. One tests again whether the boundary is correct and finds that 6 is on the wrong side. This now leads to a_3. Successive trials may lead to a_4 and eventually to a_5 which separates correctly all the dots from all the crosses. The procedure gradually "learns" the correct answer to the task of finding a decision surface and is therefore called the learning machine.

In the m-dimensional case a decision surface must be found with a lower dimensionality than the pattern space. For reasons of computational convenience [and, more specifically, to be able to use eqn. (9)], it is preferable that the decision surface should be linear and pass through the origin of the pattern space. This is not possible in Fig. 1. It becomes possible, however, if the two-dimensional space is augmented by the addition of a third dimension, z. This extra dimension is usually given a value of unity and it is added to all pattern vectors. The vector $\mathbf{x} = (x_1, x_2)$ now becomes $\mathbf{x} = (x_1, x_2, z)$. It is now possible to separate the two classes by a linear decision surface (here a plane) (see Fig. 7). Class K is above the plane and class L below.

This plane can conveniently and unambiguously be described by an orthogonal or normal vector in the origin, called the weight vector and represented by $\mathbf{w}$.

This can be generalized to the d-dimensional case. To all pattern vectors a $(m+1)$th component is added so that they are now given by $\mathbf{x} = (x_1, x_2, \ldots, x_m, z)$ and a linear decision surface can be sought that will be represented by its normal vector.

The normal vector not only permits a specification of the surface but also allows easy classification. The scalar product of the normal vector and the pattern vector is given by

$$\mathbf{w} \cdot \mathbf{x} = |\mathbf{w}|\,|\mathbf{x}| \cos \theta \tag{9}$$

where $|\mathbf{w}|$ and $|\mathbf{x}|$ are the magnitudes of vectors $\mathbf{w}$ and $\mathbf{x}$ and θ is the angle

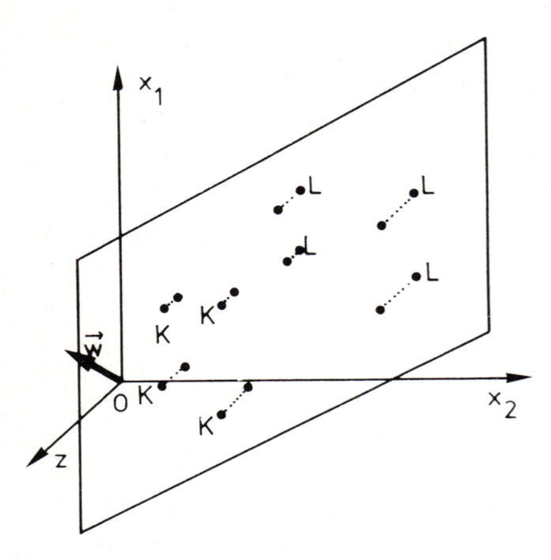

Fig. 7. Separation of two classes by a plane through the origin.

between the two. When **w** and the pattern vector lie on the same side of the plane $\cos\theta > 0$ and, as $|\mathbf{w}|$ and $|\mathbf{x}|$ are positive quantities, $\mathbf{w}\cdot\mathbf{x}$ must be positive. When the pattern vector of a sample of class L is considered, **x** and **w** do not lie on the same side of the plane and therefore $90 < \theta < 270$, so that $\cos\theta < 0$. The scalar product of **w** and **x** is now negative.

In other words, one verifies whether the sign of the scalar product is positive to decide whether the sample is part of K. When it is not, it should be classified in L.

In practice, the calculations are not carried out by using eqn. (9). In the same way as **x** is represented by its components along the axes, **w** can also be decomposed in its components $w_1, w_2, \ldots, w_m, w_{m+1}$ along the same axes. The scalar product $\mathbf{w}\cdot\mathbf{x}$ is then equal to the sum of the products of the components

$$\mathbf{w}\cdot\mathbf{x} = w_1x_1 + w_2x_2 + \ldots + w_mx_m + w_{m+1}z \tag{10}$$

The coefficients $w_1, \ldots$ can be considered as weights of the variables $x_1, \ldots$ which is why **w** is called the weight vector.

The determination of **w** leads to a simple classification rule. Before a classification is possible it is necessary, however, to find a decision surface (or its associated weight vector) that permits the separation of classes K and L. This is accomplished during the training or learning step, using an iterative procedure. It is initiated by selecting, sometimes arbitrarily, an initial weighting vector and investigating whether all the pattern vectors fall on the correct side of the associated surface. When an object u characterized by a pattern $\mathbf{x}_u$ is found to be misclassified because the product

$$\mathbf{w}\cdot\mathbf{x}_u = s \tag{11}$$

produces the wrong sign, a new decision surface is obtained by reflecting it about the misclassified point. This means that one determines $\mathbf{w}'$, so that

$$\mathbf{w}'\cdot\mathbf{x}_u = -s \tag{12}$$

This process is repeated whenever a misclassified sample is found until a completely successful **w** is obtained. In this case the process is said to converge.

The learning machine explained here is the simplest of a large class of methods called threshold logic unit (TLU) methods. In the method discussed here, s is compared to zero (the threshold) to make a binary decision (K or L). Non-zero TLU methods also exist and the learning machine can be adapted for multi-category decisions by splitting them into sequences of binary decisions.

4. The *k*-nearest neighbour method

A mathematically very simple non-parametric classification procedure is the nearest neighbour method. In this method one computes the distance between an unknown, represented by its pattern vector, and each of the pattern vectors of the training set. Usually one employs the Euclidian distance. If the training set consists of a total of n samples, then n distances are calculated and one selects the lowest of these. If this is D_{ul}, where u represents the unknown and l a sample from learning group L, then one classifies u in group L. A three-dimensional example is given in Fig. 8.

In a more sophisticated version of this technique, called the k-nearest neighbour method (often abbreviated to the KNN method), one selects the k nearest samples to u and classifies u in the group to which the majority of the k samples belong.

Figure 9 gives an example of a 3-NN method. One selects the three nearest neighbours (A, B and C) to the unknown u. Since A and B belong to L one classifies u in category L.

The choice of k can be determined by optimization: one determines the prediction ability with different values of k. Usually it is found that small values of k (3 or 5) are to be preferred.

If the training set is sufficiently large, the mathematical simplicity of this method does not prevent it from yielding classification results as good as and often better than the much more complex TLU methods discussed in the preceding section. It

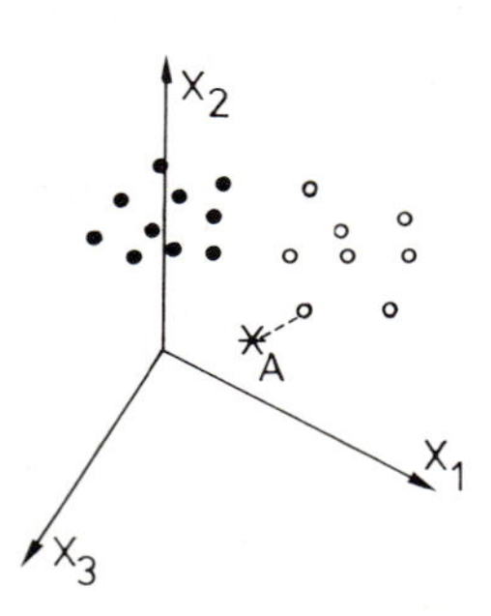

Fig. 8. 1NN classification of object A.

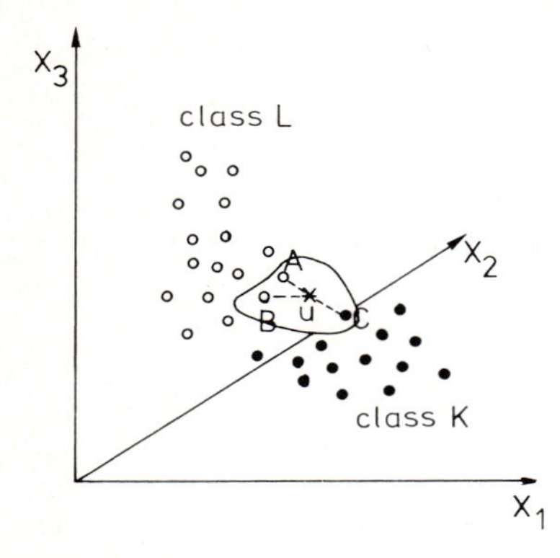

Fig. 9. 3NN classification of object u.

also has the advantage of being a multi-category method whereas most TLU methods are fundamentally binary decision methods.

The mathematical simplicity of the method also leads to some problems. They can be partially overcome, however, at the price of greater mathematical sophistication.

A first problem is that the method is sensitive to the fact that in the training set the number of samples of each class is not the same. Figure 10 gives an example of two overlapping classes which differ considerably in sample size. The unknown is classified into the class with the largest membership, because in the zone of overlap between classes more of its members are present. In fact, the unknown is rather closely situated to the center of the other class, so that its classification is at least doubtful. This can be overcome by not using a majority criterion but an alternative

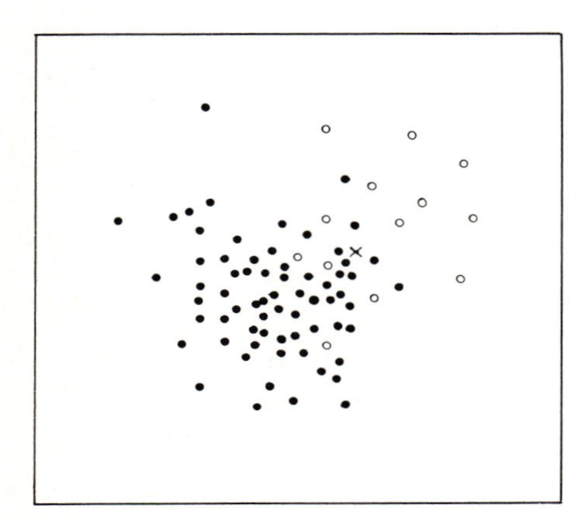

Fig. 10. A situation which necessitates classification by alternative KNN criteria.

one. In the example of Fig. 10, this might for instance be: classify the object in class K if for $k = 10$ at least 9 neighbours (out of 10) belong to K, otherwise classify the test object in L. The selection of k and of the alternative criterion value should be determined by optimization.

5. Density methods

In these methods, one imagines a potential field around the objects of the learning set. For this reason these methods have also been called potential methods.

The shape of the potential field depends on the choice of a potential function. Many functions can be used for this purpose, but for practical reasons it is recommended that a simple one such as a triangular or a gaussian function is selected. The function is characterized by its width. This is important for its smoothing behaviour (see below).

Figure 11 illustrates the triangular and the normal function for a learning class K in a one-dimensional space. From the potential functions arising from the individual points, the cumulative potential function is determined by adding the heights of the individual potential functions in each position along the x axis. Figure 11 shows that the cumulative function constitutes a continuous line which is never zero within a class. This is done for each learning class separately.

By dividing the cumulative potential function of a class by the number of samples contributing to it one obtains the (mean) potential function of the class. In this way, the potential function assumes a probabilistic character and, therefore, the potential method permits probabilistic classification. The classification of an object from the test set into one of the learning classes is determined by means of the potential of the learning class in the position of the test object. The test object is then classified into the class which gives rise to the largest potential. The boundary between two classes is given by those positions where the potentials caused by these two classes have the same value.

A one-dimensional example is given in Fig. 12. Object u is considered to belong to K, because at the location of u the potential of K is larger than the potential of L.

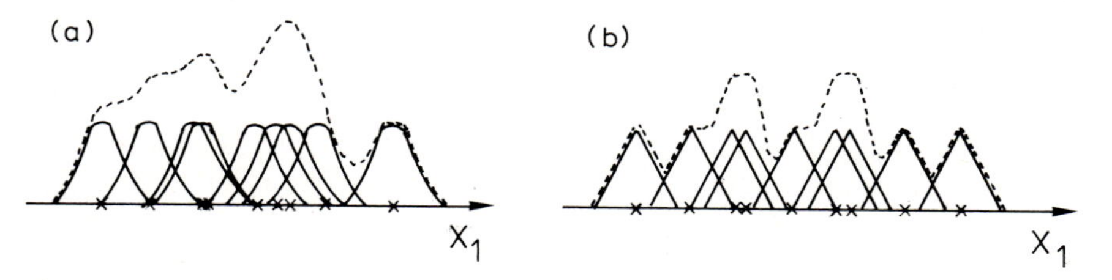

Fig. 11. Density estimation for a learning class using (a) normal and (b) triangular potential functions. The broken line indicates the cumulative potential.

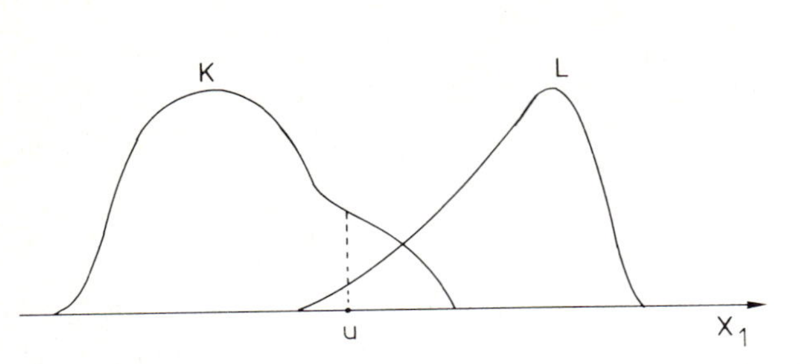

Fig. 12. Classification of an unknown object u. f(K) and f(L) indicate the potential functions for classes K and L.

One must also choose the smoothing parameter. This depends on the data set investigated. When the smoothing parameter is too small [Fig. 13(a)] most of the individual potential functions of a learning class do not overlap with each other and consequently the continuous potential surface of Fig. 12 is not obtained. In such a situation, a test object [u in Fig. 13(a)], which clearly belongs to a particular learning class, may seem to have a low degree of membership because the potential is low or even zero. An excessive smoothing parameter gives rise to potential surfaces that are much too flat [Fig. 13(b)], so that discrimination between adjoining classes is difficult. The major task of the learning procedure is the selection of a suitable smoothing parameter.

When the pattern space is extended to two dimensions, the cumulative function of each class constitutes a potential surface, such as in Fig. 14. Figure 14 depicts a 2-dimensional potential surface and shows again the influence of the smoothing parameter.

For each position in the 2-dimensional pattern space, the degree of membership of a class is given by the height of the potential surface above the plane of the pattern space. Clearly, the model can be extended mathematically to higher dimensional pattern spaces, developing so-called potential hypersurfaces.

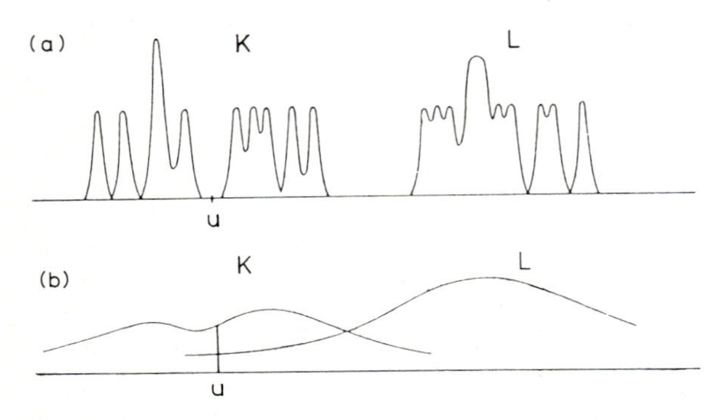

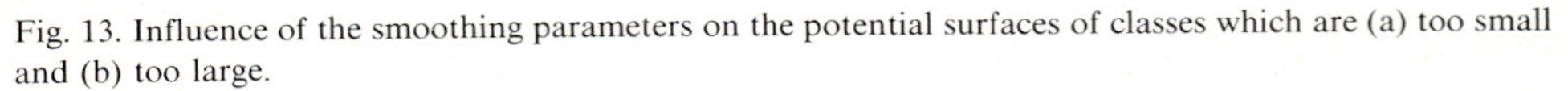

Fig. 13. Influence of the smoothing parameters on the potential surfaces of classes which are (a) too small and (b) too large.

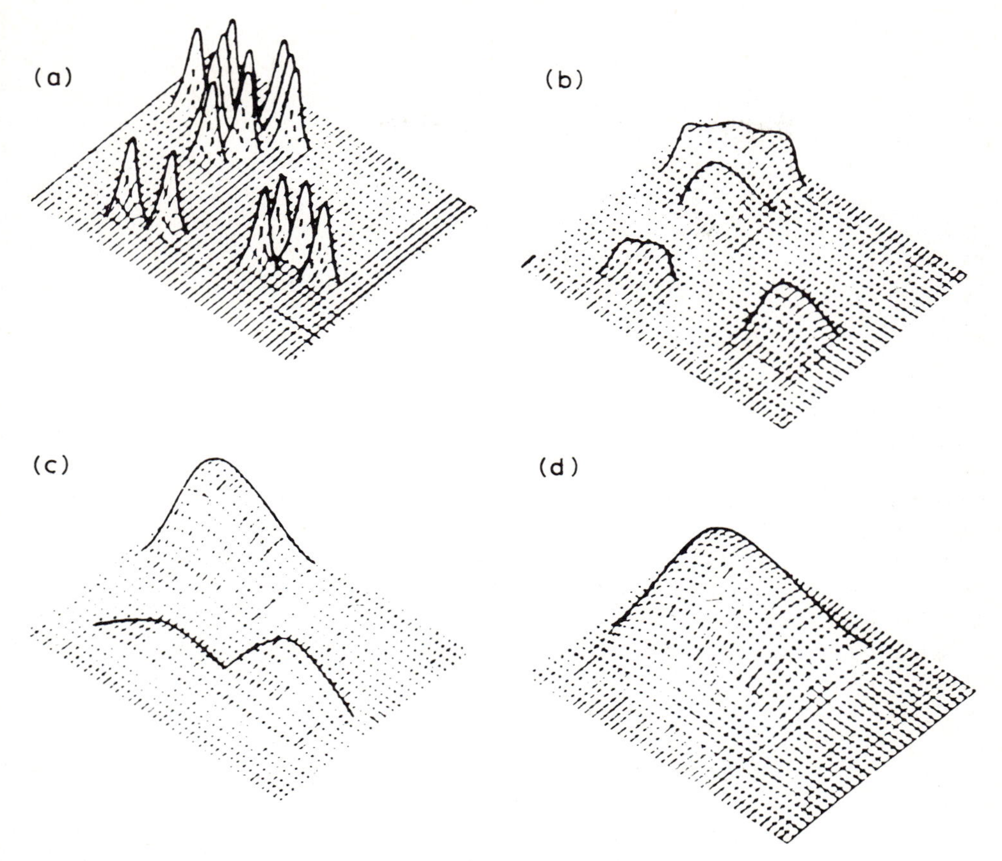

Fig. 14. Potential surfaces obtained in a two-dimensional pattern space [2].

Advantages of these methods are that (a) no a priori assumptions about distribution are necessary and (b) probabilistic decisions can be taken easily (in contrast with KNN). In chemistry, the best known method is ALLOC [3].

6. Modelling methods

Modelling methods have many advantages compared to discrimination methods. These advantages are illustrated in Chap. 20. Modelling techniques develop a separate mathematical description for each class, independently from other classes in the training set. Geometrically this corresponds with the construction of an envelope, a "class box", around each class. A new object is classified according to its position in the pattern space with relation to the class boxes. Three kinds of decision are possible, namely

(1) the object is not situated in any of the class boxes: it is not part of any class and is therefore considered to be an outlier;

(2) the object is situated within the boundaries of only one class: it can uniquely be assigned to that particular class; and

(3) the object is situated inside the boundaries of more than one class box. The class boxes overlap so that the classes are not completely separable. The object is classified as belonging to the overlapping classes.

The simplest class modelling method is based on the multivariate normal distribution and is called UNEQ [4]. A more elaborate one, based on principal components is SIMCA [5] (Sect. 6.2).

6.1 Methods based on the multivariate normal distribution

In Chapt. 20, Sect. 1, we explained that there is a difference between what clinical chemists call normal regions obtained by considering two variables individually or by combining them in a bivariate way. In this section, we will consider first bivariate and later, more generally, multivariate distributions and how they can be used to model classes.

If the objects are characterized by two variables then the data are distributed according to a bivariate distribution.

The probability function of a bivariate normal distribution is given by

$$f(x_1, x_2) = \frac{1}{2\pi\sigma_1\sigma_2\sqrt{1-\rho^2}} \exp\left[-\frac{1}{2(1-\rho^2)}\left\{\left(\frac{x_1-\mu_1}{\sigma_1}\right)^2 + \left(\frac{x_2-\mu_2}{\sigma_2}\right)^2 - 2\rho\left(\frac{x_1-\mu_1}{\sigma_1}\right)\left(\frac{x_2-\mu_2}{\sigma_2}\right)\right\}\right] \quad (13)$$

where μ_j and σ_j^2 are the mean and variance of x_j ($j = 1, 2$) and ρ is the correlation coefficient.

The surface represented by the equation can be compared with a bell-shaped hill (Fig. 15). Cross-sections of this surface with planes drawn perpendicular to the (x_1, x_2) plane yield gauss curves. Cross-sections with planes drawn parallel to the (x_1,

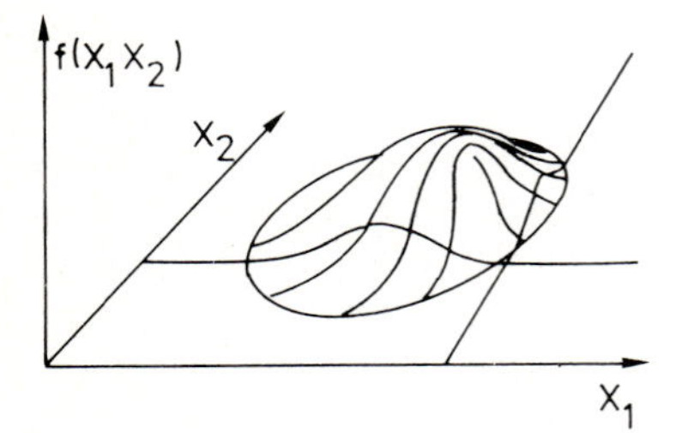

Fig. 15. Bivariate normal distribution surface.

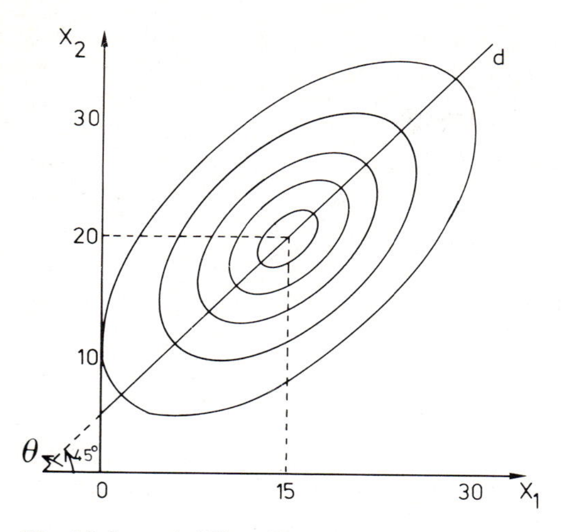

Fig. 16. Isoprobability ellipses for a bivariate normal distribution with $\mu_1 = 15$, $\mu_2 = 20$, $\sigma_1 = \sigma_2 = 5$, $\rho = 0.6$.

x_2) plane yield ellipses representing all x_1, x_2 combinations with the same probability density. They are called isoprobability ellipses. Generally the bivariate normal distribution is represented as a set of isoprobability ellipses (Fig. 16) with equations

$$a = \frac{1}{1-\rho^2}\left\{\left(\frac{x_1-\mu_1}{\sigma_1}\right)^2 + \left(\frac{x_2-\mu_2}{\sigma_2}\right)^2 - 2\rho\left(\frac{x_1-\mu_1}{\sigma_1}\right)\left(\frac{x_2-\mu_2}{\sigma_2}\right)\right\} \qquad (14)$$

a is a positive constant; the smaller the constant the higher up the hill of Fig. 15 the cross-section is performed. For $a = 5.99$ an ellipse is obtained containing 95% of the datapoints.

The shape of the ellipses and their position in the (x_1, x_2) plane are determined by the values of σ_1, σ_2 and ρ. The center of the ellipse is the point with coordinates (μ_1, μ_2); the major axis of the ellipse (d) and the minor axis perpendicular to it, are common to all isoprobability ellipses.

The major axis d makes an angle θ with the x_1 axis, depending on the ratio σ_1/σ_2 and on ρ.

$$\theta = \tfrac{1}{2}\arctan\frac{2\rho\sigma_1\sigma_2}{\sigma_1^2 - \sigma_2^2} \qquad (\text{for } \sigma_1 > \sigma_2)$$

$$\theta = 45° \quad \text{when } \sigma_1^2 = \sigma_2^2 \quad (\text{and } \rho > 0) \qquad (15)$$

The last equation shows that with scaled variables, i.e. after having performed the z-transform (see also Chap. 22), the position of the ellipse in the (x_1, x_2) plane is always the same ($\theta = 45°$). For scaled variables, the ratio of the lengths of the major and minor axes is given by

$$\sqrt{\frac{1+|\rho|}{1-|\rho|}}$$

One observes that for $\rho = 0$, this ratio is 1 and a circle is obtained. For $\rho = 1$, the ellipse as computed with eqn. (14) is undefined and, in fact, it is found that all datapoints lie on a straight line.

The major axis of the ellipse is the orthogonal regression line, i.e. the line which is obtained by minimizing the sums of squares of the residuals measured perpendicularly to the line (see also Chap. 3). It is also called the principal component, defined in Chap. 21. The equation of this principal component is

$$y - \bar{y} = \frac{\sigma_2^2 - \sigma_1^2 + \sqrt{\left(\sigma_2^2 - \sigma_1^2\right)^2 + 4\rho^2\sigma_1^2\sigma_2^2}}{2\rho\sigma_1\sigma_2}(x - \bar{x}) \tag{16}$$

More generally, for m normally distributed variables a multivariate distribution can be written for which the probability function is given by

$$\mathrm{f}(x) = \frac{1}{(2\pi)^{m/2}|\Gamma|^{1/2}} \mathrm{e}^{-1/2\{(\mathbf{x}-\boldsymbol{\mu})'\Gamma^{-1}(\mathbf{x}-\boldsymbol{\mu})\}} \tag{17}$$

where μ is the population mean vector, indicating the centroid of the population in pattern space and Γ is the variance–covariance matrix.

For $m = 3$ isoprobability ellipsoids (see Fig. 15, Chap. 20) and for $m > 3$ isoprobability hyperellipsoids are obtained. The square root of the expression between brackets in the exponent of eqn. (17) is called the generalized or Mahalanobis distance, D.

$$D^2 = (\mathbf{x} - \boldsymbol{\mu})'\Gamma^{-1}(\mathbf{x} - \boldsymbol{\mu})$$

For a bivariate distribution this reduces to the a of eqn. (14). To better understand the meaning of this equation, it can be rewritten as

$$\left(\frac{x_1 - \mu_1}{\sigma_1}\right)^2 + \left\{\frac{\dfrac{(x_2 - \mu_2)}{\sigma_2} - \rho\dfrac{(x_1 - \mu_1)}{\sigma_1}}{\sqrt{1 - \rho^2}}\right\}^2$$

One then notes that the part of the second variable that is explained by correlation with the first variable is substracted. It is a distance, corrected for correlation. When there is no correlation or when correlation is not taken into account, it reduces to the Euclidian distance (for scaled variables) of Chaps. 20 and 22. An important property of the generalized distance is that it is scale invariant. This means that the z-transform does not influence the result and is therefore not necessary (see also Sect. 8.1).

As explained in Sects. 1 and 6 of Chap. 20, one can model each class separately with multivariate normal distributions. In this method, which has been called UNEQ, one assumes for each class separately a multivariate normal distribution and one computes the Mahalanobis distance for each object in the learning class. It can be shown that the Mahalanobis distance follows a χ^2 distribution and it is thus possible to compute a 95% confidence limit for each class. Once this has been done, one computes the distance between object u to be classified to each centroid of a class. For object u and class K this is given by

$$D^2_{Ku} = (\mathbf{x}_u - \bar{\mathbf{x}}_K)' \mathbf{C}_K^{-1} (\mathbf{x}_u - \bar{\mathbf{x}}_K) \tag{18}$$

where $\mathbf{C}_K$ is the estimated variance–covariance matrix of class K and $\bar{\mathbf{x}}_K$ the observed vector of means for class K.

This method requires the estimation of the variance–covariance matrix. For this reason enough data must be available and one estimates that the number of data should exceed the number of variables by a factor of 6.

6.2 SIMCA

The modelling properties of the principal components technique (see Chap. 21) have been used to develop a modelling technique called SIMCA (soft independent modelling of class analogy), which has been applied successfully to solve many chemical pattern recognition problems.

SIMCA considers each class separately, as does UNEQ. For each class separately a principal component analysis is performed which leads to a PC model for each class (so-called disjoint class models). Such a model is given by

$$x_{ij}^K = \bar{x}_{Kj} + \sum_{r=1}^{p_K} v_{jr}^K u_{ir}^K + e_{ij}^K \tag{19}$$

where

$\bar{x}_{Kj}$ = mean of variable j in class K

p_K = number of significant principal components in class K

v_{jr}^K = loading of variable j on component r in class K

u_{ir}^K = score of object i on component r in class K

This is equivalent to eqn. (25): the additional symbol K means that only the data of class K are considered to construct the model. As many models as there are classes are obtained.

It should be remembered from Chap. 21 that one determines first a new origin given by the means of each variable. Then, one determines the number of significant components and, finally, one estimates the loadings and the scores on the components. This is represented by Fig. 17 for 3 original variables for which 2 principal components were found to be necessary for a single class.

The plane defined by u_1 and u_2 describes the major variation of the data points of the class; u_{k1} and u_{k2} then define the projection of point k on this plane. One observes that point k itself is usually not found in the plane, but is somewhat distant from it. The distance e_{kj} between object k and the plane along variable j is, as explained in Chap. 21, the variation due to k and is unexplained by the modelling components. The larger e_{kj} therefore the larger the unexplained variation due to object k. If this variation becomes too large (in other words if k is too far

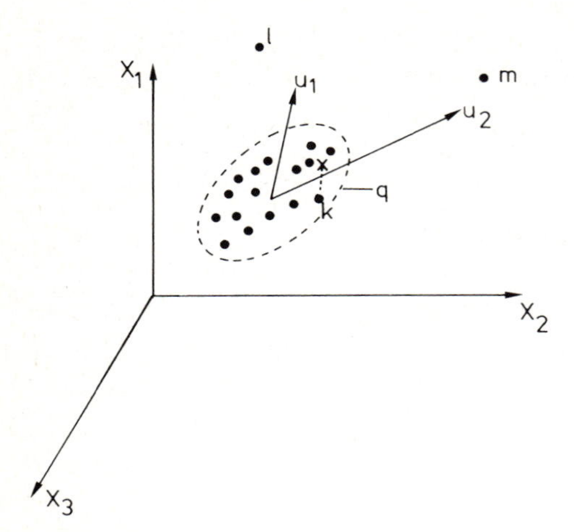

Fig. 17. SIMCA: principal components u_1 and u_2, model class K. k is a member of K; l and m are not.

away from the plane), this means that k is not well described by the PC model. This leads to the geometrical representation in Fig. 18. One defines planes a and b situated at a critical distance from the u_1–u_2 plane. A point k situated within the sandwich of planes a and b can be considered as well described by the u_1–u_2 plane and therefore as a member of the class considered, while points situated outside the sandwich are considered to be non-members. Thus nearly all points of Fig. 18 are considered to be part of the model with the exception of point l situated outside the a–b sandwich.

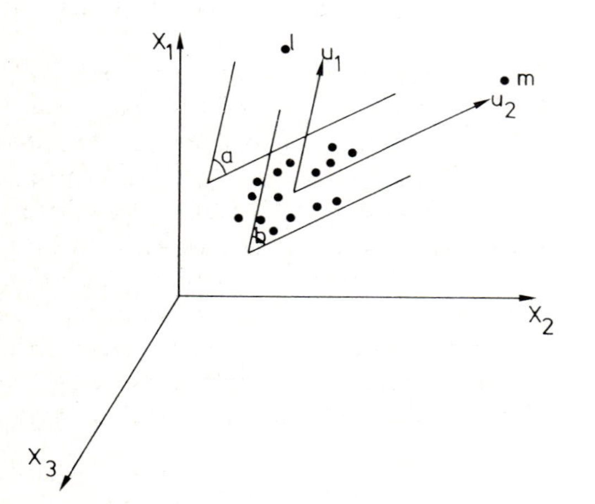

Fig. 18. SIMCA: the planes a and b, equidistant from u_1 and u_2 bound the space in which objects are considered to belong to K (see Fig. 17). l is now excluded.

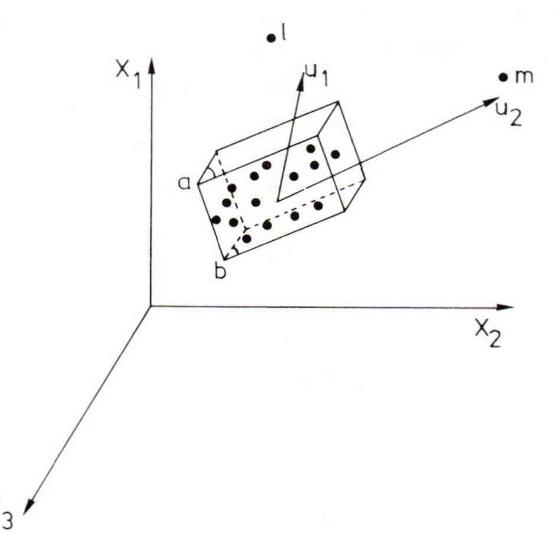

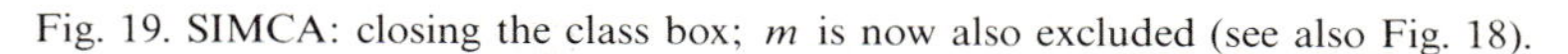

Fig. 19. SIMCA: closing the class box; m is now also excluded (see also Fig. 18).

There is, however, still a problem with object m. Although this is clearly not a member of the class it is situated within the a–b sandwich and therefore it would be considered as a member. This is avoided by closing the class model along the principal axes, leading to the class box of Fig. 19. The resulting box now delimits a portion of the 3-dimensional space which contains class K. If only one principal component, i.e. a line had been found sufficient to describe the data, the class box would have been a cylinder around this line. It is perfectly possible that in a more-class pattern recognition problem, different types of boxes should be used for different classes, since the models are disjoint and therefore the model of one class does not depend on that of another. Such a situation is depicted in Fig. 20. The form of the boxes should be compared with the rugby balls obtained by EQ and UNEQ.

Let us now reconsider the process of constructing the class box which we described intuitively with the aid of Figs. 17–19 in a more formal way.

Once the PC model of the classes K have been derived using training sets consisting of known n_K members, objects i belonging to the class are fitted to each of the disjoint models by means of multiple regression. This means that one considers the v_{jr}^K values as independent variables, $x_{ij}^K - \bar{x}_j^K$ as the dependent variable and determines the scores u_{ir}^K and residuals e_{ij}^K for object i.

Once the PC model of the classes K have been derived using training sets consisting of n_K known K-members, one derives the sandwiching planes. The distance at which these planes are situated depends clearly on the spread of the data around the plane. This is given by the standard deviation of s_0^K of the e_{kj}^K values for class K

$$s_0^K = \left[\sum_{j=1}^{m} \sum_{i=1}^{n_K} \left(e_{ij}^K \right)^2 / (n_K - p_K - 1)(m - p_K) \right]^{1/2} \tag{20}$$

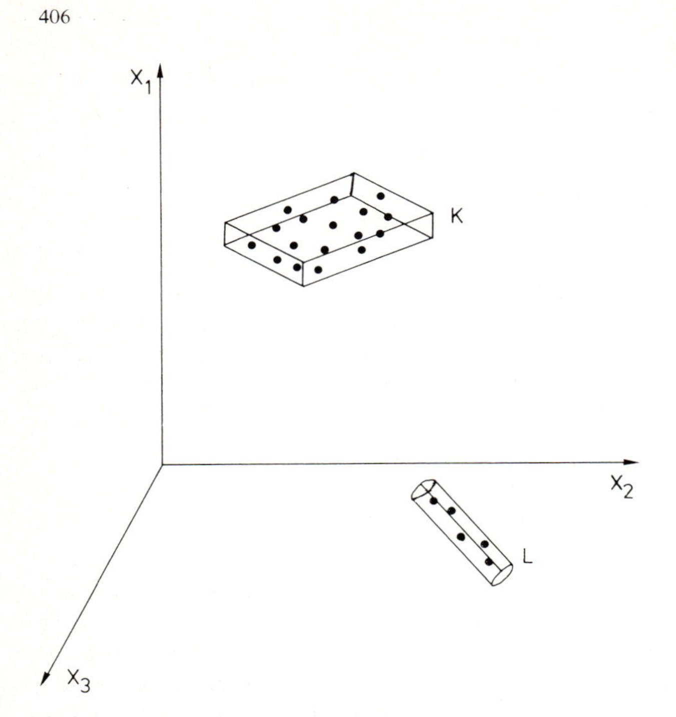

Fig. 20. SIMCA: a two-class classification. K is described by a two-component model and L by a one-component model.

where the numerator is the number of degrees of freedom. If one now wants to classify an unknown object k one first fits the object to each of the disjoint models of the K classes. This means that one determines the point of projection on the PC planes (or on the PC lines). Mathematically this is done by computing the u_{kr} coefficients and the e_{kj} residuals from multiple regression with $x_{ki} - \bar{x}_j^K$ as the dependent variable and the v_{jr}^K values as the independent ones. The e_{kj} variable determines the distance from the plane. By writing this as the standard deviation s_k^K for object k

$$s_k^K = \left[\sum_{j=1}^{m} \left(e_{kj}^K \right)^2 / (m - p_K) \right]^{1/2} \tag{21}$$

one obtains a quantity which can be compared to the variance of the class using the F-ratio

$$F = \frac{\left(s_k^K \right)^2}{\left(s_0^K \right)^2} \tag{22}$$

If the F-value is larger than the critical F-value at a given level of significance, one can conclude that the distance of k from the plane is significantly larger than that of the class K as a whole. One now defines

$$\left(s_{lim}^{K}\right)^{2}=\left(s_{0}^{K}\right)^{2}\cdot F^{0.05} \tag{23}$$

and decides that k is a member of K if

$$\left(s_{k}^{K}\right)^{2}<\left(s_{lim}^{K}\right)^{2} \tag{24}$$

One must also close the class model along the PC axes. Let us consider a 2-dimensional PC model. Closing the model along the PC axes means then deriving a normal region in a plane and using the u values. Since principal components, by definition are not correlated, each principal component may be considered separately. The reader is referred to the literature [5] for details on how to do this.

7. Feature selection

7.1 Univariate strategy

One general way of selecting features is to compare the means and the variances of the different variables before pattern recognition is applied. Intuitively, a variable for which the mean is the same for each class is of little use for discriminating the classes in question. In the same way, variables with widely different means for the classes and small intraclass variance should be of value and one therefore selects those variables for which the expression

$$\frac{\bar{x}_{Kj}-\bar{x}_{Lj}}{\sqrt{s_{Kj}^{2}+s_{Lj}^{2}}} \tag{25}$$

is highest. In this expression, derived for the case where the number of class members is the same in both classes, $\bar{x}_{Kj}$ is the mean of variable x_j for class K and s_{Kj} is the standard deviation of the same variable for this class. If the object of the selection is to find an optimal combination of variables for discrimination it should be noted that, as the correlation between variables is not taken into account, one selects in this way the best individual variables, but not necessarily the best combination of variables (see Chap. 9). The same is true when one uses a method called variance-weighting. This method permits weights to be given to the variables on the basis of their power to discriminate between the training sets. These weights are measures of the ratio of between-class variance to within-class variance for the learning groups.

For two classes, K and L, the weights are obtained by using the equation

$$w_{j}=\frac{\dfrac{n_{K}\cdot n_{L}}{n^{2}}\displaystyle\sum_{k=1}^{n_{K}}\sum_{l=1}^{n_{L}}\left(x_{kj}-x_{lj}\right)^{2}}{\dfrac{n_{K}}{n}\displaystyle\sum_{k=1}^{n_{K}}\sum_{k'=1}^{n_{K}}\left(x_{kj}-x_{k'j}\right)^{2}+\dfrac{n_{L}}{n}\displaystyle\sum_{l=1}^{n_{L}}\sum_{l'=1}^{n_{L}}\left(x_{lj}-x_{l'j}\right)^{2}} \tag{26}$$

References p. 412

where n_K and n_L are the number of individuals that are members of classes K and L $(n_K + n_L = n)$, x_{kj} is the concentration of the jth variable or the measurement value of this variable (k and k' are part of K, l and l' of L) and w_j is the weight of the jth variable.

7.2 Multivariate strategy

In pattern recognition based on discrimination one often prefers to select sets of variables on the basis of a stepwise procedure. In the so-called forward selection procedure one starts by selecting one variable according to a certain criterion, then one adds at each step the parameter that optimizes the criterion. Backward selection is also possible: in each step one eliminates the variables which have the least effect on the criterion and one proceeds in this way until a compromise between as small a number of variables as possible and a sufficiently good discrimination is obtained. What is meant by sufficiently depends on the object of the analytical study: how to make decisions is discussed in Chap. 26.

The selection of variables is best carried out with, as criterion, the classification result, for instance expressed as correct prediction percentage. This means that to select d variables out of m one evaluates the classification obtained with each such combination of variables. This may be very time consuming and is only feasible for forward selection where d is smaller than in backward procedures. In our experience it is used with advantage for methods such as ALLOC and KNN. For other methods such as SLDA of LLM, it is too time-consuming. One then often uses a statistical criterion such as the F-test. This is designed to test whether the additional variable has a significant influence on the discrimination.

In some instances sets of variables can be selected directly. If the variables have been standardized prior to SLDA, one can select those variables that have the largest weight coefficients. Another direct method is to determine the contribution percentage of each variable j to the total distance in the discriminant space between the centroids of classes K and L. The contribution of variable j is proportional to its weight w_j and to the difference of the values for the centroids of both classes, i.e. the difference between the mean values for classes K and L, $\bar{x}_{Kj} - \bar{x}_{Lj}$. The "discrimination power" of each variable may therefore be expressed as a percentage by

$$100 \frac{|w_j(\bar{x}_{Kj} - \bar{x}_{Lj})|}{\sum_{j=1}^{m} |w_j(\bar{x}_{Kj} - \bar{x}_{Lj})|} \qquad (27)$$

7.3 Mixed multivariate–univariate strategy

The multivariate strategy is to be preferred to the univariate one except when the number of variables is too large compared with the number of objects. This occurs frequently. Suppose one wants to classify wines on the basis of their capillary GLC chromatograms. At least 100 variables will then be obtained and it is improbable that more than a few tens of wines will be analyzed.

It was shown that if the number of dimensions, m, is relatively large compared with the sample size of the training set, n, non-significant classifications are obtained. It has been shown that it is possible to separate with 100% success two sets of numbers (sample size 25 for each set, number of dimensions 30) obtained from a random number generator and for which there was therefore no genuine reason for separability! In general, n/m should exceed 3 and in SIMCA it must be at least 2.

A feasible strategy when $m > 3n$ is to make a cluster analysis of the variables over all objects. One then looks for a grouping of the variables indicating some kind of similarity in their behaviour over the data set. Thus, one hopefully finds that the variables cluster into a "small" number of groups. One can then take one variable from each group.

7.4 Conclusions

Clearly, feature selection can have economical consequences since it eliminates the need for the determination of certain variables. In some instances, the number of features remaining is surprisingly low. In an application concerning air quality, 35 features could be reduced to 2 [6]. In such cases simple graphical procedures can then replace the rather complex methods described in this chapter for further classification.

An important aspect of feature selection is also that it is often found that a few irrelevant variables introduce so much noise that a good classification cannot be obtained. When these irrelevant variables are deleted, however, a clear and well separated class structure is often found. The deletion of irrelevant variables is therefore an important aim of feature selection (see also Chap. 20).

Some methods are more sensitive to the presence of noise than others. For instance when using ALLOC feature selection is highly recommended while it is of little use with SIMCA. On the other hand when a feature selection is considered necessary ALLOC often gives the best results, while SIMCA is not very easy to apply.

8. Recommended methods and strategy

This section discusses a recommended strategy which can be followed in almost every instance, at least if one has all the computer programs available. If one does not, other methods must of course be used. In that case, SIMCA and UNEQ may be replaced in order of preference by SLDA, KNN and LLM. The flow sheet of Fig. 21 summarizes the strategy to be recommended in most cases.

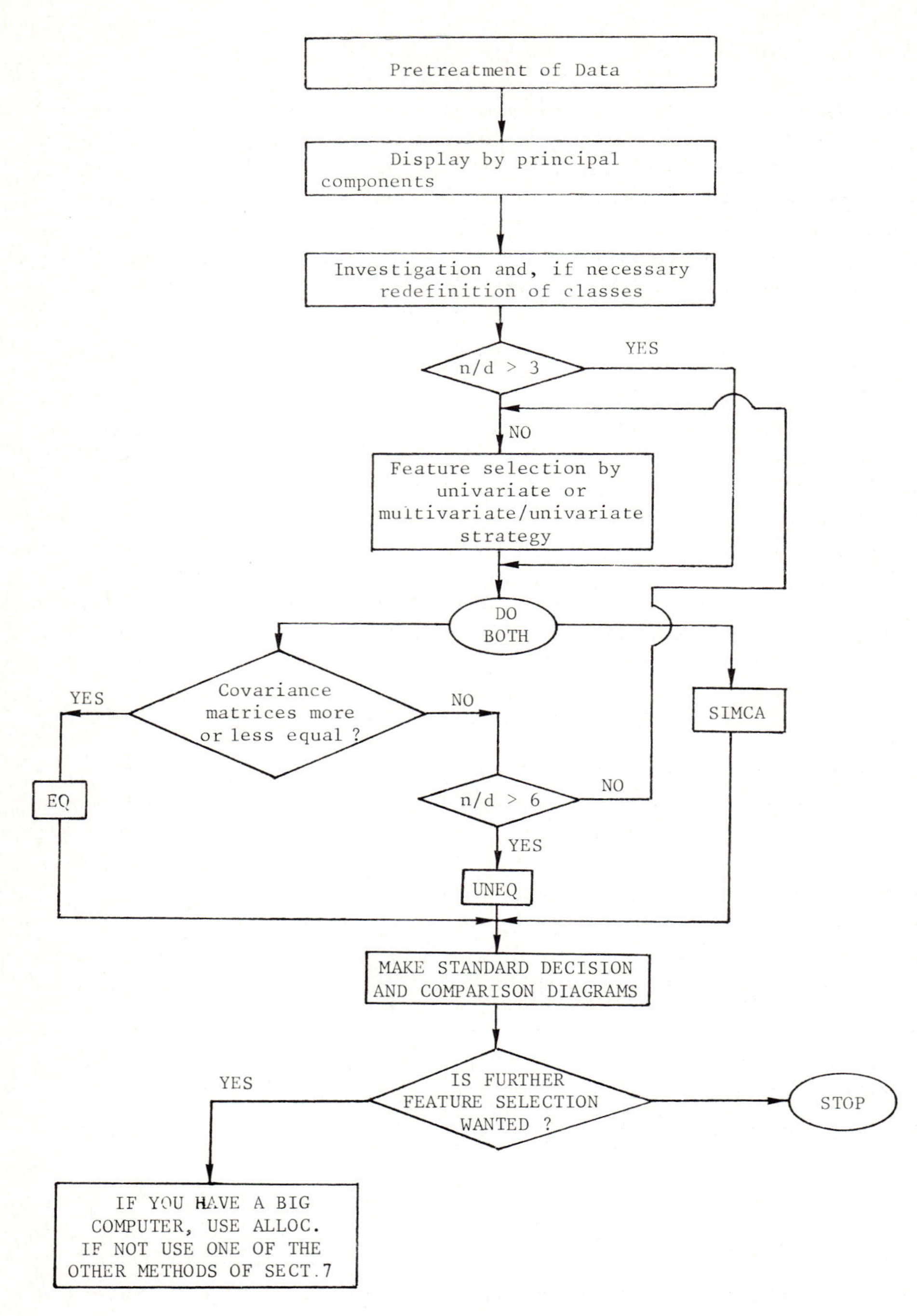

Fig. 21. Flow scheme to be followed for a new pattern recognition application.

8.1 *Pretreatment of the data*

One starts with a pretreatment of the data. It is preferable to work with data with a more or less normal distribution. To observe whether this distribution is present, one makes a histogram for each of the variables and, if this leads to a more normal distribution, one transforms the data, for instance by taking the log of the variables. This log transformation normalizes tailed distributions.

Scaling by the z-transform is usually also carried out to avoid scale effects of the units in which the variables were measured. When using SIMCA it is preferable to scale each class separately. UNEQ does not require scaling at all.

8.2 *Looking at the data*

One then proceeds by looking at the data, i.e. by making a principal components plot (see Chap. 21). This usually gives a good idea of how difficult the discrimination will be and consideration of the loadings of the principal component usually gives a good preliminary idea of the importance of the variables. It should be stressed here again that principal components does not have the purpose of discriminating between classes. Surprises cannot therefore be excluded. Principal components also shows whether certain methods may be used or not. For instance a compact class next to a disperse one excludes SLDA and necessitates a particular variant of KNN.

8.3 *Redefinition of classes*

In the next step one investigates the structure of the classes separately again using the principal components method. If the class is inhomogeneous and consists of subclasses, it is usually recommended that the subclass be used for the modelling methods. Outliers have an extreme influence on most methods and one must investigate whether they may be deleted. If possible, one should do so.

8.4 *Feature selection*

The next step depends on the number of objects and the number of variables. It is preferable that n(objects)/m(variables) should exceed 3 and for UNEQ it is even to be preferred that it should exceed 6. If these conditions are not met, eliminate the least promising variables using the univariate or mixed univariate/multivariate strategies of Sect. 7.

8.5 *Selection of method*

Since each pattern recognition method is subject to certain conditions and performs better in some situations than in others, one should always start out with more than one method. We recommend SLDA as a first approach, SIMCA and UNEQ or EQ for the final classification.

References p. 412

If the original problem was to discriminate between classes one may stop here. If feature selection was wanted one will proceed to do this, preferably with ALLOC.

References

1 D. Coomans, D.L. Massart and I. Broeckaert, Potential methods in pattern recognition. Part 4: A combination of ALLOC and statistical linear discriminant analysis, Anal. Chim. Acta, 133 (1981) 215.

2 D. Coomans and D.L. Massart, Potential methods in pattern recognition. Part 2: CLUPOT, an unsupervised pattern recognition technique, Anal. Chim. Acta, 133 (1981) 225.

3 J.D.F. Habbema, Some useful extensions of the standard model for probabilistic supervised pattern recognition, Anal. Chim. Acta, 150 (1982) 1.

4 M.P. Derde and D.L. Massart, UNEQ: a disjoint modelling technique for pattern recognition based on normal distribution, Anal. Chim. Acta, 184 (1986) 33.

5 S. Wold and M. Sjostrom, SIMCA: a method for analyzing chemical data in terms of similarity and analogy, in B.R. Kowalski (Ed.), Chemometrics, Theory and Application, ACS Symposium Series, No. 52, 1977.

6 J. Smeyers-Verbeke, J.C. Den Hartog, W.H. Dekker, D. Coomans, L. Buydens and D.L. Massart, The use of principal component analysis for the investigation of an organic air pollutants data set, Atmos. Environ., 18 (1984) 2741.

Recommended reading

General

L. Domokos, I. Frank, G. Matolcsy and G. Jalsovszky, Pattern recognition applied to vapour-phase infrared spectra, Anal. Chim. Acta, 154 (1983) 181.

A.M. Harper, D.L. Duewer and B.R. Kowalski, ARTHUR and experimental data analysis. The heuristic use of a polyalgorithm, in B.R. Kowalski (Ed.), Chemometrics: Theory and Applications, American Chemical Society, Washington, DC, 1977, p. 14.

J.F.K. Huber and G. Reich, Extraction of information on the chemical structure of monofunctional compounds from retention data in gas–liquid chromatography by pattern recognition methods, Anal. Chim. Acta, 122 (1980) 139.

P.C. Jurs and T.L. Isenhour, Chemical Applications of Pattern Recognition, Wiley, New York, 1975.

M. Sjostrom and B.R. Kowalski, A comparison of five pattern recognition methods based on the classification results from six real data bases, Anal. Chim. Acta, 112 (1979) 11.

Linear discriminant analysis and related methods

D. Coomans, D.L. Massart and L. Kaufman, Optimization by statistical linear discriminant analysis in analytical chemistry, Anal. Chim. Acta, 112 (1979) 97.

A.G. Jacobs and M.R. Holland, Discriminant functions for the differential diagnosis of hypercalcemia, Lancet, (ii) (1980) 322.

P.A. Lachenbruch, Discriminant Analysis, Haffner Press, New York, 1975.

H.E. Solberg, Discriminant analysis, Crit. Rev. Clin. Lab. Sci., 9 (1978) 209.

Linear learning machine

M. Bos, The learning machine in quantitative chemical analysis. 2. Potentiometric titrations of mixtures of three bases, Anal. Chim. Acta, 112 (1979) 65.

M. Ichise, H. Yamagishi and T. Kojima, Analog feedback linear learning machine applied to the peak height analysis in stair-case polarography, J. Electroanal. Chem., 113 (1980) 41.

N.J. Nilsson, Learning Machines, McGraw-Hill, New York, 1965.

Nearest neighbour methods

D. Coomans and D.L. Massart have published several articles in Anal. Chim. Acta, the first of which is D. Coomans and D.L. Massart, Alternative *k*-nearest neighbour rules in supervised pattern recognition, Anal. Chim. Acta, 138 (1982) 15.

W.A. Byers, B.S. Freiser and S.P. Perone, Structural and activity characterization of organic compounds by electroanalysis and pattern recognition, Anal. Chem., 55 (1983) 620.

Potential methods

D. Coomans and I. Broeckaert, Potential Pattern Recognition in Chemical and Medical Decision Making, Research Studies Press, Letchworth, 1986.

The same authors have published, with various co-authors, a series of articles in Anal. Chim. Acta, the first of which is D. Coomans, D.L. Massart, I. Broeckaert and A. Tassin, Potential methods in pattern recognition, Anal. Chim. Acta, 133 (1981) 215.

D.J. Hand, Kernel Discriminant Analysis, Research Studies Press, Chichester, 1982.

Modeling methods, multivariate distributions, and Mahalanobis distance

C. Albano, W.J. Dunn, III, U. Edlund, E. Johansson, B. Nordén, M. Sjöström and S. Wold, Four levels of pattern recognition, Anal. Chim. Acta, 103 (1978) 429.

C. Albano, G. Blomquist, W.J. Dunn, III, U. Edlund, B. Eliasson, E. Johansson, B. Norden, M. Sjöström, B. Söderström and S. Wold, Characterization and classification based on multivariate data analysis, in A. Varmavuori (Ed.), Proceedings 27th IUPAC, Pergamon Press, Oxford, 1980.

J.C. Boyd and D.A. Lacher, The multivariate reference range: an alternative interpretation of multi-test profiles, Clin. Chem., 28 (1982) 259.

W.J. Dunn, D.L. Stalling, T.R. Schwartz, J.W. Hogan, J.D. Petty, E. Johansson and S. Wold, Pattern recognition for classification and determination of polychlorinated biphenyls in environmental samples, Anal. Chem., 56 (1984) 1308.

B. Kagedal, A. Sandstrom and G. Tibbling, Determination of a trivariate reference region for free thyroxine index, free triiodothyronine index, and thyrotropin from results obtained in a health survey of middle-aged women, Clin. Chem., 24 (1978) 1744.

O.M. Kvalheim, K. Oygard and O. Grahl-Nielsen, SIMCA multivariate data analysis of blue mussel components in environmental pollution studies, Anal. Chim. Acta, 150 (1983) 145.

H.L. Mark and D. Tunnell, Qualitative near-infrared reflectance analysis using Mahalanobis distances, Anal. Chem., 56 (1985) 1449.

S. Wold, Pattern recognition by means of disjoint principal components models, Pattern Recognition, 8 (1976) 127.

Chapter 24

Decisions in the Analytical Laboratory

In the preceding chapters, we discussed how signals can be converted into analytical results and how analytical results can be combined into knowledge about the object or system that has been sampled. This knowledge is then used to solve analytical problems and to make decisions. Analytical chemistry helps to decide which actions should be taken. Analytical results from the clinical laboratory, for instance, are usually followed by a diagnosis of the disease, leading to a therapy. In industrial quality control, results will be of help in guaranteeing a certain product quality, or in deciding on actions to improve this quality. Until now, we have considered the analytical procedure as a separate system, but of course it shows interactions with its environment. A general model for these interactions is not available. Many analytical procedures are part of a control loop and thus help to regulate processes, whether these processes are a therapy of patients or quality control. The major part of analytical procedures is carried out in the analytical laboratory. An analytical recipe can be considered to be a subsystem of the analytical laboratory. Other subsystems are the instruments of different kinds, personnel with different tasks and skills, and the laboratory information management system. Between the procedures and people, several interactions (relationships) exist. For instance, a manual procedure does not produce information without a technician. The whole set of elements and relationships, i.e. the organization of men and machines, influences the performance of the laboratory to a large degree. Studies of this performance or attempts to optimize it, require the construction of a laboratory model, in particular a model that can predict the effect of a change in the organization on the production of knowledge. Experiments with a real laboratory are too costly and can lead to disappointments. Models comprising procedures and instruments with their characteristics have been recently published. These models are based on digital simulation, which is a discipline from Operations Research. In a paper on analytical laboratory organizational design Cook [1] used the following definition of an analytical laboratory organization: "An analytical chemistry laboratory organization is the rational coordination of the activities of a number of people for the achievement of some common explicit goals, through division of labour and function and through a hierarchy of authority and responsibility." It is evident that the optimal laboratory organization strongly depends on the environment of the laboratory (industry, government, university). There are as many different laboratory organizations as there are analytical laboratories, although it is possible to distinguish between some main types (routine, research, etc.).

Properties of the analytical procedure which define the amount of knowledge which can be acquired from an object or process are accuracy, precision, analysis time, and the sampling strategy. These features together define the quality of an analytical result, which determines the uncertainty left after analysis. In Chap. 9, we have seen that the information yield of an analysis is the difference between the uncertainty before and the uncertainty after analysis. Because complete certainty is not attainable, decision making is necessarily based on incomplete knowledge.

Two forms of decision making influence the laboratory operation: the decisions which follow from analytical results, and the decisions in the analytical laboratory itself, or laboratory strategies. A strategy is a plan to reach a defined goal in an optimal way. Laboratory strategies or decisions deal with

(a) decisions on sampling schemes for the description of an object or for the monitoring and control of a process,

(b) decisions on the method selection. This also includes decisions on combinations of analytical procedures, and

(c) decisions on the sample and information flow in the laboratory.

As said before, decision making has to be based on incomplete knowledge. Four levels of knowledge can be distinguished.

(1) A set of data without explicit indication of the relationships between the data.

(2) A statistical or other reduction of the data into characteristic figures: mean value, standard deviation, coefficient of correlation, peak positions, etc.

(3) A descriptive model which describes the relationships between the characteristic figures, or blocks of figures.

(4) A predictive model which forecasts the effect of decisions on the value of the characteristic numbers.

Of course, levels (3) and (4) are the most advanced tools for decision making because they provide direct information on the consequences of alternative decisions. We will now have a closer look at the two types of decision made in an analytical laboratory.

1. Decisions based on analytical results

In general, analytical results are used to describe an object, or to monitor or control a process. Sampling and analysis of heterogeneous (bulk) material must lead to a representative value of the average composition of the material or lot. In some other instances, decisions have to be made about individuals. This is a diagnostic problem, which occurs, among others, in clinical chemistry, environmental chemistry, and food chemistry. The judgement on whether or not a certain food sample contains a dangerous amount of mycotoxins can be considered as a diagnostic problem. The value of a test with a certain precision for diagnostic purposes depends to a large extent on the variability among the individuals investigated. It is intuitively clear that tests with a very low precision will contribute considerably to the observed variability. They will therefore lead to a high uncertainty in the diagnosis. On the other hand, it is also evident that when a very precise method is used, a further increase in precision will hardly contribute to the reduction of the uncertainty in the diagnosis. Therefore, a compromise has to be found between the cost of analytical precision and the return obtained by diminishing the uncertainty about the analyzed object. A formal analysis of the effect of the analytical precision on the diagnostic value of a test can be carried out using a statistical evaluation, known as Bayes' theorem, which is explained in Chap. 26.

When the object to be analyzed is a process, a stream of material with properties varying in time, the sampling strategy together with the analytical parameters (precision, analysis time) should be derived from the dynamic behaviour of the process parameters. In the case of process control, process values have to be forecast ahead of the last analytical result until the next analytical result is received. Here, the autocorrelation function (see Chap. 14) is an optimal predictor of future process values when based on the last-obtained analytical result. An extrapolation is therefore made from the time of the last analytical result to the next moment an analytical result will be obtained. In Chap. 27, a quantitative decision model is derived which relates the sampling scheme and the analytical parameters to the quality of the attainable control. Of course, the more samples are analyzed, the shorter the analysis time is and the better the precision is, the better becomes the quality of the control loop. Again, cost–benefit considerations must decide about the analytical conditions. There is a point at which the additional analysis costs caused by a better quality of the analytical result are no longer justified by the additional returns, obtained by a better quality.

2. Decisions about the analytical process

In the case of process control, the analysis time is an important performance parameter. When the analysis is carried out in a laboratory, the analysis time is the sum of the times required for the analysis of the sample, transportation of the sample and the results, and delays. At higher sampling rates, the utilization factor of the available analysis capacity increases. As a result, the delays may also increase and because of these longer delays, the quality of process control may be worse. Thus, the negative effect of longer delays may override the positive effect of the higher sampling rate. In many instances, the relationships between the performance characteristics (e.g. delays) and the corresponding characteristics of the laboratory are complex or even obscure. The performance characteristics of a laboratory are strongly influenced by the organization of the laboratory. It is evident that decisions on analytical laboratories, which can be considered to be men–machine systems, are best based on a descriptive or predictive model [levels (3) and (4) of knowledge] of the system. These models are derived from applied mathematics and typical problems such as allocation, inventory, queueing, etc. are studied in a science called operations research. Applications of operations research in analytical chemistry are discussed in Chap. 25. The behaviour of queues of samples and analytical results can be approximated by models from queueing theory. More exact models with a higher predictive accuracy and precision can be derived by the application of digital simulation. Simulation is based on a simulation model of the laboratory. Such a model consists of classes of objects with given characteristics, e.g. instruments,

Reference p. 418

analytical procedures, and classes of strategies, e.g. priorities, scheduling of samples in batches, etc. When the characteristics (e.g. analysis times, arrival times, etc.) of the objects and the strategies in the simulation model are sufficiently close to those of the laboratory, one can calculate delays, utilization factors of equipment, and queue sizes, which are close to the actual ones. An example is given in Chap. 25. The application of operations research in analytical chemistry is not limited to the study of laboratory organizations, but can also be used to optimize strategies for carrying out combinations of separate analytical procedures. A preparative chromatographic separation scheme for multicomponent samples, for instance, can be executed along various pathways. The end of a pathway, or a node, represents a situation in which some of the compounds remain on the column and others are eluted. A system of nodes and pathways constitutes a network or graph. One of the purposes of studying graphs is to find the shortest path between the initial situation, having no compounds separated, and the final situation in which all compounds are separated. Graphs can be studied by graph theory, which is a mathematical method from operations research. In Chap. 25, graph theory is applied to optimize the ion-exchange separation of samples containing several ions. Graph theory can also be applied to detect clusters in unsupervised pattern recognition (Chap. 25).

Reference

1 C.F. Cook, Analytical laboratory organizational design in a changing environment, Anal. Chem., 48 (1976) 724A.

Chapter 25

Operations Research

Ackoff and Sasieni [1] defined operations research (OR) as "the application of scientific method by interdisciplinary teams to problems involving the control of organized (man–machine) systems so as to provide solutions which best serve the purposes of the organization as a whole".

The term "organization" is important in this context. As stated by Goulden [2] in his article entitled "Management studies and techniques for application in analytical research, development and service", there are three essential components of much human endeavour: the work to be undertaken, the organization necessary to carry out that work, and the people by whom the work will be done. Analytical chemists tend to pay more attention to the work to be accomplished (and the tools with which to do it) than to the other two components. This becomes evident when one considers the clinical laboratory. Thousands of articles have been published on how to determine a biochemical parameter in an efficient way, but only a few on how to design an optimal configuration for a clinical laboratory!

It is a characteristic of organizations that they are complex systems and the optimization therefore usually consists of a comparison of many different possibilities. OR techniques are used to find the optimal solution for problems in which many combinations are possible. This is a very common situation in analytical chemistry and for this reason OR techniques should be of considerable value in this field.

Many of the problems discussed are problems of organization in the true sense. The queueing example given in Sect. 1 concerns the organization of a laboratory. Other examples, such as the shortest path application, concern the development of methods. In this case, there is, however, always an organizational analogue.

It may appear surprising to find several chapters devoted to techniques from the management or economic sciences in a book such as this. These techniques are, however, certainly relevant. The manager's job is to make the best possible use of the resources at his disposal to achieve a certain goal (usually commercial). This formula, however, describes equally well the task of most analytical chemists, even of those who are involved only with research. Hence, it is reasonable that the techniques used by modern managers to help them in their decisions will also be useful to analytical chemists.

It is very important to note here that we have written "to help them with their decisions" and not "to make their decisions". Although OR methods are mathematical methods, they rarely offer exact and ready-made solutions for real-life problems. OR methods use models which can rarely be sufficiently precise to cover all relevant factors. Therefore, the solutions obtained should be understood more as

References p. 434

a guide for evaluating solutions. This difficulty in applying OR results is not restricted to analytical chemistry but is also encountered in more classical applications. For instance, the optimal solution of a job allocation problem in industry may be rejected by management on the grounds of possible difficulties with trade unions.

As indicated above, OR consists of a collection of mathematical techniques. Some of these are linear programming, integer programming, queueing theory, dynamic programming, graph theory, game theory, and simulation. The prototype problems that can be solved are [1]

(1) allocation,
(2) inventory,
(3) replacement,
(4) queueing,
(5) sequencing and coordination,
(6) routing,
(7) competition, and
(8) search.

Several, but not all, of these prototype problems have been applied in analytical chemistry. In this book, we have grouped the applications into three categories, namely a queueing problem, a routing problem (shortest path) and an application which has been discussed earlier but is really based on OR models. We have added to it a section on simulation.

1. A queueing problem

The time between the arrival of a sample and the communication of a result may be a performance characteristic for an analytical laboratory. In practice, one often finds that a large part of the time is spent waiting. In fact, there are two waiting times that are of importance to a laboratory, viz.

(a) the time a sample is waiting because the apparatus or personnel are occupied with previous samples and

(b) the time that apparatus or personnel wait because no samples are available.

If these waiting times are too long, there is clearly insufficient agreement between the proposed work load and the analysis capacity and it is therefore obviously of interest to investigate the relationships between these two quantities. This is done with the use of queueing theory.

The waiting time for a determination depends on

(a) the mean rate of arrival of samples, λ;

(b) the mean rate of analysis or, to use queueing theory language, the mean service rate per channel, μ. $\mu = 1/\bar{t}_a$, where $\bar{t}_a$ is the average service time; and

(c) the number, m, of pieces of apparatus (or technicians) that are available to carry out the analysis; in queueing theory language, the number of channels.

The objective of the queueing analysis is to determine parameters such as $\bar{w}$, the mean waiting time before commencement of the actual determination (queueing theory language: mean waiting time in the queue), or $\bar{n}_q$, the mean queue length.

The analysis is primarily of interest when $\lambda/m < \mu$, i.e. when the mean number of samples submitted for analysis is smaller than the analysis capacity. If this is not the case, the queue will grow indefinitely.

The quantity $\lambda/m\mu = \rho$ is called the traffic intensity or utilization factor and plays an important role in computations, as will be shown later. When $\rho < 1$, queueing analysis can be carried out and it has been shown that, in this instance, a steady state is reached. This means that, after a certain initial time needed to establish the steady state, a situation is reached in which one is able to predict the values of the parameters $\bar{t}_a$ and $\bar{n}_q$. For the calculation of $\bar{t}_a$ and $\bar{n}_q$, certain assumptions about the distributions of inter-arrival time and service time are necessary. Queues are generally described with a shorthand notation such as A/B/m, where A and B represent the distributions for inter-arrival time and service time and m is the number of channels. For example, in the M/M/1 system, both the inter-arrival time and the service time are exponentially distributed, which in this field is denoted by M, and there is only one service channel. This system is the simplest possible.

The following assumptions are made concerning the distribution of the arrival rate and the service times.

(a) The number of arrivals during a given time interval has a Poisson distribution. This hypothesis implies that the probability of n arrivals in an interval $(0, t)$ is given by

$$P_n(t) = \frac{e^{-\lambda t}(\lambda t)^n}{n!} \tag{1}$$

where λt is the average number of arrivals during this interval.

The result for the probability of n arrivals in a Poisson distribution can be obtained from the three basic assumptions of this distribution.

(i) The probability of a single arrival $P_1(\Delta t)$ during a short time interval, Δt, is proportional to the length of the interval. It is given by $\lambda \Delta t + o(\Delta t)$, where λ is the parameter of the Poisson distribution and $o(\Delta t)$ is a small value which becomes negligible for small Δt. In terms of the probability, $P_1(t)$, it can be seen that eqn. (1) implies

$$\lim_{\Delta t \to 0} \frac{P_1(\Delta t)}{\Delta t} = \lambda \tag{2}$$

(ii) The probability of more than one arrival during the small interval Δt is negligible for small Δt. As a result of these two assumptions

$$\lim_{\Delta t \to 0} \left[P_1(\Delta t) + P_0(\Delta t) \right] = 1 \tag{3}$$

(iii) The numbers of arrivals during non-overlapping time intervals are statistically independent.

(b) The service time has an exponential distribution. This hypothesis means that the probability that the service time equals t is given by $\mu\, e^{-\mu t}$, where μ is the parameter of the exponential distribution. Further, it is assumed that service times and times between successive arrivals are independent.

To describe the state of the system at a given time t, we shall use the following concepts.

The number of elements present in the system at an instant t is called $N(t)$. The study of queueing systems is mainly concerned with the behaviour of the system in a state of equilibrium. At this point, the probabilities P_n do not depend on t. In the single-server queue, the condition for reaching such a state is given by

$$\rho = \frac{\lambda}{\mu} < 1 \tag{4}$$

The main results which can be obtained from these assumptions concern a number of parameters which describe the expected way the system will behave. For the simple model described above, we shall give formulas for the following parameters.

P_n = probability for n samples to be present in the system;
$\bar{n}$ = mean number of samples present in the system;
$\bar{n}_q$ = mean number of samples present in the queue;
$\bar{w}$ = mean waiting time; and
$P(W \leqslant A)$ = probability that the waiting time does not exceed A.

$$P_n = \rho^n(1-\rho) \qquad n = 0, 1, 2, \ldots \tag{5a}$$

$$\bar{n} = \frac{\rho}{1-\rho} \qquad \bar{n}_q = \frac{\rho^2}{1-\rho} \qquad \bar{w} = \frac{\rho}{\mu(1-\rho)} \tag{5b}$$

$$P(W \leqslant A) = 1 - \rho\, e^{-\mu A(1-\rho)} \tag{5c}$$

A laboratory usually consists of several service posts. For example, the samples have to be centrifuged (service post 1) before being distributed over several pieces of apparatus for determination of the concentration (service posts 2, ..., m). It can be shown that if a given service point receives samples from various sources (i), each with a Poisson input rate λ_i, in general the total input will not be a Poisson process, but this node still behaves as if it is an M/M/n system with input rate $\Sigma\lambda_i$. Therefore, many systems can be investigated using introductory queueing theory.

It remains to be shown that the rate of arrival of samples in an analytical laboratory follows a Poisson distribution. It has been shown that this is the case in at least some laboratories. Figure 1 gives the distribution of arrival rates in a laboratory for structural analysis. The actual distributions can be represented by Poisson distributions.

Therefore, one may expect that some analytical laboratories can be studied with queueing theory. For some other laboratories, this is not the case. The sample input of clinical and some industrial control laboratories is time-dependent. In the early morning, the laboratory is almost empty and by the evening all samples received are completed. In terms of queueing theory, such laboratories never reach a "steady state". The solution of non-stationary queues requires complex mathematics and, to

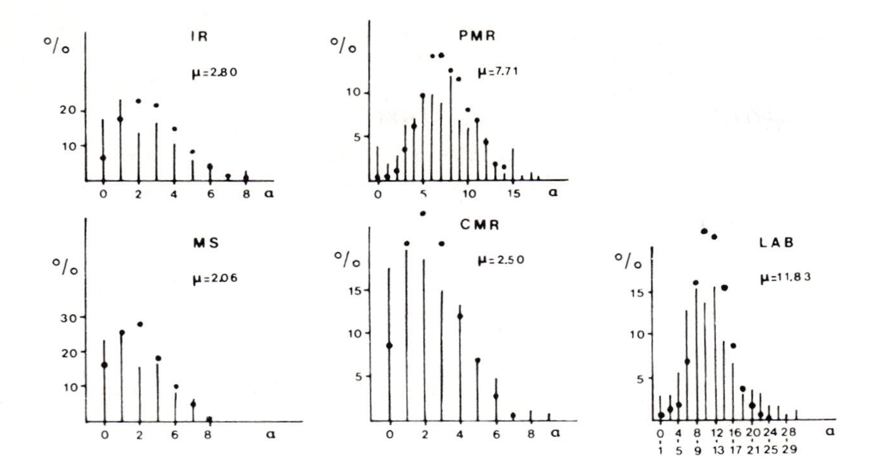

Fig. 1. Histograms of the number of samples per day arriving at the various sections of a laboratory for structural analysis. ●, Fitted Poisson distribution.

our knowledge, in such laboratories only simulation (see Sect. 4) has been applied. However, in such laboratories, the waiting time usually is not important, provided that the analysis is carried out on the day of arrival of the sample: the waiting time is not a performance characteristic.

Other laboratories, for instance analytical departments in research laboratories, receive a more or less continuous flow of samples. This flow can be described in terms of statistical parameters such as the mean and the variance of the inter-arrival time of the samples. If these parameters, together with the parameters describing the statistical behaviour of the analysis time, remain constant for a sufficiently long period, a steady state will be observed, allowing the application of queueing theory. However, this is only true for fairly simple systems and it is obvious that such complex systems as real analytical laboratories cannot be described easily by the rather simple models on which queueing theory is based. In that case, digital simulation can be considered as an alternative method for handling more complex models. Nevertheless, queueing theory provides some interesting conclusions about delay times in analytical laboratories and some examples of general interest will be discussed.

(a) Fluctuations of the analysis time

From eqn. (5c), one concludes that, for exponentially distributed inter-arrival times of the samples and analysis times, the waiting time shows an exponential distribution. This means that the waiting time for the results of some samples is much longer than the average waiting time. The statistical nature of both times causes the delay of some samples to be much longer than the average.

A good way to control the average delay is to control the probability distribution of the analysis time. It was demonstrated that systems without statistical fluctuations of the analysis time show half the waiting time of an M/M/1 system. The magnitude of the fluctuations of the analysis time can be expressed as the coefficient of variation, C, which is the ratio of the standard deviation and the mean of the probability distribution function. The influence of this coefficient on the waiting time is given by the Pollaczek–Khinchin mean-value equations

$$\frac{\overline{w}}{\overline{t}_{\mathrm{a}}} = \frac{\rho(1 + C_{\mathrm{b}}^2)}{2(1-\rho)} \tag{6}$$

$$\frac{\overline{T}}{\overline{t}_{\mathrm{a}}} = 1 + \frac{\rho(1 + C_{\mathrm{b}}^2)}{2(1-\rho)} \tag{7}$$

where $\overline{T}$ is the delay time ($\overline{T} = \overline{w} + t_{\mathrm{a}}$).

Statistical fluctuations in the analysis time can be eliminated or reduced by standardizing the manipulations or by introducing automated procedures. The effect of doing this can be predicted by using eqns. (6) and (7).

(b) Influence of the analysis time

It is obvious from eqn. (5b) that the utilization factor has an important effect on the mean delay in the laboratory. As ρ approaches unity, the average time in the laboratory grows in an unbounded fashion. In some laboratories, it is common practice to analyse all samples twice in order to detect analytical errors. Assuming that a duplicate analysis requires double the time of a single analysis, a considerable decrease in the average delay is observed if the second analysis is omitted. For example, for a system with $\rho = 0.9$ and an average analysis time of 0.5 h, the average delay time decreases from 10 to 0.9 h. The analyst should then decide whether the decrease in the average delay is worth the increased probability of delivering faulty results.

(c) The number of channels of the apparatus

Consider an instrument with m service channels. When the instrument becomes available, it will accept m samples from the queue and analyse them simultaneously. If the analyst finds less than m samples in the queue, then this number of samples will be analysed. For this particular instance, the change in delay time on increasing the number of channels or the capacity of the instrument, assuming a constant analysis time, can be calculated. Figure 2 demonstrates the effect of increasing the instrument capacity for a system with $\rho = 0.9$.

(d) The number of analysts

Suppose that two analysts perform the same analytical procedure. From the equations of an M/M/m system, the effect of admitting a third analyst can be

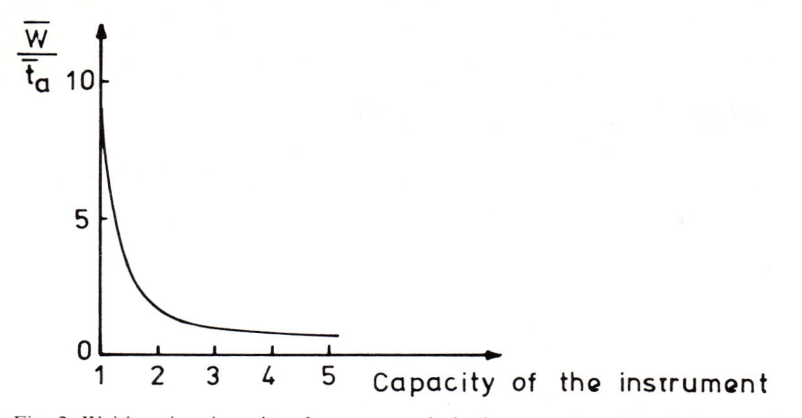

Fig. 2. Waiting time in units of average analysis time as a function of the capacity of the instrument ($\rho = 0.9$).

calculated as a function of ρ. From Fig. 3, one can see that for $m\rho = 1.5$, that is both analysts are 75% employed, the admittance of a third analyst reduces the delay time to 50% of the initial value.

(e) Priorities

The samples submitted to a laboratory do not necessarily have the same priority. For example, urgent samples in a clinical laboratory are positioned at the top of the queues and are analysed before samples of lower priority, irrespective of their delay time.

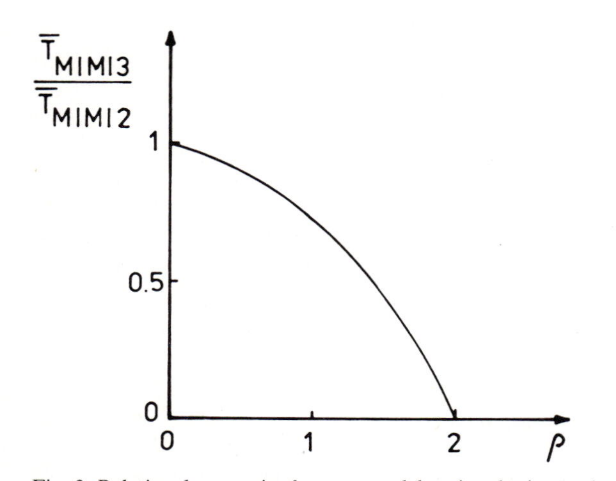

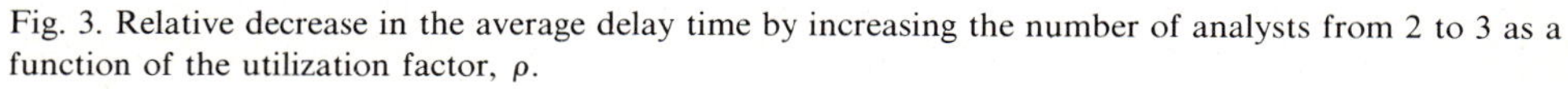

Fig. 3. Relative decrease in the average delay time by increasing the number of analysts from 2 to 3 as a function of the utilization factor, ρ.

References p. 434

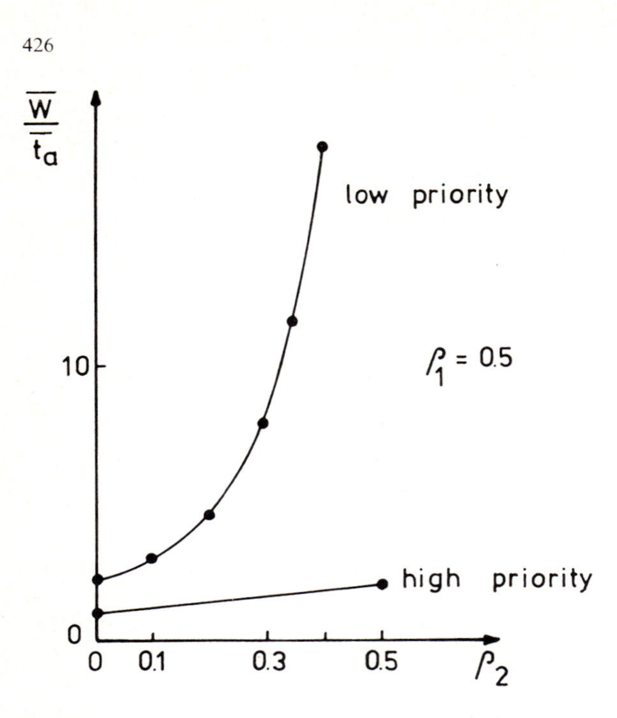

Fig. 4. Waiting time in units of average analysis time for the two priority classes under absolute priority discipline as a function of the utilization factor, ρ_2, for the samples of low priority. The utilization factor, ρ_1, for the samples of high priority is kept constant at 0.5.

For equal average analysis times of both priority classes, the total average delay time of the samples is not affected by a priority rule, but a great difference in average delay may be observed between the classes. It can be shown that, in an M/M/1 system, the delay time of the high- and low-priority classes, the high priority class having absolute priority over the low priority class, are $(\rho_1\bar{t}_{a1} + \rho_2\bar{t}_{a2})/(1-\rho_1)$ and $(\rho_1\bar{t}_{a1} + \rho_2)/(1-\rho_1)(1-\rho)$, respectively, where subscript 1 indicates the high priority class and $\rho = \rho_1 + \rho_2$. From Fig. 4, it can easily be seen that an increased delivery of samples of the low-priority class has a very small effect on the delay time of the high-priority samples. It can also be shown that an increased delivery of high-priority samples affects the delay time of both classes.

All of the examples presented above are calculated for open systems. These are systems for which the inter-arrival times of the samples do not depend on the delay times. However, analytical laboratories often interact with their environment and form a closed system with it. When the investigator receives the analytical results, he starts new experiments and sends new samples to the analytical laboratory. The time lag between two samples (or sample series) consists of the delay time of the result and the time needed to react on the result. Decreasing the delay time of the result in a closed system diminishes the average inter-arrival time of the samples. As a consequence, the expected reduction in the delay times is not obtained. However, the throughput of the laboratory is increased.

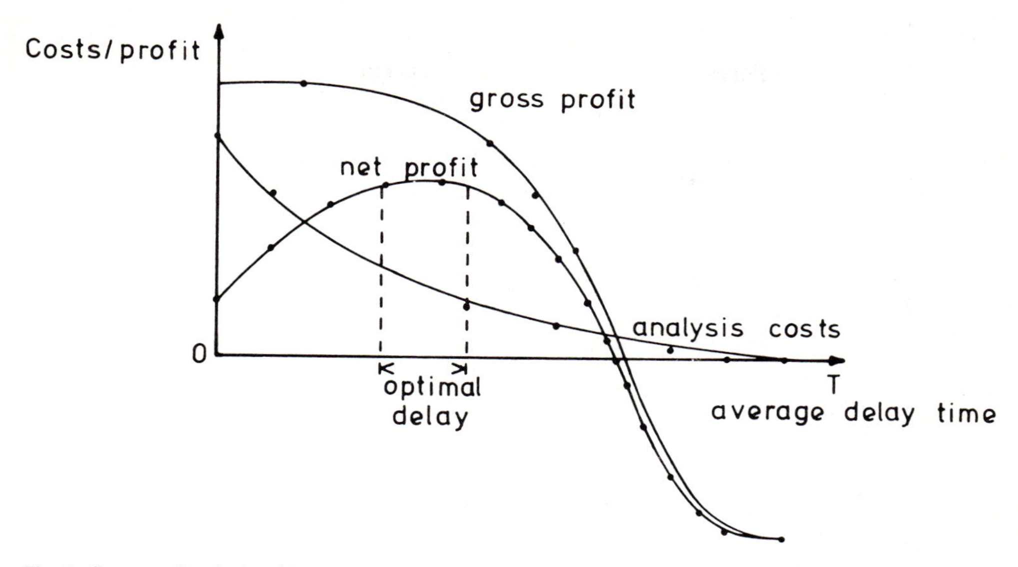

Fig. 5. Cost–profit relationships.

Delay times can be considered to be performance characteristics of the analytical laboratory. However, this criterion is interrelated with another criterion, namely cost. Enhancement of the equipment (technicians and instruments) decreases the delay times of the samples. As a result, the operation of the laboratory is more expensive, but the gross profit increases. A maximal net profit is then obtained for some mean delay time (Fig. 5).

2. A shortest path problem

A network or graph consists of a set of points (nodes) connected by lines (edges or links). These links can be one-way (one can go from point A to B, but not vice versa) or two-way. When the edges are characterized by values, it is called a weighted graph. In the usual economic problems for which one applies networks, these values are cost, time, or distance.

In this section, a routing problem is considered in which one must go from one node (the origin) to another (the terminal node). There are many ways by which this is possible and the routing problem consists in finding a path that minimizes the sum of the values of the edges that constitute the path. This is easily understood if one supposes that the nodes are towns and the values of the edges between neighbouring towns are the distances between the towns. The routing problem then consists in choosing the shortest way of going from one town to another. This type

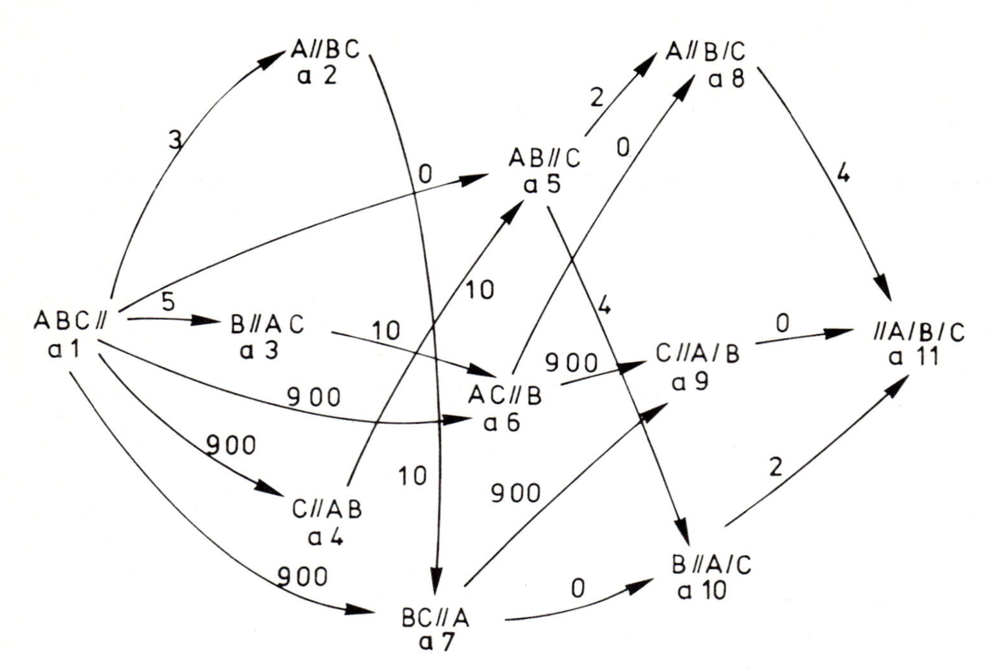

Fig. 6. Network describing the separation of three ions, A, B, and C.

of problem is called a shortest or minimal path problem. It is applied here to the optimization of chromatographic separation schemes for multicomponent samples and more particularly to the ion-exchange separation of samples containing several different ions.

In this type of application, one usually employs more or less rapid and clear-cut separation steps. This can be explained best by considering the simplest possible case, namely the separation of three ions, A, B, and C. The original situation is that the three ions have been brought together (not separated) on a chromatographic column and the final situation should be that they are eluted and separated from each other. These two situations constitute the initial and the terminal nodes of the network. They are denoted by ABC// and //A/B/C. The elements which remain on the column are given to the left of symbol // and the symbol / means that the ions to the left and to the right of it are separated. There are many ways in which one can go from situation ABC// to situation //A/B/C, as shown in the network in Fig. 6.

Step 1. There are two possibilities, viz.

(a) one can elute one element and retain the other two on the column. This leads to nodes AB//C, AC//B, BC//A; or

(b) one can elute two elements and retain the other. This leads to nodes A//BC, B//AC and C//AB.

Step 2.

(a) Following step 1a, one elutes one of the two remaining ions. For example, if in the first step A was eluted, one now elutes B or C. In step 2a one can reach the situations A//B/C, B//A/C or C//A/B.

(b) Following step 1b, two ions are eluted together and are therefore not separated; they have to be adsorbed first on another column. In the meantime, one can elute the single ion left on the first column. One then has two elements adsorbed on a column and one eluted. The different possibilities are AB//C, AC//B, BC//A, i.e. the situations reached also after step 1a. From there one proceeds to step 2a.

Step 3.

Step 3 follows step 2a. Only one ion remains on the column. It is now eluted, so that the terminal node is reached.

These different possibilities and their relationships can be depicted as a directed graph (Fig. 6). Directed graphs are graphs in which each edge has a specific direction. To find the shortest path one has to give values to the edges of the graph. As the problem is to find the procedure that permits one to carry out the separation in the shortest time possible, these values should be the times necessary to carry out the steps symbolized by the links in the graph or a value proportional to the time. We shall not detail the manner in which these times were derived. Essentially, there are three possibilities.

(a) If the separation depicted by a particular link is possible, the time is considered to be equal to the distribution coefficient of the ion which is the slowest to be eluted in this step (the distribution coefficient as defined in ion exchange chromatography is proportional to the elution time). There is a very large literature on distribution coefficients, particularly for metal ions (it is probable that at least 1000 such coefficients can be found for each ion) and a computer program was used to select the best eluting agent (from nearly 400 possible substances) for each separation step depicted by a link.

(b) If the separation depicted by a particular link is not possible, one gives a very high value to that link.

(c) If the link contains a transfer from one column to another, a value corresponding to the estimated time needed for this transfer is given.

One may question whether the application of graph theory is really necessary as no doubt a separation such as that described above can be investigated easily without it. However, when more ions are to be separated, the number of nodes grows very rapidly.

The calculation of the number of nodes is rather complicated. It is simpler in the special case where transfers from one column to another are not allowed (all ions eluted from the column must be completely separated from all of the other ions). In this instance, and for three elements, the following nodes should be considered: ABC//, A//B/C, AB//C, AC//B, BC//A, B//A/C, C//A/B, and //A/B/C. If one considers only the stationary phase (to the left of //), one notes that all the combinations of zero, one, two, and three ions out of three are present.

Calling n the total number of ions and p $(0 \leqslant p \leqslant n)$ the number of ions in a particular combination taken from these n, the total number of combinations is equal to $\sum_{p=0}^{n} C_n^p$, where C_n^p is the symbol used for the number of combinations of p elements out of a set of n. This can be shown to be equal to 2^n. In this particular instance, a separation scheme for eight elements would contain 256 nodes. For the general case (transfers allowed), no less than 17008 nodes would have to be considered! Even for 4 elements, 38 nodes are obtained and it begins to be difficult to consider all of the possibilities without using graph theory. The shortest (cheapest) path in a graph can be found, for example, with a very simple algorithm due to Ford [3]. Let us suppose that one has to construct a highway from town a_1 to a town a_{11}. There are several possible layouts, which are determined by the towns through which one must pass, and these must be selected from a_2–a_{10}. The values of the links in the resulting graph are given by the estimated costs. The problem is, of course, to find the cheapest route. A value $A_1 = 0$ is assigned to town (node) a_1 and the value of all the nodes a_n directly linked to a_1 is computed by using the equation $A_n = A_1 + l(a_1, a_n)$, where $l(a_1, a_n)$ is the length of edge (a_1, a_n). In this way, one assigns the values 3, 5, 900, 0, 900, and 900 to the nodes a_2, a_3, a_4, a_5, a_6 and a_7, respectively. One repeats this procedure for the nodes a_m linked directly to one of the nodes a_n by using the equation $A_m = A_n + l(a_n, a_m)$. One continues to do this until a value has been assigned to each of the nodes in the graph. One would, for example, assign the values 15, 1800, 900, and 902 to a_8, a_9, a_{10} and a_{11}, respectively.

In the first stage, one has assigned a possible value to all of the nodes, but not necessarily the lowest possible value. For example, the value A_7 of node a_7 is now 900. The value 900 is artificially high, indicating that for practical reasons it is impossible to go from a_1 to a_7. In the highway example, this could mean a mountain ridge and in the ion-exchange case, it could be a separation that cannot be carried out in a reasonable time. This value is derived here from edge (a_1, a_7) using the equation $A_7 = A_1 + l(a_1, a_7)$. Town a_7, however, can also be reached from town a_2. The value of A_7 is then given by $A_2 + l(a_2, a_7)$ and is equal to 13; this replaces the original value 900. In this way, all of the nodes are checked until one is satisfied that each town is reached in the cheapest possible way. The optimal path is then found by retracing the steps that led to the final value for A_{11}. In the graph in Fig. 6, this is the path a_1, a_5, a_8, a_{11} with a total value of 6.

The graph in Fig. 6 is the graph obtained for the separation of Ca, Co, and Th on a cation-exchange column. One arrives at this graph by replacing nodes a_1–a_{11} by CaThCo//, Ca//Th/Co, Th//Ca/Co, Co//Ca/Th, Ca Th//Co, Ca Co//Th, Th Co//Ca, Ca//Th/Co, Co//Ca/Th, Th//Ca/Co, and //Ca/Th/Co, respectively. The weights of the edges are distribution coefficients obtained from the literature as explained above. The conclusion in this particular instance is that one must first reach situation a_5, i.e. Ca Th//Co, meaning that one must first elute Co. As the weight of the edge is 0, this means that at least one solvent has been described in the literature that permits one to elute Co with a distribution coefficient of 0 without eluting Ca and Th. The following steps are the elution of Th (distribution coefficient = 2) and Ca (distribution coefficient = 4).

3. Operations research models used in clustering

Operations research models can be used in many different fields of chemometrics. One field in which there are several applications is clustering. In fact, the minimal spanning tree method described in Sect. 3 of Chap. 22 is a graph–theoretical method.

Another method, Jardine and Sibson's method [4], is also based on graph–theoretical concepts. The method starts by considering each object as a node and by linking those objects which are more similar than a certain threshold. If a Euclidian distance is used, this means that only those nodes are linked for which the distance is smaller than a chosen threshold distance. When this is done, one determines the so-called maximal complete subgraphs. A complete subgraph is a set of nodes for which all the nodes are connected to each other. Maximal complete subgraphs are then the largest of these complete subgraphs. In Fig. 7, they are (B, G, F, D, E, C), (A, B, C, F, G), (H, I, J, K), etc. Each of these is considered as the kernel of a cluster. One now joins those kernels that overlap to a large degree with, as criterion, the fact that they should have at least a prespecified number of nodes in common (for instance, 3). Since only the first two kernels satisfy this requirement, one considers A, B, C, D, E, F, G as one cluster and H, I, J, K as another.

Another clustering method based on operations research is MASLOC, which has already been discussed in Chap. 22.

4. Digital simulation

As an illustration of an application of digital simulation, consider a relatively simple laboratory consisting of 4 small spectroscopic sections (one or perhaps two analysts per section): IR, PMR, ^{13}C-NMR, and MS, which are operating independently of each other. Incoming samples are distributed over the four sections on the basis of given criteria. A small percentage of samples is also routed to several

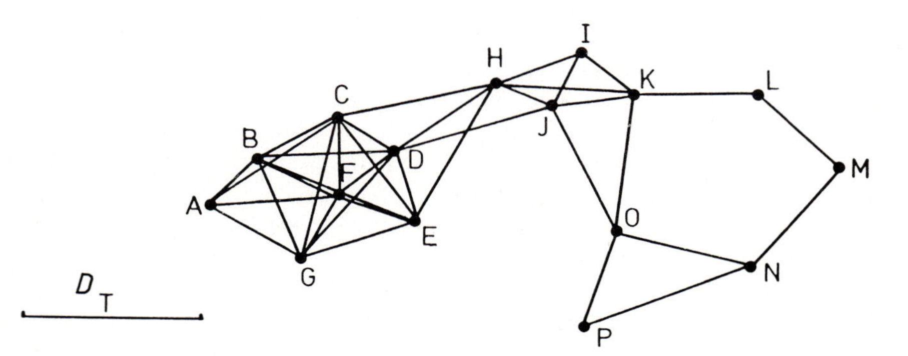

Fig. 7. Step 1 of the Jardine and Sibson method [4]. Only objects less distant than D_T are linked.

References p. 434

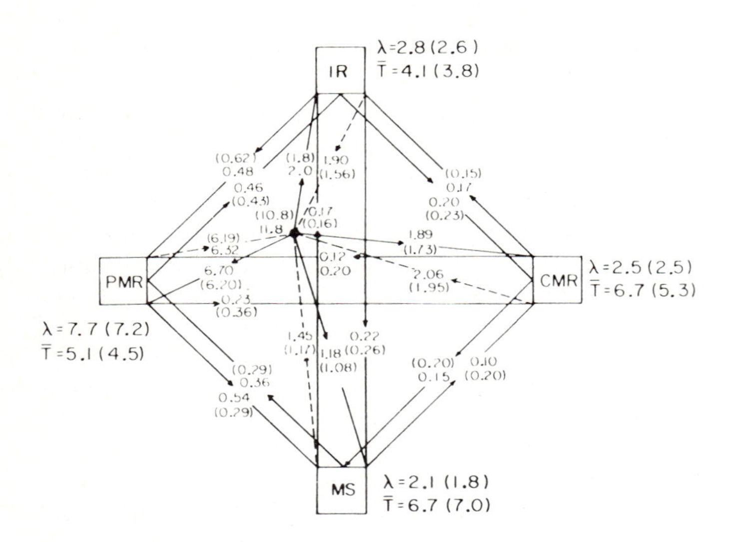

Fig. 8. Sample flows (samples day^{-1}) and delays (days) in a laboratory for spectroscopic analysis. (The numbers in parentheses are simulated data.)

sections sequentially (Fig. 8). Every analytical method consists of two steps: the measurement of the sample and the interpretation of the spectra.

Samples awaiting measurement, spectra waiting for interpretation, and the results waiting to be communicated are the queues in the laboratory. For this simple laboratory, this already means 12 queues in total. Four types of operation characteristics or strategies can be distinguished: strategies which control the priority of different types of sample, strategies which govern the path or route of the sample through the laboratory; strategies for the assignment of personnel to a certain type of analysis; strategies which regulate the priority between measurement and interpretation. To give an impression of the complexity of this system, consider the possible strategies which can be used to regulate priorities between samples.

(1) The sample in line with the earliest laboratory arrival date is selected first for analysis.

(2) The sample in line with the earliest arrival date at a particular spectroscopic section is analyzed first.

(3) The samples in each section are partitioned in groups according to the number of unsuccessful methods tried before. Priority is assigned to the samples which have been routed either along the largest number of sections or the smallest number of sections.

(4) Samples get priority according to the analysis time (difficulties) expected. There are again two possibilities: the shortest-expected analyzing time first, or the samples are partitioned in two groups, e.g. "easy" and "difficult" samples, with different priority.

Realizing that priorities may range from absolute priority to equal priority and that a similar list of strategies can be produced for the three other types of strategies, it becomes obvious that decision making in analytical laboratories is a difficult and complex matter to formalize. A set of fixed operating characteristics defines the performance of the laboratory, e.g. the statistical parameters (distribution, mean, standard deviation, time constant) of the sample arrivals, analysis times, down time of instruments, time between down times, the non-analyzing activities of the analysts.

Sometimes, initially fixed operating characteristics may become variable, e.g. when one has to decide whether an instrument should be replaced or not, or whether the number of instruments should be changed.

It is clear now that, in many instances, queueing theory cannot provide sufficient information for actual decision making for such complex systems. As mentioned in Sect. 1, queueing theory indicates roughly the importance of variables. It can be reasonably admitted that some results obtained for simple systems may be extrapolated to more complex systems. For instance, the delay of low-priority samples is more sensitive for the work load than the high-priority samples, regardless of the complexity of the laboratory.

It is obvious that experimentation with a real laboratory system cannot be considered because this would be too expensive and time-consuming and might even lead to chaos. A good alternative way to study a laboratory system may be to simulate its behaviour by means of a simulation model. Computer simulation experiments and modelling in general are usually conducted in several stages. After the formulation of the questions to be answered, laboratory data are collected and processed, e.g. the inter-arrival times of the samples, the mean down time of the instruments, and the delay times of various groups of samples. Some data may be directly obtained from the book-keeping of the laboratory. Others have to be gathered by interviewing the analysts, for instance, to find out which priority policies are used.

The most precarious and time-consuming stage of computer simulation is the formulation of the mathematical model. In this stage all variables, parameters, and relationships are quantified. Variables are selected on the basis of an estimate of their relative importance. This is important because the inclusion of too many variables renders the simulation model unnecessarily complex. On the other hand, when one or more important variables are missed, the simulation result will be far from reality.

Furthermore, it is necessary to build the model in as efficient a way as possible in order to obtain good results with a minimum effort. For this purpose, various computer languages have been developed, especially for the programming of simulation models. GPSS, SIMULA, and GASP are typical languages suitable for the

References p. 434

simulation of queueing and scheduling systems. In order to be able to obtain any meaningful result from experimentation with the model, the model should pass a validation test. Such a validation consists of proving that the model produces results that are consistent with the known performance of the real system. Often, the validity will not be satisfactory at a first trial. It will then be necessary to carry out additional observations in the laboratory in order to be able to refine the model properly. Once the computer model has been proved to be valid, actual simulation experiments can be conducted. Because a large number of variables are involved in a simulation model, a good experimental design (Chap. 17) is very important in order to obtain the requested information with a minimum of simulation experiments. As a result, simulation is a tedious and difficult process which necessitates the effort of a multidisciplinary team. Therefore, the decision to program a simulation model of the laboratory should not be undertaken lightly. The answer to the ultimate question whether the effort to build a model is useful depends on the profit obtained by avoiding wrong decisions. A necessary condition for a model to be useful is that the credibility with the policy makers is high enough. Building a model that is not used is a waste of time. Decision making in analytical laboratories is usually limited to a few alternative policies. These alternatives generally differ incrementally from current policies, which means that the advances are made in small steps. Drastic changes or complete reorganizations of the laboratory are usually not acceptable to policy makers. This technique of small incremental changes of policies makes it possible to compare the actual response of the laboratory with the predictions obtained with the model. When both agree during an extended period of time, the credibility of the model grows and the model will become a more decisive tool in decision making. When reality and prediction disagree, adaptations of the model are imperative, which hopefully will lead to a better tool for future decisions.

References

1 R.L. Ackoff and M.W. Sasieni, Fundamentals of Operations Research, Wiley, New York, 1968.
2 R. Goulden, Management studies and techniques for application in analytical research, development and service, Analyst, 99 (1974) 929.
3 A. Kaufmann, Introduction à la Combinatorique en vue des Applications, Dunod, Paris, 1968.
4 J. Jardine and R. Sibson, Comput. J., 11 (1986) 177.

Recommended reading

Books

Good first introductions to operations research are ref. 1 and

C.F. Palmer, Quantitative Aids for Management Decision Making (with Applications), Saxon House, Farnborough, U.K., 1979.

A reference work is

H.M. Wagner, Principles of Operations Research. With Applications to Managerial Decision, Prentice-Hall, London, 1975.

Specific aspects are discussed in

A. Battersby, Network Analysis for Planning and Scheduling, Macmillan, London, 1970.

T.H. Naylor, J.L. Balintfy, D.S. Burdick and K. Chu, Computer Simulation Techniques, Wiley, New York, 1966.

N. Roberts, D. Andersen, R. Deal, et al., Introduction to Computer Simulation. A System Dynamics Modeling Approach, Addison-Wesley, Reading, MA, 1983.

M. Zeleny, Multiple Criteria Decision Making, McGraw-Hill, New York, 1982.

Articles

D.L. Massart, C. Janssens, L. Kaufman and R. Smits, Application of the theory of graphs to the optimalisation of chromatographic separation schemes for multicomponent samples, Anal. Chem., 44 (1972) 2390.

D.L. Massart, L. Kaufman and D. Coomans, An operational research model for pattern recognition, Anal. Chim. Acta, 122 (1980) 347.

K. Vaananen, S. Kivirikko, S. Koskenniemi, J. Koskimies and A. Relander, Laboratory simulation. A study of laboratory activities by a simulation method, Methods Inform. Med., 13 (1974) 568.

B.G.M. Vandeginste, Digital simulation of the effect of dispatching rules on the performance of a routine laboratory for structural analysis, Anal. Chim. Acta, 122 (1980) 435.

B.G.M. Vandeginste, Strategies in molecular spectroscopic analysis with application of queueing theory and digital simulation, Anal. Chim. Acta, 112 (1979) 253.

P. Winkel, Operational research and cost containment. A general mathematical model of a work station, Clin. Chem., 30 (1984) 1758.

Chapter 26

Decision Making

In 1973, the editor of *Analytical Chemistry* stated that an analytical method is a means to an end and not an end in itself. In Chap. 10, we have already studied the relationship between utility and cost and we came to the conclusion that analytical chemists must restrict the information they produce (and therefore the costs that this entails) to that quantity for which the utility of obtaining it is considered higher than the cost. This involves a decision about the quantity of information to be obtained (or about precision, number of samples to be analysed, limit of detection, or any other of the performance characteristics determining the quantity of information).

One has not only to make decisions about the analytical process but also about the results obtained. In many (and really in most) instances, analytical chemistry is action-oriented and leads to decisions such as to declare a patient ill or healthy, a batch of manufactured goods as being of sufficient quality or not, a lot of animal feed free from mycotoxins or not, etc. These are simple examples and, in general, one is not restricted to binary decisions. For instance, a clinical analysis will not usually lead to the conclusion that a patient is ill or healthy but that he suffers from a particular disease rather than from another disease. In any event, one observes that decisions have to be made, so that it is relevant to study the problem of decision making.

Decision making has essentially to do with uncertainty. Under complete certainty, it is not difficult to make decisions. However, this is rarely the case. Medical doctors do not claim that their diagnoses are 100% certain, it is not possible to analyze each individual element from a batch of manufactured goods, and since mycotoxins are present in so-called "hot spots" (i.e. they occur only in a few particular places in a lot) one is never really sure that no mycotoxins are present in a certain lot because certainty can be obtained only by using up the whole lot for analysis. Therefore, in practice, decisions are made under uncertainty. Of course, one wants to reduce the uncertainty to an acceptable level and this is, in many instances, the job of the analytical chemist. We will now describe how additional information influences decisions made under uncertainty. A very simple problem will be used to introduce some mathematical foundations of decision making.

1. A decision matrix

Consider first an industrial laboratory which has to decide whether to use quality control or not. If quality control is carried out, faulty samples will be detected and thereby one can assume that customers will be satisfied, but one incurs the costs of

TABLE 1

Table of possible outcomes

	H_1 acceptable sample	H_2 deficient sample
d_1 : quality control	O_{11}	O_{12}
d_2 : no quality control	O_{21}	O_{22}

quality control. If one does not use quality control, money is saved in the laboratory but there will be some dissatisfied customers. One can summarize the possible outcomes using Table 1.

To each outcome, one can attach a utility. The best possible outcome is, of course, O_{21} to which we attach maximal utility (U_{21}). This utility is given an arbitrary value 1 ($U_{21} = 1$). The worst possible outcome is sending undetected faulty goods to a customer. This is given utility 0 so that $U_{22} = 0$. Outcome O_{11} leads to a satisfied customer but the cost of quality control has to be taken into account and O_{12} also leads to a satisfied customer but the additional cost of replacing the sample by a new one must now also be considered. O_{11} and O_{12} are less desirable than O_{21}, but more acceptable than O_{22}, so that $0 < U_{12} < U_{11} < 1$. The exact values depend on the cost of quality control and production and commercial costs. If it is possible to evaluate these, an exact value can be given to both utilities. In some instances, this will not be possible and one will have to tolerate a certain measure of subjectivity. Anyway, let us suppose that one has arrived at the decision that $U_{11} = 0.9$ and $U_{12} = 0.5$. To be able to make a decision, one also needs to know the probabilities of H_1 and H_2. These are obtained from past experience and given together with the utilities in Table 2.

Since the utility of d_2 (not to carry out quality control) is higher than d_1, d_2 is preferred.

TABLE 2

Decision matrix with utilities and probabilities

	H_1	H_2
d_1	$U_{11} = 0.9$	$U_{12} = 0.5$
d_2	$U_{21} = 1.0$	$U_{22} = 0.0$
Probability	0.9	0.1

The utilities for d_1 and d_2 are given by

$U(d_1) = 0.9 \times 0.9 + 0.1 \times 0.5 = 0.86$

$U(d_2) = 0.9 \times 1 + 0.1 \times 0.0 = 0.9$

This result depends on two factors.

(a) The probability. When the probability of delivering acceptable samples decreases, the advantage of d_2 also decreases. It is easy to compute that for $p = 0.833$ (i.e. 17% defective samples) d_1 and d_2 become equally attractive.

(b) The cost of quality control. When the cost of quality control decreases, this leads to higher utilities U_{11} and U_{12}. Carrying out quality control costs 0.1 in utility units; when this is reduced to 0.06, so that $U_{11} = 0.94$ and $U_{12} = 0.54$, d_1 and d_2 again become equally attractive.

2. Probability theory

Since, in this chapter, we will have to apply probability theory, some basic definitions and laws must now be given.

(1) The probability of an event A is a positive number smaller than or equal to 1

$0 \leqslant P(\mathrm{A}) \leqslant 1$

(2) The sum of the probabilities of all possible simple events $\mathrm{A}_1, \ldots, \mathrm{A}_n$ belonging to the same experiment is 1

$$\sum_{i=1}^{n} P(\mathrm{A}_i) = 1$$

(3) The union (A or B) of two events A and B is the event which consists of all simple events that belong to A or to B or to both A and B. It is noted as $\mathrm{A} \cup \mathrm{B}$ or as $\mathrm{A} + \mathrm{B}$.

The intersection (A and B) of two events A and B is the event which consists of all simple events that belong both to A and to B. It is noted as $\mathrm{A} \cap \mathrm{B}$ or as A.B or, simply, as AB

$P(\mathrm{A} \cup \mathrm{B}) = P(\mathrm{A}) + P(\mathrm{B}) - P(\mathrm{A} \cap \mathrm{B})$

This is called the addition law.

(4) Two events A and B are said to be exclusive if no simple event belongs to both A and to B. For exclusive events, one notices that

$P(\mathrm{A} \cup \mathrm{B}) = P(\mathrm{A}) + P(\mathrm{B})$

(5) A and $\overline{\mathrm{A}}$ (read "not A") are complementary events

$P(\overline{\mathrm{A}}) = 1 - P(\mathrm{A})$

This is the law of complementation.

(6) Two events A and B are said to be independent if

$P(\mathrm{AB}) = P(\mathrm{A})P(\mathrm{B})$

(7) $P(\mathrm{A}/\mathrm{B})$ is understood as the probability of an event A when the event B has occurred.

$P(\mathrm{A}/\mathrm{B}) = P(\mathrm{A})$

where $P(\mathrm{A}/\mathrm{B})$ is the conditional or posterior probability, if $P(\mathrm{A})$ and $P(\mathrm{B})$ are independent. $P(\mathrm{A}/\mathrm{B})$ can then be shown to be

$$P(\mathrm{A}/\mathrm{B}) = \frac{P(\mathrm{AB})}{P(\mathrm{B})} \qquad (1)$$

References p. 451

3. The operating characteristic curve and the power function in quality control

In Sect. 1, we considered an extremely simple example of decision making. Let us now investigate more generally the basics of acceptance sampling quality control. One assumes that the process is under statistical control in the sense that a certain constant rate of defectives is produced throughout a lot. How, then, does one establish a sampling plan that will assure both the buyer and the producer of a fair decision? Consider a sampling plan in which five samples are taken from a lot. Using the binomial distribution (Chap. 2), one can now determine the probability of finding no defectives in these five samples in function of the true defective rate. This is shown in Fig. 1. When one accepts the batch if no defectives are found for the 5 samples, then the curve relates probability of acceptance to the true defective rate. It is called the operating characteristic curve, a term which comes from statistical decision theory.

Statistical decision theory is an important aspect of quality control. Returning to Fig. 1, one observes that, under a sampling plan which includes acceptance of the lot only when 0 samples out of 10 are found defective, the consumer or buyer still has about an even change ($p = 0.43$) of accepting lots with 8% defectives. In fact, both the producer and the consumer run a risk. This is shown more explicitly in Fig. 2.

A sample of 20 objects is taken and the operating characteristic curves are determined for no defectives ($c = 0$), for one defective ($c = 1$) and for two defectives ($c = 2$). Consider, for instance, the $c = 1$ situation. When 1 sample out of 20 is found to be defective on inspection, a lot with a true defective rate of 2.2% has a

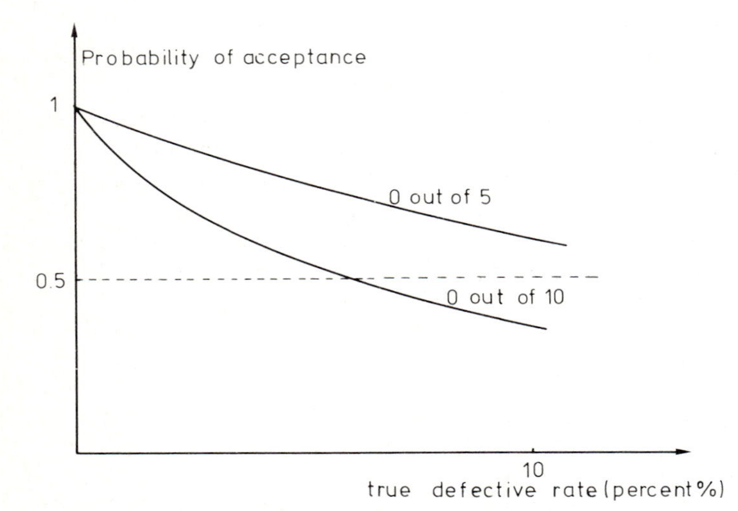

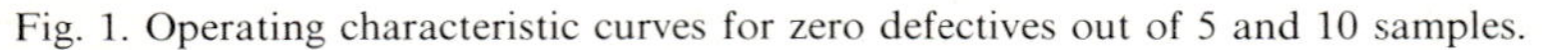
Fig. 1. Operating characteristic curves for zero defectives out of 5 and 10 samples.

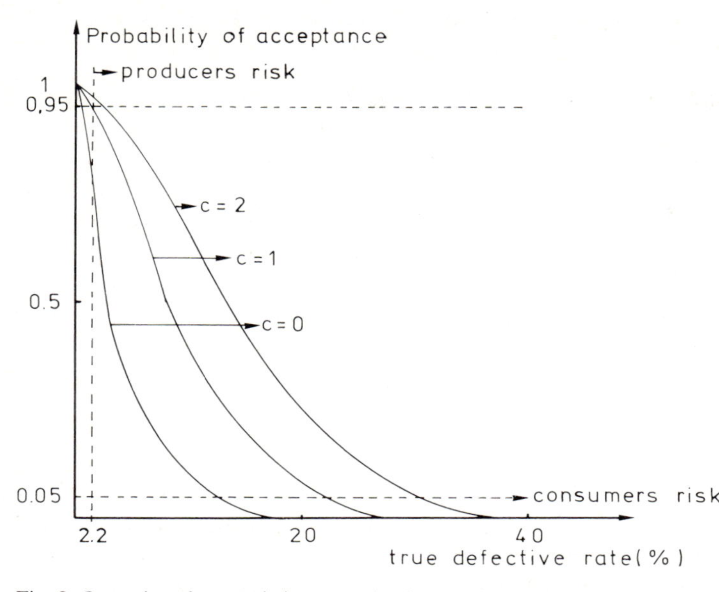

Fig. 2. Operating characteristic curves for 0, 1, and 2 defectives, c, and sample size, $n = 20$.

probability of acceptance of 95% or, in other words, the producer runs the risk that, 5 times out of 100, a lot with a true defective rate of 2.2% is rejected. This means that a lot which, in reality, is better than the measured defective rate is rejected: this is a producer's risk. On the other hand, the consumer has a 0.05 probability of accepting a lot with somewhat more than 20% true defectives! Consumer's and producer's risk are related to the type I and type II errors of Chap. 2.

The type I error is the producer's risk of wrongly rejecting an acceptable lot, while the type II error is the consumer's risk of wrongly accepting a bad lot.

In fact, as shown in the next example, the use of the operating characteristic curves in statistical decision theory is not confined to acceptance sampling.

A commercial firm produces a reference serum for clinical laboratories. The certified mean value is 300, reached by carrying out analytical determinations. The standard deviation of the determination is known to be 24. Suppose a re-evaluation is carried out: 64 determinations gave a higher mean. Should the certified mean value of 300 be replaced by the higher mean? The two hypotheses are

H_0 : the true value is 300, $\mu = 300$

H_1 : the true value is higher than 300, $\mu > 300$

Suppose that the new mean is 310.

The distributions around the two means 300 and 310 are given in Fig. 3. One can now consider a decision limit as given in this figure.

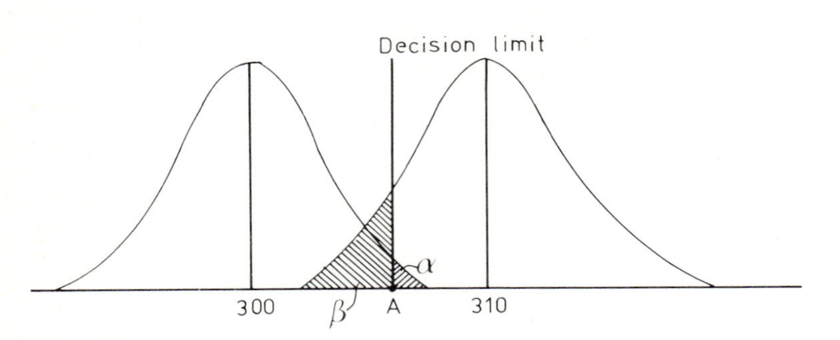

Fig. 3. Effect of decision limit on type I (α) and type II (β) errors.

The probability of accepting H_0 when it is not true (retaining 300 when it should be 310) is given by the area β, while the probability of rejecting H_0, when it is true (changing to 310 when it should be 300) is given by α. α is the type I error (also called the significance level), while β is the type II error.

Clearly, α and β for a given decision limit depend on the distance between the two means. If the mean found on re-evaluation was found to be 305, β, the probability of retaining 300 when it should be 305, will be higher: it has become more difficult to discriminate between the two means. The curve relating β to the distance between the two means for a given α is called the operating characteristic curve, while the curve relating $1-\beta$ to the distance for a given α is called the power curve. Suppose that α is set equal to 0.01. Then, one finds from Table 1 of the Appendix that $z = 2.33$ and since

$$z = \frac{\bar{x} - 300}{\sigma/\sqrt{N}} = \frac{\bar{x} - 300}{24/\sqrt{64}}$$

then $\bar{x} = 307$.

For this and higher values of the decision limit, we will replace the old mean by the new. We can now also compute β at any value of the new mean. For a mean of 310

$$\frac{307 - 310}{24/\sqrt{64}} = -1$$

and the area β can then be derived from Table 1 of the Appendix, i.e.

$$\beta = 0.15$$

Computations of β at the same α for other values of the new means can now be carried out. They yield Fig. 4.

One concludes that, at the given level of significance, the probability of keeping the old mean is 1 up to 300 and that it then starts dropping. At 315, it is practically

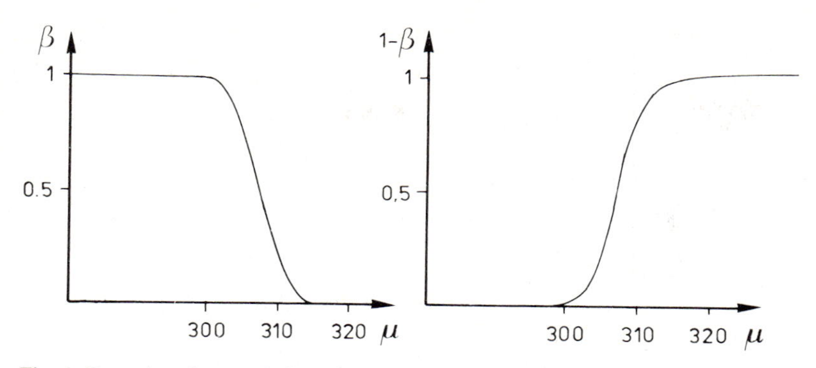

Fig. 4. Operating characteristic and power curves for the worked example at $\alpha = 0.01$.

certain that the new mean will replace the old. One can now make the connection with Fig. 2. This is called an operating curve [1].

4. Prior and posterior probabilities. Bayes' rule

We consider again the example of Sect. 1 but make it more complex by assuming a given distribution of defectives over the lots instead of a constant defective rate. We suppose now that the product is made in batches of 10 000 and that one knows that some batches are excellent while some contain many faulty individual samples. Experience has led to the conclusion that there is an equal probability of producing batches with 0, 10, and 20% defective samples. Note that there is still a 10% over-all probability that a sample is defective.

It could be worthwhile for further decisions about carrying out quality control on the whole batch to have a better idea of what kind of batch is present. To achieve this, the laboratory responsible for quality control proposes, as a rule, to select 5 samples out of each batch of 10 000 on a random basis and to analyze these samples.

On the basis of the result, further decisions about inspecting all the samples will be taken. The preliminary inspection can yield the result that 0, 1, . . . , 5 samples are defective. What do these results mean? Let us consider the situation that not a single sample is found defective. The chance of finding 5 acceptable samples out of 5 in batches with 0, 10%, or 20% defective samples can be obtained from a binomial distribution. It is given by 1, $(0.9)^5 = 0.59$, and $(0.8)^5 = 0.33$, respectively. The probability of finding 1 defective sample is 0, 0.32, and 0.41.

Let us now summarize the available information about the batches of goods. We know that

(a) a batch can have no defectives (event G1), 10% defectives (event G2), or 20% defectives (event G3). These three events are disjoint, which means that they are mutually exclusive (a batch must belong to only one of the three categories) and exhaustive (there are no other possibilities);

References p. 451

(b) the batches are equally probable. In other words, one knows before carrying out the sampling programme that the probability of encountering a batch with no defectives is 0.33. One can also say that, prior to the additional information, the probability of G1, called $P(\text{G1})$, is equal to 0.33. $P(\text{G1})$, $P(\text{G2})$, $P(\text{G3})$ are therefore called the a priori or the prior probabilities;

(c) the sampling programme leads to a number of possible findings X (X = A, B, C, D, E or F) where A means that one finds 0 defective samples and F that one finds 5. Again, one can calculate the probability that one obtains A in the case of events G1, G2, or G3. This is symbolized as $P(\text{A}/\text{G1})$, $P(\text{A}/\text{G2})$, $P(\text{A}/\text{G3})$, respectively and is called the conditional probability of A given G1, G2, and G3. For instance, P (zero defectives in samples batch/10% defectives in lot) = 0.59.

What we do not know yet is the probability of having a batch with 10% defective samples when no samples are found to be defective in the sampling program. This is symbolized by $P(\text{G2}/\text{A})$ and is called the probability of G2 posterior to finding that X = A or, more shortly, the a posteriori or the posterior probability of G2.

To obtain $P(\text{G2}/\text{A})$, one applies a rule called Bayes' rule. From eqn. (1), one obtains

$$P(\text{G2}/\text{A}) = \frac{P(\text{G2.A})}{P(\text{A})}$$

and

$$P(\text{A}/\text{G2}) = \frac{P(\text{G2}.A)}{P(\text{G2})} \tag{2}$$

so that

$$P(\text{G2}/\text{A}) = \frac{P(\text{A}/\text{G2})\,P(\text{G2})}{P(\text{A})} \tag{3}$$

Furthermore, $P(\text{A})$ can be rewritten as

$$P(\text{A}) = P(\text{A.G1}) + P(\text{A.G2}) + P(\text{A.G3})$$

and by analogy with eqn. (2)

$$P(\text{A}/\text{G1}) = \frac{P(\text{G1.A})}{P(\text{G1})}$$

$$P(\text{A}/\text{G3}) = \frac{P(\text{G3.A})}{P(\text{G3})}$$

so that

$$P(\text{A}) = P(\text{A}/\text{G1})\,P(\text{G1}) + P(\text{A}/\text{G2})\,P(\text{G2}) + P(\text{A}/\text{G3})\,P(\text{G3})$$

and

$$P(\text{G2}/\text{A}) = \frac{P(\text{G2})\,P(\text{A}/\text{G2})}{P(\text{G1})\,P(\text{A}/\text{G1}) + P(\text{G2})\,P(\text{A}/\text{G2}) + P(\text{G3})\,P(\text{A}/\text{G3})} \tag{4}$$

This is called Bayes' rule for the specific case of three possible events G1, G2, and G3. For this example, the result is

$$P(\mathrm{G2/A}) = \frac{0.33 \times 0.59}{0.33 \times 1 + 0.33 \times 0.59 + 0.33 \times 0.33} = 0.31$$

$P(\mathrm{G2/A}) = P$ (10% defectives in lot/zero defectives in sampled batch)

while

$P(\mathrm{G1/A}) = 0.52$

and

$P(\mathrm{G3/A}) = 0.17$

Finding no defective sample out of five trials has little effect on the a priori probability of 33% that a batch with 10% defectives is present, it increases the probability that a no-defectives batch is present, and decreases the probability of a 20% defectives batch.

If one wishes, a decision matrix can now be constructed as in Sect. 1 with this information.

The example considered in this chapter is an extremely simple example of the importance of designing optimal sampling programs. Chapter 27 gives some idea about how to do this and also discusses a particular kind of sampling problem, namely how to space sampling or measurements in time to obtain adequate amounts of information.

5. The diagnostic value of a test

Determining the diagnostic value of a test is very much related to the problem discussed in the preceding sections.

In general, the more sophisticated or elaborate a procedure, the more information one expects and the more it costs. A common question in economics is: "If more money is spent, a better (or more of a) product will be obtained. How are additional costs and benefits related?" This kind of question is answered by cost–benefit calculations (see Chap. 10). In the same way, an analytical chemist might ask, "given a certain procedure for determining lead in drinking water, spending an amount can yield a faster method with a better precision. Is it worthwhile to spend that amount?"

It appears that the only analytical chemists to have asked regularly and tried to answer this question are clinical chemists. Therefore we will consider here medical decision making and do this in three steps, viz. the evaluation of the technical value of the test, the evaluation of its diagnostic value, and the analysis of costs and benefits.

In clinical chemistry, the simplest possible objective is to separate two classes, namely healthy and ill people. The values of a chemical test follow different (often Gaussian or near-Gaussian) distributions for ill and healthy persons. This is

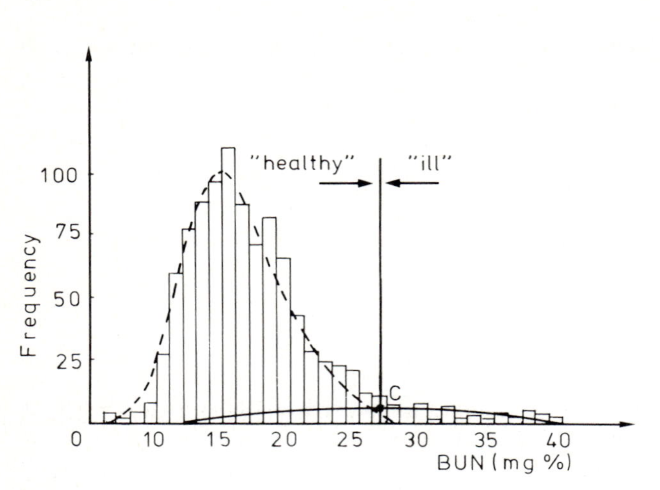

Fig. 5. A "normal" and an "abnormal" population. Adapted from ref. 2.

illustrated by Fig. 5. When values higher than C are found, one assumes that the person is probably ill, and when it is lower, that the person is probably healthy. Even without analytical errors, classification errors are made in doing so. In Fig. 5, about 5% of the healthy persons are classified as ill and these 5% therefore constitute false-positive results. On the other hand, 50% of ill persons are not detected. There are, therefore, 50% false-negatives.

The "technical value" of the test can therefore be summarized in the evaluation matrix of Table 3 (it should be noted that, in the medical literature, this is usually called a decision matrix, since it does not include costs or utilities, it is, however, quite different in scope from Table 2).

TABLE 3

Evaluation of a clinical test (technical value)

Correct diagnosis	Test result		
	Ill = "positive"	Healthy = "negative"	
Healthy	FP	TN	TN + FP
Ill	TP	FN	FN + TP
	TP + FP	TN + FN	

TN = true negatives; number of healthy patients diagnosed as healthy, i.e. number of correctly diagnosed healthy patients.
FN = false negatives; number of ill patients diagnosed as healthy, i.e. number of ill patients falsely diagnosed as healthy.
TP = true positives; number of ill patients diagnosed as ill, i.e. number of correctly diagnosed ill patients.
FP = false positives; number of healthy patients diagnosed as ill, i.e. number of healthy patients falsely diagnosed as ill.

One observes that the evaluation Table 3 contains four values. It is preferable to try and condense the information in a smaller number of values. For this purpose, one uses criteria such as the sensitivity and specificity. These terms are those used in medical terminology; readers should be aware that their meaning is very different from their usual meaning in analytical chemistry.

$$\text{Sensitivity} = \frac{\text{Diseased persons positive to the test}}{\text{All diseased persons tested}} \times 100$$

or

$$\text{Sensitivity} = \frac{\text{TP}}{\text{TP} + \text{FN}} \times 100$$

$$\text{Specificity} = \frac{\text{Non-diseased persons negative to the test}}{\text{All non-diseased persons tested}} \times 100$$

or

$$\text{Specificity} = \frac{\text{TN}}{\text{TN} + \text{FP}} \times 100$$

The definition of the symbols TP, TN, FN, and FP are given in Table 3.

These two terms together describe a test. When one prefers a single performance characteristic, one can use

(a) the efficiency

$$\text{EF} = \frac{\text{TP} + \text{TN}}{\text{TP} + \text{TN} + \text{FP} + \text{FN}} \times 100$$

(which is equal to the % correct diagnoses)

(b) the Youden index

$$\text{YI} = \text{sensitivity} + \text{specificity} - 100$$

The diagnostic value of the test depends not only on its technical value, but also on the a priori probabilities of both classes (ill or healthy). Let us suppose that one knows from prior experience that 10% of the people are ill and 90% are not. This now leads to Table 4. This knowledge must also be taken into account in calculating the diagnostic value of the method, which can be done using Bayes' theorem.

As an example, let H_1 = the fact that a person is healthy, H_2 = the fact that a person is ill, A = the finding of a negative value, and B = the finding of a positive

TABLE 4

Evaluation of the clinical tests of Fig. 5 (technical value)

	B	A	$P(H_n)$
H_1	0.05	0.95	0.90
H_2	0.50	0.50	0.10

References p. 451

value. Bayes' theorem here becomes

$$P(H_1/A) = \frac{P(H_1)P(A/H_1)}{P(H_1)P(A/H_1) + P(H_2)P(A/H_2)}$$

where $P(H_1/A)$ is the probability that H_1 is true when event A occurs or

$$P(H_1/A) = \frac{0.90 \times 0.95}{0.90 \times 0.95 + 0.10 \times 0.5} = 0.945$$

Therefore, finding a value less than C (see Fig. 5) indicates that there is 94.5% chance of a person being healthy. In this particular situation, the additional information produced by the test has increased the probability of being healthy from the a priori probability value of 90% to 94.5%.

By appropriate assignation of the symbols in eqn. (4), one can also answer other questions, such as [2]

(a) what is the probability $P(H_1/B)$ that the patient is healthy when a value higher than C is obtained? The result is 0.474 and gives the fraction of false-positive values;

(b) what is the probability $P(H_2/A)$ that the patient is ill with a value below C? The result is 0.055 and gives the fraction of false-negative values; and

(c) what is the probability $P(H_2/B)$ that the patient is ill with a value higher than C? The result of 0.526, meaning that only about half of the ill population in detected with this test.

Besides determining the technical and the diagnostic value of the test, one can determine the utility of the test by considering the cost and the benefit or value of the result. These are summarized in Table 5. The cost–benefit analysis is carried out by comparing cost and benefits, taking into account the various probabilities. For instance, the cost and benefit for H_1 must be multiplied by $P(H_1)$ and those for the upper left case must again be multiplied with the probability of finding B when H_1 is true. This is done systematically in Table 5.

TABLE 5

Cost–benefit analysis of a test

	Positive (B)	Negative (A)
Healthy (H_1)	Benefit: none	Benefit:constatation of health (VA_1)
	Cost:unnecessary treatment (CB_1)	Cost: cost of the test (CA_1)
Ill (H_2)	Benefit:illness is treated (VB_2)	Benefit: none
	Cost:cost of treat-ment (CB_2)	Cost: untreated illness (CA_2)

Utility = $P(A/H_1)P(H_1)(VA_1 - CA_1) + P(B/H_2)P(H_2)(VB_2 - CB) - P(A/H_2)P(H_2)CA_2 - P(B/H_1)P(H_1)CA_1$

6. The ROC curve

Until now, we have supposed more or less implicitly that the decision limit is situated at the crossing of two distributions (point A in Fig. 6). However, this is not necessarily so. One might decide that all persons who could be ill should undergo a therapy and that therefore the cut-off point should be B. On the other hand, with a high risk (or costly) therapy, one might prefer to treat only those people who definitely have the disease. The cut-off point should then be C.

Clearly, the sensitivity and the specificity of the test depend on this decision point. For instance, if point C is chosen, the sensitivity is smaller but the specificity is higher. In fact, sensitivity and specificity are related and therefore so are true and false positive rates. The former is the sensitivity divided by 100 and the latter is equal to [1 − (specificity/100)]. In general, the more sensitive a test is, the less specific it becomes. To analytical chemists, this conclusion will certainly not come as a surprise. On the other hand, it appears that analytical chemists have not tried to describe the phenomenon in a formal way.

It is not difficult to think of immediate analogies in analytical chemistry. These are evident, for example, in practical courses in analytical chemistry, where the first exercises are often of a qualitative nature and students are confronted with the difficult decision of whether a certain colour, indicating a particular metal ion, is present or not. Everyone who has taught such a course knows that there are students who record the presence of the colour only when it is undeniable. Such a student has a very small score of false positives and a large score of false negatives. His overscrupulous colleague, who always detects a hint of the colour to be observed, rarely misses one of the ions present but detects many ions that are, in fact, absent. The same argument can now be repeated for tests where a decision limit has been established. A decision limit such as C in Fig. 6 causes many false negatives but few true positives.

The relationship between false positives and true positives is given by the so-called receiver operating characteristic curve (ROC curves), a term introduced in medical decision making from signal detection theory. Three hypothetical ROC curves are given in Fig. 7. One observes that, at one end of the curve, one finds zero true positives and zero false positives, i.e. a situation of absolute specificity and zero sensitivity.

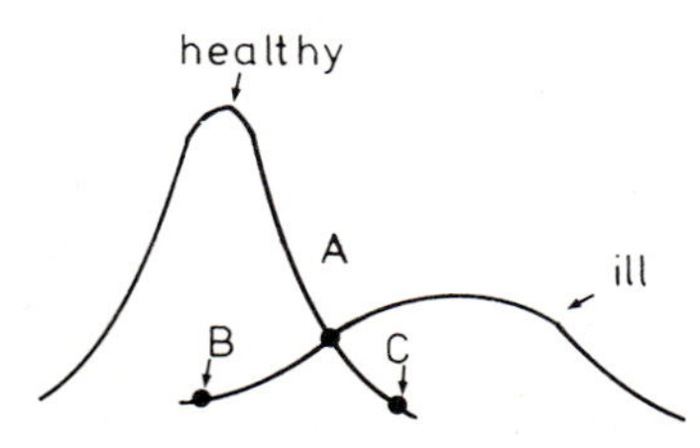

Fig. 6. The selection of a cut-off point.

References p. 451

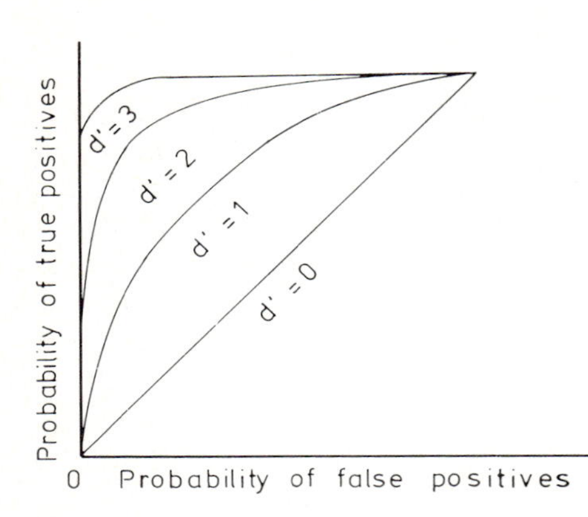

Fig. 7. The ROC curves for $d' = 0, 1, 2,$ and 3 (and equal variance). Reproduced with permission of John Wiley and Sons.

The name given to this curve indicates a relation to the operating curve of Sect. 3. To perceive this relationship, it is sufficient to note that

FP rate $= \alpha$ (type I error)

FN rate $= \beta$ (type II error)

TN rate $= 1 - \alpha$

TP rate $= 1 - \beta$

The ROC curve then relates α to $1 - \beta$ as a function of the decision rule and the ROC curve really is a different way of picturing the quantities investigated in the acceptance sampling problem.

The shape of a ROC curve gives an indication of the value of a test. If the test is to have any value, the fraction of true positives must always be higher than the fraction of false positives.

This is not the case under the diagonal of Fig. 7 so that this region of the diagram must not be considered further. The diagonal itself describes a situation where true positives and false positives are equally probable. This occurs only when the two distributions overlap completely (case of total identity). Clearly, the test will be better when the ROC curve is more concave. The region of interest is therefore confined to the region above the diagonal.

One can use the ROC curve for two purposes, to compare the technical value of different tests and to find the best cut-off point for a particular test.

The technical value of a test is determined by the extent of overlap between the distributions obtained for the test values for the two populations ("ill" and "not ill"). The extent of overlap itself is determined by the difference between the maxima of the two distributions and by the variance of the distributions. In signal detection theory, one defines the parameter d' as the ratio of the difference between maxima and the common standard deviation. This parameter can be considered to be a quality criterion for a test. Figure 7 gives the ROC curves for $d' = 0, 1, 2,$ and 3

TABLE 6

Cost calculations for a test for occult blood in stool to detect colon cancer

Rate FP	Rate FN	Cost of FP	Cost of FN	Total cost
0.11	0.03	54.83	3.96	58.79
0.02	0.20	9.97	26.40	36.37
0.01	0.34	4.98	44.88	49.86

Example of calculation:
Cost of FP: $0.11 \times 500 \times (1-0.003) = 54.83$.
Cost of FN: $0.03 \times 44000 \times 0.003 = 3.96$.

(and equal variance).

As discussed above, the cut-off point must depend on the relative prevalence of the two populations to be discriminated and cost–benefit relations.

The problem can be rephrased as: find the minimum of [FN × cost of a FN × prevalence of illness + FP × cost of a FP × prevalence of health]. As an example, consider occult blood testing in stool to detect colon cancer. Only costs (no benefits) are considered.

Cost of FN: decrease of the 5-year survival rate from 95 to 40% which in ref. 3 is considered to constitute a mean economical cost of $44 000.

Cost of FP: additional testing which will eventually convert the FP into a TN $500.

Prevalence: in the population investigated, the prevalence is estimated to be 3 per 1000.

From the ROC curve for the current testing procedure, a few FP/FN rates are derived. For instance when a 0.11 FP rate is tolerated, the ROC curve indicates that the corresponding FN rate is 0.03. This then leads to the consequences depicted in Table 6. Clearly the 0.02/0.20 solution is to be preferred.

References

1 M.R. Spiegel, Schaum's Outline Series, Theory and Problems of Statistics, Schaum, New York, 1961, pp. 178–179. (The worked example of Sect. 3 has been paraphrased from this reference.)
2 H.F. Martin, B.J. Gudzinowicz and H. Fanger, Normal Values in Clinical Chemistry, Dekker, New York, 1975.
3 S.N. Finkelstein and M.M. Kristein, in E.S. Benson, D.P. Connelly and M.D. Burke (Eds.), Clinics in Laboratory Medicine, Vol. 2, Saunders, Philadelphia, 1982.

Recommended reading

Books

A.J. Duncan, Quality Control and Industrial Statistics, Irwin, 1975.
R.S. Galen, S.R. Gambino. Beyond Normality: The Predictive Value and Efficiency of Medical Diagnoses, Wiley, New York, 1975.

M. Gy, Sampling of Particular Materials: Theory and Practice, Elsevier, Amsterdam, 1979.
D.V. Lindley, Making Decisions, Wiley-Interscience, London, 1971.
J. Lesourne, Cost–Benefit Analysis and Economic Theory: Studies in Mathematical and Management Economics, North-Holland, Amsterdam, 1975.
E.J. Mishan, Cost–Benefit Analysis, Allen and Unwin, London, 1971.
E.A. Patrick, Decision Analysis in Medicine: Methods and Applications, CRC Press, Boca Raton, FL, 1979.
M.C. Weinstein and H.V. Fineberg, Clinical Design Analysis, Saunders, Philadelphia, 1980.

Articles

J.R. Beck and E.K. Shultz, The use of relative operating characteristic (ROC) curves in test performance evaluation, Arch. Pathol. Lab. Med., 110 (1986) 13.
J.C. Boyd, Perspectives on the use of chemometrics in laboratory medicine, Clin. Chem., 32 (1986) 1726.
J.E. Goin, ROC curve estimation and hypothesis testing: applications to breast cancer detection, Pattern Recognition, 15 (1982) 263.
G.J. Hahn, Random samplings, Chem. Tech. (Leipzig), (5) (1982) 286.
K. Linnet and E. Brandt, Assessing diagnostic tests once an optimal cutoff point has been selected, Clin. Chem., 32 (1986) 1341.
C.E. Metz, Basic principles of ROC analysis, Semin. Nucl. Med., 8 (1978) 283.
J.P. Nolan, N.J. Tarsa and G. DiBenedetto, Case-finding for thyroid disease: costs and health benefits, Am. J. Clin. Pathol., 83 (1985) 346.
A.R. Prest and R. Turvey, Cost–benefit analysis: a survey, Econ. J., 75 (1965) 683.

Chapter 27

Process Control

Analytical information can be used to control a process and industrial processes have to be controlled in order to manufacture a product of a desired quality. The analytical chemist is often faced with the development of analytical procedures or analyzers to be used, either on-line or off-line, for this purpose. The requirements for such procedures or analyzers are set by the nature of the process fluctuations, the quality of the analytical result, and the cost–benefit relation between costs of analysis and profit returned by improved control.

The information which is obtained by sampling and analyzing a process essentially depends on three factors: (1) the sampling frequency per unit of time (hour, day), (2) the precision of the analytical result, and (3) the time lag between the moment of sampling and the moment that the analytical result is available (result delay). If a control action has to be based on the result of an analytical measurement, the quality of the control will depend on the quality of the measurement. Imprecise results will lead to imprecise actions and a result which is too late is useless. However, the more samples that are analyzed with better precision and the shorter the delay, the more expensive the analytical process is. When one has to decide which analytical procedure will be used and which sampling scheme will be applied, one has to find the point where the marginal costs for the analytical activities are no longer compensated by the marginal returns when using that information for process control. Expressed in economical terms (introduced in Chap. 10), this means that the marginal costs must be lower than the marginal returns.

Let us clarify this with an example. Suppose that a chemical plant produces a nitrogen fertilizer which has product requirements per unit of product sold of mean nitrogen content ($\overline{N}$) 23.0% and minimal specified amount 22.3%. Figure 1(a) shows the natural (uncontrolled) fluctuations of the N content (standard deviation = s_x) over the units of product sold. It is assumed that subsequent mixing of the fertilizer to homogenize the delivered product is not feasible. When the fluctuations of the N content in the product are too large, the average N content should be higher than the specified amount, to be sure to sell at least the minimal required content of 22.3% [Fig. 1(b)]. When a probability of, for example, 1% is tolerated for selling fertilizer of a too-low content, then the following data are obtained (assuming a normal distribution).

When $s_x = 1.2\%$ N; $\overline{N}$ to be sold $= 22.3 + (2.33 \times 1.2) = 25.1\%$ N

When $s_x = 0.6\%$ N; $\overline{N}$ to be sold $= 22.3 + (2.33 \times 0.6) = 23.7\%$ N

2.33 is the one-sided value for the z distribution at the level of 99% (see Table 1 of the Appendix)

References p. 466

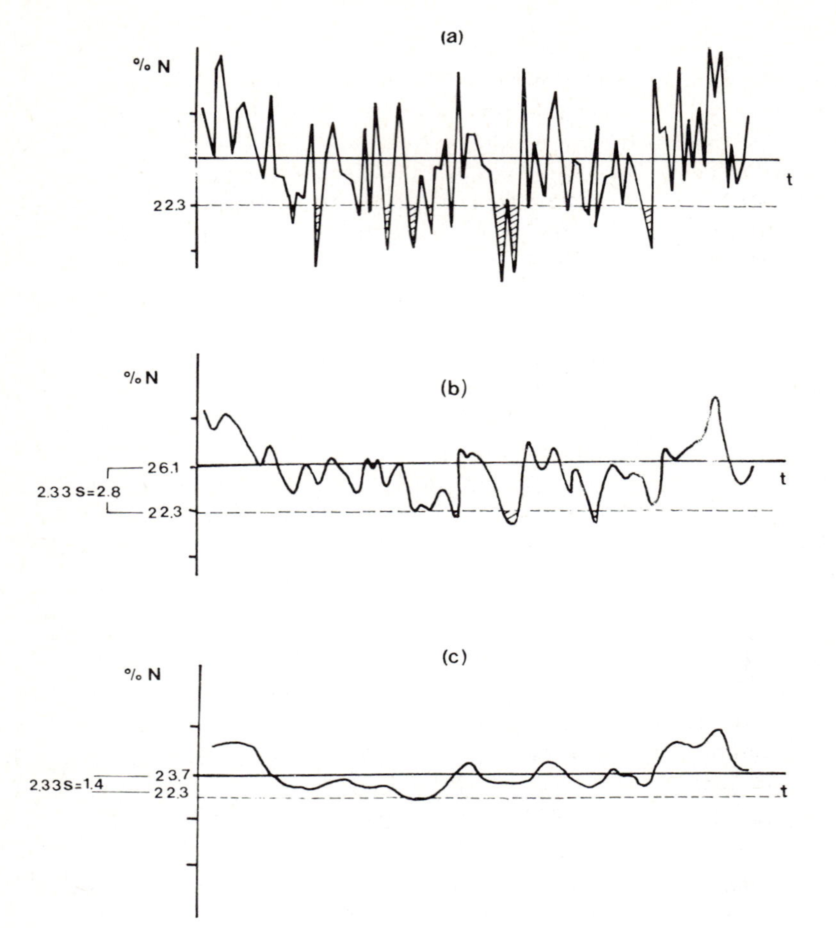

Fig. 1. (a) Natural (uncontrolled) fluctuations of the nitrogen content as a function of time (product units) with the lower limit for the nitrogen content per unit sold equal to 22.3%. (b) Residual fluctuations of the nitrogen content of a controlled process with $s_x = 1.2\%$. (c) As (b) but with $s_x = 0.6\%$.

In order to meet the specifications of average and minimal contents, s_x should be reduced to $(23.0 - 22.3)/2.33 = 0.30\%$. Reduction of the standard deviation can be achieved by control of the process. For that purpose, it is necessary to know the process state at regular intervals. The subsequent analytical problem is that a wide variety of analytical methods is available to determine the amount of N in the product stream, but all have different characteristics: i.e. analysis time, precision and costs per analysis (Table 1). Two basic decisions have to be taken: (1) which analytical method should be chosen and (2) which analysis (or sampling) frequency is optimal.

TABLE 1

Analysis time and precision of analytical methods for the determination of the nitrogen content of nitrogen fertilizers

Method	Analysis time, T_d (min)	Precision, s_a (% N)
Total N: distillation	75	0.17
Total N: automated	12	0.25
NO_3-N: autoanalyzer	15.5	0.51
NO_3-N: ion SE	10	0.76
$NH_4NO_3/CaCO_3$: X-ray	8	0.8
Total N: neutron activation	5	0.17
γ-Ray absorbtion	1	0.64

This is quite a complex problem because meaningful answers can only be obtained when one is able to balance the costs for analysis against the process profit. The process profit in our example is negatively influenced by two factors, the cost of selling too much N and the costs associated with selling below the specifications.

1. Derivation of a process model

The amount of information which is obtained about a process by a chemical analysis is the difference between the uncertainty concerning the process before analysis, H_{before}, and the uncertainty after analysis, H_{after}, (see Chap. 9).

$$I = H_{\text{before}} - H_{\text{after}}$$

The more samples that are taken and the more precise the analytical results, the better one knows the state of the process or the smaller the uncertainty will be after analysis. In other words, the amount of information, I, depends on the sampling frequency, the method characteristics, and the process characteristics. Of course, control actions are not limited to the times when results are received, but control may also be needed between two consecutive analytical results [R_1 and R_2 in Fig. 2].

In order to be able to control a process in a continuous way, one needs a process model by which process values can be estimated or forecast at any time ahead of the receipt of the last analytical result. Control actions are thus based on a prediction function. The simplest prediction function is to assume that the process state remains unchanged between two measurements. In that case, control manifolds are adjusted every time new process information has been obtained (new measurement). In terms of process control, this is called a hold circuit.

Van der Grinten [1–4] concluded that many chemical processes behave like a

References p. 466

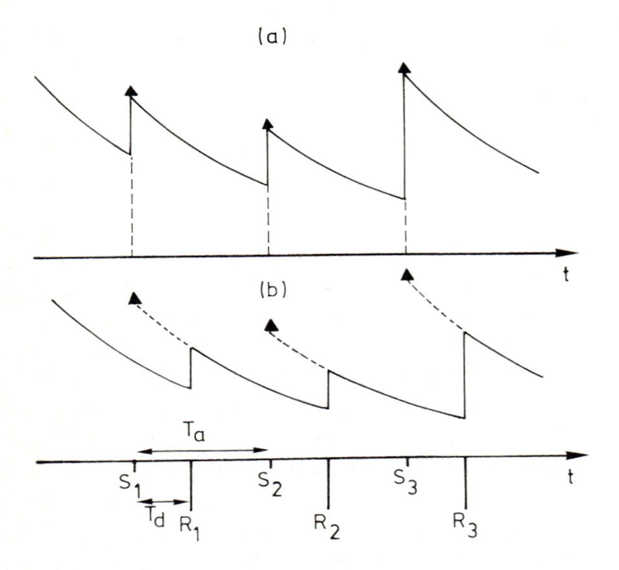

Fig. 2. Sampling characteristics. $T_a = S_{i+1} - S_i$ is the sampling interval (or interval between analytical results); T_d is the analysis (or dead) time. (a) Interpolation between sampling times S_1, S_2, S_3,..... in process reconstruction. (b) Extrapolation between result times R_1, R_2, R_3,.... in process control.

first-order stochastic process (see Chap. 14), which can be described with the autoregressive model discussed in Sect. 5.2 of Chap. 14.

$$x(t+\tau) - \bar{x} = r(\tau)[x(t) - \bar{x}] + e(t+\tau) \tag{1}$$

where $x(t+\tau)$ is the process value at time $(t+\tau)$, $\bar{x}$ is the mean process value, e is the difference between model and process values or residual, $x(t)$ is the process value at time t, and $r(\tau)$ is the autocorrelation between process values separated by a time τ (= autocorrelation for a time lag τ). Consequently, the forecast of process states at a time $1\Delta t, 2\Delta t, \ldots, \tau\Delta t$ ahead (where Δt is the sampling interval) is given by

$$x(t+1) - \bar{x} = r(1)[x(t) - \bar{x}] + e(t+1)$$

$$x(t+2) - \bar{x} = r(2)[x(t) - \bar{x}] + e(t+2)$$

$$x(t+\tau) - \bar{x} = r(\tau)[x(t) - \bar{x}] + e(t+\tau)$$

The values $r(1), r(2), \ldots, r(\tau)$ represent values of the autocorrelation between process values separated by times $\Delta t, 2\Delta t, \ldots, \tau\Delta t$, respectively (see Chap. 14). Consequently, the autocorrelation function can be used as a predictor (or interpolator) for unobserved process values.

In Chap. 14, it has been derived that the autocorrelation function of a first-order

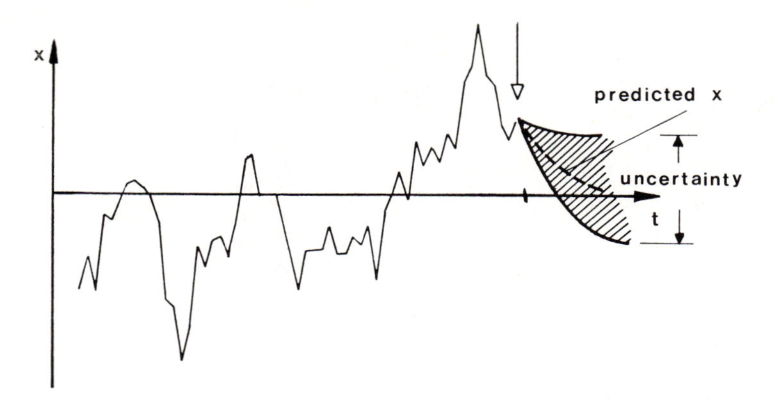

Fig. 3. - - -, The autocorrelation function as a predictor for future process values. ——, Two-sided confidence interval of the predicted process value.

process is an exponentially decreasing function. The prediction function based on a first-order stochastic model therefore becomes

$$x(t+\tau) - \bar{x} = [x(t) - \bar{x}] \exp(-\tau/T_x)$$

where T_x is the time constant of the process. For large τ values, $x(t+\tau) = \bar{x}$. This means that, for long times after the last measurement, one predicts that the most probable state of the process is its mean value. The prediction function for a process with $T_x = 10$ is shown in Fig. 3.

Predictions are only meaningful if one can attribute confidence intervals to the predicted value. These confidence intervals depend on the standard deviation of the difference between the predicted and true process value (s_ϵ) or prediction error.

For a large τ value, $r(\tau)$ is equal to zero and therefore the predicted process value, $x(t+\tau)$, is equal to the process mean ($\bar{x}$). Because the standard deviation of the natural fluctuations of the process about the mean is equal to s_x, $s_\epsilon = s_x$. Obviously, for $\tau = 0$, one knows exactly the course of the process and therefore $s_\epsilon = 0$. In general, it can be derived that the standard deviation of the differences between predicted and true values as a function of τ is given by

$$s_\epsilon(\tau) = s_x \sqrt{[1 - r^2(\tau)]}$$

As an example, the standard deviation of the prediction error as a function of the time over which the process state is predicted by the autocorrelation function is indicated in Fig. 3 by the width of the shaded area.

2. Quality of a control loop

One can use the autocorrelation function to interpolate process values between consecutive sampling points [S_1, S_2, S_3,.... in Fig. 2(a)]. This is called process reconstruction and the result is shown in Fig. 2(a).

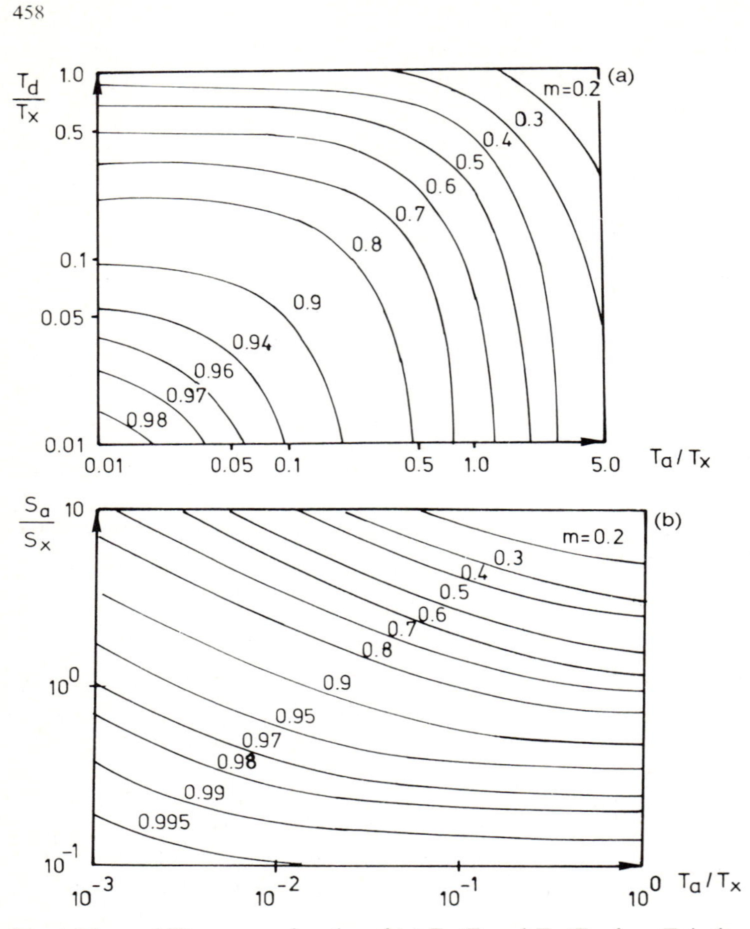

Fig. 4. Measurability, m, as a function of (a) T_a/T_x and T_d/T_x where T_a is the sampling interval and T_d is the analysis time and (b) s_a/s_x and T_a/T_x where s_a is the standard deviation of the analytical procedure, s_x is the standard deviation of the uncontrolled process and T_x is the time constant of the uncontrolled process.

Because a process reconstruction is usually carried out at the end of a period of time during which the measurements have been collected (e.g. a year), one should not take into account the fact that the analytical result of the sample has been delayed for a time T_d after the sampling action: the process is thus interpolated between S_i and S_{i+1} (i.e. two consecutive sampling times) and not between R_i and R_{i+1} (i.e. two consecutive reports of analytical results), as is the case for process control. The sum of squares of the difference between the true process values and the reconstructed process divided by the number of observations represents the remaining uncertainty about the process, called the residual variance (s_ϵ^2).

In process control, the autocorrelation function is used to predict the process

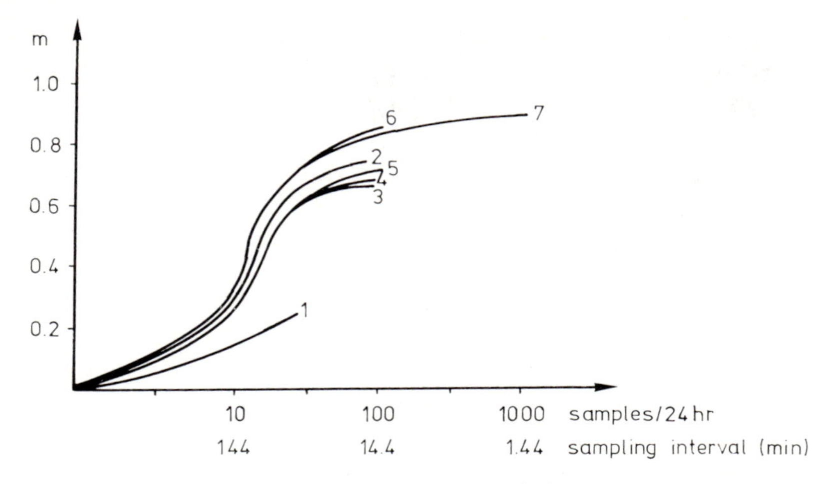

Fig. 5. Measurability of candidate methods for the analysis of the nitrogen content of a nitrogen fertilizer as a function of the sampling interval.

state (value) until the next analytical result is available. Here, one should therefore take the analysis time into account as is shown in Fig. 2(b).

The quality of a control loop can be expressed as the relative decrease of the process variance by control. This is called the controlability or control efficiency and is given by $(s_x^2 - s_\epsilon^2)/s_x^2$ where s_x^2 is the variance of the uncontrolled process and s_ϵ^2 is the variance of the controlled process. For a stationary process, $0 < s_\epsilon^2 < s_x^2$. Therefore, the controlability is bounded by zero and one. If it is one, the process variance after control is reduced to zero; if it is zero, control has no effect on the process variance. The controllability is a product of two factors: the first, f, is related to the imperfection of the controller and the second, m, is related to the imperfection of the information which is obtained by the measuring process. m is called the measurability and is given by

$$m^2 = \frac{s_x^2 - s_\epsilon^2}{s_x^2} \tag{2}$$

It is the maximum control that can be achieved by using an ideal control system. When the controller is perfect ($f = 1$), the controlability is equal to the measurability.

Van der Grinten [1–4] developed criteria that can serve as a guide for the analytical chemist who has to decide how precisely, how rapidly, and how frequently to analyze in order to achieve an optimal process control, or at least to select the best procedure or analyzer for the process control. An important relationship has been derived between the properties of the analytical method and the

References p. 466

measurability of a process when employing the autocorrelation function as a predictor.

$$m = \exp(-T_d/T_x) \cdot \exp(-T_a/2T_x) \cdot \exp(-T_s/3T_x) \cdot \left(1 - \frac{s_a}{s_x}\sqrt{\frac{T_a}{T_x}}\right) \tag{3}$$

or

$$m = m_d \cdot m_a \cdot m_s \cdot m_n$$

where T_x is the time constant of the process (see Chap. 14), T_a is the time span between the samples, T_d is the analysis time, T_s is the sampling time, s_a^2 is the analysis variance, s_x^2 is the process variance, m_a is the measurability due to the sampling interval, m_d is the measurability due to the analysis time, m_s is the measurability due to the sampling collection time, and m_n is the measurability due to precision.

From eqn. (3), it follows that the effects of the analysis time, sampling time and sampling interval can be considered separately. The effect of a decrease of the analysis time, T_d, by 10% is given by

$$\frac{m'}{m} = \frac{\exp(-0.9T_d/T_x)}{\exp(-T_d/T_x)} = \exp(0.1T_d/T_x)$$

Equally, the effects of a decrease of the time between the samples, T_a, and the sample collection time, T_s, by 10% are respectively given by

$$\frac{m'}{m} = \exp(0.1T_a/2T_x)$$

when $s_a \ll s_x$ and

$$\frac{m'}{m} = \exp(0.1T_s/3T_x)$$

Because, usually, $T_s \ll T_a$ and $T_s \ll T_d$, the smallest effect is expected for the time during which the sample is collected. The effects of the analysis time, T_d, compared with the time between the samples, T_a, depend upon their relative magnitude. A good indication of the expected effects on the measurability by varying the characteristics of the analytical methods is obtained when considering the plots of m (Fig. 4) as a function of T_a/T_x, T_d/T_x, and s_a/s_x.

The variance of the controlled process can be calculated as a function of the measurability.

$$s_\epsilon^2 = s_x^2(1 - m^2) \tag{4}$$

s_ϵ/s_x is tabulated (Table 2) as a function of the measurability, m. This shows that the standard deviation of the controlled process is less than half the original value when $m \geqslant 0.80$. Consequently, measurabilities lower than 0.80 are of no practical interest.

The residual variance when a process is reconstructed is found by substituting $T_d = 0$ in eqn. (3).

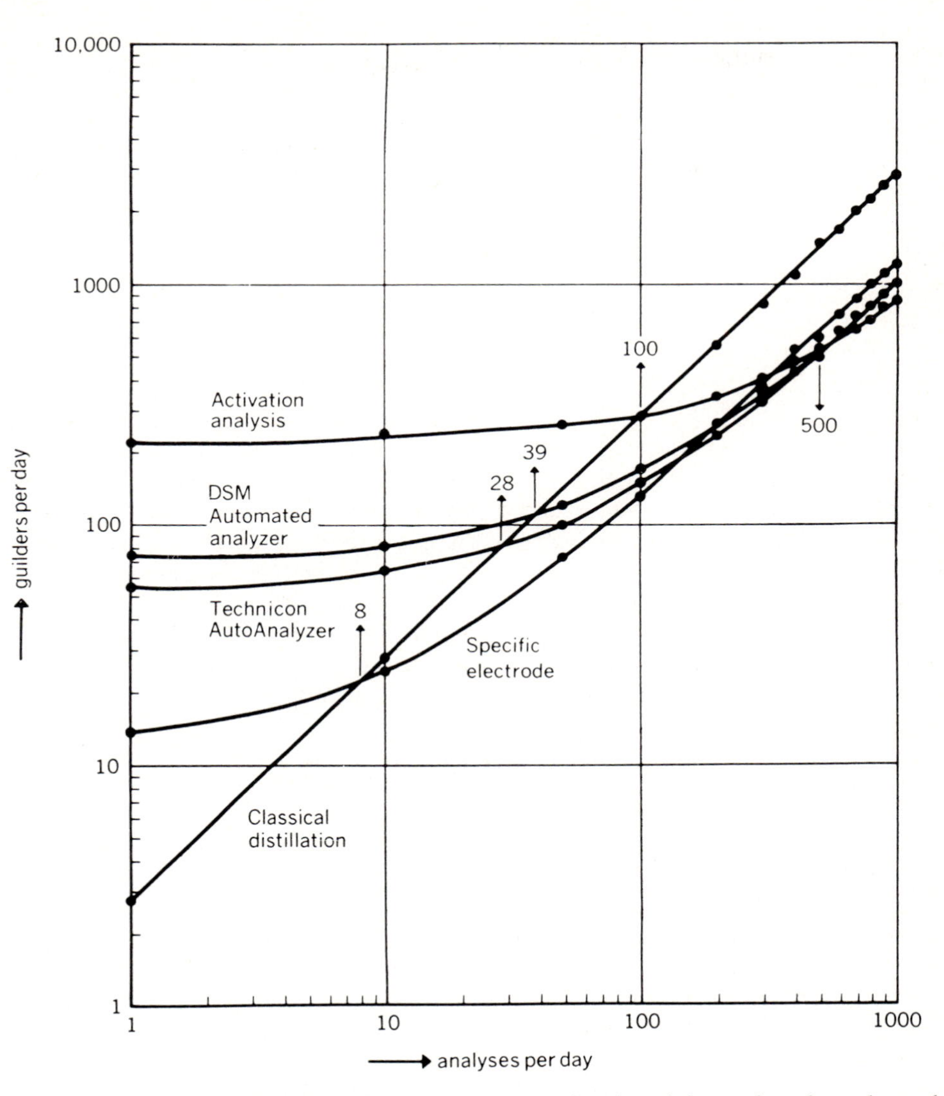

Fig. 6. Analysis costs of candidate analytical methods as a function of the number of samples analysed per day.

The (T_a, T_d) space in Fig. 4(a) can be divided in two regions: an upper triangle, where T_a is smaller than T_d, and a lower triangle, where T_a is larger than T_d. In the lower triangle, the analysis times, T_d, are shorter than the sampling interval, T_a. Here, control will improve when the sampling interval is shortened until the sampling interval becomes equal to the analysis time. However, when the analysis

TABLE 2

Standard deviation of the controlled process as a function of the measurability

Measurability, m	Standard deviation of the controlled process (fraction of the value for uncontrolled process)
1	0
0.9	0.43
0.8	0.6
0.7	0.71
0.6	0.8
0.5	0.86

time is already shorter than the sampling interval, no effect should be expected from a further shortening of the analysis time. If insufficient process control can be obtained (too small m), measures have to be taken to sample the process more frequently and to choose a shorter analysis method at the same time. It should be remarked at this point that, in real situations, the dead time is the sum of the analysis time and the waiting time. All measures which augment the work load (longer analysis times or shorter sampling intervals) introduce an additional upward shift in Fig. 4(a) because the waiting times, and thus the dead times, become longer. Conversely, measures which decrease the work load introduce an additional downward shift in the figure to smaller dead times.

An example of the expected effects is given below. Suppose that one controls a chemical process with a time constant, T_x, of 1 day. One sample is taken every day ($T_a = 1$) and the analysis time is 0.5 days (from sampling to the report of the analytical result). Thus, $T_d/T_x = 0.5$ and $T_a/T_x = 1$. It is read from Fig. 4(a) that $m_a m_d = 0.42$. Assuming that the other measurability factors are equal to one, then the overall measurability equals 0.42.

From eqn. (4), it follows that s_ϵ/s_x is equal to 0.92, meaning that the standard deviation of the controlled process is still 92% of its original value. To improve that figure, the analytical chemist has several options. Let us consider three of them: (1) to increase the sampling frequency, (2) to choose a faster analytical method, and (3) a combination of both.

With Fig. 4(a) in hand, the solution of the problem is quite obvious: a decrease of T_a/T_x from 1 to 0.25 (4 samples per day) while keeping T_d constant has a minor effect on the measurability because the maximal attainable measurability for $T_d/T_x = 0.5$ is equal to 0.6. This means that doubling the sampling frequency does not augment the available information about the process. The selection of a method which is twice as fast as the first (thus $T_d/T_x = 0.25$) also has a minor effect on the measurability (an increase to 0.6, which means a standard deviation of 80%). The third option is the best, namely to choose a faster method when increasing the sampling frequency to 4 samples a day. The measurability for $T_d/T_x = 0.25$ and $T_a/T_x = 0.25$ is 0.8, which means a reduction of the standard deviation to 60%.

3. An example

Let us recall the example of the fertilizer plant studied by Leemans [5]. The problem of selecting the best analytical method for the control of this plant can be assessed in a quantitative way. It has been found experimentally that the time constant of the fertilizer process is equal to 66 min. When the time span between two samples, T_a, is 30 min, the measurability factors listed in Table 3 are found for a number of candidate methods.

For each method, $T_a = 30$ min and therefore $m_a = \exp(-30/60) = 0.80$. As an example, we calculate the measurability factors of the Kjeldahl method.
Analysis time = 75 min
Precision = 0.17
Standard deviation of uncontrolled process = 1.2

$$m_d = \exp(-T_d/T_x) = \exp(-75/66) = 0.32$$

$$m_n = 1 - \frac{s_a}{s_x}\sqrt{\frac{T_a}{T_x}}$$

$$= 1 - \frac{0.17}{1.2}\sqrt{\frac{30}{60}}$$

$$= 0.99$$

The overall measurability, m, of each method is then $m = 0.8 m_n \cdot m_d$. The analytical method with the highest measurability in Table 3 is neutron activation. This result is independent of the chosen sampling frequency. By changing the sampling frequency, T_a, one can calculate the overall measurability as a function of the sampling frequency. This yields the plot given in Fig. 5.

When it is assumed that the time between the samples, T_a, should not fall below the total analysis time, including the sample preparation, a maximal measurability

TABLE 3

Measurabilities of the analytical methods for the determination of N in a fertilizer

Sampling rate: 2 samples per hour
Time constant of the process (T_x): 66 min
Measurability due to sampling (m_a): 0.8

Method	Measurability due to analysis time, m_d	Measurability due to precision	$m_a m_d m_n$
Total N: distillation	0.32	0.99	0.24
Total N: automated	0.84	0.97	0.65
NO_3-N: autoanalyzer	0.79	0.92	0.58
NO_3-N: ion SE	0.86	0.85	0.58
$NH_4NO_3/CaCO_3$: X-ray	0.89	0.83	0.59
Total N: ν-activation	0.93	0.99	0.74
γ-Ray absorption	0.98	0.88	0.69

References p. 466

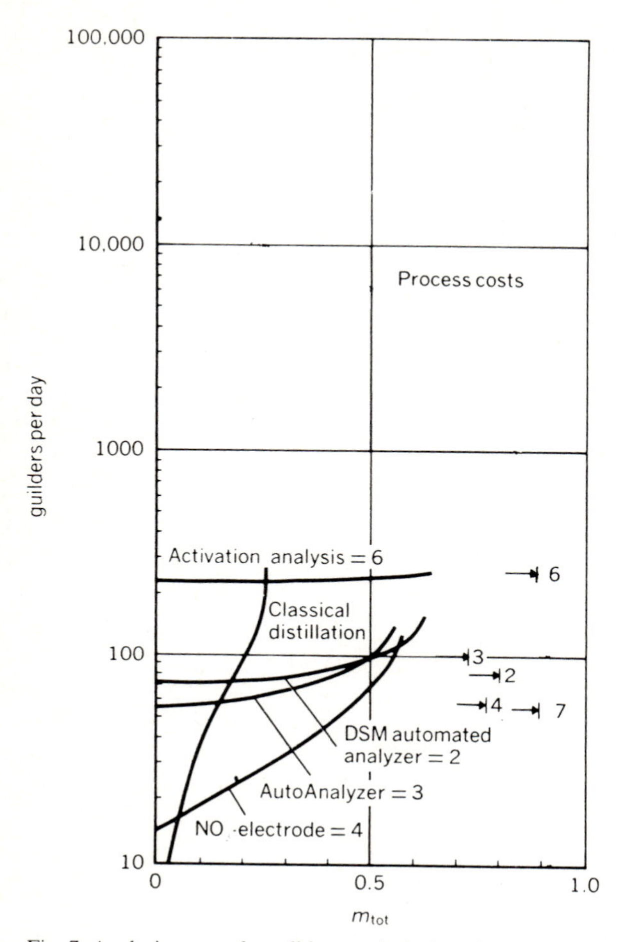

Fig. 7. Analysis costs of candidate analytical methods versus the measurability.

of only 0.6 is attainable. In this particular situation, it was found that the manual sampling constituted the bottle-neck of the analysis time. By switching to on-line methods without a manual sampling step, both the analysis time and time between the samples could be lowered considerably, yielding a measurability equal to 0.9 for some of the methods.

Some conclusions of Leemans can now be summarized as follows.

(1) The classical distillation yields the smallest measurability factor in spite of the high precision. This is caused by the long analysis time of the method.

(2) The measurability is increased by using the faster automatic distillation procedure, although its precision is somewhat worse.

(3) The best performance in this case is obtained with a non-specific and imprecise but fast procedure.

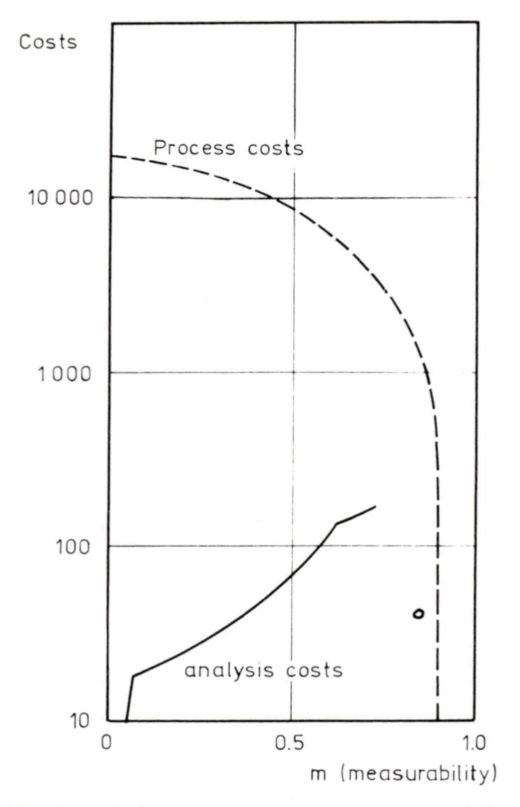

Fig. 8. Analysis and process costs versus measurability.

By the condition derived in the previous section, namely that the time between the samples should not drop below the analysis time, one finds the maximal attainable measurability for each analytical method, by substituting $T_a = T_d$ in eqn. (3). The sampling frequency in turn defines the number of samples which have to be analyzed by the laboratory, and thus the analysis costs. An example of such cost functions is given in Fig. 6. A procedure which requires a low capital investment in comparison with the labour costs shows a linear relationship between costs and sample load (samples/day). On the other hand, when the capital investment is the most important part of the costs, the costs are fairly independent from the sample load (within a certain range). Most of the procedures will behave between these two extremes. The plots which relate the sampling frequency to the measurability (Fig. 5) and the sampling frequency to the analysis costs (Fig. 6) can be combined to give the relationship between analysis costs and measurability (Fig. 7).

This example clearly illustrates the importance of the measurability concept for the analytical chemist. Because the measurability is a direct measure for the residual

References p. 466

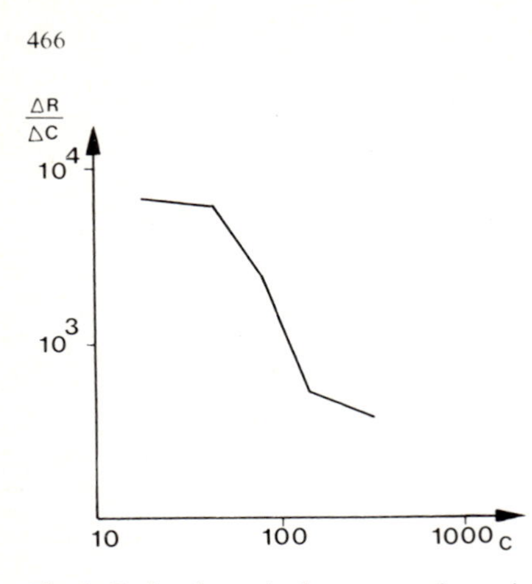

Fig. 9. Ratio of marginal returns and marginal costs versus the cost level.

standard deviation of the controlled process, a relationship can be derived between the marginal returns and the marginal costs. Therefore plots are made of the analysis costs and process costs versus the measurability (Fig. 8). The analysis costs are found by taking the bottom line in Fig. 7.

From Fig. 8, a plot can be constructed of the decrease of the process costs (marginal returns) for a given increase of the analysis costs (Fig. 9). This gives the optimal level of analysis costs of the laboratory. The method with the highest measurability at that costs level is the in situ determination of the specific gravity by γ-ray absorption.

A complication mentioned before (Chap. 26) is that the total time between sampling and the analytical result may be considerably longer than the actual analysis time because of the delays in the laboratory. Consequently, the laboratory organisation influences the measurability. Thus, the cost effectiveness of different laboratories may differ because of a different laboratory organisation. Therefore one should know the influence of various strategies on the delays of various groups of samples. A laboratory organisation is of such complexity that theoretical models (such as those provided by queueing theory) can only give a rough indication of the importance of the parameters. Experiments with the real life laboratory are, for obvious reasons, impossible. Hence the solution is to design a substitute for reality (a simulation model) (Chap. 25) which one hopes will mimic reality in all the variables of interest. The model then becomes a tool for practical decision making, where the outcome of various strategies and decisions is simulated before actual implementation.

References

1 P.M.E.M. van der Grinten, Finding optimum controller settings, Control Eng., 10 (1963) 51.
2 P.M.E.M. van der Grinten, Determining plant controllability, Control Eng., 10 (1963) 87.

3 P.M.E.M. van der Grinten, Control effects of instrument accuracy and measuring speed. I, J. Instrum. Soc. Am., 12 (1965) 87.
4 P.M.E.M. van der Grinten, Control effects of instrument accuracy and measuring speed. II, J. Instrum. Soc. Am., 13 (1966) 58.
5 F.A. Leemans, The selection of an optimum analytical method, Anal. Chem., 43(11) (1971) 36A.

Recommended reading

G. Kateman, Sampling, in B.R. Kowalski (Ed.), Chemometrics, Mathematics and Statistics in Chemistry, Reidel, Dordrecht, 1984.
G. Kateman and F. Pijpers, Quality Control in Analytical Chemistry, Wiley, New York, 1981, Chap. 2.5.

Appendix

TABLE 1

The normal distribution

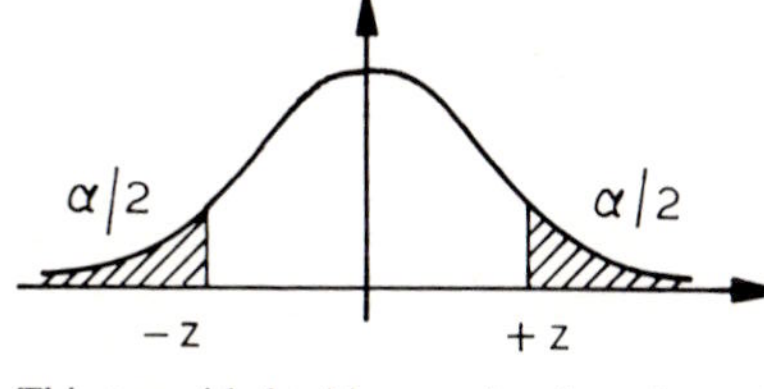

This two-sided table contains the values of z such that the probability of a standardized normally distributed variable to lie outside the interval $[-z, z]$ is a given number α. The values of α are given to two decimal places. The first decimal place is located at the left of the table and the second one above it. Beneath the table, z values are given that correspond to a few extreme α numbers. (Reproduced with permission from R.A. Fisher and F. Yates, Statistical Tables for Biological Agricultural and Medical Research, Oliver and Boyd, Edinburgh, 1963.)

α	0.00	0.01	0.02	0.03	0.04	0.05	0.06	0.07	0.08	0.09
0.0	∞	2.575829	2.326348	2.170090	2.053749	1.959964	1.880794	1.811911	1.750686	1.695398
0.1	1.644854	1.598193	1.554774	1.514102	1.475791	1.439531	1.405072	1.372204	1.340755	1.310579
0.2	1.281552	1.253565	1.226528	1.200359	1.174987	1.150349	1.126391	1.103063	1.080319	1.058122
0.3	1.036433	1.015222	0.994458	0.974114	0.954165	0.934589	0.915365	0.896473	0.877896	0.859617
0.4	0.841621	0.823894	0.806421	0.789192	0.772193	0.755415	0.738847	0.722479	0.706303	0.690309
0.5	0.674490	0.658838	0.643345	0.628006	0.612813	0.597760	0.582842	0.568051	0.553385	0.538836
0.6	0.524401	0.510073	0.495850	0.481727	0.467699	0.453762	0.439913	0.426148	0.412463	0.398855
0.7	0.385320	0.371856	0.358459	0.345126	0.331853	0.318639	0.305481	0.292375	0.279319	0.266311
0.8	0.253347	0.240426	0.227545	0.214702	0.201893	0.189118	0.176374	0.163658	0.150969	0.138304
0.9	0.125661	0.113039	0.100434	0.087845	0.075270	0.062707	0.050154	0.037608	0.025069	0.012533

α	0.002	0.001	0.0001	0.00001	0.000001	0.0000001	0.00000001	0.000000001
z	3.090232	3.29053	3.89059	4.41717	4.89164	5.32672	5.73073	6.10941

TABLE 2

Student's t-distribution

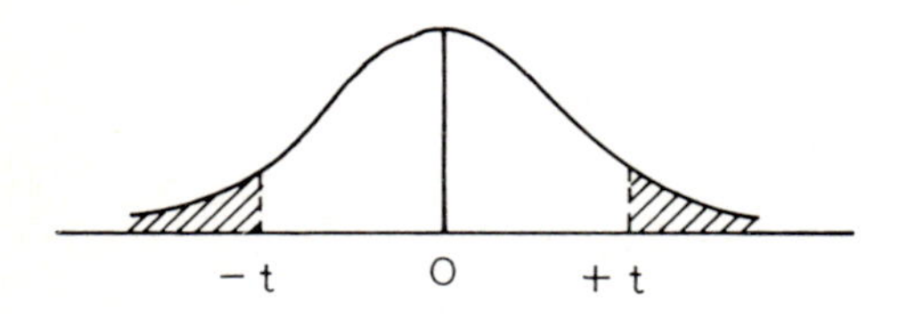

In the table for the t-distribution, the rows correspond to the numbers of degrees of freedom (d.f.) and the columns to values of α. The table is two-sided, meaning that the probability for the variable to lie outside the interval $[-t, t]$ equals α. Note that for increasing d.f., the t-distribution converges to the standardized normal distribution. This can be seen by comparing the last row of the table (for d.f. $= \infty$) with the corresponding values of the normal distribution. (Reproduced with permission from D. Schwarz, Méthodes Statistiques à l'Usage des Médecins et des Biologistes, Flammarion, Paris, 1963, p. 287.)

	α								
d.f.	0.90	0.50	0.30	0.20	0.10	0.05	0.02	0.01	0.001
1	0.158	1.000	1.963	3.078	6.314	12.706	31.821	63.657	636.619
2	0.142	0.816	1.386	1.886	2.920	4.303	6.965	9.925	31.598
3	0.137	0.765	1.250	1.638	2.353	3.182	4.541	5.841	12.924
4	0.134	0.741	1.190	1.533	2.132	2.776	3.747	4.604	8.610
5	0.132	0.727	1.156	1.476	2.015	2.571	3.365	4.032	6.869
6	0.131	0.718	1.134	1.440	1.943	2.447	3.143	3.707	5.959
7	0.130	0.711	1.119	1.415	1.895	2.365	2.998	3.499	5.408
8	0.130	0.706	1.108	1.397	1.860	2.306	2.896	3.355	5.041
9	0.129	0.703	1.100	1.383	1.833	2.262	2.821	3.250	4.781
10	0.129	0.700	1.093	1.372	1.812	2.228	2.764	3.169	4.587
11	0.129	0.697	1.088	1.363	1.796	2.201	2.718	3.106	4.437
12	0.128	0.695	1.083	1.356	1.782	2.179	2.681	3.055	4.318
13	0.128	0.694	1.079	1.350	1.771	2.160	2.650	3.012	4.221
14	0.128	0.692	1.076	1.345	1.761	2.145	2.624	2.977	4.140
15	0.128	0.691	1.074	1.341	1.753	2.131	2.602	2.947	4.073
16	0.128	0.690	1.071	1.337	1.746	2.120	2.583	2.921	4.015
17	0.128	0.689	1.069	1.333	1.740	2.110	2.567	2.898	3.965
18	0.127	0.688	1.067	1.330	1.734	2.101	2.552	2.878	3.922
19	0.127	0.688	1.066	1.328	1.729	2.093	2.539	2.861	3.883
20	0.127	0.687	1.064	1.325	1.725	2.086	2.528	2.845	3.850
21	0.127	0.686	1.063	1.323	1.721	2.080	2.518	2.831	3.819
22	0.127	0.686	1.061	1.321	1.717	2.074	2.508	2.819	3.792
23	0.127	0.685	1.060	1.319	1.714	2.069	2.500	2.807	3.767
24	0.127	0.685	1.059	1.318	1.711	2.064	2.492	2.797	3.745
25	0.127	0.684	1.058	1.316	1.708	2.060	2.485	2.787	3.725
26	0.127	0.684	1.058	1.315	1.706	2.056	2.479	2.779	3.707
27	0.127	0.684	1.057	1.314	1.703	2.052	2.473	2.771	3.690
28	0.127	0.683	1.056	1.313	1.701	2.048	2.467	2.763	3.674
29	0.127	0.683	1.055	1.311	1.699	2.045	2.462	2.756	3.659
30	0.127	0.683	1.055	1.310	1.697	2.042	2.457	2.750	3.646
∞	0.126	0.674	1.036	1.282	1.645	1.960	2.326	2.576	3.291

TABLE 3

The F-distribution

In the following table, values of the inverse cumulative frequency distribution of the F-distribution are given for several values of n_1 and n_2, which are respectively, the number of degrees of freedom of the numerator and denominator. All values of the table correspond to a cumulative function equal to 0.975. This is equivalent to a two-sided table with α equal to 5%. (Reproduced with permission from D.L. Massart, A. Dijkstra and L. Kaufman, Evaluation and Optimization of Laboratory Methods and Analytical Procedures, Elsevier, Amsterdam, 1978.)

n_2 \ n_1	1	2	3	4	5	6	7	8	9	10	12	15	20	24	30	40	60	120	∞
1	647	779	864	899	922	937	948	956	963	968	976	985	993	997	1001	1005	1010	1014	1018
2	38.51	39.00	39.17	39.25	39.30	39.33	39.36	39.37	39.39	39.40	39.41	39.43	39.45	39.46	39.46	39.47	39.48	39.98	39.50
3	17.44	16.04	15.44	15.10	14.88	14.73	14.62	14.54	14.47	14.42	14.34	14.25	14.17	14.12	14.08	14.04	13.99	13.87	13.90
4	12.22	10.65	9.98	9.60	9.36	9.20	9.07	8.98	8.90	8.84	8.75	8.66	8.56	8.51	8.46	8.41	8.36	8.31	8.26
5	10.01	8.43	7.76	7.39	7.15	6.98	6.85	6.76	6.68	6.62	6.52	6.43	6.33	6.28	6.23	6.18	6.12	6.06	6.02
6	8.81	7.26	6.60	6.23	5.99	5.82	5.70	5.60	5.52	5.46	5.37	5.27	5.17	5.12	5.07	5.01	4.96	4.90	4.85
7	8.07	6.54	5.89	5.52	5.29	5.12	4.99	4.90	4.82	4.76	4.67	4.57	4.47	4.41	4.36	4.31	4.25	4.20	4.14
8	7.57	6.06	5.42	5.05	4.82	4.65	4.53	4.43	4.36	4.29	4.20	4.10	4.00	3.95	3.89	3.84	3.78	3.73	3.67
9	7.21	5.71	5.08	4.72	4.48	4.32	4.20	4.10	4.03	3.96	3.87	3.77	3.67	3.61	3.56	3.51	3.45	3.39	3.33
10	6.94	5.46	4.83	4.47	4.24	4.07	3.95	3.85	3.78	3.72	3.62	3.52	3.42	3.37	3.31	3.26	3.20	3.14	3.08
12	6.55	5.10	4.47	4.12	3.89	3.73	3.61	3.51	3.44	3.37	3.28	3.18	3.07	3.02	2.96	2.91	2.85	2.79	2.72
15	6.20	4.77	4.15	3.80	3.58	3.41	3.29	3.20	3.12	3.06	2.96	2.86	2.76	2.70	2.64	2.58	2.52	2.46	2.40
20	5.87	4.46	3.86	3.51	3.29	3.13	3.01	2.91	2.84	2.77	2.68	2.57	2.46	2.41	2.35	2.29	2.22	2.16	2.09
24	5.72	4.32	3.72	3.38	3.15	2.99	2.87	2.78	2.70	2.64	2.54	2.44	2.33	2.27	2.21	2.15	2.08	2.01	1.94
30	5.57	4.18	3.59	3.25	3.03	2.87	2.75	2.65	2.57	2.51	2.41	2.31	2.20	2.14	2.07	2.01	1.94	1.87	1.79
40	5.42	4.05	3.46	3.13	2.90	2.74	2.62	2.53	2.45	2.39	2.29	2.18	2.07	2.01	1.94	1.88	1.80	1.72	1.64
60	5.29	3.93	3.34	3.01	2.79	2.63	2.51	2.41	2.33	2.27	2.17	2.06	1.94	1.88	1.82	1.74	1.67	1.58	1.48
120	5.15	3.80	3.23	2.89	2.67	2.52	2.39	2.30	2.22	2.16	2.05	1.94	1.82	1.76	1.69	1.61	1.53	1.43	1.31
∞	5.02	3.69	3.12	2.79	2.57	2.41	2.29	2.19	2.11	2.05	1.94	1.83	1.71	1.64	1.57	1.48	1.39	1.27	1.00

Values for 0.025 can be found in the same table using the formula

$$F_{\alpha/2,k,m} = \frac{1}{F_{1-(\alpha/2),m,k}}$$

Together, these values can be used to find the following interval containing 95% of the total probability

$$(F_{0.025,k,m};\ F_{0.975,k,m})$$

or

$$\left(\frac{1}{F_{0.975,m,k}};\ F_{0.975,k,m}\right)$$

TABLE 3 (continued)

The second table for the *F*-distribution contains values for which the cumulative function is equal to 0.95. This is equivalent to a two-sided table with α equal to 10%.

n_2 \ n_1	1	2	3	4	5	6	7	8	9	10	12	15	20	24	30	40	60	120	∞
1	161	199	215	224	230	234	237	239	240	242	244	246	248	249	250	251	252	253	254
2	18.51	19.00	19.16	19.25	19.30	19.33	19.35	19.37	19.38	19.40	19.41	19.43	19.45	19.45	19.46	19.47	19.48	19.49	19.50
3	10.13	9.55	9.28	9.12	9.01	8.94	8.89	8.85	8.81	8.79	8.74	8.70	8.66	8.64	8.62	8.59	8.57	8.53	8.53
4	7.71	6.94	6.59	6.39	6.26	6.16	6.09	6.04	6.00	5.96	5.91	5.86	5.80	5.77	5.75	5.72	5.69	5.65	5.63
5	6.61	5.79	5.41	5.19	5.05	4.95	4.88	4.82	4.77	4.74	4.68	4.62	4.56	4.53	4.50	4.46	4.43	4.40	4.36
6	5.99	5.14	4.76	4.53	4.39	4.28	4.21	4.15	4.10	4.06	4.00	3.94	3.87	3.84	3.81	3.77	3.74	3.70	3.67
7	5.59	4.74	4.35	4.12	3.97	3.87	3.79	3.73	3.68	3.64	3.57	3.51	3.44	3.41	3.38	3.34	3.30	3.27	3.23
8	5.32	4.46	4.07	3.84	3.69	3.58	3.50	3.44	3.39	3.35	3.28	3.22	3.15	3.12	3.08	3.04	3.00	2.97	2.93
9	5.12	4.26	3.86	3.63	3.48	3.37	3.29	3.23	3.18	3.14	3.07	3.01	2.94	2.90	2.86	2.83	2.79	2.75	2.71
10	4.96	4.10	3.71	3.48	3.33	3.22	3.14	3.07	3.02	2.98	2.91	2.84	2.77	2.74	2.70	2.66	2.62	2.58	2.54
12	4.75	3.89	3.49	3.26	3.11	3.00	2.91	2.85	2.80	2.75	2.69	2.62	2.54	2.51	2.47	2.43	2.38	2.34	2.30
15	4.54	3.68	3.29	3.06	2.90	2.79	2.71	2.64	2.59	2.54	2.48	2.40	2.33	2.29	2.25	2.20	2.16	2.11	2.07
20	4.35	3.49	3.10	2.87	2.71	2.60	2.51	2.45	2.39	2.35	2.28	2.20	2.12	2.08	2.04	1.99	1.95	1.90	1.84
24	4.26	3.40	3.01	2.78	2.62	2.51	2.42	2.36	2.30	2.25	2.18	2.11	2.03	1.98	1.94	1.89	1.84	1.79	1.73
30	4.17	3.32	2.92	2.69	2.53	2.42	2.33	2.27	2.21	2.16	2.09	2.01	1.93	1.89	1.84	1.79	1.74	1.68	1.62
40	4.08	3.23	2.84	2.61	2.45	2.34	2.25	2.18	2.12	2.08	2.00	1.92	1.84	1.79	1.74	1.69	1.64	1.58	1.51
60	4.00	3.15	2.76	2.53	2.37	2.25	2.17	2.10	2.04	1.99	1.92	1.84	1.75	1.70	1.65	1.59	1.53	1.47	1.39
120	3.92	3.07	2.68	2.45	2.29	2.18	2.09	2.02	1.96	1.91	1.83	1.75	1.66	1.61	1.55	1.50	1.43	1.35	1.25
∞	3.84	3.00	2.60	2.37	2.21	2.10	2.01	1.94	1.88	1.83	1.75	1.67	1.57	1.52	1.46	1.39	1.32	1.22	1.00

TABLE 4

Tables for the Mann–Whitney test

The following tables contain critical values of the U statistic for significance levels α equal to 5% and 10% for a two-sided test. If an observed U value is less than or equal to the value in the table, the null hypothesis may be rejected at the level of significance of the table. (Reproduced with permission from W. Beyer (Ed.), Handbook of Tables for Probability and Statistics, CRC Press, Boca Raton, FL, 1966.)

	Critical values of U for α equal to 5%																			
n_1	$n_2=1$	2	3	4	5	6	7	8	9	10	11	12	13	14	15	16	17	18	19	20
1																				
2								0	0	0	0	1	1	1	1	1	2	2	2	2
3					0	1	1	2	2	3	3	4	4	5	5	6	6	7	7	8
4				0	1	2	3	4	4	5	6	7	8	9	10	11	11	12	13	13
5			0	1	2	3	5	6	7	8	9	11	12	13	14	15	17	18	19	20
6			1	2	3	5	6	8	10	11	13	14	16	17	19	21	22	24	25	27
7			1	3	5	6	8	10	12	14	16	18	20	22	24	26	28	30	32	34
8		0	2	4	6	8	10	13	15	17	19	22	24	26	29	31	34	36	38	41
9		0	2	4	7	10	12	15	17	20	23	26	28	31	34	37	39	42	45	48
10		0	3	5	8	11	14	17	20	23	26	29	33	36	39	42	45	48	52	55
11		0	3	6	9	13	16	19	23	26	30	33	37	40	44	47	51	55	58	62
12		1	4	7	11	14	18	22	26	29	33	37	41	45	49	53	57	61	65	69
13		1	4	8	12	16	20	24	28	33	37	41	45	50	54	59	63	67	72	76
14		1	5	9	13	17	22	26	31	36	40	45	50	55	59	64	67	74	78	83
15		1	5	10	14	19	24	29	34	39	44	49	54	59	64	70	75	80	85	90
16		1	6	11	15	21	26	31	37	42	47	53	59	64	70	75	81	86	92	98
17		2	6	11	17	22	28	34	39	45	51	57	63	67	75	81	87	93	99	105
18		2	7	12	18	24	30	36	42	48	55	61	67	74	80	86	93	99	106	112
19		2	7	13	19	25	32	38	45	52	58	65	72	78	85	92	99	106	113	119
20		2	8	13	20	27	34	41	48	55	62	69	76	83	90	98	105	112	119	127

TABLE 4 (continued)

n_1	Critical values of U for α equal to 10%																			
	$n_2=1$	2	3	4	5	6	7	8	9	10	11	12	13	14	15	16	17	18	19	20
1																			0	0
2					0	0	0	1	1	1	1	2	2	2	3	3	3	4	4	4
3			0	0	1	2	2	3	3	4	5	5	6	7	7	8	9	9	10	11
4			0	1	2	3	4	5	6	7	8	9	10	11	12	14	15	16	17	18
5		0	1	2	4	5	6	8	9	11	12	13	15	16	18	19	20	22	23	25
6		0	2	3	5	7	8	10	12	14	16	17	19	21	23	25	26	28	30	32
7		0	2	4	6	8	11	13	15	17	19	21	24	26	28	30	33	35	37	39
8		1	3	5	8	10	13	15	18	20	23	26	28	31	33	36	39	41	44	47
9		1	3	6	9	12	15	18	21	24	27	30	33	36	39	42	45	48	51	54
10		1	4	7	11	14	17	20	24	27	31	34	37	41	44	48	51	55	58	62
11		1	5	8	12	16	19	23	27	31	34	38	42	46	50	54	57	61	65	69
12		2	5	9	13	17	21	26	30	34	38	42	47	51	55	60	64	68	72	77
13		2	6	10	15	19	24	28	33	37	42	47	51	56	61	65	70	75	80	84
14		2	7	11	16	21	26	31	36	41	46	51	56	61	66	71	77	82	87	92
15		3	7	12	18	23	28	33	39	44	50	55	61	66	72	77	83	88	94	100
16		3	8	14	19	25	30	36	42	48	54	60	65	71	77	83	89	95	101	107
17		3	9	15	20	26	33	39	45	51	57	64	70	77	83	89	96	102	109	115
18		4	9	16	22	28	35	41	48	55	61	68	75	82	88	95	102	109	116	123
19	0	4	10	17	23	30	37	44	51	58	65	72	80	87	94	101	109	116	123	130
20	0	4	11	18	25	32	39	47	54	62	69	77	84	92	100	107	115	123	130	138

TABLE 5

Table for the Wilcoxon test

The following table contains approximate 10%, 5%, 2%, and 1% points for the T statistic for various sample sizes. If an observed T value is less than or equal to the value in the table, the null hypothesis may be rejected at the corresponding level of significance. (Reproduced with permission from W. Beyer (Ed.), Handbook of Tables for Probability and Statistics, CRC Press, Boca Raton, FL, 1966.)

	n														
α	5	6	7	8	9	10	11	12	13	14	15	16	17	18	19
0.10	1	2	4	6	8	11	14	17	21	26	30	36	41	47	54
0.05		1	2	4	6	8	11	14	17	21	25	30	35	40	46
0.02			0	2	3	5	7	10	13	16	20	24	28	33	38
0.01				0	2	3	5	7	10	13	16	19	23	28	32

	n														
α	20	21	22	23	24	25	26	27	28	29	30	31	32	33	34
0.10	60	68	75	83	92	101	110	120	130	141	152	163	175	188	201
0.05	52	59	66	73	81	90	98	107	117	127	137	148	159	171	183
0.02	43	49	56	62	69	77	85	93	102	111	120	130	141	151	162
0.01	37	43	49	55	61	68	76	84	92	100	109	118	128	138	149

	n															
α	35	36	37	38	39	40	41	42	43	44	45	46	47	48	49	50
0.10	214	228	242	256	271	287	308	319	336	353	371	389	408	427	446	466
0.05	195	208	222	235	250	264	279	295	311	327	344	361	379	397	415	434
0.02	174	186	198	211	224	238	252	267	281	297	313	329	345	362	380	398
0.01	160	171	183	195	208	221	234	248	262	277	292	307	323	339	356	373

TABLE 6

The chi-square distribution
(Reproduced with permission from D.L. Massart, A. Dijkstra and L. Kaufman, Evaluation and Optimization of Laboratory Methods and Analytical Procedures, Elsevier, Amsterdam, 1978.)

	χ^2_k								
k	0.01	0.05	0.10	0.25	0.50	0.75	0.90	0.95	0.99
1	0.00016	0.0039	0.0158	0.102	0.455	1.32	2.71	3.84	6.63
2	0.0201	0.103	0.211	0.575	1.39	2.77	4.61	5.99	9.21
3	0.115	0.352	0.584	1.21	2.37	4.11	6.25	7.81	11.3
4	0.297	0.711	1.06	1.92	3.36	5.39	7.78	9.49	13.3
5	0.554	1.15	1.61	2.67	4.35	6.63	9.24	11.1	15.1
6	0.872	1.64	2.20	3.45	5.35	7.84	10.6	12.6	16.8
7	1.24	2.17	2.83	4.25	6.35	9.04	12.0	14.1	18.5
8	1.65	2.73	3.49	5.07	7.34	10.2	13.4	15.5	20.1
9	2.09	3.33	4.17	5.90	8.34	11.4	14.7	16.9	21.7
10	2.56	3.94	4.87	6.74	9.34	12.5	16.0	18.3	23.2
11	3.05	4.57	5.58	7.58	10.3	13.7	17.3	19.7	24.7
12	3.57	5.23	6.30	8.44	11.3	14.8	18.5	21.0	26.2
13	4.11	5.89	7.04	9.30	12.3	16.0	19.8	22.4	27.7
14	4.66	6.57	7.79	10.2	13.3	17.1	21.1	23.7	29.1
15	5.23	7.26	8.55	11.0	14.3	18.2	22.3	25.0	30.6
16	5.81	7.95	9.31	11.9	15.3	19.4	23.5	26.3	32.0
17	6.41	8.67	10.1	12.8	16.3	20.5	24.8	27.6	33.4
18	7.01	9.39	10.9	13.7	17.3	21.6	26.0	28.9	34.8
19	7.63	10.1	11.7	14.6	18.3	22.7	27.2	30.1	36.2
20	8.26	10.9	12.4	15.5	19.3	23.8	28.4	31.4	37.6
21	8.90	11.6	13.2	16.3	20.3	24.9	29.6	32.7	38.9
22	9.54	12.3	14.0	17.2	21.3	26.0	30.8	34.0	40.3
23	10.2	13.1	14.8	18.1	22.3	27.1	32.0	35.2	41.6
24	10.9	13.8	15.7	19.0	23.3	28.2	33.2	36.4	43.0
25	11.5	14.6	16.5	19.9	24.3	29.3	34.4	37.7	44.3

Index

F

J

K

L

M

R

S